Preserving the Legacy

Industrial Processes and Waste Stream Management

Howard H. Guyer
Lead Author and Editor

JOHN WILEY & SONS, INC.

New York / Chichester / Weinheim / Brisbane / Singapore / Toronto

Technical Illustration: Richard J. Washichek, Graphic Dimensions, Inc.
Photo Research: Beatrice Hohenegger

This material is based on work supported by the National Science Foundation under Grant No. DUE94-54521. Any opinions, findings, and conclusions or recommendations expressed in this material are those of the author(s) and do not necessarily reflect those of the National Science Foundation.

This publication is designed to provide accurate and authoritative information in regard to the subject matter covered. It is sold with the understanding that the publisher is not engaged in rendering legal, accounting, or other professional services. If legal advice or other expert assistance is required, the services of a competent professional person should be sought.

Library of Congress Cataloging-in-Publication Data:

0-471-29984-7

Table of Contents

To the Student

This book has been written for students of environmental technology. It is also an excellent resource for anyone who wishes to learn more about how the many different items we use daily are produced. *Industrial Processes and Waste Stream Management* is the fourth volume in the *Preserving the Legacy* textbook series developed by INTELECOM Intelligent Telecommunications in association with the Partnership for Environmental Technology Education (PETE).

Industrial Processes and Waste Stream Management is an introductory college-level textbook, which assumes the study of the *Introduction to Environmental Technology* text as well as prerequisite skills in and knowledge of basic chemistry. The text has been divided into two main sections: the first seven chapters are considered foundational to the remaining twenty-one chapters, which are industry-specific. Within the foundational chapters, the tools and regulations for waste stream management are presented, followed by chapters describing the four EPA-recognized treatment technologies, including physical, chemical, thermal, and biological methods. Chapter 7 presents the concepts that must be known to implement a successful pollution prevention program.

The following twenty-one chapters have an industry-specific focus and are presented cafeteria-style. It is assumed that most courses will not have sufficient time to cover all the chapters, but will rather select the ones most appropriate for specific needs and local employment opportunities. By providing an extensive range of industries, it is expected that students seeking employment in an unfamiliar industry can later use the text as a resource to inform themselves about the processes and waste streams of concern.

To assist the student in mastery of the material, several learning devices have been provided. Chapters start with a list of learning objectives to help the reader focus on the broad concepts to be mastered. An introductory section in each chapter provides an overall view of the topics included, before launching into detailed explanations. Each technical word that is critical to the understanding has been bolded the first time it is used and has been included in an extensive glossary of more than one thousand terms. Each section within the chapter is followed by several questions that give students the opportunity to check their understanding. Finally, at the end of each chapter several critical thinking questions or classroom activities are included. These questions and activities are intended to be of assistance to the teacher in finding ways for the students to demonstrate their understanding; they also suggest ways for individuals or groups of students to continue their learning beyond the classroom. The text also includes a bibliography to guide further study and an index to assist in the quick location of topics of interest.

In what is a unique learning opportunity, many of the topics presented in this text can also be visually experienced by the student using the integrated *Preserving the Legacy* educational video series. These fifteen videos take the students across the nation and into normally restricted areas. They experience engineers describing how a process works, professors and other specialists analyzing the underlying scientific principles, and environmental engineers explaining how a waste treatment process works. EPA specialists are interviewed to explain the regulations and how they apply; animation is used to visually describe processes that cannot otherwise be observed. Through all of this, students are provided with many role models of working environmental technicians.

Acknowledgements

The development of this integrated set of learning tools has been no small undertaking. The project would have never been completed, were it not for a network of extremely dedicated individuals. I would like to first acknowledge the contributions of each of the members of the National Academic Council (NAC). As PETE regional representatives, they assisted in the overall *Preserving the Legacy* project design and, in particular, the design and review of this text. They are:

—David Y. Boon, Front Range Community College

—Ann Boyce, Bakersfield College

—Eldon Enger, Delta College

—William T. Engle, Jr., PhD, University of Florida, TREEO Center

—Douglas A. Feil, Kirkwood Community College

—Steven R. Onstot, Esq., Fullerton College

—Douglas Nelson, SUNY, Morrisville, New York

—Ray Seitz PhD, Columbia Basin College (Retired)

—Andrew J. Silva, South Dakota School of Mines

I would like to acknowledge each of the contributing authors and co-authors. They are:

—Christopher J. Biermann PhD, Oregon State University

—David Y. Boon, Front Range Community College

—Ann Boyce, Bakersfield College

—Jack Cavanaugh, Concurrent Technical Corporation

—Walter Cordell III, Hunt Wesson/ConAgra

—Lisa Cummins, Safety Clean Corporation

—Douglas A. Feil, Kirkwood Community College

—Howard H. Guyer, Fullerton College

—Margaret W. F. Lee, Lee Consulting Company

—Larry McGaughey PhD, Radian International LLC

—Richard Marty PhD, S.M. Stoller Corporation

—Sylvia Medina, North Wind Environmental

—William "Pat" Miller, University of Washington (Retired)

—Jeff Morrell PhD, Oregon State University

—Luis Nunez PhD, Argonne National Laboratory

—David M. Olsen, Merced College

—Thomas Shahady PhD, Catawaba Valley Community College

—Andrew J. Silva, South Dakota School of Mines

—John Simonsen PhD, Oregon State University

—Melinda Trizinsky PhD, Clean Cities Inc.

—Azita Yazdani, Pollution Prevention International, Inc.

The development, production, and scripts for the integrated videos are a credit to the skills of David Weisman, producer/director. With his staff, they researched each topic, found the best locations and subject matter experts, and carefully edited the elements into a flexible instructional tool that visually extends the text.

Once the words were committed to paper, it was the effort of the INTELECOM staff, led by Sally V. Beaty, INTELECOM President and project Principal Investigator, and her staff, in particular Judy Sullivan, that provided the central coordination and support to hold the various parts of the project together. I would particularly like to thank the special and tireless efforts of Beatrice Hohenegger, INTELECOM's Senior Editor. Her countless hours of organizing and checking every detail, following the often circuitous permission routes, as well as her polishing of the final manuscript to remove ambiguities and to improve student understanding have contributed to the production of this textbook.

For their ability to juggle creativity and conformity, and to meet timelines again and again with each round of proofing and polishing, I extend thanks to Dick and Mary Ann Washichek of Graphic Dimensions, Inc.

For the initial efforts, recognition is due to Paul R. Dickinson, Director of the Partnership for Environmental Education (PETE), to Sally V. Beaty in orchestrating the writing and management of a complex grant, and to the vision and funding of the National Science Foundation's Department of Undergraduate Education (DUE) that made this project possible.

Howard H. Guyer,
Lead Author and Editor

PRESERVING
THE LEGACY

1

Introduction to Waste Streams

by Douglas A. Feil and Andrew J. Silva

Chapter Objectives

Upon completing this chapter, the student will be able to:

1. **Explain** why it is in the best interest of everyone to minimize or eliminate waste production.

2. **Differentiate** between pollution prevention and minimization of waste.

3. **Analyze** the material balance concept as it relates to manufacturing.

4. **Distinguish** between the four categories of waste treatment technologies and give an example of each.

5. **Recognize** the hierarchy of hazardous waste management and explain EPA's rationale for the rankings.

6. **Explain** waste minimization as it applies to RCRA hazardous waste management.

7. **Explain** what constitutes pollution prevention and minimization of waste.

Chapter Sections

1-1 Introduction

Efforts to regulate health and environmental safety in the United States started more than a century ago. These first efforts, however, were only intended to protect human health by removing such things as narcotics and poisons from patent medicines and by preventing transportation accidents and explosions. It was not until the 1970s that the public became aware of the growing detrimental effects of pollution, making the regulation of health and environmental safety a national priority. The newly formed **Environmental Protection Agency (EPA)** was intended to control the amount and cleanup of waste existing in the environment. However, by the time legislation came into effect, it was already inadequate. The increased automobile usage, the growth of industry, and the beginning of the "throw-away" society demanded updated measures and renewed nationwide efforts. The failure of the initial legislation, and its subsequent amendments, was largely due to the fact that the general focus was on treat-

ment and disposal of generated waste. It is only since the beginning of the 1990s that we have realized that the only effective way to deal with waste is simply not to generate it. The focus, therefore, has shifted to front-of-pipe measures or pollution prevention.

This chapter is intended to provide an introduction to waste stream management with the pollution prevention philosophy in mind. After a brief history, the chapter introduces the sources of pollution and some of the federal laws and regulations that apply to waste generation. This is followed by brief descriptions of the four general categories of waste stream treatment technologies recognized by EPA. The minimization of waste through recycling and the pollution prevention technology of source reduction are the two subjects that follow. The final part of the chapter examines two of the tools – material flow and balance diagrams – used by environmental professionals to determine the point of origin and amount of waste generated.

1-2 Historical Perspective

The concern for nature and the effect of industrial pollution are not new subjects. The philosopher and essayist Henry David Thoreau, naturalist and first president of the Sierra Club John Muir, and others voiced concerns about these issues early in the 19th century.

In 1962 biologist Rachel Carson published *Silent Spring*, a book that focused our attention on the use of pesticides and their alarming effects on natural systems. In the late 1960s events such as fires on the Cuyahoga River in Cleveland, massive fish kills in the Great Lakes, and a four million gallon oil spill from a platform off the coast of Santa Barbara in California brought the reality of environmental contamination in America into sharp focus.

Waste that for years had been put "out-of-sight out-of-mind" came clearly into the public's view. Population growth and industrialization had overwhelmed nature's ability to assimilate human waste. Finally, the first views of earth from the moon in 1969 made the nation realize the fragile nature of our small blue planet.

In 1965 the Solid Waste Disposal Act (SWDA) was passed; it was intended to turn our nation's open smoldering dumps into soil-covered sanitary landfills. In 1970 Congress amended the Act calling for a comprehensive investigation – through the administration of surveys and studies – of the nation's hazardous waste management practices. As can be seen in Table 1-1, this was the beginning of what is sometimes called the decade of environmental regulations.

Shortly after its enactment, the responsibility for administrating the amended Act was passed to the newly created EPA. However, its preoccupation with air and water pollution control laws of the time caused it to simply ignore the mandates of the 1970 amendment.

In 1975 Congress was again holding hearings to update the 1970 amended SWDA and the issues surrounding the hazardous waste management studies. By then pro-environment feelings had reached new heights in the United States. After nearly a year of hearings – on October 21, 1976 – Congress enacted the Resource Conservation and Recovery Act (RCRA), giving the United States a new statute to replace the much amended Solid Waste Disposal Act.

Year	Legislation
1970	**National Environmental Policy Act** – Federal branches or agencies proposing projects had to prepare environmental impact statements. **Clean Air Act** – Amended the original CAA passed in 1955. Made prevention and control of harmful substances being discharged into the air a national goal. **Occupational Safety and Health Act** – Protected workers from health hazards. **Resource Recovery Act** – Encouraged refuse recycling and better waste management practices.
1972	**Federal Water Pollution Control Act** (Amendment) – Became the **Clean Water Act** designed to make all U.S. waters safe for swimming by 1983. **Federal Insecticide, Fungicide, and Rodenticide Act** – Controlled use and safety for all pesticides.
1974	**Safe Drinking Water Act** – Established uniform federal drinking water standards.
1976	**Resource Conservation and Recovery Act** – Encouraged recycling and hazardous waste control. **Toxic Substance Control Act** – Evaluated and regulated both new and existing chemicals not until then regulated under any other environmental laws.
1977	**Clean Air Act** (1977 Amendment) – Established the **National Ambient Air Quality Standards** (NAAQS) and hazardous air pollutants (HAPs).
1980	**Comprehensive Environmental Response, Compensation, and Liability Act - Superfund** – Required cleanup of abandoned and uncontrolled hazardous waste sites.

Table 1-1: The decade of environmental legislation.

There was little indication initially of the sweeping effects it would eventually have on the American public and industry, but no longer could waste be dumped into rivers, the air, or on the ground. The public demanded – through government intervention – safe sanitary landfills, breathable air, and swimmable and fishable waters.

As can be seen from Figure 1-1, the number of federal hazardous substance laws passed during the 1970s is unprecedented. The focus of the nation, however, had changed from regulation of waste disposal to regulation of waste treatment prior to disposal. In the 1980s the nation again became alarmed. Many treatment technologies of the 1970s turned out only to move the pollutant from one **environmental compartment** (air, water, land, or biota) to an adjacent compartment. Toxins that were removed from the air and the waterways were being placed into or on the land. Landfills were filling fast. New landfill sites were hard to establish and the cries for recycling echoed across the land. Pleas were made to reduce, recycle, and reuse resources. Continued pressure on the government resulted in the 1984 congressional amendment to RCRA, whereupon the new federal strategy would be to make waste minimization – especially its reduction – the priority in waste management.

The 1990s brought yet a new direction. Pollution prevention (P2) became the battle cry. Why produce a waste, if its production can be prevented? Pollution prevention changed the way we looked at waste management. Disposal, treatment, and recycling are all reactive measures required after waste is produced. Pollution prevention, on the other hand, is a proactive measure that focuses on preventing the production of waste.

As the focus of waste stream management has changed from reactive **end-of-pipe** treatment technologies to front-of-pipe pollution prevention, so do the topics in this chapter progress from disposal, through pollution control, to today's pollution prevention technologies. Each year more industries are finding that pollution prevention not only pays in the form of reduced treatment costs and a smaller regulatory burden, but it also increases production economics.

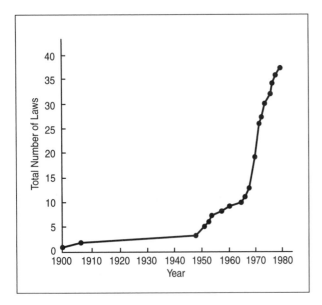

Figure 1-1: Number of federal laws enacted per year regulating the use, manufacture, transportation, and disposal of hazardous substances.

Checking Your Understanding

1. Name two historical figures that expressed concern about our treatment of the environment.

2. When EPA was created, what were two of its primary goals?

3. What is the focus of the environmental movement in the 1990s?

1-3 Waste Generation and Distribution

Since the start of the industrial revolution, industries have been plagued with all types of waste problems. Only since public pressure brought about the formation of the EPA, however, has there been any major effort by the federal government to clean up our environment and force industries to meet emission standards. Industrial emissions tend to fall into three basic categories: those affecting the air, the water, and the land.

Air

Air pollutants can be defined as any emissions that affect air quality. For regulatory purposes these pollutants are subdivided into primary and secondary categories. **Primary pollutants** are those substances that enter the atmosphere directly from a source. **Secondary pollutants** – like ozone – are new substances formed in the atmosphere by the interaction of primary pollutants and other substances.

The major primary air pollutants include such things as unburned hydrocarbons (HCs), carbon monoxide (CO), nitrogen oxides (NOx), sulfur oxides (SOx), and particulates smaller than 10 mm in diameter (PM 10). The major sources of primary air pollutants are transportation; domestic heating; electric power generation; the burning of refuse, forests, agricultural wastes, and industrial fuels; and industrial process emissions. The secondary pollutant ozone, for example, is not directly released into the atmosphere, but rather is the product of a complex set of photochemical reactions involving unburned HCs and NOx in the presence of ultraviolet light.

In passing the 1970 amendments to the Clean Air Act (CAA), Congress authorized the establishment of stringent uniform National Ambient Air Quality Standards (NAAQS). This established a third category of air pollutants, called **hazardous air pollutants (HAPs)** or air toxics. These HAPs include all materials that may be reasonably anticipated to result in health problems. Examples include such things as arsenic, asbestos, benzene, beryllium, mercury, radionuclides, radon-222, and vinyl chloride.

Water

Water pollution often presents industry with a double problem. Frequently it is necessary to treat process water prior to use as well as after use to remove contaminants. In either case, the contaminants tend to fall into three broad groups: floating materials, suspended matter, and dissolved impurities.

Floating materials include such things as greases and oils that are toxic to aquatic life and may pose fire hazards, as in the case of the fires on the Cuyahoga River. Suspended matter may be of mineral origin, such as silt and mining tailings; when suspended it reduces light penetration and when it settles on the bottom it may change the ecology of the stream. It may also be made up of organic substances that rob the water of its dissolved oxygen content. In either case, these materials can smother microorganisms and ruin fish spawning grounds. Dissolved impurities include acids, bases, pesticides, and soluble metal salts. Their effects can range from slight taste and odor problems in drinking water, to accumulation of toxic concentrations in food chains, to acute toxic conditions that result in massive plant and animal kills, as in the case of the Lake Superior fish kills.

Some of the earliest legislative attempts to control environmental discharges had to do with water pollution. The Clean Water Act of today is a complete revision of federal legislation dating back to the Refuse Act of 1899. This act, and its major modifications made in 1948, 1956, 1965, 1966, and 1970, failed to control major and increasing pollution of waterways across the United States. The much stronger amendments in 1972 mandated that the nation's waters once again become fishable and swimmable and that by 1985 discharges to waterways be eliminated. To achieve these ambitious goals, effluent limitations were set and a permitting system called the National Pollutant Discharge Elimination System (NPDES) was established. The CWA was further strengthened by the 1987 amendments, which included the protection of wetland estuaries.

Solids

Solid wastes are of two types; nonhazardous and hazardous. Nonhazardous solid wastes include such things as fly ash, coal dust, unburned mineral matter from coal, metallic foundry dusts, sand from foundry castings, and some sludges. Hazardous wastes are defined as those that are toxic, flammable, reactive, or corrosive. There are two basic laws that govern the handling and disposal of hazardous wastes. The Resource Conservation and Recovery Act (RCRA) of 1976 and the Comprehensive Environmental Response, Compensation, and Liability Act (CERCLA) of 1980. Both deal with the dangers posed to the environment by chemicals, but from two different perspectives. CERCLA is designed to remedy past mistakes, and RCRA is concerned with avoiding future mistakes through proper waste management practices.

From the above discussion it is apparent that legislative action has always been directed either toward a particular type of pollutant or to protect a specific environmental compartment. But one of the vexing problems about contamination is that it frequently crosses from one compartment into another – a process called **cross-media transfer**. Examples of cross-media transfer include the improper disposal on the soil of a water insoluble, volatile, organic

contaminant like trichloroethylene (TCE). TCE is an excellent solvent that was commonly used in cleaning and drying electronic parts and degreasing operations. Through evaporation into the air and percolation into the soil it can reach each of the other compartments and it can also incorporate into living organisms. Acid rain is another example of an air pollutant that is transferred into the water compartment – rain and lakes – and later becomes both a soil and biota contaminant through cross-media transfer.

Too often in the past waste treatment methods have actually turned out to be little more than cross-media transfers. Pollutants were removed from one environmental compartment only to be introduced into another.

Checking Your Understanding

1. What is the difference between a primary and a secondary air pollutant?

2. Name five hazardous air pollutants (HAPs).

3. Which act, RCRA or CERCLA, attempts to remedy past environmental mistakes?

4. What did the 1972 amendments to the CWA set as the nation's goals?

5. What is the term given to the movement of a contaminant from one environmental compartment into another?

1-4 End-of-pipe Waste Treatment Technologies

It is standard throughout industry to separate the end-of-pipe technologies into four waste treatment categories – **physical**, **chemical**, **biological**, and **thermal**. There are numerous examples, variations, and combinations of these methods. Some – like screening or sedimentation – are unquestionably physical. Others, such as chemical precipitation or pH adjustment are obviously chemical in nature. Biological treatment methods rely on the action of living organisms to accomplish the task. Thermal treatment methods rely on the use of heat or combustion to transform the waste by oxidation. However, not all combinations are clearly definable. There is no full agreement, for example, on whether activated carbon adsorption and ion exchange are chemical or physical processes. Incineration and bioremediation both result in chemical changes, but they have been placed in separate categories.

It should also be understood that the various types of treatment processes do not necessarily operate in isolation. It is very common to have a sequence of different treatment categories operating in tandem on a given waste stream. For example, in a wastewater treatment plant the effluent is put through a series of physical separation methods before biological methods are employed. At the end of the biological treatment, the effluent undergoes more physical treatment – and sometimes even a chemical precipitation – before it is released to the environment.

The real issue presented by end-of-pipe technology is not how it should be classified, but how to best make use of the chemical and physical properties of the pollutant to effectively concentrate, reduce, or eliminate its hazard prior to discharge. The answer is not always straightforward. Scientific knowledge must provide the foundation for the method selected, but it must also be tempered by such considerations as the economic and risk assessments as well as the regulatory and sociopolitical realities. For example, although EPA approved the use of incinerators and established stringent performance standards, some states still refuse – largely because of political pressure – to issue permits for the use of hazardous waste incinerators.

Physical Treatment Technologies

Physical treatment technologies make use of the waste's physical properties to separate and/or concentrate the waste, but do nothing to detoxify or change its chemical characteristics. Sedimentation and filtration are common examples. Depending on the pore size of the filtering medium, objects as large as bowling balls to no larger than 0.2 micron can be removed using various types of screen, fabric, or membrane filters.

Chemical Treatment Technologies

Chemical treatment methods make use of the waste's chemical characteristics to manipulate, concentrate, transform and/or detoxify the pollutant. Liquid waste streams are often subjected to methods such as neutralization, electrolysis, chemical precipitation and flocculation, oxidation and reduction, and even combustion.

Biological Treatment Technologies

Biological treatment uses living organisms to separate, concentrate, or detoxify waste, and is the most commonly employed technology. Most municipal wastewater treatment rely on aerobic, anaerobic, and facultative bacteria to reduce the amount of organic materials in wastewater. Many biological treatment technologies result in chemical transformations; however, the chemical change is performed by the metabolism of a living organism.

Use of intrinsic (native) bacteria and other microorganisms to remove unwanted hydrocarbons from contaminated soil and groundwater is also an application of biological treatment technology, which is referred to as **bioremediation**. Biological treatment methods have even been developed that can be used to remove unwanted pollutants from a gaseous waste stream.

Thermal Treatment Technologies

Finally, thermal treatment technologies involve the use of heat to separate, concentrate, or destroy waste. Incineration is undoubtedly the oldest waste treatment method used by man. Modern furnaces have names like fixed-bed, fluidized-bed, reciprocating grate and multiple hearth incinerators. Rotary kilns, secondary combustion chambers, and plasma arc incinerators are all used to destroy wastes by heating them to high temperatures.

Due to its tendency to form small amounts of harmful by-products, incineration remains the most controversial method used for the destruction of wastes. It is, however, the most effective way to reduce the waste's volume, while at the same time transforming its character.

Checking Your Understanding

1. What are the four recognized waste treatment categories?

2. What are two things that a physical treatment technology will do?

3. What is the controversy surrounding the use of thermal treatment technologies?

4. What is the term used to describe the use of intrinsic bacteria to clean contaminated soil?

1-5 Environmental Laws

Several federal laws and regulations that affect industries who generate and treat hazardous wastes were introduced earlier in this chapter. In this section and in the next chapter we will expand on some of them and include others that are directed more and more toward pollution prevention. Most laws currently on the books relate either to controlling substances as they are used, or to hazardous wastes after they are generated. Generally, these laws can be grouped into the following four broad categories.

Chemical Use, Handling, and Evaluation

In this category we find laws such as the Toxic Substances Control Act (TSCA); the Federal Insecticide, Fungicide, and Rodenticide Act (FIFRA); and the Federal Food, Drug, and Cosmetic Act (FFDCA). Each of these laws requires testing or evaluation as well as control of chemicals used in commerce.

Each law targets chemicals by their specific uses, but focuses primarily on raw materials and products, not wastes. Industrial facilities that use many raw materials and products are subject to these types of laws. In particular, the Occupational Safety and Health Act (OSHA), which regulates employee safety in the workplace, deals extensively with the hazards of working with chemicals, equipment, energy, and common hazards in industrial facilities.

By-products of Processes

This category includes laws such as the Clean Air Act (CAA), the Clean Water Act (CWA), and the Safe Drinking Water Act (SDWA). These laws protect the quality of the air and water compartments by regulating emissions, discharges, and releases of substances deleterious to these compartments. The discussion of mass balance later in this chapter will point out the importance of accounting for all emissions and discharges in industrial facilities.

Cleanup and Disposal

This category includes the Resource Conservation and Recovery Act (RCRA) and the Comprehensive Environmental Response Compensation and Liability Act (CERCLA), also known as Superfund. These two laws cover the current and past "sins" of hazardous and solid waste management as well as emergency response to contamination. Since industrial facilities generate both solid and hazardous wastes and must be prepared for emergencies involving the hazardous substances, these two laws have a tremendous impact on the average facility. RCRA, for instance, requires tracking of hazardous wastes from "cradle-to-grave," thus making accurate recordkeeping inevitable.

Transportation of Chemicals and Hazardous Substances

This last category includes the Hazardous Materials Transportation Act (HMTA), which requires specific procedures and extensive recordkeeping to properly ship hazardous substances by highway, sea, air, or rail. Every facility in this country that ships hazardous raw materials or products must be familiar with the provisions of this law to insure that its operations are in compliance.

"Pollution prevention makes economic sense. We'll save money on raw materials, we'll have less waste to dispose of, and we'll protect American citizens and our own environment."

Carol Browner, USEPA

The Pollution Prevention Act of 1990 (PPA) declared pollution prevention to be the national policy of the United States. Pollution prevention, also called source reduction, aims to both conserve finite natural resources and prevent waste and harmful substances from contaminating the environment.

In November 1994 the EPA clarified its goals for **waste minimization** by the further release of the Waste Minimization National Plan. Its two important components are 1) to ensure that certain hazardous wastes are reduced at the source prior to treatment, storage, or disposal under RCRA, whenever possible, and 2) to avoid transferring these constituents across environmental media.

Reducing the amount of waste produced at the source is the only permanent and practical solution to the myriad of problems caused by environmental pollutants. Remember, pollution prevention makes economic sense because today's generators are tomorrow's potentially responsible parties with legal liability – even when they have followed current regulations!

Checking Your Understanding

1. What is the meaning of the acronym SDWA?

2. Which act applies to employee health and safety?

3. Which two laws apply to the cleanup and disposal of hazardous wastes?

4. Which act controls the transportation of hazardous materials?

5. Which act declared pollution prevention to be this nation's policy?

1-6 Hazardous Waste Stream Management

Hierarchy of Hazardous Waste Management

As suggested earlier, prior to the passage of legislation regulating the discharge and disposal of waste, most businesses did not concern themselves with tracking or measuring the amount of waste they produced. The November 1984 amendments to RCRA – known as the Hazardous and Solid Waste Amendments (HSWA) – established a new federal strategy to make waste minimization – especially its reduction – the priority in waste management. The strategy would be accomplished through the application of the **hierarchy of hazardous waste management**.

In this hierarchy, source reduction was given highest priority followed by recycling, treatment, and disposal to the land. In short, this policy was in-tended to encourage industry to reduce hazardous waste at its source, rather than use end-of-pipe treatment and/or disposal.

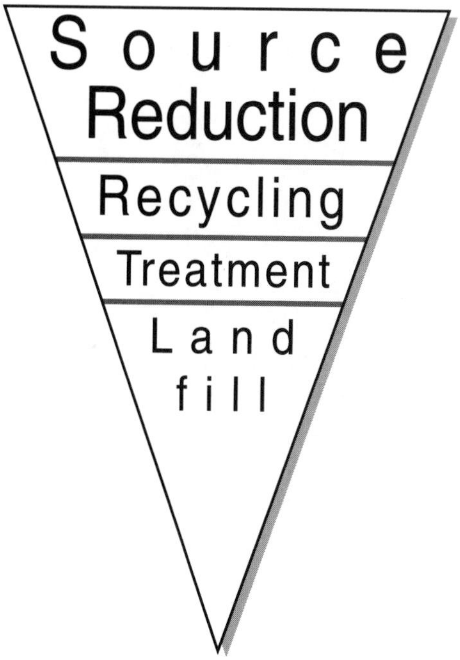

Since it is often impractical to completely eliminate the production of hazardous waste from an industrial process, reducing the volume generated lessens the environmental impact while at the same time lowering the operating costs, decreasing the complexity of waste management, and reducing the potential liability. Volume reduction can be accomplished through such strategies as inventory control, improved housekeeping, production/process modifications, product substitution or reformulation, waste segregation, and new uses for wastes.

For nearly the first twenty years (1970-1990) of its existence, EPA placed its primary focus on end-of-pipe waste treatment to reduce environmental damage. Congressional mandates now place the emphasis on preventing waste. **Minimization of waste** is an umbrella term that includes **source reduction**, recycling, and waste treatment processes. It is important, however, to understand the difference between minimization of waste versus waste treatment, recycling, and source reduction.

Figure 1-2: EPA's hierarchy of hazardous waste management.

> "Pollution prevention is not the only strategy for reducing risk, but is the preferred one. Where prevention or recycling are not feasible, treatment followed by safe disposal as a last resort will play an important role in achieving environmental goals."
>
> Carol Browner, USEPA

Minimization of waste results in reducing the weight, volume, or toxicity – to the extent feasible – of a waste prior to any off site treatment, storage, or disposal. Source reduction, on the other hand, is preventing the formation of waste. In practice, source reduction and pollution prevention have the same result. Minimization of waste can be broken down into three basic activities:

1. Source Reduction – In-plant changes that either reduce or eliminate the generation of waste. Pollution prevention means source reduction prior to production of a waste;

2. Recycling – Reuse, recovery, or reclamation of a waste stream for use as an ingredient in a production process or recovery of a reusable product; and

3. Treatment – The application of physical, chemical, biological, and thermal technologies to minimize volume, mass, or toxicity of a waste.

From an environmental perspective, source reduction is preferable to recycling and treatment. Recycling is less desirable because the waste is still generated, and the recycling process results in waste residues. Disposal is the final discharge of the waste either into the air via a smokestack; into the water through a discharge pipe leading to a river, lake, or ocean; or onto the land. This is typically regulated by government permit.

Checking Your Understanding

1. Which amendment to RCRA established source reduction as the nation's highest priority?

2. Which one of the following terms: recycling, waste treatment, minimization of waste, or source reduction includes all the others?

3. What are the four steps in the waste management hierarchy?

1-7 Tools for Waste Stream Analysis

Prior to the passage of legislation regulating waste discharge and disposal, most businesses did not concern themselves with tracking or measuring the amount of wastes they generated. There were usually no costs associated with air and liquid discharges, and only insignificant costs associated with hauling and paying local landfill fees to dispose of solid wastes.

The passage of the Clean Air Act and Clean Water Act, however, required companies to start treating most liquid and gaseous wastes prior to discharge. RCRA restricted the kinds of waste that could be disposed of in landfills. It also required local governments to design and build landfills under increasingly stringent specifications with corresponding increases in disposal charges. Generators of solid hazardous waste were required to treat it prior to disposal in the landfills. The cost for disposing of all types of hazardous waste soared. To maintain profitability, companies became concerned about identifying and quantifying the wastes they were producing.

Many tools are used by the environmental professional to evaluate and analyze the nature of waste streams: flow measurements; data on concentration and mass obtained from laboratory analysis; data on product composition; and measurements from wastewater, air, and solid waste discharges. The movement of materials through the process can be displayed on a **flow diagram**. Through the use of boxes, icons, and arrows the diagram shows all of the points where materials enter and where wastes, by-products, and products exit the process.

After a flow diagram is prepared, the amount, or mass, of each material and waste involved in each step of the process can be analyzed using **material balance** calculations. Comparing the total amount of material entering and exiting the process allows for the determination of fugitive losses or data errors.

Material Flow Diagrams

To understand the steps involved in the development of a flow diagram, an example is helpful. On the following pages, a material flow diagram and a portion of a material balance sheet will be presented for the hypothetical Kola® Bottling Company located in Somewhere, USA.

Kola's® soft drinks are sold worldwide through a chain of distributors. Its bottling operation is described as follows:

The flavoring syrup, sweetener, caps, and prelabeled plastic bottles are all purchased from outside vendors, according to Kola's® specifications.

At the Mixing Station the flavoring, water, and sweetener are mixed. At the Bottling Station, the mixture is injected into bottles. Just before the cap is applied, pressurized carbon dioxide – CO_2 – is injected to make the mixture effervescent.

The capped bottles then move to the Packaging/Shipping Station where they are placed in six-packs, and then four six-packs are put into a case. The cases of Kola® are then shipped to the soft drink distributors.

Oregon requires all bottling companies to pay a refund on returned empty bottles. Bottles collected from its network of distributors are returned to Kola® for recycling. At the Receiving/Sorting Station, the bottles are sorted to remove trash, sent to the Shredding Station to be shredded, and then sent to the Loading Station to be sent by rail to a furniture manufacturer, who converts them into upscale patio furniture.

A simple flow diagram for Kola's® bottling operation is presented in Figure 1-3.

The flow diagram is constructed to depict the major processes occurring in the plant. For clarity the processes have been divided into two groups, each depicting one of the major activities performed at the plant: soft drink production and bottle recycling. Groups of processes that perform a major function are often called **process circuits**. In this example, the bottling circuit consists of Kola® mixing, bottling, CO_2 injection, and packaging/shipping. The bottle recycling circuit consists of empty bottle receiving/sorting, shredding, and loading/shipping. Arrows ($\rightarrow$) are used to show the route materials take through the circuit, and points at which raw materials enter the process. This flow diagram is a qualitative diagram. The term qualitative is used because it identifies the materials entering and leaving the plant, but it does not specify the quantity of those materials.

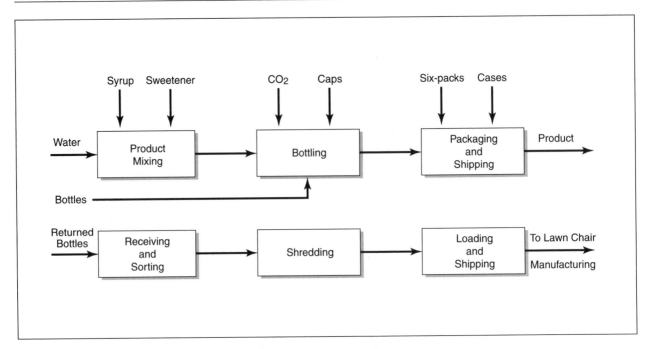

Figure 1-3: Simple flow diagram for Kola's® bottling and return operations.

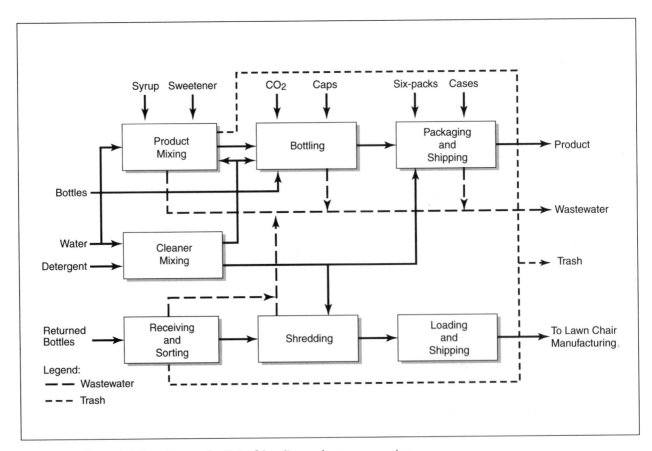

Figure 1-4: Expanded flow diagram for Kola's® bottling and return operation.

The steps involved in this example are few when compared with the steps in an oil refinery, pulp and paper mill, or nuclear power plant. Obviously, more complicated processes require more complicated diagrams than the one presented here. Although the complexity of flow diagrams varies, the purpose for constructing the diagram remains the same: to depict, as simply as possible, the steps in the process and the flow of material from one step to the next.

What Figure 1-3 does not show are all of the other processes and materials that are used or generated in operating the plant. For example, federal and state regulations require that mixing equipment, bottling machines, packaging machines, recycling equipment, pipe lines, conveyors, floors, walls, etc. be regularly sanitized. Occasionally, defective caps, bottles, paperboard six-packs and cases, or off-spec sweeteners and syrups are received from the suppliers. These defective materials may be identified before they enter the process, or they may be used and fail during the process. In either case, the operation results in the generation of waste. Also, the bottles being returned to the site for recycling often contain non-recyclable trash and residue in the form of unconsumed liquid. All of these materials must be accounted for if a flow diagram is to be used later to conduct a waste audit. These additional materials have been included on the flow diagram shown in Figure 1-4.

As can be seen, the intended use of a flow diagram determines the amount of detail it must contain. If its purpose is to simply track raw materials for inventory control purposes, then the flow diagram depicted in Figure 1-3 is adequate. If the diagram is to be used to identify and track process wastes, then it must include the detail included in Figure 1-4. If, however, the intent is to identify and track all of the wastes produced by a facility, then neither diagram is adequate. To be complete, Figure 1-4 would have to include all of the other wastes, such as trash generated by the custodial and front office staff, and even the parking lot run-off.

Such a detailed material flow diagram is a useful tool for the environmental technician, because it identifies the areas, processes, and types of wastes generated, as well as the raw materials from which the wastes were generated. This information is essential in determining:

—what processes produce waste;

—what chemicals may be in the waste;

—what type of permit or regulations apply to the waste;

—what process machines may need repair or replacement; and

—what raw material substitutions would result in reduced waste or complete determination of it.

Such diagrams provide a template for conducting environmental audits and for gathering the plant's waste inventory.

Material Balance

Law of Conservation of Mass

Mass In = Mass Out

After a detailed material flow diagram has been constructed, material balances can be determined. Material balances are the calculations that compare the amounts of material entering and exiting the process. These calculations are firmly rooted in the scientific principle known as the Law of Conservation of Mass. Simply stated, the sum of the weights of all materials entering and exiting the plant must be equal.

If the calculations demonstrate a material balance in which the mass entering equals the mass exiting, then all raw materials, products, and wastes at the plant have been identified. If, however, calculations fail to demonstrate equal masses in and out, then either a mistake has been made in the measurements and calculations, or some raw material, waste, or product has not been identified. Since raw materials are typically purchased and products are sold, their quantities are usually well known and easily documented. When an imbalance occurs, therefore, it is usually due to faulty measurements or a fugitive waste stream that has not been identified.

Because of the details included in the Figure 1-4 flow diagram, it can be used to perform a material balance calculation for the Kola® Bottling plant. Consider the following additional information:

Each of Kola's® 12 fl. oz. bottles contains 1 fl. oz. syrup, 2 fl. oz. of sweetener, and 9 fl. oz. of water.

Daily Detergent/Disinfectant Usage	
Mixing/Bottling Equipment	300 gal
Receiving/Sorting Area	50 gal
Packing/Shipping Area	25 gal
Shredding Area	25 gal
Mixture TOTAL	**400 gal**

The mixing equipment and bottling machines are cleaned after each of the three shifts per day. The floors, walls, and remaining equipment are cleaned at the end of each production day. The cleaner used is a low sudsing detergent/disinfectant. It is mixed according to a formula using one part detergent to three parts of water. Each day 400 gallons of the cleaning mixture are required to clean the facility and equipment.

Item	Quantity
Water Consumption	5,925.0 gal
Bottles of Kola® Produced	80,000 bottles
Bottles Used	81,500 bottles
Caps Used	81,400 caps
Weight of Bottles Returned for Shredding	20,000 lbs
Weight of Shredded Bottles Shipped	18,800 lbs
Weight of Trash from Returned Bottle Area	150 lbs
Detergent Concentrate Consumed	100 gal
Kola® Syrup Used	634.4 gal
Sweetener Used	1,268.8 gal
Six-Pack Cartons Used	13,500 cartons
Case Cartons Used	3,350 cases

Table 1-2: Daily inventory for Kola® bottling plant.

Potable water is obtained from the local water district. The company records its daily water usage. The bottling and packaging machines record the number of bottles, six-packs, and cases entering each machine each day. At the end of a normal production day the readings obtained are indicated in Table 1-2.

Using the material flow diagram developed for Figure 1-4, a new diagram can be created (Figure 1-5) that includes the quantities of materials entering and leaving each of the various steps in the daily process. With this information it should be possible to determine if the material flow diagram has accounted for all of the facility's wastes. This is accomplished by comparing the amounts of all the raw materials entering the plant, and the wastes and product leaving the plant. Table 1-3 has been prepared to help organize the information.

As careful examination of Table 1-3 reveals, the material balance calculation cannot be completed

Materials In	Quantity	Materials Out	Quantity
Kola® Syrup Used	634.4 gal	Product	80,000 bottles
Sweetener Used	1,268.8 gal	Shredded Plastic	18,800 lbs
Bottles, New	81,500 bottles	Liquid Waste	???
Caps	81,400 caps	Trash	???
CO_2	Insignificant		
Six-Pack Cartons Used	13,500 cartons		
Case Cartons Used	3,350 cases		
Returned Bottles	20,000 lbs		
Detergent Concentrate	100 gal		
Water	5,925.0 gal.		
TOTALS			

Table 1-3: Daily materials balance tabulation.

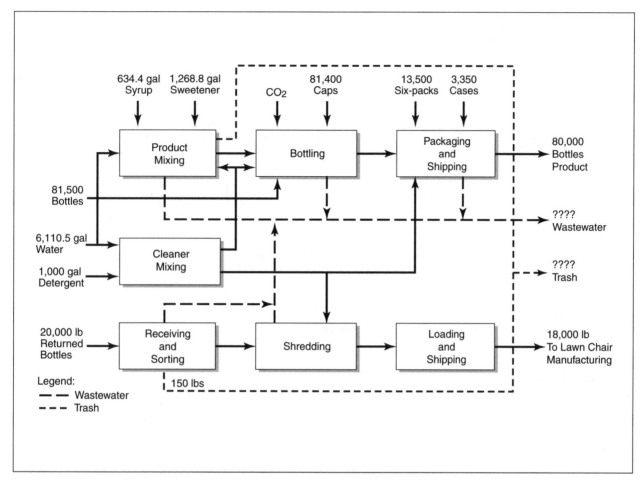

Figure 1-5: Quantities of materials entering and leaving the bottling plant.

for either the liquid or solid waste streams because the amount for several of the items is not known (see question marks). It is not an uncommon situation for an environmental technician to encounter! However, by using known information for items isolated in "sub-circuits," missing information can still often be determined.

Consider, for example, just the liquids entering and leaving the plant daily. This part of the flow diagram has been isolated in Figure 1-6.

To make comparisons and perform calculations, the quantities need to have common units. It is also helpful to place that information into a table. Table 1-4 has been constructed to help identify incoming and exiting liquids, and do the necessary calculations to convert all measurements to common units.

Note that the data contained in Table 1-4 still contains two unknowns. It will not be possible to

determine the amount of liquid waste produced by the plant, unless we can first determine how much waste liquid is contained in the returned bottles. This can be accomplished by performing a material balance just around the receiving/sorting station in the return bottle circuit. Figure 1-7 is a flow diagram for the receiving/sorting station circuit that contains weights of material entering and leaving the station.

The daily material balance for just the receiving/sorting station is given in Table 1-5.

The 18,800 lbs. of shredded bottles and the 150 lbs. of trash leaving this station are solids. They enter the plant as part of the 20,000 lbs. of plastic bottles and liquid waste shown on the flow diagram. The remainder of these 20,000 lbs. entering the plant is the liquid waste. Therefore, we can perform the calculations shown in Table 1-6.

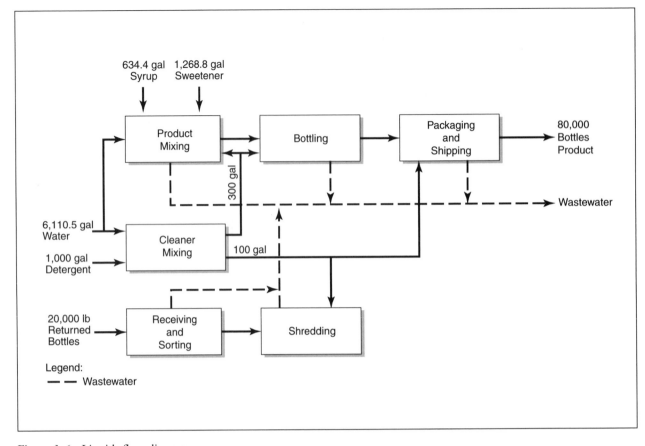

Figure 1-6: Liquids flow diagram.

Liquid In	Quantity (gallons)	Liquid Out	Quantity (gallons)
Water	5,925.0	Kola® 80,000 btls × 12 fl. oz./btl × 1 gal/128 fl. oz.	7,500.0
Kola® Syrup	634.4	Plant Liquid Waste	???
Sweetener	1,268.8		
Detergent	100.0		
Returned Liquid in Bottles	???		
TOTALS	???		

Table 1-4: Daily quantities of liquids in and out of the plant.

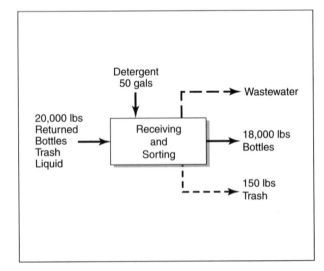

Figure 1-7: Material balance for receiving/sorting station.

Materials In	Quantity	Materials Out	Quantity
Returned Bottles and Liquid	20,000 lbs	Trash	150 lbs
Cleaning Solutions	50 gal	Cleaning Solutions	50 gal
		Bottles, Shredded	18,800 lbs
		Returned Liquid in Bottles	???

Table 1-5: Daily materials balance for receiving/sorting station.

Materials In	Quantity	Materials Out	Quantity
Returned Bottles and Liquid	20,000 lbs	Trash	150 lbs
		Bottles, Shredded	18,800 lbs
Totals	**20,000 lbs**	**Totals**	**18,950 lbs**
		Returned Liquid in Bottles (Total In – Total Out)	**1,050 lbs**
		Volume of Returned Liquid in Bottles 1,050 lb × 16 fl. oz./lb × 1 gal/128 fl. oz. =	**131.3 gal**

Table 1-6: Daily materials balance for solids at the receiving/sorting station.

Substance	IN	OUT	
	Liquid	Kola®	Waste
Detergent 400 gal × 1 part/4 parts	100.0 gal	0	100.0 gal
Liquid Waste in Returned Bottles (From Table 1-5)	131.3 gal	0	131.3 gal
Syrup 1 fl. oz./bottle × 80,000 bottles × 1 gal/128 fl. oz.	634.4 gal	625 gal	9.4 gal
Sweetener 3 fl. oz./bottle × 80,000 bottles × 1 gal/128 fl. oz.	1,268.8 gal	1,250 gal	18.8 gal
Subtotals	**2,134.5 gal**	**1,875 gal**	**259.5 gal**
Total Water Purchased	5,925.0 gal		
Water in Kola® 9 fl. oz./bottle × 80,000 bottles × 1 gal/128 fl. oz.		5,625.0 gal	
Water in Cleaner 400 gal × 3 parts/4 parts			300 gal
Total Liquids	**8,059.5 gal**	**7,500.0 gal**	**559.5 gal**
Material Balance	**8,059.5 gal**	**8,059.5 gal**	

Table 1-7: Liquid mass balance calculation.

As can be seen from the last calculation, 131.3 gallons of liquid waste entered the plant as a part of the weight of the returned bottles. We have now accounted for all liquids entering and leaving the plant. With this calculated information, we can now complete the two previous tables by adding the missing information.

In both Table 1-4 and 1-5, the amount of liquid returned with the bottles is 131.3 gallons. By adding this information to Table 1-4, it is now possible to calculate the sum of all the liquids entering the plant. Since the sum of all liquids entering and exiting the plant must be equal, then the total of all liquids exiting must also be 8,050.5 gallons.

Since the volume of liquid contained in 80,000 bottles of Kola® is 7,500 gallons, the difference of 559.5 gallons must be the plant's total liquid waste. The material balance for only the liquid portion of the flow diagram has now been completed. To further analyze the liquid flow portion of the diagram, and the generation of the liquid wastes, Table 1-7 has been prepared.

Developing material flow diagrams and calculating the material balances provides companies with an accurate picture of the overall waste production. This information is essential for developing strategies to eliminate, reduce, recycle, or dispose of waste. When considering the company's profitability, the most obvious waste management strategy is waste avoidance. If waste cannot be completely avoided, then reduction or resource recovery, through reuse and recycling, remain as valuable alternatives. If use of these alternatives still results in waste, then end-of-pipe treatment may be employed. This can include any single or combination of biological, chemical, thermal, or physical treatment technologies – the topics of the following chapters.

Summary

The laws passed by Congress and implemented by EPA in the 1970s resulted in an initial focus on attempts to control the disposal of wastes into the environment. This focus underwent a gradual evolution resulting in the passage of the Pollution Prevention Act of 1990, which set pollution prevention as a national policy. In the early "command-and-control" era, a large number of federal laws and regulations were passed to control the practices of industry. The significant laws can be subdivided into those that affect chemical use, handling, and evaluation (TSCA, FIFRA & FFDCA), by-products of processes (CCA, CWA & SDWA), cleanup and disposal (RCRA & CERCLA), and transportation (HMTA).

Many of these laws mandated industry to account for the amount of waste it was generating. Two tools that are commonly used for this task are the materials flow diagram and mass balance calculations.

To deal with the identified waste streams, EPA recognized four different treatment technologies – physical, chemical, thermal, and biological. Recycling, reuse, recovery, and new uses for the wastes were also encouraged. As a result of the passage of the PPA, the emphasis shifted away from end-of-pipe treatment methods and attempts to recover and reuse wastes, to the front-of-pipe practices that prevent the formation of pollution. The role of pollution prevention is, therefore, to focus on the significant opportunities that industry has to reduce or prevent pollution at the source, through cost-effective changes in production, operation, and raw material usage.

Checking Your Understanding

1. What type of diagram shows the movement of materials through a process or plant?

2. What two masses must be equal in a mass balance calculation?

3. What is a process circuit?

4. Can you complete Table 1-4 using the calculated information from Table 1-5?

Critical Thinking Questions

1. What is the difference between pollution prevention and minimization of waste activities?

2. What is EPA's rationale for its hierarchy of hazardous waste management?

3. In what significant aspects do the four recognized EPA treatment technologies differ?

4. Balance Calculation Problem

Using the information provided here, as well as the flow diagrams and data presented earlier in the chapter, complete the following tables and mass balance for Kola® Bottling Company's solid waste stream.

By conducting a solid waste audit and taking a number of measurements, Kola's® environmental technician determined that empty 12 fl. oz. plastic bottles weigh 0.9 oz., the bottle caps weigh 0.3 oz., and both the 5 gallon syrup and 5 gallon sweetener containers weigh 25 oz. each. It was also determined that the paperboard six-packs and cases weigh 3.2 oz. and 12.5 oz., respectively. It was also learned that Kola® receives its detergent/sanitizer concentrate in metal containers that are returned to the manufacturer, resulting in no solid waste generation at the Cleaner/Mixing Station.

Weight of Kola® Packaging Materials	
Bottles, 12 fl. oz.	0.9 oz.
Syrup Container, 5 gal.	25.0 oz.
Sweetener Container, 5 gal.	25.0 oz.
Bottle Caps	0.3 oz.
Paperboard Six-packs	3.2 oz.
Paperboard Cases	12.5 oz.
Bottle and Cap	1.2 oz.

Help the technician by completing the following tables and calculating the amount of solid waste being produced by Kola® on a daily basis.

Mass of Solids Entering/Exiting by Station			
Station	Number	Weight Each	Total
Bottles Entering			
Syrup Containers, 5 gal.			
Sweetener Containers, 5 gal.			

Mass of Solids Entering/Exiting by Station (cont.)			
Station	Number	Weight Each	Total
Caps			
Six-packs			
Cases			
Bottle and Cap of Kola®			
Trash	???	???	????
Shredded Plastic Shipped			
Returned Plastic & Trash Received			

Solids Mass Balance		
Station	Mass In	Mass Out
Bottles Entering		
Syrup Containers, 5 gal.		
Sweetener Containers, 5 gal.		
Caps		
Cases		
Bottle and Cap of Kola®		
Trash		???
Shredded Plastic Shipped		
Returned Plastic & Trash Received		
Totals for Mass In and Mass Out		

Do the Mass In and Mass Out columns equal? _____ Yes _____ No

How many pounds of Trash would have to be added to make them equal? _____ lb.

After analyzing the two (liquid and solid) material balances, if you were Kola's® environmental technician what minimization of waste strategies would you recommend, and why?

2

Waste Stream Regulations

by Ann Boyce

Chapter Objectives

Upon completing this chapter, the student will be able to:

1. **Describe** basic requirements of the environmental, health and safety, and transportation laws that affect the operations of an industrial facility.

2. **Explain** how each of these laws may impact industrial facilities.

3. **Recognize** the value of pollution prevention and waste minimization and the advantages of being classified as a small quantity generator.

4. **List** ways to acquire more information on current and proposed laws and regulations.

Chapter Sections

2-1 Introduction

2-2 Chemical Use, Handling, and Evaluation Laws

2-3 Laws Relating to By-products of Processes

2-4 Cleanup and Disposal of Chemicals

2-5 Transportation of Chemicals and Hazardous Substances

2-6 Getting Information on Environmental, Health and Safety, and Transportation Laws

2-1 Introduction

In the first chapter we learned that starting in the 1970s a large number of laws were enacted to control the day-to-day operations and waste stream disposal activities of industrial facilities. In this chapter we will look at each of these laws and explore in some detail how their provisions impact various types of industrial operations. In general, the activities of industry that are most highly regulated include: the use, handling, and storage of chemicals; air emissions; oil spills; effluent discharges; underground injection; hazardous waste management; site clean-up; emergency planning and notification; chemical testing and reporting; employee safety; and hazardous materials transportation. We shall also see how pollution prevention (P2) activities decrease the regulatory burden of most of these laws on industry. Lastly, we will learn how to locate current information regarding laws, regulations, and governmental policies affecting environmental, health, and safety practices. In order to increase the reference value of this chapter, the citation for implementing regulations is provided in brackets after the titles.

2-2 Chemical Use, Handling, and Evaluation Laws

Toxic Substances Control Act (TSCA) – [40 CFR 700-799]

Imagine you are head of a company that has just discovered a revolutionary new chemical that, when used as a key ingredient in a furniture stripper, will remove most finishes with ease and effectiveness – unlike any other on the market. Your staff has also begun to explore other potential uses for this marvelous chemical. You know this substance is a real money-maker and are eager to get it on the market, but you will have to wait – because the **Toxic Substances Control Act (TSCA)** will not allow you to produce and distribute the product until it has been properly evaluated for its effect on human health and the environment.

TSCA, which was enacted in 1976 and is administered by the EPA, has four major goals:

1. Screen new chemical substances to determine if they pose a risk;

2. Require testing of chemical substances that are initially identified by the screening process for possible risks; require testing of chemicals that are on the market, but which new data suggest may be a risk;

3. Gather information on existing chemicals;

4. Control chemicals, both new and existing, that are determined to pose a risk.

Screening New Chemicals

The provisions of TSCA that require the screening of new chemicals are known as **premanufacture notification (PMN)** requirements. With few exceptions, companies intending to manufacture a new chemical must notify EPA at least 90 days before manufacture. This notification must include basic information about the chemical and available data on its risk to public health.

The exceptions to the PMN requirements include: chemicals not intended to enter commerce, those the EPA determines do not present an unreasonable risk, chemicals that are manufactured in very small annual volumes, and those covered by other chemical assessment and control laws. An exemption is made, for instance, for chemicals used only for research and development or produced in volumes of less than 1,000 kilograms per year. Examples of other chemical assessment and control laws are the Food, Drug, and Cosmetic Act and the Federal Insecticide, Fungicide, and Rodenticide Act. Chemicals regulated by these acts are exempt from the provisions of TSCA.

All chemicals manufactured after July 1, 1979 are considered new chemical substances for the purpose of PMN. Chemicals manufactured and used in commerce before this date were used to build an inventory of chemical substances. New chemicals are automatically added to this inventory as they complete PMN and begin to be manufactured.

A complete PMN must contain the following:

1. The name of the substance (this can be the common or trade name);

2. The chemical identity and molecular structure;

3. The estimated production levels, typically for the first three years of production;

4. Proposed use(s) of the chemical;

5. Method(s) of disposal of the chemical;

6. Estimated levels of exposure in the workplace and the number of workers who potentially could be exposed;

7. A description of by-products, impurities, and other related products;

8. Available test data on health and environmental effects related to manufacture if the manufacturer has possession and control of these data;

9. A description of available test data.

During EPA's 90-day review period, the risks posed by the chemical are evaluated by conducting a literature search and looking at the effects caused by chemicals of similar composition. EPA is required by TSCA to take regulatory action if it has information that the chemical presents or will present "…a significant risk of serious or widespread harm to human beings from cancer, gene mutations, or birth defects." If the EPA review concludes that the chemical presents unreasonable risks, it can take regulatory action through the courts and rulemaking process to ban its manufacture. Administratively, it can issue an order banning production until more information is provided.

If the EPA review concludes that there are no unreasonable risks, the manufacturer is notified that it can begin production of the chemical, and the substance is added to the growing inventory of TSCA chemical substances. Once approved, a chemical can undergo additional review if its use changes. The changes in use may require that certain restrictions be placed on the chemical under the "significant new use" rule (SNUR). These restrictions may include disallowing its use in consumer products or requiring additional precautions such as personal protective equipment (PPE) or engineering controls when it is used.

Testing

EPA can require testing of new chemicals and chemicals already on the market. This testing requirement can be imposed under two circumstances:

1. If the chemical may pose an unreasonable risk to public health or the environment and proper evaluation cannot be conducted due to insufficient data;

2. If large quantities of the chemical are capable of entering the environment or of coming into contact with humans in substantial amounts, and data to evaluate its effects are insufficient.

Testing involves scientifically evaluating the chemical to determine whether it poses health or environmental hazards. In order to require testing, the EPA must show that existing data are inadequate to evaluate its risk. Testing is a long, involved, and costly process that must be done in accordance with specific guidelines published by EPA.

Gathering Information

EPA has the authority to require chemical manufacturers and processors to collect, record, and/or submit information regarding their chemicals. Since TSCA mandates the evaluation of chemical risk, it is imperative for EPA to have access to the information it needs to make proper determinations.

For instance, the manufacturer of a newly approved chemical can be required to provide additional data. EPA may direct the manufacturer to provide reports on effects of the chemical, exposure levels experienced in the workplace or environment, uses, or production levels in order to increase confidence in the chemical data.

EPA can impose these data gathering requirements on manufacturers or processors of chemicals it is considering for regulation because EPA suspects they may pose hazards. They may require that records be kept for up to 30 years.

Manufacturers, processors, and distributors have an additional obligation to report "substantial risks" to EPA within 15 days. Examples of "substantial risks" are studies indicating that a chemical may have a carcinogenic or mutagenic effect, or the aftermath of spills or releases indicates harmful effects not previously known.

Impact of TSCA on Industrial Facilities

PMN/Testing – If a company plans to manufacture a new chemical for a specific process, or use an existing chemical in a new application, it must expend time, effort, and money to meet the PMN and potential testing requirements of TSCA. Testing, in particular, is very expensive and may take years to complete. Therefore, the substitution of a less hazardous material – a pollution prevention (P2) strategy – will pay off doubly because of reduced risk and avoiding the cost of TSCA's testing requirements.

Banning and Phase-out of Certain Chemicals Under TSCA – Under TSCA's chemical regulation provisions, chemicals in common industrial use can be banned. Examples include polychlorinated biphenyls (PCBs), chlorofluorocarbons (CFCs) used as aerosol propellants, easily crumbled asbestos in schools, and certain substances in metalworking fluids.

The provisions of TSCA were used by EPA to ban the manufacture of PCBs and require the phaseout of existing uses of the chemical. TSCA also enabled the EPA to ban the manufacture and processing of CFCs for use as aerosol propellants. This action was taken because CFCs are believed to deplete the stratospheric ozone layer.

In the late 1980s further concern about the effects of CFCs on the stratospheric ozone as well as their contribution as greenhouse gases led to the U.S. ratification of the Montreal Protocol, which went into effect on January 1, 1989. This treaty called for freezing production of CFCs at 1986 levels. Although TSCA could have been used as a means to control CFCs, it was decided that since CFCs are air emissions, they could be controlled under the Clean Air Act.

TCDD, better known as dioxin, represents an example of the use of TSCA authority to regulate a chemical where an imminent hazard exists. In the late 1970s, it was discovered that dioxin was in waste generated during the manufacture of certain herbicides. EPA was very concerned about disposal of these wastes and starting in 1980 required that it be notified by facilities 60 days before such disposal occurred. EPA's justification was that disposal of these highly toxic wastes in municipal landfills or incinerators could pose a serious and imminent health hazard. Currently, the disposal of dioxins is covered under RCRA.

Recordkeeping and Reporting – The strict requirements for reporting and recordkeeping related to toxic chemical substances impact many industrial operations. Importers, manufacturers, or processors of chemical substances identified by EPA must report information on production volume, environmental releases, and/or chemical uses every four years. Small businesses are exempt from these requirements but are still required to report this information in some situations. This information is used by EPA to develop the inventory (the TSCA Inventory) of all chemical substances manufactured or processed in the United States. Table 2-1 is an excerpt from the 1985 Edition of the TSCA Chemical Inventory.

27642-27-9	Benzamide, 4-hydroxy-N-methyl- $C_8H_9NO_2$
27668-52-6	1-Octadecanaminium, N,N-dimethyl-N-[3-(trimethoxysilyl)propyl]-,chloride $C_{26}H_{58}NO_3Si.Cl$
27676-62-6	1,3,5-Triazine-2,4,6(1H,3H,5H)-trione, 1,3,5-tris[[3,5-bis(1,1-dimethylethyl)-4-hydroxyphenyl] methyl]- $C_{48}H_{69}N_3O_6$
27685-51-4	Cobaltate(2-), tetrakis(thiocyanato-N)-, mercury(2+) (1:1), (T-4)- $C_4CoN_4S_4.Hg$
27685-83-2	Benzamide, N,N-1,5-anthraquinonylenebis[2,3,4,5-tetrachloro- $C_{28}H_{10}C_{18}N_2O_4$
27686-84-6	1,2-Benzenediol, 4,4-(2,3-dimethyl-1,4-butanediyl)bis-, (R*,S*)- $C_{18}H_{22}O_4$
27690-02-4	1,5-Anthracenedisulfonic acid, 9,10-dioxo-, dipotassium salt- $C_{14}H_8O_8S_2.2K$
27692-91-7	1,2-Ethanediamine, N-ethyl-N,N-dimethyl-N-phenyl- $C_{12}H_{20}N_2$
27697-00-3	1,2-Benzenedicarboxylic acid, mono[2-[(2-methyl-1-oxo-2-propenyl)oxy]ethyl]ester $C_{14}H_{14}O_6$
27697-51-4	Ethanaminium, N,N,N-trimethyl-, chloride $C_5H_{14}N.Cl$
27790-37-0 XU	Sulfuric acid, Potassium tin (2+) salt (2:2:1) $H_2O_4S.K. 1/2Sn$
27791-59-9 XU	2-Propenoic acid, polymer with N,N-methylenebis[2-propenamide] and 2-propenamide $(C_7H_{10}N_2O_2.C_3H_5NO.C_3H_4O_2)_x$
27791-81-7 XU	2-Propenoic acid, 2-methyl-, polymer with butyl 2-propenoate, methyl 2-methyl-2-propenoate and 2-propenoic acid $(C_7H_{12}O_2.C_5H_6O_2.C_3H_4O_2)_x$
27794-93-0	Phosphonic acid, [nitrilotris(methylene)] tris-,potassium salt $C_3H_{12}NO_9P_3.xK$
27796-66-3 XU	Diphosphoric acid, monammonium salt $H_4O_7P_2.H_3N$
27813-02-1	2-Propenoic acid, 2-methyl-,monoester with 1,2-propanediol $C_7H_{12}O_3$
27816-23-5 P	2-Propenoic acid, 2-cyano-, 2-methoxyethyl ester $C_7H_9NO_3$
27822-88-4	2-Oxazolinium, 2-[p-(dimethylamino)styryl]-4-(hydroxymethyl)-3,4-dimethyl-,iodide $C_{16}H_{23}N_2O_2.I$
27829-72-7	2-Hexenoic acid, ethyl ester, (E)- $C_8H_{14}O_2$
27831-63-6	Benzenesulfonic acid, 2,4-dihydroxy-3,5-bis[(p-nitrophenyl)azo]- $C_{18}H_{12}N_6O_9S$

Table 2-1: An excerpt from the 1985 edition of the TSCA Chemical Inventory.

Records of significant adverse effects on health or the environment must be kept as well. Any manufacturer, importer, processor, or distributor of a chemical substance or mixture who obtains information that a substance presents a substantial risk of injury to health or the environment must promptly report this information to EPA.

Companies that plan to export chemical substances regulated under TSCA must notify EPA within seven days of export. This notification is required for each country each calendar year, starting with the first export. Importers of chemical substances must comply with an import certification requirement of the U.S. Customs Service authorized by TSCA.

Federal Insecticide, Fungicide, and Rodenticide Act (FIFRA) – [40 CFR 162-180]

If you were a manufacturer planning to produce a new pesticide that is intended to kill or control bacteria, fungi, insects, rodents, weeds, or other living organisms, your proposal would not fall under the purview of TSCA but of FIFRA. The two laws have similar goals of requiring evaluation, control, and information gathering on chemical substances. However, the intended use of a pesticide is to kill organisms, which puts it into a different category of hazard from those regulated under TSCA.

Considering the definition of a pesticide, it is easy to imagine how many types of chemicals this law regulates. For instance, disinfectants intended to kill or retard the growth of bacteria in bathrooms, hospitals, on clothing, and other materials, are regulated under FIFRA. However, if the chemical's use involves killing microorganisms within humans or animals, as is the case with antibiotic drugs, then it falls under the purview of the Food and Drug Act (FDA).

Under the provisions of FIFRA, a company must register a new pesticide with EPA before beginning production. This process involves submitting test results, information on proposed uses, and suggested labeling. FIFRA also requires accurate labeling of the product and descriptions of the protective measures to be used by workers who use the product. As with TSCA, FIFRA enables EPA to ban, restrict, and control pesticides.

Impact of FIFRA on Industrial Facilities

The impact of FIFRA on industrial facilities parallels that of TSCA, but applies only to facilities specifically involved in the manufacture, formulation, or application of pesticides. These facilities are required to gather data, report to EPA, and train workers.

Occupational Safety and Health Act (OSHA) – [29 CFR]

The goal of the Occupational Safety and Health Act of 1970 (OSHA) is to ensure safe and healthful working conditions for workers in firms with greater than 10 employees. The Occupational Safety and Health Administration within the Labor Department administers the law by developing and enforcing regulations. A number of provisions within the OSHA regulations in 29 CFR affect a broad range of industrial facilities. They include: General Industry Safety Standards, Hazard Communication, Hazardous Waste Operations and Emergency Response, and Confined Space Entry.

General Industry Safety Standards

While the OSHA regulations have specific parts for industries such as construction, shipyards, marine terminals, and longshoring, all other industries fall under 29 CFR Part 1910, the General Industry Safety Standards. These standards regulate a broad range of industries and cover a multitude of workplace health and safety hazards including fire suppression, blood-borne pathogens, asbestos, respiratory protection/PPE, air contaminants, chemical hazards, lockout/tagout, and control of physical hazards such as moving parts, noise, radiation, and explosions. Several of the key standards applicable to general industry will be highlighted in this section.

Hazard Communication Standard

This standard requires employers to assess the hazards of all chemicals in the workplace and inform workers of the hazards associated with the chemicals in their work areas. In addition to hazard assessment, the standard requires labeling of all containers, making **Material Safety Data Sheets (MSDS)** available to all employees, employee training, and extensive recordkeeping.

Hazardous Waste Operations and Emergency Response

This standard is designed to protect workers engaged in hazardous waste operations and emergency response (HAZWOPER) activities. This requires implementation of programs in each of the following areas: training, health and safety, medical surveillance, decontamination, materials handling, and emergency response.

For the purposes of this standard, OSHA defines hazardous waste operations as:

— Cleanup operations required by a government entity (federal, state, local) involving hazardous substances that are conducted at "uncontrolled hazardous waste sites." Uncontrolled hazardous waste sites are defined as areas where an accumulation of hazardous waste creates a threat to the health and safety of individuals or the environment or both;

— Corrective actions involving cleanup operations at sites covered by RCRA;

— Voluntary cleanup operations at sites recognized by government entities as uncontrolled hazardous waste sites;

— Operations involving hazardous wastes that are conducted in treatment, storage, and disposal facilities; and

— Emergency response operations related to release, or substantial threats of release, of hazardous substances.

Required training must enable employees to perform their assigned tasks and functions in a safe and healthful manner that prevents endangering themselves or other employees. Employees at a site covered by HAZWOPER regulations must have appropriate training before being allowed to work onsite. For general site workers the standard requires a minimum of 40 hours of instruction offsite and a minimum of three days (24 hours total) of actual field experience under the direct supervision of a trained supervisor. Workers onsite only occasionally must receive a minimum of 24 hours of offsite instruction and one day (eight hours) of onsite supervised instruction. Supervisors must receive an additional eight hours of specialized training before supervising workers. In addition, all employees with this type of training must receive eight hours of refresher training annually to be acquainted with new training examples, incidents, and relevant topics.

Employers covered under this standard must develop and implement a written plan addressing health and safety for onsite employees who are involved in hazardous waste operations. This program must be designed to identify, evaluate, and control safety and health hazards at the facility and provide for emergency response procedures.

A medical surveillance program helps assess and monitor the health and fitness of employees working with hazardous substances. Medical surveillance is required for all employees who: 1) are exposed above **Permissible Exposure Limits (PELs)** for more than 30 days, 2) wear respirators for more than 30 days a year, 3) are injured due to overexposure, or 4) are members of a HazMat team (emergency responders). Medical surveillance records must be kept for a minimum of 30 years. Employers must inform each employee of the existence of these records and make them available upon written request. Table 2-2 lists some PELs for common industrial materials.

Medical examinations are required:

—As a baseline before beginning work;

—Annually or biannually;

—When determined by a specific exposure (e.g. involved in a high-risk exposure); and

—As an exit exam when leaving the company or changing job assignments.

Employers are required to develop, communicate, and implement decontamination procedures before any equipment or employees are allowed to enter areas at a facility where the potential exists for exposure to hazardous substances. During decontamination procedures large volumes of waste can be, and generally are, generated. Therefore:

—Each drum or other container must be inspected for leaks, damage, and proper labeling prior to handling;

—Damaged drums or containers must have their contents emptied into an appropriate container;

—Procedures must be used that minimize the risks involved in opening, sampling, or moving drums or containers.

The emergency response component of this standard is designed to prepare for and handle anticipated emergencies. An Emergency Response Plan must be written to include: pre-emergency planning, personnel roles, emergency recognition, evacuation routes, decontamination procedures, emergency medical treatment, personal protective equipment (PPE), emergency equipment, and procedures for reporting incidents. The employer is not required to train all employees in emergency response procedures. However, all employees must have sufficient training to recognize an emergency situation and know how to summon the emergency response team. Emergency response personnel must have completed training before being required to perform actual emergency procedures. Employees trained for emergency response must be annually certified for competency. This training must include:

—Elements of the emergency response plan;

—Job-specific standard operating procedures;

—Selection and use of PPE;

—Emergency incident procedures;

—Methods to minimize risk; and

—Symptoms of and response to overexposure.

Confined Space Entry

The Confined Space Entry Standard became effective in 1993 and requires the issuing of permits for entry into those confined spaces that pose special dangers for workers. Confined spaces may contain potential hazards such as toxic, explosive, or asphyxiating atmospheres. They may also have limited openings, affecting the ease of entry to and/or exit from the space. Spaces of concern include process vessels (brewing and food processing industries), water towers, bulk material hoppers, silos, auger-type conveyors, mixers, tank cars, and utility conduits. The regulations describe:

—Acceptable entry conditions;

Chemical Name, Structure/Formula, CAS and RETCS Nos., and DOT ID and Guide Nos.	Synonyms, Trade Names, and Conversion Factors	Exposure Limits (TWA unless noted otherwise) (PELs)	IDLH	Physical Description
Acetaldehyde CH_3CHO 75-07-0 AB 1925000 1089 26	Acetic aldehyde, Ethanal, Ethyl aldehyde 1 ppm = 1.83 mg/m³	NIOSH Ca See Appendix A See Appendix C (Aldehydes) OSHA* 200 ppm (360 mg/m³)	Ca [2,000 ppm]	Colorless liquid or gas (above 69° F) with a pungent, fruity odor. Sp. Gr: 0.79 Class IA Flammablel
Acetic acid CH_3COOH 64-19-7 AF1225000 2790 60 (10-80% acid) 2789 29 (>80% acid)	Acetic acid (aqueous), Ethanoic acid, Glacial acetic acid (pure compound), Methanecarboxylic acid [Note: Can be found in concentrations 5-8% in vinegar.] 1 ppm = 2.50 mg/m³	NIOSH 10 ppm (25 mg/m³) ST 15 ppm (37 mg/m³) OSHA* 10 ppm (25 mg/m³)	50 ppm	Colorless liquid or crystals with a sour, vinegar- like odor. [Note: Pure compound is a solid below 62° F. Often used in an aqueous solution.]
Acetic anhydride $(CH_3CO)_2O$ 108-24-7 AK1925000 171539	Acetic acid anhydride, Acetic oxide, Acetyl oxide, Ethanoic anhydride	NIOSH C 5 ppm (20 mg/m³) OSHA* 5 ppm (20 mg/m³)	200 ppm	Colorless liquid with a strong, pungent, vinegar-like odor.
Acetone $(CH_3)_2CO$ 67-64-1 AI3150000 1090 26	Dimethyl ketone, Ketone propane, 2-Propanone 1 ppm = 2.42 mg/m³	NIOSH 250 ppm (590 mg/m³) OSHA* 1000 ppm (2400 mg/m³)	2,500 ppm [LEL]	Colorless liquid with a fragrant, mint-like odor.

Table 2-2: An excerpt from a NIOSH Pocket Guide showing various PELs.

—Isolating the permit space;

—Purging and ventilating;

—Testing and monitoring the air;

—Proper permit information;

—Worker training;

—Knowledge of the duties of the authorized entrants, their attendants and supervisors; and

—Rescue.

Impact of OSH Act on Industrial Facilities

Many industrial processes involve chemical interactions, chemical cleaning, and various light and radiation exposures. Hazards include acute and chronic exposures to toxic chemicals and radiation, electric shock, stress, and fatigue. In addition, other potential hazards that employees and employers should be aware of include falls, overexertion, sprains/strains, and injuries from falling objects or from being caught in, under, or between objects. Production workers are those most frequently exposed to hazardous conditions. Nearly every production job involves the use of chemicals for cleaning, stripping, or degreasing parts and equipment. Maintenance personnel who enter enclosed or confined spaces are also exposed to toxic substances.

Solvents are used to dissolve various materials. Those that have commonly been used include trichloroethylene, toluene, acetone, methylene chloride, perchlorethylene, glycol ether, isopropyl alcohol, chloroform, and xylene. Exposure occurs by skin absorption and by inhalation. Skin exposure may result in dermatitis or skin rash, edema or swelling, and blistering. These exposures can result from chemical splashes and spills, from directly immersing one's hands into solvents and chemicals, from contact with solvent-soaked clothing or solvent-wet objects, and from the use of improper personal protective equipment. Solvents can dissolve the body's natural protective barrier of fats and oils leaving the skin unprotected against further irritation.

In addition, the inhalation or absorption of solvents may affect the central nervous system, acting as a depressant and an anesthetic, or causing headaches, nausea, drowsiness, dizziness, complaints of irritation, abnormal behavior, general ill feeling, and even unconsciousness. Excessive and/or continued exposure to certain solvents may result in liver, lung, kidney, and reproductive damage, as well as cancer.

Acids and alkalis are a second category of chemicals that are frequently used by industry. These substances may cause serious burns if they are splashed into the eyes or onto the skin. If vapors or mists are inhaled, they may result in a burning of the linings of the nose, mouth, throat, and lungs.

Other common materials that employees may come into contact with include metals, toxic gases, plastics/resins, PCBs, fiberglass, and asbestos. Radiation, noise, and occupational stress are also hazards to which workers can be exposed. Job stress may result from prolonged repetitive and monotonous detail work, overtime and work speedups, as well as from lifting, improper sitting, and prolonged standing.

It is clear that industry is impacted a great deal by issues related to worker health and safety. Not complying with the regulations designed to control or minimize workplace hazards can result in stiff fines and penalties as well as the expense of lost work time and production shutdowns from accidents and injuries. Workplace injuries and illnesses can also increase the costs of workers' compensation and medical insurance. Potential lawsuits are yet another concern.

Checking Your Understanding

1. What are two uses for which premanufacture notifications (PMNs) are not required?

2. Name the four major goals of TSCA.

3. Describe two circumstances in which EPA can require testing of new chemicals or chemicals already on the market.

4. What is the name of the act that regulates the manufacture and use of pesticides?

5. Which standard of the OSH Act requires making MSDSs available to all employees?

6. How many hours of offsite training are required for employees working at a site covered by HAZWOPER?

7. Workers exposed to hazardous substances above permissible exposure limits for more than 30 days per year must be covered by what type of program?

8. In what year did the Confined Space Entry Standard become effective?

2-3 Laws Relating to By-products of Processes

Clean Air Act (CAA) – [40 CFR 50-80]

The goal of the Clean Air Act (CAA) – and its amendments of 1990 – is to reduce air pollution. To achieve this goal it has several components:

—Establishment of **National Ambient Air Quality Standards (NAAQS)**;

—Regulation of hazardous air pollutants;

—Establishment of pollution performance standards; and

—Approved State Implemented Programs (SIPs).

The CAA requires EPA to identify air pollutants that are potentially hazardous to human health and the environment. The act also requires EPA to establish national primary air quality standards for pollutants that are harmful to human health as well as secondary air quality standards for those pollutants that are harmful to the environment.

NAAQSs establish levels of air quality that are to be applied uniformly throughout regions in the United States. An air quality control region is classified as a "nonattainment" area if a NAAQS is violated anywhere in the region. The EPA has set NAAQSs for six pollutants: ozone, carbon monoxide, particulate matter (PM-10), sulfur dioxide, nitrogen dioxide, and lead.

The CAA also established standards for regulating hazardous air pollutants. These pollutants are known to produce adverse health effects at low levels of exposure. **National Emissions Standards for Hazardous Air Pollutants (NESHAPs)** have been developed for a small group of substances including: arsenic, asbestos, benzene, beryllium, coke oven emissions, mercury, radionuclides, and vinyl chloride.

Under the CAA, for example, EPA has listed asbestos as a hazardous air pollutant and subsequently established emission standards for the manufacture, fabrication, spray application, waste packaging, labeling, and disposal of asbestos. The act also established standards for asbestos emissions during renovation and demolition projects.

The 1990 Amendments to the CAA – a massive legislative effort with nearly 1,000 pages – have a significant impact on how companies do business. The amendments impose more than 175 new regulatory standards, include more than 30 guidance documents, and established six scientific panels and numerous research projects.

Impact of CAA on Industrial Facilities

The CAA, as amended, addresses five general areas that impact businesses – Mobile Sources, Air Toxics, Acid Rain/Acid Deposition, CFCs/Global Warming, and Permitting. These areas have the following general requirements:

—Mobile Sources – The focus is to reduce carbon monoxide and other pollutants generated by mobile sources. This is accomplished through the use of alternative fuels, changing diesel fuels, and increasing diesel and automobile inspections. Requirements for mobile sources also affect refueling locomotives and heavy-duty trucks as well as the rebuilding of heavy-duty engines.

—Air Toxics – This section regulates specified hazardous air pollutants (HAPs). Some commonly listed HAPs are: benzene, chlorine, ethylene glycol, hexane, methyl ethyl ketone, PCBs, and metal compounds. Rules are developed under this section to control emissions from various industries. Any changes required to reduce air toxics are made by making permits more stringent.

—Acid Rain/Acid Deposition – This section requires reducing the amounts of sulfur dioxide and nitrogen oxide emissions from fossil fuel-fired facilities. For example, the overall goal for reducing SO_2 emissions is by 10 million tons below 1980 national levels.

—CFCs/Global Warming – Requires complete production phase-out of certain CFCs (chlorofluorocarbons) by the end of the decade. These include halons, except when used for fire suppression; methyl chloroform, except for small amounts used for metal testing; and carbon tetrachloride. HCFCs production is to be frozen in 25 years, and recycling and proper disposal is required. Motor vehicle air conditioners and all other refrigeration unit repairs are heavily regulated.

—Permitting – The annual cost of operating permits for air pollution sources continues to go up and enforcement provisions are continually increased. Enforcement officers can issue spot fines for air pollution violations. The trend for the future is to increase the requirements for testing and monitoring.

The NAAQSs of the CAA can have significant impact on many industries. For instance, many industries are emission sources of volatile organic compounds (VOCs), which react in the atmosphere to produce ozone. Thus, although the NAAQS is for ozone, the relevant emissions for monitoring purposes are VOCs.

In both attainment and nonattainment areas, whenever new plants are built or emissions from existing sources increase as a result of expansion, a New Source Review (NSR) is triggered. Special rules apply in attainment areas. These are called Prevention of Significant Deterioration (PSD) requirements and include the following:

—Installation of **Best Available Control Technology (BACT)**;

—A detailed air quality analysis showing that there will be no violation of PSD increments;

—Prediction of future air quality standards; and

—Possible monitoring of air quality for one year prior to the issuance of the permit.

Restrictions in nonattainment areas are even more severe. The principal requirements of NSR in nonattainment areas are:

—Installation of **Lowest Achievable Emission Rate (LAER)** technology;

—Provision for offsets, allowing reduced emissions from other sources to compensate for increased emissions from new sources; and

—Demonstrating that NAAQSs are being met by doing an analysis of air quality.

Ozone nonattainment areas are classified as Marginal, Moderate, Serious, Severe, or Extreme. Attainment deadlines are based on a sliding scale that reflects the severity of the pollution. See Table 2-3 for areas EPA has classified as Extreme, Severe, or Serious in the late 1990s.

A source defined as "major" must install Reasonable Available Control Technology (RACT) as prescribed in the State Implementation Plan (SIP).

A major source is defined both by the size of the source's emissions and the category of the nonattainment area. In addition, if a firm has the potential to emit more than 100 tons per year (TPY), it is also considered to be a major source. Potential to emit means the maximum capacity of a stationary source to emit a pollutant under its physical and operational design.

A determination of the necessary RACT requirements is made on the basis of a case-by-case review of each facility. In an attempt to issue uniform guidelines, EPA has begun to issue guidance on controls for industrial categories. Some examples of industrial applications for which control guidance is available include: miscellaneous metal parts and products, plastic parts, and solvent cleaning.

The NAAQSs apply to a small number of the most common pollutants. Additional controls that directly restrict the emission of 189 Hazardous Air Pollutants (HAPs) are also established by the CAA. EPA is authorized to establish Maximum Achiev-

Extreme (1 Area)
Los Angeles-Anaheim-Riverside, CA

Severe (8 Areas)
Baltimore, MD Chicago, IL-IN-WI Houston-Galveston-Brazoria, TX Milwaukee-Racine, WI Muskegon, MI New York, NY-NJ-CT Philadelphia, PA-NJ-CT San Diego, CA

Serious (16 Areas)
Atlanta, GA Bakersfield, CA Baton Rouge, LA Beaumont-Port Arthur, TX Boston, MA El Paso, TX Fresno, CA Hartford, CT Huntington-Ashland, WV-KY-OH Parkersburg-Marietta, WV-OH Portsmouth-Dover-Rochester, NH-ME Providence, RI Sacramento, CA Sheboygan, WI Springfield, MA Washington, DC-MD-VA

Table 2-3: Nonattainment areas of the United States identified by EPA in the late 1990s.

Hazardous Air Pollutant	
Antimony compounds	Hydrochloric acid
Arsenic compounds	Hydrofluoric acid
Arsine	Methanol
Carbon tetrachloride	Methyl isobutyl ketone
Catechol	Nickel compounds
Chlorine	Phosphine
Chromium compounds	Phosphorus
Ethyl acrylate	1, 1, 1-trichloroethane
Ethyl benzene	Trichloroethylene
Ethylene glycol	Xylene

Table 2-4: Chemicals listed for the semiconductor manufacturing industry that are scheduled for Maximum Achievable Control Technology (MACT) Standards.

able Control Technology (MACT) standards for source categories that emit at least one of the pollutants on the HAPs list. As an example, Table 2-4 shows chemicals that have been listed for the manufacture of semiconductors.

In addition, EPA is in the process of identifying categories of industrial facilities that emit substantial quantities of any of these 189 pollutants. Regulations to meet these standards will require pollution prevention (P2) strategies to achieve a 75 – 90 percent reduction in emissions below current levels and can include:

— Installation of control equipment;

— Process changes;

— Material substitution;

— Work practice changes; and

— Operator training or certification.

A pollution source will receive a six-year extension of the date for compliance with the MACT standard if it achieves a 90 percent reduction in its air toxic emissions prior to the date the standard is proposed for its industry category. Demonstration of emissions reduction must be made before the standard is proposed.

The CAA and its implementing regulations define the minimum standards and procedures required for state operated permit programs. The permit system is a new approach established by the CAA Amendments that is designed to define the requirements each source must meet and to facilitate enforcement. In addition, permit fees will generate revenue to fund implementation of the program.

Any facility defined as a "major source" is required to secure a permit. A source is defined as a single point from which emissions are released or as an entire industrial facility that is under the control of the same person(s), and a major source is defined as a source that emits or has the potential to emit:

— 10 TPY or more of any hazardous air pollutant;

— 25 TPY or more of any combination of hazardous air pollutants; or

— 100 TPY of any air pollutant.

For ozone nonattainment areas, major sources are defined as sources with the potential to emit:

— 100 TPY or more of volatile organic compounds (VOCs) in areas defined as marginal or moderate;

— 50 TPY or more of VOCs in areas classified as serious;

— 25 TPY or more of VOCs in areas classified as severe; and

— 10 TPY or more of VOCs in areas classified as extreme.

In addition to major sources, all sources that are required to undergo New Source Review, that are subject to New Source Performance Standards, or that are identified by Federal or State regulations, must obtain a permit.

Once a source submits an application, it may continue to operate until the permit is issued. When issued, the permit will include all air pollution control requirements applicable to the facility. Among these are compliance schedules, emissions monitoring, emergency provisions, self-reporting responsibilities, and emissions limitations. The maximum permit term is five years.

States are required to develop fee schedules sufficient to cover permit program costs. The CAA sets a minimum annual fee of $25 per ton for all regulated pollutants (except carbon monoxide), but individual states can set higher or lower fees as long as they collect sufficient revenues to cover program costs.

The CAA Amendments provide for a phase-out of the production and consumption of CFCs and other chemicals that are causing the destruction of the stratospheric ozone layer. The requirements apply to any individual, corporate, or governmental entity that produces, transforms, imports, or exports these controlled substances.

Clean Water Act (CWA) – [40 CFR 100-140 and 400-470]

The Clean Water Act (CWA) is the basic Federal law governing water pollution control in the United States. It regulates discharges into surface waters from all types of sources (municipal, industrial, and non-point sources). Many industries produce a number of pollutants regulated under the CWA. The most applicable provisions of the CWA affecting industry include: Spills of Oil and Hazardous Substances, the **National Pollutant Discharge Elimination System (NPDES)** Permit Program, and the Storm Water Program.

Discharge of Oil – (40 CFR 110)

The regulations in this part apply to the discharge of oil, which is prohibited by the CWA. Prohibited discharges include certain discharges into or upon the navigable waters of the United States or adjoining shorelines.

These regulations define the term "discharge" as including but not limited to any spilling, leaking, pumping, pouring, emitting, emptying, or dumping into the marine environment of quantities of oil that:

1. Violate applicable water quality standard; or

2. Cause a film or sheen upon or discoloration of the surface of the water or adjoining shorelines, or cause a sludge or emulsion to be deposited beneath the surface of the water or upon the adjoining shorelines.

Oil Pollution Prevention – (40 CFR 112)

This section establishes procedures, methods, equipment, and other requirements to prevent the discharge of oil from non-transportation-related onshore and offshore facilities into or upon the navigable waters of the United States or adjoining shorelines. A business that owns or operates above-ground oil storage tanks having capacities greater than 1,320 gallons, and that could reasonably be expected to discharge oil to the navigable waters of the United States, must prepare a written **Spill Prevention Control and Counter-Measure (SPCC)** Plan. This plan must include information on storage, drainage and transfer operations, loading and unloading, oil drilling, inspections and recordkeeping, security, personnel training, and spill prevention procedures. The SPCC Plan must be reviewed and certified by a Registered Professional Engineer in order to satisfy the requirements of this part.

Reportable Quantities of Hazardous Substances Under the Federal Water Pollution Control Act – (40 CFR 116 and 40 CFR 117)

The Federal Water Pollution Control Act (FWPCA) designates hazardous substances under the CWA and establishes reportable quantities (RQs) for each listed substance. When an amount equal to or in excess of the RQ is discharged, the facility must provide notice to the Federal government of the discharge, following Department of Transportation requirements. This requirement does not apply to facilities

Hazardous Substance	RQ in Pounds
Acetic acid	5,000
Ammonia	100
Ammonium fluoride	100
Ammonium hydroxide	1,000
Antimony trichloride	1,000
Antimony trioxide	1,000
Arsenic trioxide	1
Carbon tetrachloride	10
Chlorine	10
Chromic acid	10
Ethyl benzene	1,000
Ethylenediamine	5,000
Ferric chloride	1,000
Ferric nitrate	1,000
Hydrochloric acid	5,000
Hydrofluoric acid	100
Isoprene	100
Nickel compounds	10 to 100
Nitric acid	1,000
Phosphoric acid	5,000
Phosphorus oxychloride	1,000
Phosphorus trichloride	1,000
Potassium cyanide	10
Potassium hydroxide	1,000
Sodium hydroxide	1,000
Sulfuric acid	1,000
Trichloroethylene	100
Xylene	1,000

Table 2-5: Reportable quantities (RQs) that may apply to the semiconductor manufacturing industries.

that discharge the substance under an NPDES permit, a Dredge and Fill permit, or to a **publicly owned treatment works (POTW)**, as long as any applicable effluent limitations or pretreatment standards have been met. An example of the RQs that may apply to the semiconductor manufacturing industries is shown in Table 2-5.

NPDES

The CWA authorized the establishment of regulations and the issuance of permits to control the discharge of pollutants to the waters of the United States. The National Pollutant Discharge Elimination System (NPDES) permit program includes regulations governing these discharges. Most states are authorized to administer NPDES programs that are at least as stringent as the federal program; EPA administers the program in states that are not authorized to do so. Remember that individual states may have more stringent requirements.

The NPDES program requires permits for the discharge of pollutants from any point source into navigable waters, except those covered under dredge and fill permits. The CWA defines the term pollutant as encompassing almost anything that a source might discharge, including dredge spoil, solid waste, incinerator residue, sewage, garbage, sewage sludge, munitions, chemical wastes, biological materials, radioactive materials, heat, wrecked or discarded equipment, rock, sand, cellar dirt, and industrial, municipal, and agricultural wastes discharged into water. The term point source means any discernible, confined, and discrete conveyance, such as a ditch or a pipe. Navigable waters are defined as "…waters of the United States." Courts have construed this term very broadly and include wetlands and ephemeral streams as navigable.

Thus, a source will be required to obtain an NPDES permit if it discharges almost anything directly to surface waters. A source that sends its wastewater to a publicly owned treatment works (POTW) is not required to obtain an NPDES permit, but may need to comply with pretreatment requirements and be required to obtain an industrial user permit from the POTW to cover its discharge. Even if the source does not produce any wastewater, it may still be subject to the NPDES permit process if it discharges "storm water associated with an industrial activity," including some construction activities.

Storm Water Permits – (40 CFR 122.26)

Storm water permits are required for facilities where material handling equipment or activities, raw materials, intermediate products, final products, waste materials, by-products, or industrial machinery are exposed to storm water that drains to a separate municipal storm sewer system or directly to a receiving water. Storm water permits are not required where the runoff flows through a combined sewer to a POTW and is treated prior to discharge.

Impact of the CWA on Industrial Facilities

The effluent discharge, storm water provisions, and NPDES permitting are the standards within the CWA that are most typically applicable to a broad cross-section of industrial facilities. Each of these provisions requires recordkeeping, monitoring and/or testing, and reporting.

For instance, an application for an NPDES permit for process wastewater must include:

— Information on the location of the outfall(s);

— A line drawing showing the water flow through the facility (including a water balance diagram and calculation);

— A description of average flows;

— The method of treatment of wastewater before discharge; and

— An estimate of the facility's actual production of wastewater if an effluent limitation guideline applies.

In addition, the applicant must report the amounts of the following pollutants for every outfall:

— Biochemical Oxygen Demand (BOD);

— Chemical Oxygen Demand (COD);

— Total Organic Carbon (TOC);

— Total Suspended Solids (TSS);

— Ammonia (measured as N);

— Temperature (both winter and summer); and

— pH.

The application must also report the results of any biological toxicity tests that may have been conducted on the effluent within the previous three years. Finally, the facility must provide information on its effluent characteristics. Industrial facilities will need to test for those priority pollutants listed in 40 CFR 122, Appendix D that the applicant knows or has reason to believe will be discharged in greater than trace amounts. Some facilities will need to test for all 126 priority pollutants. Table 2-6 lists some priority pollutants, for example, that are typically present in discharges from the semiconductor packaging industries.

Each applicant must also indicate whether he/she knows or has reason to believe the facility in question discharges any of the other hazardous substances, or non-conventional pollutants listed in Appendix D. Quantitative testing is not required for the other hazardous pollutants; however, the applicant must describe why the pollutant is expected to be discharged and provide the results of any quantitative data about discharge of that pollutant. Quantitative testing is required for the non-conventional pollutants if the applicant expects them to be present in the discharge. For example, Table 2-7 shows the typical hazardous and non-conventional pollutants from this list for the semiconductor manufacturing industries.

If the facility discharges only non-process wastewater, the application must include information on the location of the outfall(s), a description of the type of waste, and values for each of the following:

— Biochemical Oxygen Demand (BOD);

— Total Suspended Solids (TSS);

— Fecal Coliform;

— Total Residual Chlorine;

— Oil and Grease;

— Chemical Oxygen Demand (COD);

— Total Organic Carbon (TOC);

— Ammonia (measured as N);

— Discharge Flow;

— pH; and

— Temperature (both winter and summer).

Standard permit conditions apply to all NPDES permits. These conditions describe the legal effect of the permit and the conditions under which it can be revoked. They notify the permittee of penalties that may be assessed if the permit is violated. Standard permit conditions also describe the permittee's duties and obligations during the effective period of the permit, including the duty to comply with all conditions in a current permit. The permittee must maintain records of all monitoring information for a period of at least three years from the date of the sample, and monitoring results must be reported at the intervals specified in the permit. The NPDES permitting authority (either EPA or an approved state) is allowed to enter the facility at any reasonable time to conduct an inspection or to monitor activity. The NPDES permitting authority must be notified if the discharger knows or has reason to believe that any toxic discharge has exceeded any effluent limitation in the permit. Other generic requirements are also contained in this section of the permit.

Priority Pollutants	
Antimony	Lead
Chromium	Methylene chloride
Copper	Nickel
Cyanide	Silver
Ethyl benzene	1,1,1-Trichloroethane

Table 2-6: Priority pollutants used in semiconductor packaging that may be present in discharge.

Hazardous Pollutants	Non-conventional Pollutants
Butyl acetate	Aluminum, total
Ethylenediamine	Boron, total
Isoprene	Chlorine, total residual
Xylene	Iron, total
	Nitrate/nitrite
	Phosphorus, total
	Titanium, total

Table 2-7: Hazardous and non-conventional chemicals used in semiconductor manufacturing.

Box 2-1 ■ Technology-Based Effluent Limitation Guidelines for the Semiconductor Industry

In the absence of effluent limitation guidelines for a facility category, permit writers establish technology-based controls using their Best Professional Judgment. In essence, the permit writer determines effluent guidelines for a single facility. The permit writer will use information such as permit limits from similar facilities using similar treatment technology, performance data from actual operating facilities, and scientific literature. Best Professional Judgment may not be used as a substitute for existing effluent guidelines and only applies to direct dischargers of wastewater. The following is an example of the technology-based effluent limitation guidelines for the semiconductor industry:

The provisions in Part 469 Subpart A apply to discharges resulting from all process operations associated with the manufacture of semiconductors (except sputtering, vapor deposition, and electroplating). The effluent limitations shown in Tables 2-8 to 2-11 are used as the basis for NPDES permits for the industry. As used in this Part, the term total toxic organics (TTO) means the sum of the concentrations of each of the following toxic organic compounds found in the discharge at concentrations greater than ten (10) micrograms per liter:

— 1, 1, 2-Trichloroethane

— 1, 2, 4-Trichlorobenzene

— 1, 2-Dichlorobenzene

— 1, 2-Dichloroethane

— 1, 1-Dichloroethylene

— 1, 2-Diphenylhydrazine

— 1, 3-Dichlorobenzene

— 1, 4-Dichlorobenzene

— 1, 1, l-Trichloroethane

— 2, 4, 6-Trichlorophenol

— 2, 4-Dichlorophenol

— 2-Chlorophenol

— 2-Nitrophenol

— 4-Nitrophenol

— Anthracene

— Bis (2-ethylhexyl) phthalate

— Butyl benzyl phthalate

— Carbon tetrachloride

— Chloroform

— Dichlorobromomethane

— Di-n-butyl phthalate

— Ethyl benzene

— Isophorone

— Methylene chloride

— Naphthalene

— Pentachlorophenol

— Phenol

— Tetrachloroethylene

— Toluene

— Trichloroethylene

Along with standard permitting conditions, NPDES permits contain effluent limitations that reflect economically feasible control of water quality for the receiving body of water; monitoring, reporting, and recordkeeping requirements; and may contain storm water treatment provisions. Other special conditions may be imposed on facilities through their NPDES permits, including: construction schedules, best management practices, additional monitoring, and spill prevention plans.

A principal means for attaining water quality objectives under the Clean Water Act is the establishment and enforcement of technology-based effluent limits. These limits take into account the economic feasibility of the limits along with their control capabilities. Because of differences in production processes, quantities, and composition of discharges, separate standards are established for discharges associated with different industry categories. These standards are referred to as technology-based effluent limitation guidelines. (See Box 2-1 and Tables 2-8 to 2-11 for an example.)

States determine the appropriate uses of each water body within the state (e.g., drinking water supply, fishable/swimmable waters, and irrigation water). States then establish water quality standards, or maximum pollutant levels, necessary for those bodies of water to attain or maintain the designated use. NPDES permits must also meet any more stringent permit limitations based on state water quality standards. Unlike the technology-based limitations discussed above, water quality-based controls focus on the effects of the discharge on the receiving water. Pollution control limits are necessary to protect local water quality from surface water discharges.

As mentioned above, only those facilities that discharge pollutants directly into waters of the United States need to obtain an NPDES permit. Facilities that discharge to POTWs, however, must comply with pretreatment requirements. Pretreatment requirements were developed because of concern that dischargers' waste containing toxic, hazardous, or concentrated conventional industrial wastes might "pass through" POTWs, or that pollutants might interfere with the successful operation of the POTW's biological treatment system.

Table 2-12, for example, shows national pretreatment standards for existing facilities that do chemical etching and plating and discharge more than 38,000 liters per day of effluent. Local pretreatment programs may impose additional requirements on such facilities.

Pollutant or Pollutant Property	Maximum for 1 day (mg/l)	Average of Daily Values for 30 Consecutive Monitoring Days Shall Not Exceed (mg/l)
Total Toxic Organics	1.37	Not Applicable
pH	Within the range of 6.0 to 9.0	Within the range of 6.0 to 9.0

Table 2-8: Semiconductor Best Practicable Control Technology Currently Available (BPT) effluent limitations.

Pollutant or Pollutant Property	Maximum for 1 day (mg/l)	Average of Daily Values for 30 Consecutive Monitoring Days Shall Not Exceed (mg/l)
Total Toxic Organics	1.37	Not Applicable
Fluoride (T)	32	17.4

Table 2-9: Semiconductor Best Available Control Technology Economically Available (BAT) effluent limitations.

Pollutant or Pollutant Property	Maximum for 1 day (mg/l)	Average of Daily Values for 30 Consecutive Monitoring Days Shall Not Exceed (mg/l)
Total Toxic Organics	1.37	Not Applicable
Fluoride (T)	32	17.4
pH	N/A	Within the range of 6.0 to 9.0

[1] Applies to facilities that commenced construction after April 18, 1983.

Table 2-10: Semiconductor New Source Performance Standards (NSPS) effluent limitations[1].

Pollutant or Pollutant Property	Maximum for 1 day (mg/l)	Average of Daily Values for 30 Consecutive Monitoring Days Shall Not Exceed (mg/l)
pH	Within the range of 6.0 to 9.0	Within the range of 6.0 to 9.0

Table 2-11: Semiconductor Best Conventional Pollution Control Technology (BCT) effluent limitations.

Pollutant or Pollutant Property	Maximum for 1 day (mg/l)	Average of Daily Values for 30 Consecutive Monitoring Days Shall Not Exceed (mg/l)
CN, T[1]	1.9	1.0
Cu	4.5	2.7
Ni	4.1	2.6
Cr	7.0	4.0
Zn	4.2	2.6
Pb	0.6	0.4
Cd	1.2	0.7
Total Metals	10.5	6.8
TSS	20.0	13.4
pH	[2]	[2]
TTO	2.13	–

[1] Cyanide, total
[2] Within the range of 7.5 to 10.0.

Table 2-12: Semiconductor pretreatment standards for common metals, chemical etching and milling, electroless plating, and electroplating facilities discharging 38,000 liters or more per day Pretreatment Standards for Existing Sources (PSES) limitations.

In addition to the categorical standards applicable to specific industries, general pretreatment standards apply to all facilities. These general pretreatment standards prohibit the following from being introduced into a POTW:

1. Pollutants that create a fire hazard in the POTW including, but not limited to, wastestreams with a closed cup flash point of less than 140°F or 60°C using specified test methods;

2. Pollutants that will cause corrosive structural damage to the POTW, but in no case discharges with pH lower than 5.0, unless the works is specifically designed to accommodate such discharges;

3. Solid or viscous pollutants in amounts that will cause obstruction to the flow in the POTW, resulting in interference;

4. Any pollutant, including oxygen-demanding pollutants (BOD, etc.), released in a discharge at a flow rate and/or pollutant concentration that will cause interference with the POTW;

5. Heat in amounts that will inhibit biological activity in the POTW resulting in interference, but in no case heat at such levels that the temperature at the POTW treatment plant exceeds 40°C, unless the approval authority, upon request of the POTW, approves alternate temperature limits;

6. Petroleum oil, non-biodegradable cutting oil, or products of mineral oil in amounts that will cause interference or pass through;

7. Pollutants that result in the presence of toxic gases, vapors, or fumes within the POTW in a quantity that may cause acute worker health and safety problems; and

8. Any trucked or hauled pollutants, except at discharge points designated by the POTW.

The storm water permit program requires affected facilities to submit applications that include a site map showing the topography of the facility, including:

—Drainage and discharge structures;

—The drainage area of each storm water outfall;

—Paved areas and buildings within each drainage area;

—Areas used for outdoor storage or disposal;

—Each existing structural control measure to reduce pollutants in storm water runoff, materials loading and access areas;

—Areas where pesticides, herbicides, soil conditioners, and fertilizers are applied;

—Each of the facility's hazardous waste treatment, storage, or disposal facilities;

—Each well where fluids from the facility are injected underground; and

—Springs and other surface water bodies that receive storm water discharges.

An estimation of the area of impervious surfaces, the total area drained by each outfall, and a description of the storage, handling, and disposal of "significant" materials in the three years prior to the submittal of the application must also be documented. A certification that all outfalls have been tested or evaluated for the presence of non-storm water discharges that are not covered by an NPDES permit must be made, and this certification must include a description of the method used, dates, and the observed onsite drainage points.

Quantitative data based on samples collected during storm events must be documented from all outfalls for the following:

1. Any pollutant with an established limit in an effluent guideline to which the facility is subject;

2. Any pollutant listed in the facility's NPDES permit;

3. Oil, grease, pH, BOD, COD, TSS, total phosphorus, total nitrogen, and nitrate plus nitrite nitrogen;

4. Flow measurements or estimates of the flow rate, and the total amount of discharge for the storm event(s) sampled and the method of the measurement; and

5. The date and duration of the storm event(s) sampled and rainfall measurements, and the duration between the storm event sampled and the end of the previous measurable storm event.

Pollutant or Pollutant Property	Maximum for 1 day (mg/l)	Average of Daily Values for 30 Consecutive Monitoring Days Shall Not Exceed (mg/l)
CN, T[1]	1.9	1.0
Cu	4.5	2.7
Ni	4.1	2.6
Cr	7.0	4.0
Zn	4.2	2.6
Pb	0.6	0.4
Cd	1.2	0.7
Total Metals	10.5	6.8
TSS	20.0	13.4
pH	[2]	[2]
Total Toxic Organics	2.13	–

[1] Cyanide, total
[2] Within the range of 7.5 to 10.0.

Table 2-13: Printed wiring board pretreatment standards for facilities discharging 38,000 liters or more per day Pretreatment Standards for Existing Sources (PSES) limitations.

Safe Drinking Water Act (SDWA) – [40 CFR 140-149]

The Safe Drinking Water Act (SDWA) established minimum national drinking water standards and guidelines for protecting groundwater. Under the SDWA, both primary and secondary standards were established for drinking water. Primary standards are contaminant levels that, if exceeded, have adverse human health effects. The primary drinking water standards set **maximum contaminant levels (MCLs)** for a variety of organic and inorganic substances as well as radionuclides. The MCLs for some primary drinking water standards are included in Table 2-13.

The EPA also recommends secondary standards that protect public welfare. These contaminants impart an adverse taste, smell, and color to drinking water. The standards include: chloride, copper, iron, manganese, color, foaming, acidity, and total dissolved solids.

Another provision of the SDWA is regulating the underground injection of liquids, hazardous substances, and waste. This program is designed to prevent the subsurface placement of fluids through wells (injection wells) from endangering underground sources of drinking water (USDW). The program allows states to permit and regulate underground injection of fluids to protect groundwater used for drinking water supplies.

EPA and state programs:

1. Impose minimum Underground Injection Control (UIC) standards for the siting, construction, operation, monitoring, and closure of injection wells;

2. Authorize injection by permit or by rule; incorporate hazardous waste requirements under the Resource Conservation and Recovery Act; and

3. Forbid the disposal of hazardous and radioactive wastewater into or above a USDW where the waste may endanger a USDW.

EPA has established five categories – Classes I, II, III, IV, and V – for injection of liquid wastes.

—Class I wells are those used to inject nonhazardous liquids associated with manufacturing and wastewater disposal into a formation beneath an underground source of drinking water that is located within one-quarter mile of the injection well.

—Class II injection wells are utilized by the oil and gas production industry. These may be used for disposal of liquids brought to the surface during production or for the injection of materials to enhance the recovery of oil and gas.

—Class III injection wells are used by the mining industry for onsite mining of ore bodies. These are frequently utilized for solution mining for uranium, salts, and potash.

—Class IV injection wells are utilized for injection of hazardous waste and radioactive waste into or above a formation that has an underground source of drinking water within one-quarter mile of the injection well. This class of injection well is no longer obtainable.

—Class V injection wells are those not identified in Classes I, II, III, or IV. These include injection wells utilized for a variety of remediation projects.

Impact of SDWA on Industrial Facilities

The drinking water standards (MCLs) of the SDWA are not typically a high impact portion of this law for most facilities since they apply to purveyors of drinking water. The Underground Injection Control (UIC) programs, however, apply to owners and operators of deep wells, into which trillions of gallons of hazardous and nonhazardous fluids associated with manufacturing processes and municipal wastewater disposal (Class I), oil and gas production (Class II), and solution mining (Class III) are injected annually.

UIC programs also apply to owners and operators of shallow wells, which are designed to release fluids either directly into a USDW or into the shallow subsurface that overlies USDW (Class V). Class V injection wells are generally shallow wastewater disposal wells, storm water and agricultural drainage systems, or other devices that are used to release fluids either directly into USDW or into the shallow subsurface that overlies the USDW.

State and EPA UIC program directors require that the operation of any UIC well not endanger an underground source of drinking water. If a drinking water supply is threatened, the owner or operator of the well must stop the injection practice, close the drainage system, and may face fines and expensive groundwater cleanup costs.

Checking Your Understanding

1. What is the overall goal of the CAA and its 1990 amendments?

2. List six pollutants identified by the NAAQS.

3. What is the primary concern of the CWA?

4. In most instances, who administers the NPDES program?

5. Under what conditions is a storm water permit not required?

6. What do the eight general pretreatment standards prohibit being discharged into a POTW?

7. Liquid wastes generated during oil production would fall into which of the five Underground Injection Control (UIC) standards categories?

2-4 Cleanup and Disposal of Chemicals

Resource Conservation and Recovery Act (RCRA) – [40 CFR 240-280]

The **Resource Conservation and Recovery Act (RCRA)** of 1976 (as amended by the Hazardous and Solid Waste Amendments of 1984) has three subtitles of interest to industrial facilities. Subtitle C sets up a "cradle-to-grave" system for tracking and regulating hazardous wastes. Subtitle I sets up a system for regulating underground storage tanks (USTs) containing petroleum or other hazardous substances (other than hazardous wastes). Subtitle D establishes a framework for regulating solid wastes that are not classified as hazardous wastes. The discussion in this section will focus on Subtitles C and I. Subtitle D requirements are not discussed further because their impact on most industrial facilities, though significant, is indirect.

The regulations discussed in this section are federal requirements. Many RCRA requirements are implemented through RCRA-authorized state laws, which may be more stringent than federal requirements. There are also non-RCRA state laws that set out UST and hazardous waste management requirements that may differ significantly from the Federal standards.

The primary goals of RCRA are:

— To protect human health and the environment;

— To conserve energy and natural resources;

— To reduce the amount of waste generated; and

— To ensure that wastes are managed in an environmentally sound manner.

Subtitle C

Under Subtitle C, RCRA requires proper hazardous waste management, which involves:

— Determining whether hazardous wastes are generated by classifying facility wastes.

— Establishing whether a facility is a large or small quantity generator.

— Determining whether the generator is a **Treatment, Storage, or Disposal (TSD) facility**.

— Properly accumulating and storing hazardous wastes.

— Properly shipping hazardous wastes for offsite treatment and/or disposal.

— Maintaining appropriate records and notifications.

— Reporting releases or threats of releases.

— Providing necessary training.

The generator of the waste is responsible for determining whether a waste is hazardous and what classification, if any, applies to the waste. Generators must examine the regulations and conduct any tests necessary to make this determination. In some cases, generators may use their own knowledge and familiarity with the waste to determine whether it is hazardous. It is important to note that enforcement penalties may be assessed for improperly determining that a waste is nonhazardous.

Wastes can be classified as hazardous, either because they are listed by EPA in 40 CFR or because they exhibit certain characteristics. **Listed hazardous wastes** are specifically named, for example, discarded commercial toluene, spent non-halogenated solvents, and spent cyanide plating bath solutions from electroplating operations. **Characteristic hazardous wastes**, on the other hand, are determined by a test of their properties for the following:

— Ignitability,

— Corrosivity,

— Reactivity, and

— Toxicity.

Characteristic hazardous wastes are those that fail a test such as the RCRA test for ignitability, or that fall outside (pH ≤ 2 or ≥ 12.5) the corrosivity test range.

In addition, most wastes that are derived from a listed hazardous waste, or are a mixture of a listed hazardous and nonhazardous waste are considered hazardous wastes. Environmental media (such as soil or groundwater) that contain a listed hazardous waste may also be considered hazardous.

Some RCRA requirements depend on whether the hazardous waste generator is classified as large quantity, small quantity, or conditionally exempt. **Small quantity generators (SQG)** are generators that 1) produce more than 100 kg but less than 1,000 kg of hazardous waste at a site per month and 2) accumulate less than 6,000 kilograms at any one time. **Conditionally exempt small quantity generators (CESQG)** 1) generate less than 100 kg of hazardous waste and less than one kg of acutely hazardous waste per month and 2) accumulate no more than 1,000 kg of hazardous and one kg of acutely hazardous wastes at any one time. Generators exceeding the limits for SQG and CESQG are classified as **large quantity generators (LQG)**.

Only a few hazardous waste regulations apply to firms that qualify as CESQG. They must:

— Evaluate the waste to determine whether it is hazardous;

— Not exceed accumulation limits; and

— Treat or dispose of the waste onsite or ensure delivery to a:

  – Permitted treatment, storage, or disposal facility (TSDF), or
  – State approved municipal or industrial solid waste facility, or
  – Legitimate recycling facility.

Small Quantity Generators must:

— Obtain a generator identification number;

— Store and ship hazardous waste in suitable containers or tanks;

— Manifest the waste properly;

— Maintain copies of the manifest (a shipment log covering each hazardous waste shipment) and test records;

— Comply with applicable land disposal restriction requirements; and

— Report releases or threats of releases of hazardous waste.

Both large and small quantity generators must ensure that hazardous wastes being shipped offsite are kept in areas that meet basic safety requirements. The wastes must be properly stored to prevent leaks and must be labeled with the name of the waste and the words "Hazardous Waste." The date on which accumulation began must be shown on the container and proper labels and placards must be used. Except when adding or removing waste, hazardous waste must be stored in a closed container that is in good condition, inspected at least weekly, and is compatible with the waste to be stored. Wastes stored in tanks or tank systems and waste generators who use drip pads are subject to more extensive requirements.

Large quantity generators must also submit a biennial report of their hazardous waste generation and management activity by March 1st of every even-numbered year. In the report, the generator must identify each waste transporter and each TSDF used throughout the year. The generator must also describe the hazardous waste generated and shipped, efforts made to reduce the volume and toxicity of the waste, and changes made in the volume and toxicity of the waste compared with those achieved in previous years. For generators who treat, store, or dispose of wastes onsite, additional reporting is required on methods of treatment, storage, or disposal.

Any generator (except some CESQGs) who disposes of waste onsite is classified as a treatment, storage, or disposal facility (TSDF). A small quantity generator who stores waste onsite for more than 180 days (without seeking an extension) is also classified as a TSDF, as is any large generator who stores waste onsite for more than 90 days (without seeking an extension). Extensions are granted only under very limited circumstances. Every hazardous waste TSDF must comply with specific management provisions and must apply for a permit and meet certain technical and financial responsibility requirements.

Hazardous wastes being shipped offsite must go to a RCRA-permitted facility. Large and small quantity generators must complete a **Uniform Hazardous Waste Manifest**, which can usually be obtained from state environmental agencies. The manifest must have enough copies to provide the generator, each transporter, and the owner or operator of the designated facility with one copy each for their records, and another copy to be returned to the generator by the owner or operator of the facility. Many states also require a copy of the manifest.

Hazardous wastes must be treated in accordance with EPA treatment standards before being disposed of in a hazardous waste landfill. These regulations are called the Land Disposal Restrictions and require that a written notification be transmitted to the destination facility with each shipment of hazardous waste. The notification must be:

—Signed by the generator,

—Filed with associated manifest copies, and

—Must include the EPA hazardous waste number (e.g., F002), the corresponding treatment standard(s) found in the regulations, the manifest number associated with the shipment of waste, and analytic data about the waste.

The requirements for small and large quantity generators state that, among other things, personnel must be familiar with emergency procedures to be followed in the event of spills, fires, or other releases of hazardous waste. Large quantity generators must additionally establish an appropriate hazardous waste handling training program for their employees. Small quantity generators must ensure that employees handling hazardous wastes are thoroughly familiar with proper waste handling procedures and that there is always a person on call or at the premises with responsibility for coordinating all response measures in the event of an emergency. Large quantity generators also must prepare a contingency plan for each facility that is designed to minimize hazards to human health or the environment from fires, explosions, or any unplanned release of hazardous waste or hazardous waste constituents. If there is a fire, explosion, or other release of hazardous material to the environment, the generator must immediately contact the National Response Center and supply information about the incident.

Finally, large and small quantity generators must maintain copies of each manifest, exception report, test result, and waste analysis, for at least three years. Large quantity generators must maintain copies of their biennial report for the same period of time. The generator must keep a copy of each land disposal restriction notification form for at least five years.

Subtitle I

Subtitle I of RCRA (as amended) establishes a program to prevent and clean up leaks from underground storage tanks (USTs). It covers underground storage tanks containing petroleum products and hazardous substances, except for hazardous waste storage tanks, which are regulated under Subtitle C.

A storage tank is defined as underground if ten percent or more of the volume – including the volume of underground pipes – is beneath the surface of the ground. Thus, a tank that is 90 percent aboveground is classified as an underground storage tank. Some types of underground storage tanks are not covered by Subtitle I. Some examples of exempted tanks include: underground storage tanks storing heating oil used on the premises, septic tanks and other tanks for collecting waste water and storm water, flow-through process tanks, and emergency spill tanks that are emptied immediately after use.

The Impact of RCRA on industrial Facilities

From the overview of RCRA presented above, we have noted that it may have a regulatory impact on industry in three ways. First, RCRA's cradle-to-grave system for tracking and regulating hazardous waste under Subtitle C affects most segments of industry. Second, the underground tank provisions in Subtitle I affect facilities owning an underground storage tank containing petroleum or hazardous substances. Third, the impacts of Subtitle D (solid waste) on industry are indirect; they arise as a result of the industry's use of solid waste disposal facilities, including municipal solid waste disposal facilities.

Hazardous Waste Management – The major factors that determine whether and to what extent RCRA requirements apply to a facility that generates hazardous waste are the types of hazardous wastes being produced, the volume of hazardous waste produced per month, and the length of time the hazardous waste remains onsite. Obviously, the regulatory burden to a hazardous waste generating facility is dramatically increased if it is a large quantity generator as opposed to a small quantity generator or conditionally exempt small quantity generator. This in itself is a compelling reason to promote pollution prevention (P2) activities. The savings in dollars, time, and effort to meet regulations can be enormous.

Listed wastes are found on four separate lists in 40 CFR. One list contains wastes from nonspecific sources and also includes wastes generated by industrial processes occurring in several different industries; the codes for such wastes always begin with the letter "F." The codes F001, F002, F003, and F004, which designate various types of spent solvent wastes, are examples of wastes from non-specific sources that may be generated by many types

Waste Code	Name of Description of Waste
F003	The following spent non-halogenated solvents: Xylene, acetone, ethyl acetate, ethyl benzene, ethyl ether, methyl isobutyl ketone, n-butyl alcohol, cyclohexanone, and methanol; all spent solvent mixtures/blends containing, before use, only the above spent non-halogenated solvents; and all spent solvent mixtures/blends containing, before use, one or more of the above non-halogenated solvents, and, a total of ten percent or more by volume) of one or more of those solvents listed in F001, F002, F004, and F005; and still bottoms from the recovery of these spent solvents and spent solvent mixtures.
F003	The following spent non-halogenated solvents: Toluene, methyl ethyl ketone, carbon disulfide, isobutanol, pyridine, benzene, 2-ethoxyethanol, and 2-nitropropane; all spent solvent mixtures/blends containing, before use, a total of ten percent or more (by volume) of one or more of the above non-halogenated solvents or those solvents listed in F001, F002, or F004; and still bottoms from the recovery of these spent solvents and spent solvent mixtures.
P098	Potassium cyanide
P099	Potassium silver cyanide
U002	Acetone
U080	Methylene chloride
U104	Silver cyanide
U112	Ethyl acetate
U134	Hydrofluoric acid
U154	Methanol
U161	Methyl isobutyl ketone
U226	1, 1, 1-Trichloroethane
U239	Xylene
U259	Ethylene glycol monomethyl ether

Table 2-14: Some examples of listed wastes found in semiconductor packaging.

of facilities. he second category of listed wastes includes hazardous wastes from specific sources; these wastes have codes that begin with the letter "K." Unless a manufacturing facility engages in the production of chemicals onsite, it is unlikely that it would have wastes falling into this category. The remaining lists cover commercial chemical products that have been or are intended to be discarded; waste codes beginning with "P" are considered acutely hazardous, while those beginning with "U" are simply considered hazardous. See Table 2-14 for examples of typical listed wastes from the semiconductor packaging industry.

Many waste streams in industry are characterized as hazardous because they exhibit one or more of the hazardous waste characteristics. One of the definitions of corrosivity is any aqueous waste that has a pH of 2 or less or a pH of 12.5 or more. In the printed wiring board industry, for example, an ammonia etchant used in the strip/etch process is characterized as a corrosive because of its pH.

The toxicity characteristic applies to a list of 40 substances that show toxicity, including metals, nonmetals, pesticides, and other organic chemicals. If a waste leachate (derived from putting the waste through a test called the Toxicity Characteristic Leaching Procedure, or TCLP) contains any one of these 40 constituents at levels above a specified level, the waste is considered hazardous. Table 2-15, for example, shows five of the common toxic characteristic contaminants that may be generated by the semiconductor manufacturing industry.

Most hazardous wastes generated by industry are covered by the land disposal restrictions. Spent solvents, a common hazardous waste, are banned from land disposal, for example, unless treated to appropriate levels. Many other wastes also must be treated prior to land disposal. Glass cullet from the manufacture of cathode ray tubes (CRTs), which usually fails the toxicity characteristic for lead, is an

Waste Code	Contaminant
D004	Arsenic
D007	Chromium
D011	Silver
D019	Carbon tetrachloride
D040	Trichloroethylene

Table 2-15: EPA toxic characteristic contaminants that may be found in semiconductor manufacturing waste.

example. These wastes must be treated prior to land disposal by either removing the lead or encapsulating it to prevent it from being leached.

Underground Tanks – If a facility owns or operates underground storage tanks (USTs) that are not covered by any of the allowed exemptions, the facility must comply with the federal requirements. If the facility is located in a state authorized to carry out the Underground Storage Tank program, it must comply with the provisions of the approved state program. These generally include requirements for:

— Design, construction, installation, and notification;

— General operations;

— Release detection;

— Release reporting, investigation, and confirmation;

— Release response and corrective action;

— Closure of underground storage tanks; and

— Financial assurance (for petroleum underground storage tanks).

These requirements may not seem to be affected by pollution prevention activities since they pertain to tanks containing raw materials used by a wide variety of facilities. However, proper monitoring, testing, and maintenance of USTs is a very important way to prevent pollution by keeping their contents from being released into the environment.

Comprehensive Environmental Response, Compensation, and Liability Act (CERCLA) – [40 CFR 300]

The Comprehensive Environmental Response, Compensation, and Liability Act of 1980, also known as CERCLA, or more commonly as Superfund, provides a system for identifying and cleaning up hazardous substances that are released into any part of the environment (air, water, groundwater, and land). It also provides for joint and several liability for parties responsible for a hazardous waste site. Joint and several liability means that a party may be liable for 100 percent of a cleanup

even if they did not cause the problem. For instance, a bank holding a mortgage on a property contaminated by its owner may be liable for the entire cleanup if the property owner does not have the funds to clean it up.

CERCLA was amended in 1986 by the **Superfund Amendments and Reauthorization Act (SARA)**. SARA made major changes to the law, including a reauthorization that provided $8.5 billion over five years to EPA and other federal facilities for the cleanup of abandoned and inoperative waste sites.

Title III of SARA is called the Emergency Planning and Community Right-to-Know Act (EPCRA). This part of the amendment created an emergency planning framework and established the right of local governments and members of the public to obtain information on the hazards posed by potential toxic substance releases.

CERCLA includes the following basic provisions:

— National Contingency Plan (NCP) – Establishes procedures for cleanup of hazardous materials spills.

— National Priorities List (NPL) – Annually updates the list of abandoned or uncontrolled hazardous waste sites in the country. Using the Hazard Ranking System (HRS), sites are prioritized by the degree of hazard they pose. The process for investigation and cleanup under CERCLA includes preliminary assessments, site investigations, ranking and listing on the NPL, and requirements for remediation and removal.

— Reporting and Responding to Discharges – Spills or releases of regulated amounts of hazardous substances into the environment must be reported immediately to the National Response Center. The center notifies all appropriate government agencies in order to coordinate response activities.

— Reportable Quantities – The National Response Center must be immediately notified if a hazardous substance is released in greater than the regulated reportable quantity (RQ) of that substance. The RQ may be from 1 to 5,000 pounds depending on the substance. The list of RQs is located in 40 CFR 302.

— Emergency Response Arrangements – After the National Response Center is notified of a hazardous materials release that exceeds the RQ, the lead responsibility for dealing with the release is assumed by the U.S. Coast Guard or EPA, depending on the location and type of release.

SARA Title III amendments to CERCLA include the following requirements:

—Emergency Planning – This requirement is designed to develop the emergency response and preparedness capabilities of state and local governments through better coordination and planning, especially within the local community.

—Emergency Release – Facilities must immediately notify police, fire, and emergency preparedness agencies if a release in excess of the reportable quantities (RQ) value of a hazardous substance occurs.

—Hazardous Chemical Inventory Reporting – This section requires a facility to submit a hazardous chemical inventory form to the local emergency planning committee, the state emergency planning commission, and the local fire department. This list must include those materials required to have Material Safety Data Sheets (MSDS) under the OSHA Hazard Communication Standard and those that were present at the facility during the previous calendar year if above specified **threshold planning quantities (TPQ)**. For example, the TPQ for chlorine gas is 100 pounds; therefore, any facility that keeps more than 100 pounds of gaseous chlorine onsite must file appropriate inventory forms.

—Toxic Release Inventory Reporting – Requires facilities that emit specified toxic chemicals to report the release and complete an annual Toxic Chemical Release Form (Form R). The purpose of this reporting requirement is to inform the public and regulatory agencies about routine releases of toxic chemicals into the environment. The data are used in research and in the development of regulations, guidelines, standards, and local emergency planning.

Impact of CERCLA on Industrial Facilities

The reporting requirements established by the SARA amendments to CERCLA have tremendous impact on industrial facilities. For example, certain manufacturers, processors, and users of more than 600 designated toxic chemicals must report to EPA and designated state agencies annually on emissions of those chemicals to the air, water, and land. Table 2-16 shows the TPQ and RQ for some EPCRA-Designated Extremely Hazardous Chemicals used in the semiconductor packaging industry.

Chemical Name	Threshold Planning Quantity (Lbs.)	Reportable Quantity (Lbs.)
Chlorine	100	10
Hydrogen peroxide	1,000	1
Hydrochloric acid	500	5,000
Hydrofluoric acid	100	100
Nitric acid	1,000	1,000
Potassium silver cyanide	500	1
Potassium cyanide	100	10
Sulfuric acid	1,000	1,000

Table 2-16: Threshold planning and reporting quantities for some EPCRA-Designated Extremely Hazardous Chemicals used by the semiconductor packaging industry.

Reporting releases is another important responsibility of industry under this law. Based on criteria that relate to the possibility of harm associated with the release of each substance, EPA by regulation has assigned a substance-specific reportable quantity (RQ) for most hazardous substances; RQs are either 1, 10, 100, 1,000, or 5,000 pounds (except for radionuclides). If EPA has not assigned a regulatory RQ to a hazardous substance, its statutory default RQ is one pound. Any individual in charge of a facility (or a vessel) must notify the National Response Center as soon as he or she has knowledge of a release (within a 24-hour period) of an amount of a hazardous substance that is equal to or greater than its RQ.

The requirement for reporting the presence, storage, or use of hazardous chemicals at or above their threshold planning quantities applies to the majority of industrial chemicals. Any facility that is required by OSHA's Hazard Communication Standard to have a Material Safety Data Sheet (MSDS) for an OSHA hazardous chemical, and has that chemical above the TPQ, must provide a copy of the MSDS for the substance to the State Emergency Response Commission (SERC), the Local Emergency Planning Committee (LEPC), and the local fire department.

The Form R filing requirements discussed in the overview don't affect as many facilities because the facility must use more than 10,000 pounds per year of a listed chemical before being required to submit the report. Table 2-17 lists chemicals used by the semiconductor industry that are listed on the Toxic Release Inventory.

Toxic Chemical	
Aluminum (fume or dust)	Hydrochloric acid
Ammonia	Isopropyl alcohol
Antimony	Methanol
Arsenic	Methyl isobutyl ketone
Boron trichloride	Nickel
Boron trifluoride	Nitric acid
Carbon tetrachloride	Phosphine
Catechol	Phosphoric acid
Chlorine	Phosphorus
Chromium	Silver
Copper	Sulfuric acid
Ethyl acrylate	1,1,1-Trichloroethane
Ethyl benzene	Trichloroethylene
Ethylene glycol	Xylene
Fluorine	

Table 2-17: Chemicals used by the semiconductor industry that are listed in the toxic release inventory.

Although not directed at industry, Executive Order 12856 – entitled Federal Compliance with Right-to-Know Laws and Pollution Prevention Requirements – has significant impact on the federal facilities reporting under SARA Title III requirements. Federal facilities are required to:

— Incorporate the Pollution Prevention Act of 1990 in planning, management, and acquisition actions;

— Comply with EPCRA including the Toxic Release Inventory (TRI) reporting;

— Develop voluntary goals to reduce total releases and offsite transfers of TRI toxic chemicals or toxic pollutants by 50 percent by 1999; and

— Establish a plan and goals for eliminating or reducing acquisition, manufacture, processing, or use of products containing extremely hazardous substances or toxic chemicals.

There are also acquisition and procurement of goods and services goals that include:

— Establishing a plan and goals for eliminating or reducing unnecessary acquisition of extremely hazardous or toxic products;

— Establish a plan and goals for voluntarily reducing manufactured, processed, and used toxic chemicals;

— Revising specifications and standards to eliminate or reduce acquisition and procurement of extremely toxic chemicals by 1999.

Industry should see this Executive Order as a strong indicator of the direction environmental regulation is going. Pollution prevention (P2) activities are the most logical and economical way to protect the environment while guarding against future compliance and liability costs.

Checking Your Understanding

1. Which Subtitle of RCRA establishes a system for regulating USTs?

2. State the four primary goals of RCRA.

3. What are the four properties that determine if a waste is a characteristic waste?

4. What conditions determine if a generator is listed as a small quantity generator (SQG) or a large quantity generator (LQG)?

5. Who must receive copies of the Uniform Hazardous Waste Manifest used when shipping hazardous wastes offsite?

6. To be considered as an underground storage tank, what percent must be underground?

7. State one compelling reason for promoting pollution prevention (P2) activities.

8. Which waste code denotes a hazardous waste from a specific source?

9. What does the term "joint and several liability" mean in terms of a company's liability?

10. What is the statutory default RQ for a substance prior to being assigned a regulatory RQ?

11. What direction does Executive Order 12856 suggest that industry should start moving?

2-5 Transportation of Chemicals and Hazardous Substances

Hazardous Materials Transportation Act – [49 CFR 171-180]

Hazardous materials and wastes, including radioactive materials, pose a risk to health, safety, and property during transportation. The Hazardous Materials Transportation Act (HMTA) gives authority to the U.S. Department of Transportation (DOT) to regulate the transportation of these materials by rail, water, air, and public highway. The goal of these regulations is to "protect the nation against risks to life and property which are inherent in the transportation of hazardous materials in commerce." Recent revisions to these regulations bring U.S. standards into conformance with those of the United Nations.

The HMTA governs the safety aspects of transporting hazardous materials, including:

— Classification of materials;

— Packaging (including manufacture, continuing qualification, and maintenance);

— Hazard communication (i.e. package marking, labeling, placarding, and shipping documentation);

— Transportation;

— Explaining how to properly load, transport, and unload cargo;

— Proper handling; and

— Incident reporting.

DOT defines a hazardous material as a material (including a hazardous substance) that is "...capable of posing an unreasonable risk to health, safety, and property when transported." Hazardous materials are divided into nine hazard classes:

— Explosives;

— Flammable solids;

— Radioactive substances;

— Oxidizers;

— Gases;

— Flammable liquids;

— Corrosives;

— Poisonous and infectious substances;

— Other.

Shippers must provide shipping papers with all hazardous materials shipments covered under the HMTA. Shipping papers describe the material being transported and the material's proper shipping name, which is found in the Hazardous Materials Table in 49 CFR. Shippers must select the shipping name from the Hazardous Materials Table that most appropriately describes the material. The shipping papers must also include emergency response information and an emergency 24-hour phone number.

There is an easily recognizable, color-coded, 11 inch diamond placard for each of the nine hazard classes. A placarded vehicle must have four identical placards attached to front, rear, and both sides of the vehicle. Placards alert emergency response personnel to the potential dangers of the hazardous material onboard a truck, rail car, freight container, ship, cargo tank, or portable tank. To require the hazardous materials placards, most loads must be 1,000 pounds or more.

Markings are words and numbers on a package or container used to further communicate information about the contents. Typical markings include the proper shipping name and the UN (United Nations) or NA (North American) number. Markings must also include the shipper's or receiver's address.

Warning labels are similar to placards, but are only 4 inch diamonds and are placed directly on packaging containing hazardous materials. Labels are designed to convey hazard information about the container contents quickly at a safe distance. The Hazardous Materials Table also lists the appropriate labels to use on specific hazardous materials.

Packaging has taken on a more and more important role over the years. The packaging for most hazardous materials falls into one or more of three groups: I, II, and III. The Hazardous Materials Table also identifies the type of packaging the shipper must use. Packages must be marked by the manufacturer to assure compliance with the package regulations.

DOT and EPA monitor spills of hazardous substances, which are named in the List of Hazardous Substances and Reportable Quantities (RQs) in 40 CFR. A spill of a hazardous substance that is equal to or greater than the RQs for that substance must be reported. The shipper must identify these hazardous substances by placing the letters "RQ" on the shipping paper either before or after the basic shipping description.

Impact of the Hazardous Materials Transportation Act on Industrial Facilities

The HMTA affects all industries that ship hazardous materials, which is quite a large percentage since hazardous materials can be as varied as inflatable life rafts, flares, and hazardous wastes. Table 2-18 shows an excerpt from the Hazardous Materials Table. Companies that are affected must ensure that employees have received appropriate training for han-

Symbols	Hazardous Materials Description and Proper Shipping Names	Hazard Class or Division	Identification Numbers	Packing Group	Label(s) Required (if not excepted)	Special Provisions	(8) Packaging Authorizations (§173.***)			Placards Required
							Exceptions	Non-bulk packaging	Bulk packaging	*Denotes Placard for any quantity
(1)	(2)	(3)	(4)	(5)	(6)	(7)	(8A)	(8B)	(8C)	(9)
AW	Castor beans or Castor meal or Castor pomace or Castor flake	9	UN2969	II	None		155	204	240	NONE
	Caustic alkali lquids, n.o.s.	8	UN1719	I	CORROSIVE	A7, B10, T42	None	201	242	CORROSIVE
				II	CORROSIVE	B2, T14	154	202	242	CORROSIVE
				III	CORROSIVE	T7	154	203	241	CORROSIVE
	Caustic potash, see Potassium hydroxide etc.									
	Caustic soda, (etc) see Sodium hydroxide etc.									
	Celluloid, in block, rods, rolls, sheets, tubes etc., except scrap	4.1	UN2000	III	FLAMMABLE SOLID		None	213	240	FLAMMABLE SOLID
	Celluloid, scrap	4.2	UN2002	III	SPONTANEOUSLY COMBUSTIBLE		None	213	241	SPONTANEOUSLY COMBUSTIBLE
	Cement, see Adhesives containing flammable liquid									
	Cerium, slabs, ingots, or rods	4.1	UN1333	II	FLAMMABLE SOLID	N34	None	212	240	FLAMMABLE SOLID
	Cerium, turnings or gritty powder	4.3	UN3078	II	DANGEROUS WHEN WET	A1	None	213	242	DANGEROUS WHEN WET*

Table 2-18: An excerpt from a hazardous materials regulations table.

dling hazardous materials. They must also comply with the federal standards pertaining to: classification, packaging, handling, marking/labeling/placarding, and documentation. The shipper is also responsible for verifying that carriers who transport their hazardous materials are properly registered, use safe vehicles, are properly trained for emergency response, and understand and comply with all applicable regulations.

In order to comply with DOT regulations, hazardous materials shippers must:

1. Use hazardous materials regulations to decide the material's
   — Proper shipping name
   — Hazard class and division
   — Identification number (ID)
   — Correct packaging
   — Correct label and markings
   — Correct Placard

2. Package the material, label and mark package, prepare shipping paper and emergency response information, and supply appropriate placards;

3. Certify on the shipping paper that the shipment has been prepared according to the rules, unless a private carrier is used;

4. Arrange for proper loading, blocking, and bracing;

5. Understand and follow any additional regulations (e.g., EPA/OSHA).

Carriers of hazardous materials loads have responsibilities as well. They must:

— Check that the shipper correctly named, labeled, and marked the shipment;

— Determine that proper placard(s) are affixed;

— Refuse improper shipments;

— Report collisions and incidents involving hazardous materials/wastes to the proper government agency.

These requirements are designed to prevent environmental contamination and protect public health and safety. Preventing spills is a pollution prevention (P2) activity and keeps a usable hazardous material from being turned into a hazardous waste.

Checking Your Understanding

1. What are the nine DOT hazard classes?

2. What is the DOT definition of a hazardous material?

3. What are five things that a hazardous materials shipper must do to comply with DOT regulations?

2-6 Getting Information on Environmental, Health and Safety, and Transportation Laws

Throughout this discussion of laws and regulations, we have cited the Code of Federal Regulations (CFR). In order to assure a facility is in compliance, it is necessary to be able to access the specific regulatory language contained in the regulations. There are several ways to obtain the specific CFR for the regulations being researched.

1. Purchasing the volume or volumes of CFR one needs:
    —Superintendent of Documents at the Government Printing Office;
    —Private vendors (see safety catalogs) sell them in both hardcopy and electronic formats.

2. Reviewing information at public access locations:
    —Most cities have a government repository library or law library where the CFRs are located in the reference or government documents area;
    —The House of Representatives and some environmental organizations maintain links on their Internet Web pages to the complete CFRs. Built-in search engines allow key word searches. To find these locations use your Web browser to search for Code of Federal Regulations.

Other important information that will help you learn about specific compliance issues relating to an industry and help with pollution prevention techniques and technologies can be found with the following sources:

—EPA has numerous documents related to pollution prevention and compliance that are freely distributed. Each EPA Regional Office also maintains a library staffed with knowledgeable people that can help you find the documents you need. Since it is impractical to list phone numbers in this book because they change so frequently, call your local environmental department, or look in the government pages of your phone directory for agencies such as EPA or OSHA.

—EPA, OSHA, and DOT have extensive information on their Internet pages. Use your browser to search for each by name.

—Trade associations for your industry are helpful in sharing information on new technologies and rules that affect your workplace.

—Non-profit environmental organizations often publish helpful information under government grants. Check for environmental organizations and "clearing houses" on the Internet.

Checking Your Understanding

1. What are two sources for obtaining volumes of CFR?

2. What are four other sources that can be used to obtain complieance information?

Summary

This chapter provides an overview of the major environmental, health and safety, and transportation laws that impact industry. In general they can be subdivided into those laws that regulate hazardous material storage and handling, air emissions, waste discharges, hazardous waste site management, site cleanup, emergency planning and notification, employee safety, and hazardous material transportation.

The Toxic Substance Control Act (TSCA) requires the screening of new chemical substances under premanufacturing notifications (PMNs) and review of "significant new use" for previously approved chemicals to insure they do not pose a risk. Under TSCA, several substances used in the past have been banned, including PCBs, CFCs, and easily crumbled asbestos. The Federal Insecticide, Fungicide, and Rodenticide Act (FIFRA) oversees the manufacture and use of pesticides. Under FIFRA, a company must register a new pesticide before starting manufacture.

The Occupational Safety and Health Act (OSHA) is designed to ensure safe and healthful working conditions for firms that have ten or more employees. Workers falling under OSHA regulation are protected by the following provisions: the General Industry Safety Standard; Hazard Communication; Hazardous Waste Operations and Emergency Response (HAZWOPER); and Confined Space Entry.

Laws relating to the disposal of by-products of industrial processes include the Clean Air Act (CAA), Clean Water Act (CWA) and the Safe Drinking Water Act (SDWA). Within each of these major Acts, standards are set for emissions. Within the CAA, the National Emission Standards for Hazardous Air Pollutants (NESHAPs) identifies and sets emission limits on such things as VOCs because of their direct linkage to the formation of ground-level ozone. Additional provisions falling under the CWA establish allowable oil discharge limits and the filing of a Spill Prevention Control and Counter-Measure plan by any industry that owns or operates large aboveground oil storage tanks. The Federal Water Pollution Control Act establishes reportable quantities (RQs) for hazardous materials. It also sets up the National Pollutant Discharge Elimination System (NPDES) permit program that regulates their discharge and requires storm water permits. The Safe Drinking Water Act (SDWA) establishes Maximum Contaminant Levels (MCLs) for a variety of substances to protect our drinking water supplies. It further regulates the classification and use of injection wells through the Underground Injection Control (UIC) program.

The Resource Conservation and Recovery Act (RCRA) is the major legislation that addresses the cleanup and disposal of chemical wastes. Through the use of small quantity generator (SQG), conditionally exempt small quantity generator (CESQG), and large quantity generator (LQG) classifications, RCRA sets the reporting standards and requires the use of the Uniform Hazardous Waste Manifest. RCRA Subsection I defines and regulates the use of underground storage tanks of hazardous materials.

The Comprehensive Environmental Response, Compensation, and Liability Act (CERCLA) provides a system for identifying and cleaning up wastes that are released into any part of the environment. CERCLA was amended by the Superfund Amendments and Reauthorization Act (SARA) in 1986, which establishes the Emergency Planning and Community Right-to-Know Act (called SARA Title III); SARA III established the emergency planning framework and right of local government and members of the public to obtain information on the hazards posed by potential toxic substance releases.

Finally, the Hazardous Materials Transportation Act (HMTA) covers the transportation of hazardous materials and wastes including radioactive materials that pose a risk to health and safety. HMTA requires the proper placarding of the vehicle and marking and labeling of the shipping containers that are used to transport hazardous substances.

Although there is not, at this time, a law requiring specific percentage reductions in the overall wastes generated by industry, Executive Order 12856 does mandate reducyions for federal facilities. Regulatory compliance appears to be moving in the direction of pollution prevention (P2), and given the regulatory burden that could be reduced by doing so, it makes sense for industry to take up the challenge. It will take commitment, careful planning, creative problem-solving, changed attitudes, and investments. The rewards, however, can be significant: including reduced liability, more efficient use of natural resources, reduced treatment and disposal costs, reduced regulatory involvement, and increased public relations.

Critical Thinking Activities

1. This chapter categorizes regulations into four convenient groups based on activities and sources that give rise to the hazardous material. Develop a different grouping of the laws based on an entirely different "in-common" premise.

2. Develop a chart listing the various laws on one axis and the following categories on the other: control of use, handling, and storage; emissions; spills; discharges; cleanup; planning and notification; and permitting. Use the information in the chapter to fill in the chart identifying the specific requirements of each law for each category.

3. Do a search on the Internet for information on OSHA, DOT, and EPA. From this search, find at least one piece of information that adds to the information on the laws in this chapter.

4. Obtain a copy of a few pages of regulations from the CFRs cited in the chapter. Do a critical comparison of the synopsis in this chapter with the actual regulations. Write a report addressing discrepancies or clarifications.

Checking Your Knowledge

1. Premanufacture Notification involves:
   a) Letting OSHA know before a new process for manufacturing is put online, in order for its hazards to workers to be fully investigated
   b) Communicating hazards to employees, under HAZWOPER, when they are involved in a manufacturing process
   c) Registering pesticides with EPA before they are manufactured and used in commerce
   — d) Screening a new chemical substance prior to its production and manufacture, in accordance with TSCA

2. The General Industry Safety Standards include:
   a) Pesticide registration
   — b) Control of blood-borne pathogens
   c) Bans on the use of CFCs
   d) Reporting requirements for the TSCA Inventory

3. Under the Significant New Use Rule:
   — a) Existing chemicals can be brought under review by TSCA
   b) New equipment must be tested and certified safe by OSHA
   c) Existing pesticides can be exempted from the registration process
   d) Cosmetics can be regulated under the FDCA

4. HAZWOPER has requirements for:
   a) Confined space workers to receive appropriate rescue training
   b) Pesticide workers to obtain applicator certification
   c) General industry to train workers in general safety and reading MSDSs
   — d) Hazardous waste site worker training on the hazards at their work site

5. The difference between NAAQS and NESHAPs is:
   a) NAAQS are related to industrial effluent discharges; NESHAPs are oil pollution prevention strategies
   b) NAAQS are the chemicals for which industrial facilities must test storm water runoff; NESHAPs are maximum contaminant levels for drinking water
   c) NAAQS are standards to determine whether a water body is a USDW; NESHAPs are ambient air quality standards
   — d) NAAQS are air quality standards for ambient pollutants; NESHAPs are air quality standards for higher health impact pollutants

6. Spill Prevention Control and Counter-Measure Plans are designed to:
   a) Bring facilities into compliance with the CAA by reducing emissions
   — b) Prevent discharges of oil into U.S. waters
   c) Reduce contaminants in storm water runoff
   d) Be used in lieu of NPDES permits when discharging hazardous wastes to navigable waters

7. Injection of liquid wastes is regulated under the:
   a) CAA
   — b) CWA
   c) SDWA
   d) NESHAPs

8. A nonattainment area has regionally exceeded:
   a) Appropriate drinking water standards
   b) TSCA limits for new chemical PSNs
   — c) Any of the NAAQS pollutant levels
   d) Storm water permit limits

9. MCLs are:
   a) Maximum Contaminant Levels of NAAQS pollutants
   b) Minimum Concentration Levels to achieve BACT
   c) Master Contaminant List for TSCA Inventory
   — d) Maximum Contaminant Levels for drinking water

10. Discharge of oil into navigable waters is prohibited by the:
    — a) CWA
    b) TSCA
    c) FIFRA
    d) SDWA

11. NPDES regulates discharges to:
    a) Air
    — b) Water bodies
    c) Land
    d) Above ground tanks

12. The generator subjected to the least regulatory impact is:
 — a) CESQG
 b) TSDF
 c) LQG
 d) SQG

13. All generators must do all of the following except:
 a) Evaluate waste to determine whether it is hazardous
 b) Ensure waste taken offsite is delivered to an approved TSDF
 c) Not exceed accumulation and volume limits
 — d) File a biennial report

14. Which of the following is not a characteristic for determining whether a waste is hazardous?
 a) Toxicity
 b) Lethality
 c) Corrosivity
 d) Ignitability

15. The Uniform Hazardous Waste Manifest is used to:
 a) Characterize wastes
 b) Permit underground tanks containing hazardous wastes
 c) Manage only large quantity generators
 — d) Ship hazardous wastes

16. CERCLA:
 a) Requires manifesting hazardous waste
 b) Can ban hazardous chemical substances
 — c) Lists potential contaminants to all environmental compartments
 d) Requires discharge permits

17. The Toxic Release Inventory applies to facilities that:
 a) Release any amounts of hazardous waste
 b) Have emissions related only to oil production and refining
 — c) Emit specified toxic chemicals in large quantities
 d) Discharge any TSCA Inventory chemical in any amount

18. MSDSs must be provided to state and local emergency planning committees by:
 — a) OSHA's Hazard Communication Standard
 b) SARA's requirement for TPQ chemicals
 c) CERCLA's joint and several liability standard
 d) RCRA's Subtitle D Standards

19. Which of the following is not a DOT hazard class?
 — a) Dangerous Solids
 b) Flammable Liquids
 c) Corrosives
 d) Gases

20. Shipping papers provide information on:
 a) The driver and the carrier, especially their liability insurance carrier
 — b) The name, hazards, identification, and reportable quantities of onboard materials
 c) The route, approved stopping areas, and total quantities of cargo
 d) Shipments by hazardous waste haulers going to a TSDF

3

Physical Treatment Technologies

by Howard Guyer

Chapter Objectives

Upon completing this chapter, the student will be able to:

1. **Describe** the physical properties of substances that can serve as the basis for physical treatment technologies.

2. **Explain** why physical treatment technologies do not change the chemical nature of the waste.

3. **Compare** physical treatment technologies that are particle-size dependent.

4. **Explain** why coagulation and flocculation are used as pretreatment processes prior to the application of other physical treatment technologies.

5. **Discuss** the principles involved in filtration and scrubber technologies.

6. **Explain** the principles involved in electrostatic precipitation.

7. **Compare** various physical treatment technologies that utilize vapor pressure differences.

8. **Explain** the principles involved in immobilization technologies.

Chapter Sections

3-1 Introduction

In Chapter 1 the four recognized treatment technologies (physical, chemical, biological, and thermal) and their application to wastes in the air, water, and soil media were introduced. The focus of this chapter is on **physical treatment technologies**, which refer to those processes that make use of the physical properties of a substance to separate, concentrate, or remove waste without destroying it. See Table 3-1 for a list of EPA's Physical Treatment Reporting Codes.

When the waste is in sufficient concentration, one of the many styles of filtration, distillation, air stripping, adsorption, or electrostatic precipitation techniques can be applied to remove it directly. If the substance is not of sufficiently high concentration, however, then collection of filtration back washes, coagulation, flocculation, evaporation, or sedimentation technologies are typically applied prior to employing other treatment methods.

Physical treatment methods tend to be more familiar and appear less exotic than most other treatment technologies. They are also seldom applied in isolation, but rather as steps in an overall waste stream management plan. Their typical role is to concentrate a waste stream so that other treatment methods can be applied economically. In fact, many other treatment methods would be ineffective without prior physical treatment of the input waste stream.

Separation of Components		Removal of Specific Components	
T35	Centrifugation	T48	Adsorption - molecular sieve
T36	Clarification		
T37	Coagulation	T49	Activated carbon
T38	Decanting	T50	Blending
T39	Encapsulation	T51	Catalysis
T40	Filtration	T52	Crystallization
T41	Flocculation	T53	Dialysis
T42	Flotation	T54	Distillation
T43	Foaming	T55	Electodialysis
T44	Sedimentation	T56	Electrolysis
T45	Thickening	T57	Evaporation
T46	Ultrafiltration	T58	High gradient magnetic separation
		T59	Leaching
		T60	Liquid ion exchange
		T61	Liquid-liquid extraction
		T62	Reverse osmosis
		T63	Solvent recovery
		T64	Stripping
		T65	Sand filter

Table 3-1: A list of EPA's physical treatment reporting codes.

3-2 Physical Properties

Before discussing the actual physical treatment technologies, it is useful to realize that each of us has a unique set of physical characteristics by which we can be described. These characteristics include hair, eye and skin color; our height and weight; and other distinguishing features such as the size and shape of our ears, nose, chin, or even a mole. In a similar fashion, each **pure substance** has a unique set of **physical properties**. These properties include its melting and **boiling points**, color, odor, **vapor pressure**, **density**, ability to conduct heat and electricity, **malleability** and **ductility**, and water solubility. These are all properties that can be measured without altering the nature of the substance.

Chemical properties, on the other hand, describe how the substance reacts with other substances. Important questions to ask are: is it flammable; does it react with acids, bases, strong oxidizing or reducing agents; or is it chemically inert? Chemical properties can be compared to our personalities and how we interact with others. To test a substance's chemical properties requires changing the chemical nature of the substance.

The explanation of these physical and chemical properties lies deep within the composition of the substance. Basically, it is attributed to how the atom's

Ionic Compounds	Molecular Compounds
Form hard crystals	Form soft, waxy, greasy crystals
High melting & boiling points	Low melting & boiling points
Many are water soluble	Most are not water soluble
Most are not soluble in organic solvents	Many are soluble in organic solvents
Molten form will conduct electricity	Molten form will not conduct electricity
Typically solids at room temperature	Vary from gases to solids at room temperature
Typically nonflammable	Most flammable

Table 3-2: Comparison of properties of ionic and molecular compounds.

electrons are arranged and to the type of chemical bonding (arrangements made with their neighboring atoms) holding them together. Atoms that have either lost or acquired electron(s) become ions; ionic compounds formed from these ions tend to have hard crystals with high melting and boiling points. Depending on the ions present, they may be water soluble like NaCl or not, like AgCl. Because they have strong forces holding each ion in the crystal structure, they tend to break into smaller pieces when pressure is applied and are, therefore, not malleable or ductile. Since ionic solids are composed of repeated rows of oppositely charged ions stacked one upon another, there is no single unit within the crystal. Each may be allowed to grow to almost any size.

Substances that are formed by sharing electrons, either equally or unequally, are **nonpolar** or **polar covalent** substances, respectively. Within a nonpolar covalent substance, such as CH_4, each molecule is an independent unit, like a glass marble. Because of their lack of attraction for each other, they tend to have low melting points, are electrically nonconducting, do not dissolve in water, and – if cooled sufficiently – form soft crystalline, waxy, greasy solids like color crayons.

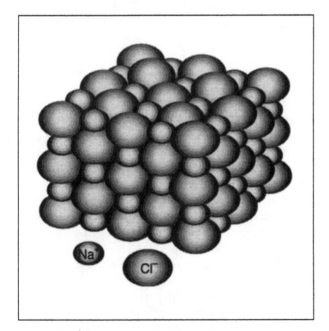

Figure 3-1: Diagram of a NaCl crystal.

Polar covalent substances, due to the unequal sharing of electrons, develop **dipoles** or partially charged sites on the molecule, that behave as small magnetic poles. The partial positive ($\delta+$) end of one polar covalent molecule is, therefore, attracted to the partially negative end ($\delta-$) of a second molecule, etc. This weak, but effective, attraction between polar covalent molecules is called **hydrogen bonding**. Polar covalent molecules, like water, are not only attracted to ions, but they are also attracted to other polar covalent molecules. Depending on the size of the molecule, the resulting hydrogen bonding gives varying degrees of water solubility to many organic substances, e.g. ethyl alcohol. Hydrogen bonding between polar molecules is also responsible for increases in their melting and boiling points, vapor pressures, viscosity, etc. The viscous, syrupy nature and water solubility of glycerin, ethylene, and propylene glycol (antifreeze), for example, is due to multiple polar hydroxide groups (–OH) that hydrogen bond not only to each other, but also with water molecules.

Using Unique Physical Properties

As previously noted, physical treatment technologies do not alter, but rather make use of the unique set of physical properties of a substance to concentrate or separate it from a waste stream. If a cup of sugar were mixed into a pail of sand, for example, using a screen in an attempt to separate them based on crystal size would likely give poor results. But, since sand and sugar differ in their water solubilities, a washing and filtering process could be used to remove the sugar from the sand. If this is then followed by evaporation of the sugar water **filtrate**, another type of physical treatment process, the crystalline sugar can be recovered. As a second example, had the sand been mixed with iron filings, the physical property of magnetism could be used to separate the iron filings from the sand.

Pretreatment Strategies

To improve the efficiency of many physical treatment procedures, a variety of pretreatment strategies have been developed. For example, when filtering and sludge-dewatering operations are employed, often the waste stream is pretreated by the addition of a **coagulant** or **flocculant**. The flocculant, typically a polymer, neutralizes the negative charges on the small (1-100 nm), colloidally suspended particles allowing them to **coagulate** or **agglomerate** into groups. The steps in the overall removal process are, therefore, charge neutralization – which allows agglomeration of the colloidal-sized particles into groups – followed by **flocculation** of these groups into larger groups or **tufts**, and finally removal by filtration.

Coagulants

Coagulants are substances that are capable of removing or destabilizing the electric charge on colloidally dispersed particles. They are typically subdivided into inorganic (lime, alum, ferric sulfate, and ferric chloride) and natural or synthetic organic polymer groups. As later discussed in Chapter 4, lime, $Ca(OH)_2$, for example, can serve as both a precipitant and a coagulant for waste streams containing heavy metal ions. Like most inorganic coagulants, lime functions by forming a metal hydroxide precipitate, destabilizing the charge on any colloidally dispersed particles by introducing positive calcium ions – Ca^{2+} – and further entrapping the colloidally dispersed particles in the precipitant-coagulant matrix.

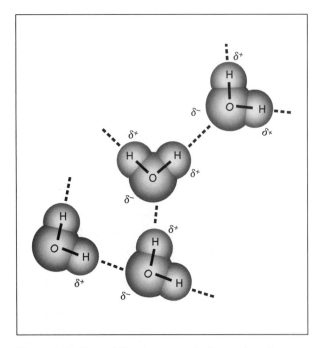

Figure 3-2: Dotted lines represent hydrogen bonding between polar covalent water molecules.

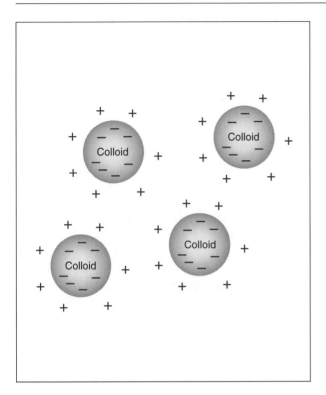

Figure 3-3: Charge distribution surrounding colloidally dispersed particles.

The organic polymers, known as **polyelectrolytes**, are high molecular weight polymers of either natural (protein, gum arabic) or synthetic origin. Polyelectrolytes function by removing or destabilizing the charge on the colloidally dispersed particles, then continuing to associate with the polymers, forming a solid known as **floc**, which can then settle.

Coagulation

Coagulation, or agglomeration, is the irreversible assembling of colloidal-sized particles from solution to form tufts. Typically, colloidal-sized particles of substances like clays or starches have the same negative surface charge, which causes them to repel each other and, therefore, resist coagulation. Through neutralization or destabilization of the charge by adding a coagulant, the particles can now agglomerate. For agglomeration to occur, the destabilized particles must collide. The most effective method to produce coagulation is, therefore, the use of high-intensity mixing to thoroughly distribute the coagulant and further destabilize the colloidal dispersion, while promoting rapid particle collisions.

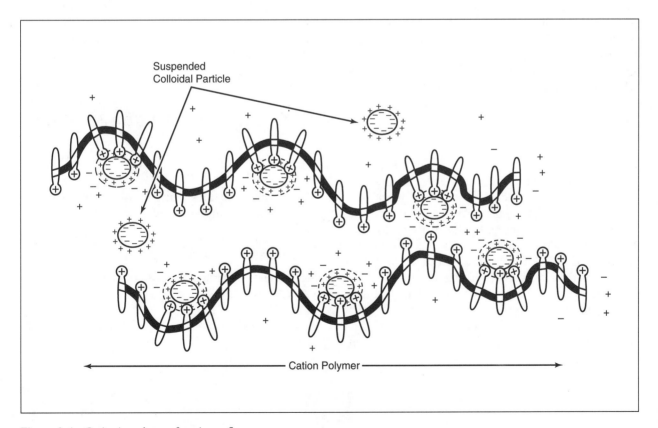

Figure 3-4: Cationic polymer forming a floc.

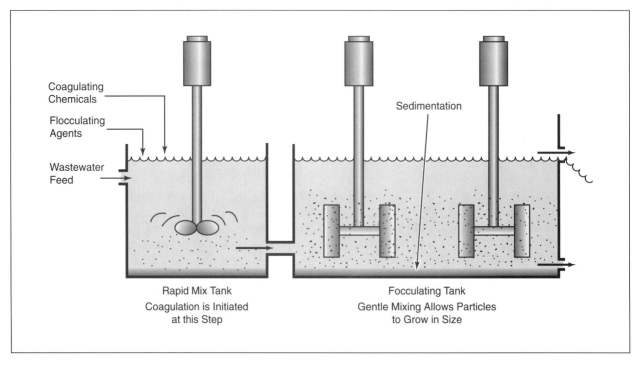

Figure 3-5: Mixing and flocculation tanks used to remove suspended wastes.

In the past, municipal water treatment processing commonly removed colloidally dispersed clay particles by adding the chemical coagulant, alum, $Al_2(SO_4)_3 \cdot 18\ H_2O$. Today, organic polymers that have the ability to act on either negatively or positively charged particles are frequently the choice when coagulating agents are needed.

Flocculation

As previously stated, the aggregation of coagulated particles into larger visible particles is called a floc. The formation of a floc is improved by gentle stirring to enhance the opportunity for coagulated particles to bump into each other. Excessive and strong stirring should be avoided, because it can overcome the relatively weak forces holding the floc together and could actually result in its destruction.

The overall goal of the pretreatment strategies – and, therefore, the efficiency of the physical treatment technology that follows – is to increase the particle size through coagulation and flocculation.

The larger floc particles tend to be easier to remove or concentrate in those waste treatment processes that rely on applied force, e.g., gravitational or centrifugal; also, they are more efficiently removed in filtering and dewatering operations.

Checking Your Understanding

1. What is the term for the weak attractive forces between polar covalent molecules?

2. What effect does being a polar substance have on the melting and boiling points?

3. Explain why colloidally dispersed particles tend to repel each other.

4. What is the main purpose of coagulation?

5. Define a floc.

6. Describe the overall goal of pretreatment methods.

3-3 Separation by Gravitational Force Technologies

Sedimentation Technologies

Sedimentation is the settling of suspended solids from a liquid suspension as the result of gravitational force. Perhaps one of the oldest and most familiar examples of sedimentation is the action of streams and rivers. For thousands of years their waters have carried silt particles toward the sea where the flow rate slows, depositing the silt particles to form the landmasses known as deltas. Sedimentation works according to the same principle: it requires that the tank, basin, or pond be sufficiently motionless to allow the particles to gradually settle. Settling rates are also influenced by physical attributes such as size, shape, and charge of the particles as well as the density difference between the suspended solid particles and the suspending liquid.

If a waste stream contains materials that are not restricted from land disposal, settling ponds constructed and managed in compliance with RCRA requirements may be used. Even though settling ponds must be constructed to meet federal regulations, require extensive regulatory oversight, and are expensive in terms of land requirements, their operations equipment is simple. Companies using settling ponds typically size them to hold several weeks of wastewater output. The holding time in a settling pond must be sufficient to allow the solid particles to settle. The liquid overflow moves from the settling pond to be discharged, recycled, or sent for further treatment. Settling tanks operate in much the same way as settling ponds and are inexpensive to purchase; however, they do require the use of large amounts of expensive floor space.

Clarifiers

Clarifiers, in general, are tanks of varying design and shape that rely on gravity sedimentation to separate solids from a flow-through liquid waste stream. Several different clarifier designs are in common use, but they retain many similarities. In all clarifiers the waste stream enters in an area known as the mixing zone and then moves into the settling zone where the solids are allowed to sink and collect as sludge. The clarified liquids are removed from above the settled solids by a process called **decantation**. The settled solids are later collected for further treatment or disposal.

As previously noted, the physical attributes of the suspended waste tend to determine the settling rate. By altering the charge, size, and/or density of the suspended solids, the settling time can be greatly reduced. For this reason, clarifier efficiency can be greatly improved when it is used in conjunction with chemical precipitation or flocculation. This means that the amount of settling time needed for the liquid to pass through the clarifier is shorter or that the physical length of the settling zone can be reduced, while still allowing sufficient time for the sludge particles to sink.

The percentage of suspended solids removed by a clarifier depends on its design and operation. In recent years, engineers have found that circular or square tanks are preferred for reasons of floor space and/or economics. As can be seen by Figure 3-6,

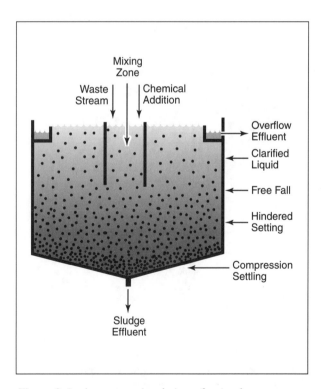

Figure 3-6: A cross-sectional view of a circular center-feed clarifier.

circular tanks, however, are less efficient due to a shorter settling zone. This deficiency has been addressed by eliminating much of the inlet flow turbulence and, therefore, reducing the size of the surrounding mixing zone. This reduction has the advantage of increasing the length of the settling zone without increasing the tank's diameter.

Slant Tube Clarifiers – An alternative design to the circular tank clarifier is the **slant tube clarifier**. A slant tube clarifier has a series of parallel tubes that are set at an angle of 45°-60° to the surface of the liquid. The tubes are housed in a tank structure with some type of overflow channel at the top and a mechanical scraper for sludge removal at the bottom. The tubes are made of plastic and are generally about two inches square. Liquid enters at the bottom and flows upward through the tubes. Solids removal takes place within the tube. Because of the design, the solid particles have only a short ver-

tical distance to fall before striking the lower side of the tube. On the tube surface, the particles agglomerate before sliding down the lower side and onto the sludge bed where they can be removed.

Inclined Plate Clarifiers – An **inclined plate clarifier** is very similar to the slant tube clarifier, except that plates are used in place of tubes. The advantage of both slant tube and inclined plate clarifiers is their large treatment capability compared to their relatively small size. These are efficient units that are available in sizes capable of handling from a few to several hundred gallons per minute. The disadvantages to both slant tube and inclined plate clarifiers are higher cost and maintenance requirements, when compared to settling tanks.

Lamella Clarifier – A more recent addition is the **lamella clarifier**, which contains inclined plates that are even more closely spaced. This creates more

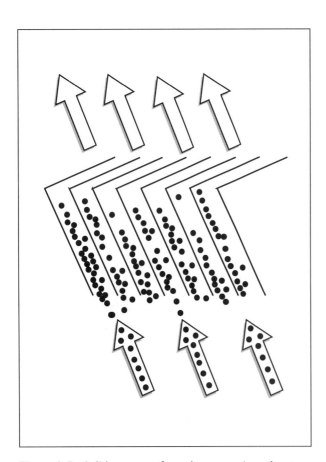

Figure 3-7: Solids separate from the contaminated water and agglomerate on the plates of the inclined plate clarifier, while the now clean water moves upward and out of the plates.

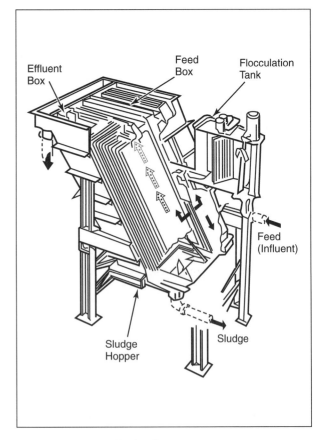

Figure 3-8: A lamella clarifier.

settling surface in a smaller volume and further reduces the unit's space requirements. The lamella plates multiply the available settling surface by greatly shortening the settling distance. The coagulation and flocculation process on the lamella concentrates the accumulated sludge as it moves down the inclined surface. The lamella unit is ideally suited for treatment of waste streams in places where space is at a premium. It is also generally less costly to purchase and install and easier to operate than most other clarifiers.

Flotation Technologies

The principle of density difference used in sedimentation is also used to float wastes that are less dense than the surrounding liquid. For example, in the settling basins at a wastewater treatment plant, while sedimentation results in grit separation, skimmers are used to remove the less dense greases and oils that are floating.

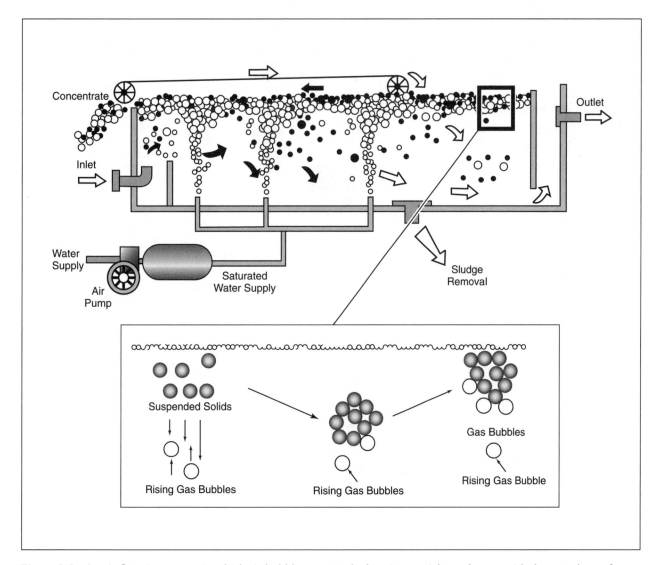

Figure 3-9: An air flotation system in which air bubbles are attached to tiny particles and move with them to the surface.

Dissolved Air Flotation

Even if a substance is denser than its surrounding liquid, **dissolved air flotation (DAF)** technologies provide a way to make the substance float. In this process finely dispersed gas bubbles are allowed to contact the suspended or emulsified waste particles to form a particle-gas composite that is less dense than the liquid. The particle-gas composite, therefore, moves to the surface where it can be easily removed by skimming.

Electroflotation – DAF technologies also include violently agitated froth flotation – used by the mining industry in the separation of ores – and **electroflotation**. The only difference between the various DAF technologies is how the gas bubbles are generated. In electroflotation, for example, the hydrogen and oxygen gases are generated by the electrolysis of the surrounding water. The waste particles, which are often charged, migrate to the electrodes where the electrolysis is taking place and coalesce on the emerging and rising bubbles.

In general, DAF can be used to treat any aqueous waste stream containing immiscible liquids, suspended solids, and other **hydrophobic** (having a low affinity for water) chemicals. It is the treatment method that is most commonly used to separate oil and greases from the wastewaters generated by the petroleum refining, meat packing, deinking, and metal working industries.

Checking Your Understanding

1. Which physical separation process is responsible for the formation of river deltas?

2. List one advantage and one disadvantage of a clarifier over a settling pond.

3. Which type of clarifier requires the least amount of floor space and why?

4. Describe the major difference between each of the DAF technologies.

3-4 Fixed Media Separation Technologies

Filtration Technologies

Filtration is one of the most widespread physical separation techniques. It serves two purposes: the removal of suspended solids from a **fluid** – gas or liquid – and the dewatering of sludges or slurries. In filtering processes, the suspended solids are separated from the fluid by passage through a **porous medium** that retains the solid on its surface or between its fibers. In filtration technology, only the physical size and shape of the contaminant solid are utilized to bring about its separation from the waste stream.

The force used to move the waste stream through the filter material is created by a pressure differential caused by either gravity, by a pump, or by centrifugal force. There are two general types of filtering media available: thin, manufactured barriers such as polymers, cloth, or metal screens, and thicker, deposited barriers formed from materials such as sand, anthracite, diatomaceous earth, or ashes.

Filtration is undoubtedly one of the oldest separation methods. In its various forms it is also the most plentiful within the physical separation technologies. The particle restraint size or **porosity** of the **filtering media** varies according to the particle size of the substance to be removed, ranging from strainers and sieves to membrane filters (see Figure 3-10). The following sections will not discuss all kinds of filters, but will consider examples of the various types of filtering techniques in each category, in order of decreasing porosity.

Strainers and Bar Screens

Strainers and **bar screens** represent the filtering media with the largest openings typically used in industry. They are generally made of metal and have openings of 1-2.5 cm (0.5-1.0 inches). A bar screen, for example, is the first filtering device used in a wastewater treatment plant. Its function is to remove only the largest types of debris – bowling balls

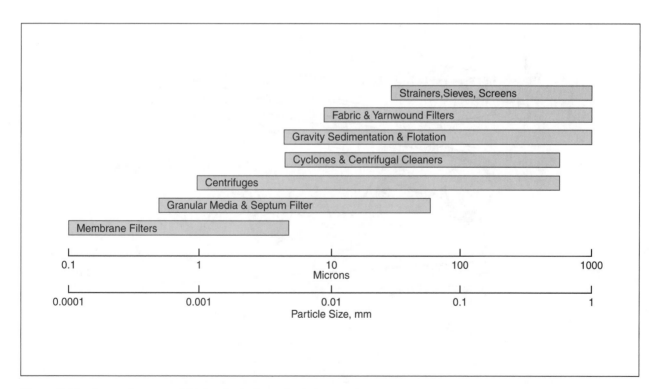

Figure 3-10: Approximate particle size for various physical separation processes.

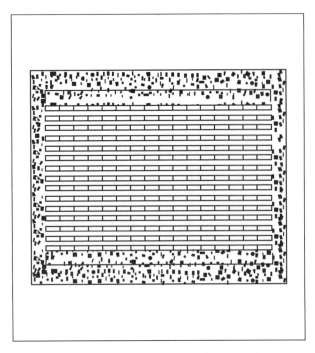

Figure 3-11: Bar screen for removing large debris.

and shopping carts – that could result in pipe blockages or pump damage. In this application, waste flows continuously through the screen; at regular intervals a rake passes across its surface to clear the bars of accumulated debris.

Belt Filter

A continuous filtering method that is often used to dewater sludges is the **belt filter**. It is composed of two belts that squeeze the waste between them as they pass through a series of rollers. The composition and pore size of the filter belt that has been selected varies according to the task. Belts are typically composed of a synthetic material with a known porosity. The belt's porosity is selected for a particular separation, based on its ability to provide the desired filtering efficiency. Once in place, sludge is evenly fed onto the lower belt and the upper belt becomes the upper edge of this belt-sludge-belt sandwich. The two belts then pass through a series

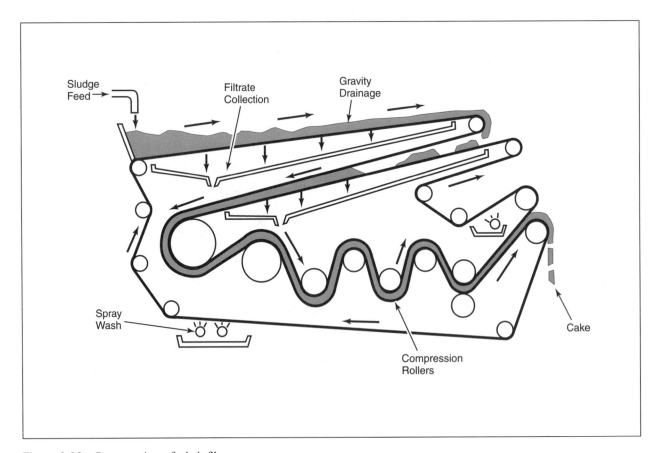

Figure 3-12: Cross-section of a belt filter.

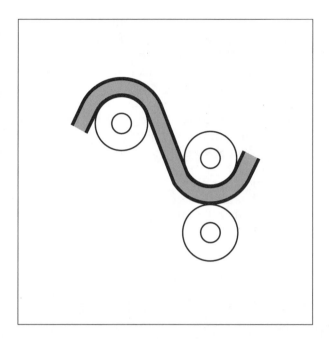

Figure 3-13: Detail of a filter belt moving through the shear zone.

of rollers that progressively apply increased pressure, through an S-shaped area known as the **shear zone**. Belt presses are particularly desirable for sludges that are difficult to dewater. The resulting dewatered sludge, containing 27-50 percent solids, is scraped from the belts for further treatment before the belt returns for its next revolution.

Rotary-drum Vacuum Filtration

Another technique that is used to dewater sludges is through the use of **rotary-drum vacuum filtration**. It is popular because it dries, or dewaters, the sludge, thereby reducing the total volume of material that will ultimately be disposed of or sent on for further treatment. In addition it is a continuous filtering process.

In a vacuum filtration unit, as shown in Figure 3-14, a compartmentalized, porous cylinder is covered by a filtering medium (cloth, coil springs, or wire mesh fabric). As the cylinder rotates through

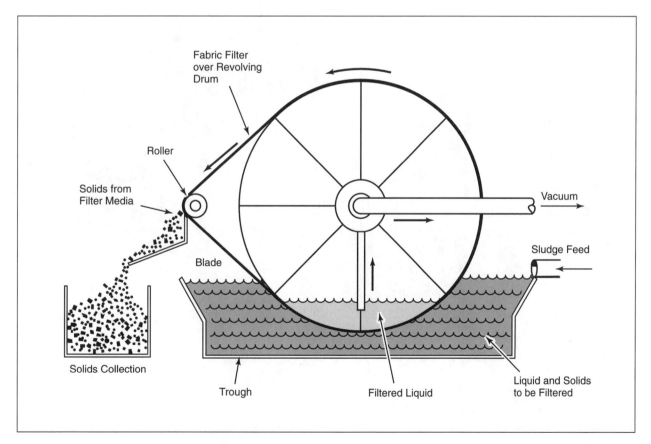

Figure 3-14: Cross-section of a rotary-drum vacuum filter.

the solution to be filtered, the vacuum inside pulls some of the sludge onto its surface forming a sludge layer. The continued application of vacuum to this segment of the compartment results in water being drawn from the sludge into the center of the cylinder. Just before the sludge is to be scrapped off the drum, the vacuum in that segment of the cylinder is reduced, improving the ability of the scraper to remove the sludge layer.

The nature of the filter medium determines the performance and life of the filter. As in most other filtering processes, the accumulated solid or **filter cake** that initially forms on the surface of the cylinder actually improves its filtering capabilities and results in the removal of increasingly smaller sized particles. Thicker sludges will generally filter and

release from the medium more readily than dilute sludges. Large solid particles are also more desirable because clogging of the filter may occur if high levels of small particles are present. Since large particles filter more readily and have less of a tendency to clog the medium, coagulation and flocculation chemicals are often used as a pretreatment step in conjunction with vacuum filtration.

The solid/water content of the filter cake may vary considerably, depending on the characteristics of the sludge or **slurry** (a dilute water and solid mixture) being dewatered. Inorganic slurries offer the potential to dewater up to 70 percent solids, while biological sludges may achieve only a 25 percent solids level.

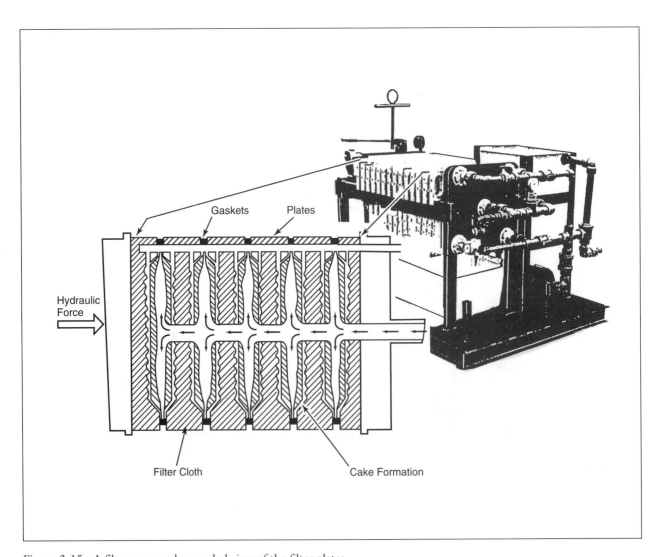

Figure 3-15: A filter press and expanded view of the filter plates.

Filter Press

Sludges and slurries can also be dewatered through the use of a **plate and frame filter press** (also known just as a **filter press**). Unlike the continuous flow process discussed above, a filter press is a batch process technology.

A filter press consists of a series of vertical plates (Figure 3-15) held on a frame and pressed together. Each plate has a corrugated or channeled surface. Between each plate is a filter medium – either cloth or woven plastic. Again, the choice for the porosity of the mesh depends on what filtering efficiency is desired and on the size of the suspended solute particles to be removed.

The sludge or slurry to be dewatered is pumped into the center of the stack of plates and works its way through the filter to the exit. Along the way the solids are trapped by the mesh of the filtering medium. As one part of the filter is filled, the waste stream must find other less restricted channels. Eventually the filter is full and there is a significant drop in the flow rate. The flow is then typically switched to a second unit, while hydraulic or air pressure is applied to the first unit to force out most of the remaining trapped water.

Once the pressure is released, the filter is opened and the individual plates are manually disassembled. After the sludge is removed by scrapping each plate, the filter is reassembled and readied for its next turn of service.

Filter presses, belt filters, and rotary-drum vacuum filters are ideally suited for treating sludges and slurries that contain 5-10 percent solids. For sludges and slurries with lower solids content, pretreatment techniques may be necessary to make these separation methods economical.

Centrifugal Filtration

Centrifugal filtration is based on the invention of Swedish scientist Gustaf de Laval, who designed a device – using centrifugal force – to separate colloidally suspended fat globules (cream) from whole milk. Although the less dense cream – acting under gravitational force – will rise to the top of a container of whole milk in several hours, this device allowed the separation to be accomplished in seconds.

Like the tub in a washing machine, its rotation around a central point results in an outward pull called centrifugal force. During centrifugal filtration operations, the waste stream is fed into a basket rotating at speeds of 1,500-2,500 rpm. Due to the resulting centrifugal force – several thousand times greater than gravitational force – the liquid is quickly passed through the basket's outer wall, where the filtering medium is located and the suspended solids are collected. In other words, it operates as a highly efficient filter.

The basket centrifuge is a batch process method that is a common choice for dewatering slurries and pumpable sludges. The sludge or slurry enters the rotating basket through a feedline. The liquid, called a **centrate**, is collected along the outside wall of the centrifuge in a process that resembles what happens during a washing machine's spin-dry cycle. Basket centrifuges are especially desirable whenever a reasonably dry cake is needed. If the sludge is to be incinerated, for example, it needs to be sufficiently dry to minimize subsequent auxiliary fuel requirements. The centrate usually contains some finely divided suspended solids that can be removed by recycling it back through the filter.

Centrifugal filtration methods are used in a variety of dewatering applications because the equipment is compact and has high throughput capacity. They are simple to operate and produce dewatered sludges containing 10-40 percent solids in a very short period of time.

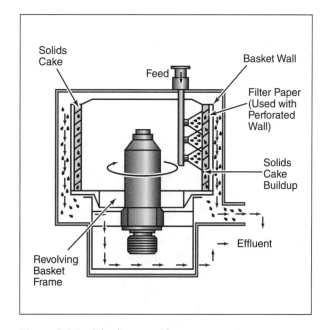

Figure 3-16: A basket centrifuge cross-section.

Checking Your Understanding

1. Identify two continuous filtering processes.

2. Which filtering process uses a shear zone to apply pressure?

3. Describe how the filter cake is removed from a filter press.

4. Which can be greater: gravitational or centrifugal force?

3-5 Other Physical Separation Technologies

Dual Media Filters

As the name implies, **dual media filters** are filters composed of two different types of filtering media. These typically consist of an upper bed of activated charcoal, anthracite (hard coal), or assorted rocks (gravel), with a lower bed of garnet or fine sand. The depth of the filter beds varies, but total bed depths of six feet are not uncommon. This technology is also a batch filtering process.

The waste stream enters at the top of the filter bed. As the solution percolates through the porous upper bed, the larger suspended particles are trapped in and on the surface of the activated carbon or coal. Smaller particles not trapped continue into the lower bed of rocks/sand. As the particle size in the lower medium layers get smaller and smaller, the space between the particles is reduced, thereby trapping increasingly smaller suspended particles.

When the filter bed becomes clogged to the point that its filtering efficiency has been lost, the flow of incoming wastewater is diverted to a second filter. **Back washing** of the first filter is accomplished by introducing a strong stream of clear water at the bottom of the filtering bed. This action frees most of the suspended material that has been trapped (concentrated) in the medium. This concentrated back wash water – heavily laden with suspended materials – is pumped off for further processing.

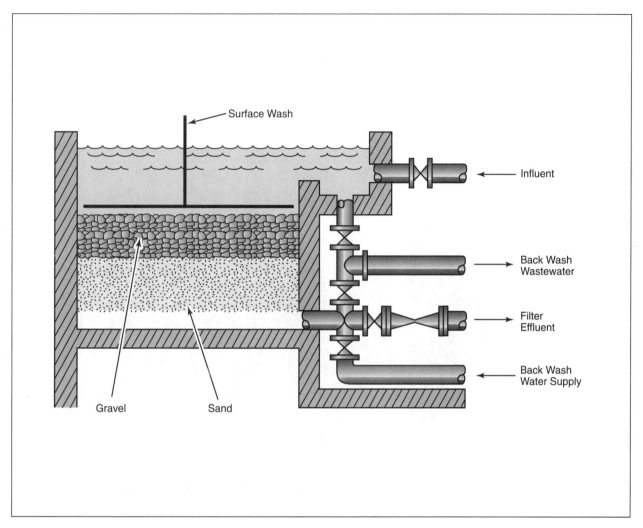

Figure 3-17: Cross-section of a dual media filter.

Once the solids are removed, back washing is stopped, which allows the filtering media to again settle. Due to the difference in densities of the media particles, the denser grains of sand settle first, followed by the less dense particles, which again form the top layer. With the integrity of the filter medium restored, the waste stream can once again be directed to this filter bed.

Baghouse Filters

Not all waste streams are liquids. One of the most common filtering technologies for removing particulate materials from a gaseous waste stream is by the use of a **baghouse**. Baghouses are basically a series of fabric bags of varying composition and porosity, stretched over wire frames.

A dirty air stream is then pulled through the filtering fabric, where the particulate matter is trapped on the exterior surface of the bag, and the clean air passes through the filter bag and on out of the stack. Wool and cotton fabric coverings were initially used. Although cotton remains popular today, many synthetic fibers, such as nylons, acrylics, polyesters, Nomex®, and Teflon®, are used because of their unique characteristics. Teflon®, for example, has good acid resistance and is capable of withstanding fairly high (500°F) temperatures. There are also many pore size options, down to sizes as small as 0.5 µm (micron).

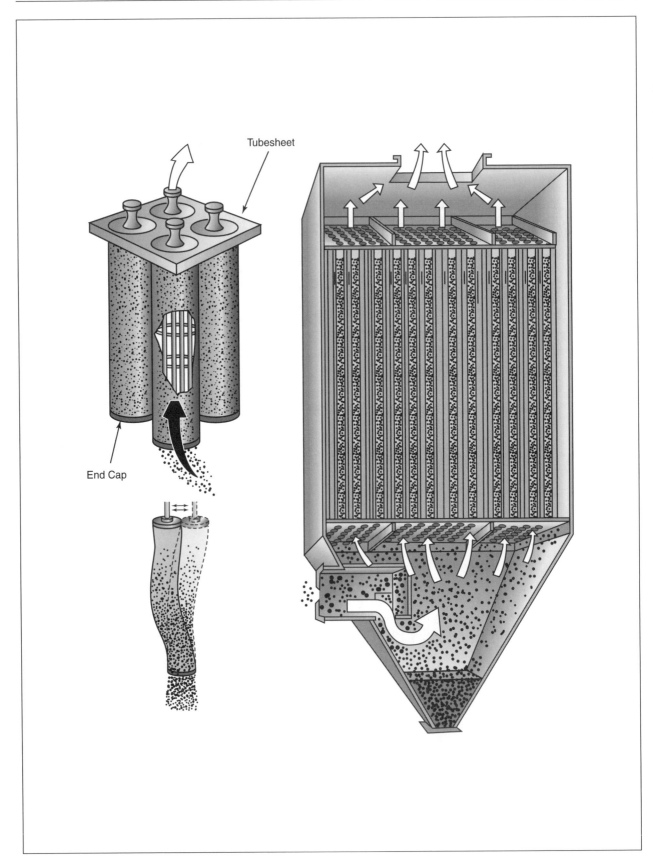

Figure 3-18: Interior of a bag house and bag shaking action.

The particulate matter or filter cake that accumulates on the external surface of the bag initially improves its filtering efficiency, until it becomes so thick that it hampers the ability of the gases to move through it. Several methods are used to remove the accumulated cake. In the intermittent cleaning method, the flow of incoming air is deflected to a second set of filter bags while the first set is cleaned. Cleaning is accomplished by either directing puffs of air into the bag or through a gentle shaking (rapping) motion. Once cleaned, the filter bags are ready to be used again.

Various methods have been devised to dislodge the accumulated cake while the baghouse remains in operation. In one of these methods, the fabric bags are supported by a flexible frame that can simply be shaken periodically, causing the cake to break apart and fall into a collection hopper below. Another option is to use high pressure traveling air streams, directed along the outside of the bag, to dislodge the cake. Use of one of these continuous methods makes it unnecessary to have a second parallel filtering system.

Cyclone Filters

A second type of filtering system designed for the removal of dry particulate matter from a gaseous waste stream is the **cyclone filter**. Cyclone filters make use of centrifugal force to remove denser par-

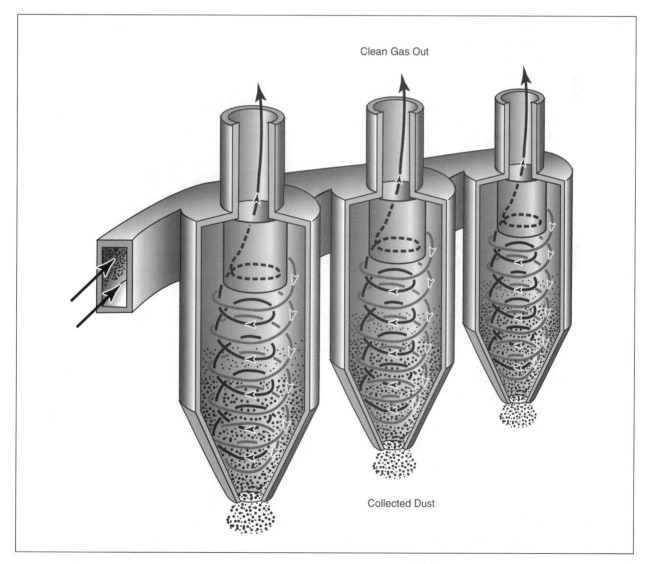

Figure 3-19: A bank of cyclone filters.

ticles from gaseous waste streams. As the dirty air stream enters the cylindrical collector (see Figure 3-19), it takes a corkscrew path downward, and the denser particles are carried to the walls where they drop into a hopper at the bottom. The clean gas rises near the center and is exhausted through the top.

Cyclone collectors are only practical when dealing with a waste stream that contains a relatively small amount of dry, reasonably large (15-50 μm) particles. They are simple to design, inexpensive to install and maintain, reliable, tolerant to high temperatures, and require a minimal amount of energy; however, they tend to be quite large and take up valuable space.

With all other factors being equal, centrifugal force is inversely proportional to the diameter of the path taken by the particle. For that reason cyclones with smaller diameters tend to have greater overall particle removal efficiencies. By placing several smaller cyclone filters in a bank, with parallel feeds, the overall particle removal efficiency can be improved over a single larger-diameter filter. Through the use of multiple smaller diameter cyclone filters, particles in the 5-20 μm size range can be removed. The one drawback to the use of smaller diameter units is that they tend to plug more readily than those with a larger diameter.

Wet and Dry Scrubbers

Physically, a scrubber may resemble and posses some of the same flow characteristics as discussed in the previous section on cyclone filters. A **scrubber**, in general, is a chamber where a dirty gas stream enters at high velocity and is suddenly slowed by expanding into the larger main chamber. Due to loss of velocity, gravity, and centrifugal forces, particles tend to settle in a downward spiraling motion in a collection area. Scrubbers come in a variety of sizes and types. They may be either wet or dry, depending on what liquid is used and in what way. Some are categorized as chemical treatment technologies because the liquid used reacts with and changes the characteristics of the waste, and will therefore be discussed in the next chapter.

Venturi Scrubbers

The venturi scrubber, which is one of the most efficient systems for particle removal, can be either wet

or dry. It consists of a converging throat or **venturi** and a diverging section into the main separator chamber. In a wet venturi scrubber the water is fed into the dust-laden gas as it enters the venturi. The

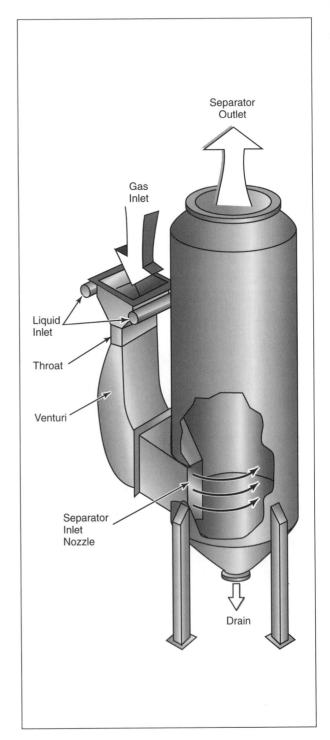

Figure 3-20: A wet venturi scrubber.

mixture exits the venturi at velocities of 100-400 mph. Due to its velocity and resulting turbulence, small dust particles have a good opportunity to come into contact with the atomized water droplets. The dust and water coalesce into larger droplets that settle more easily and are removed from the gas stream.

Even so-called dry venturi scrubbers are not totally dry. In this application, the dirty gases are dry when they move through the venturi, but encounter a water spray when they pass to the other side of the venturi.

Venturi scrubbers are very efficient and can achieve more than 99 percent removal of particles down to 0.5 μm. As such, they are typically used for removal of fine particles or – as is discussed later in Chapter 5 – as a chemical treatment technology to remove toxic vapors from incinerator stack gases. Their disadvantage is that they require a fair amount of water, which most facilities conserve by recycling.

Checking Your Understanding

1. When considering the filtering processes, which type can remove only the largest debris?

2. Explain the difference between batch and continuous flow filters.

3. Which type of filtration process uses a very rapidly rotating basket to improve filtering efficiency?

4. Which type of filtration process uses bags to remove suspended particles from a gaseous waste stream?

5. Compare the advantages and disadvantages of using several smaller diameter cyclone filters rather than only one of larger diameter.

6. What is the function of the water mist in a wet venturi scrubber?

3-6 Membrane Separation Technologies

Membrane separation technologies are expected to play an increasingly important role in the future treatment and recycling of hazardous wastes. Improved synthetic materials are providing a wide array of tough, selectively permeable membranes. These membranes are used in processes having names like **microfiltration**, **ultrafiltration**, **nanofiltration**, **hyperfiltration**, **reverse osmosis**, and **electrodialysis**. Each of these filters allows for the separation of a contaminate (solute) from a liquid phase (typically water). Membrane separation technologies can achieve several desired goals: volume reduction, recovery and/or purification of the solvent, and concentration and/or recovery of the solute.

Microfiltration

Microfiltration processes employ a microporous membrane to remove extremely small particles from an aqueous waste stream. Semipermeable plastic membranes such as Tyvek®, with pores of molecular dimensions, allow water to pass but retain suspended particles larger than 0.2 μm (0.2 micron).

Newer technologies have produced membranes with a pore size of 0.1 μm, which is called ultrafiltration. Ultrafiltration can be used to separate such things as proteins and very large carbohydrate molecules from an aqueous solution. Pore sizes of 0.001 μm or smaller allow for nanofiltration. Nano-

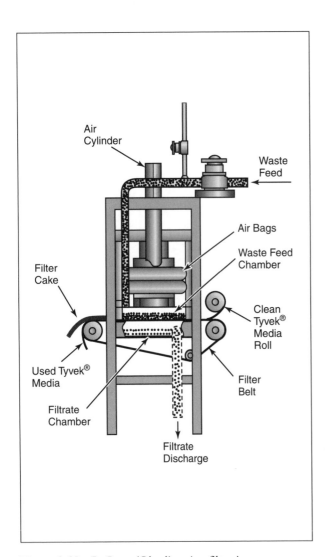

Figure 3-21: DuPont/Oberlin microfiltration system.

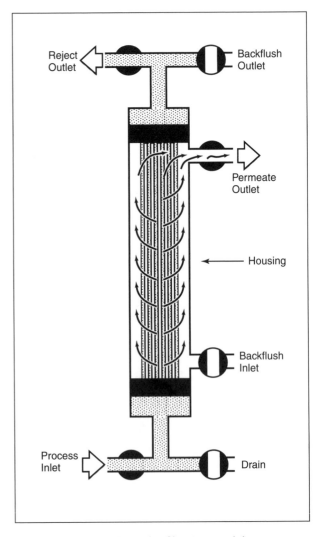

Figure 3-22: A capillary ultrafiltration module.

Filtration Method	Membrane Pore Size
Microfiltration	> 0.2 μm
Ultrafiltration	> 0.1 μm
Nanofiltration	> 0.001 μm

Table 3-3: Microfiltration ranges.

filtration, therefore, allows the removal of particles 100 times smaller than the ones captured by ultrafiltration. Due to the extremely small pore size in synthetic membranes, all of these processes are pressure driven.

In each of the above applications, water containing small suspended solids, bacteria, and/or viruses, is brought into contact with the outside of a large number of microfiltration tubes. Through the application of high external pressure, the water molecules are selectively allowed to pass into the

tubes and are collected. The materials failing to pass through the membrane are periodically removed by back flushing to restore the membrane to its original condition.

Reverse Osmosis

Reverse osmosis is the membrane technology capable of removing the smallest particles. This application functions in the reverse of nature. In nature, osmosis is described as the tendency of water to diffuse through a **semipermeable membrane** from an area of lesser **solute** (dissolved substance) concentration into an area of greater solute concentration. For example, the solute concentration in the root hairs of a tree tends to be greater than the solute concentration of the surrounding soil moisture. There is, therefore, a tendency for the water molecules to leave the moisture in the soil and diffuse into the root hair. This accumulation of water in-

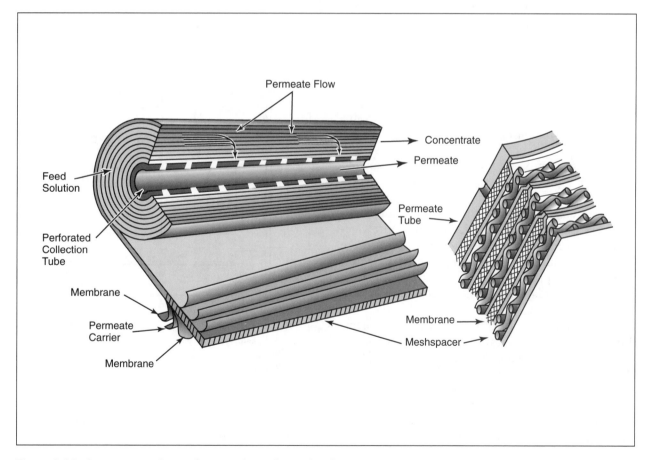

Figure 3-23: Reverse osmosis membrane and membrane detail.

side the root results in **osmotic pressure**, which, with the assistance of capillary attraction, causes the sap to rise to the top of even the tallest trees.

When reverse osmosis is used to clean an aqueous waste stream, external pressures of 100-500 psi must be applied to overcome the osmotic pressure; this causes the water to move in the opposite direction – from the area of greater solute concentration into an area of lesser concentration. Hence the name reverse osmosis.

Several different reverse osmosis configurations have been developed for commercial application. For example, the spiral wound configuration shown in Figure 3-23 consists of sheets of synthetic membrane backed by spacer material and wound up jellyroll fashion. The applied pressure forces the water molecules to pass through the membrane in the opposite direction, or onto the side that has the lower solute concentration. In this manner nearly pure water – called **permeate** – can be produced, leaving the larger molecules and ions behind.

In 1995, the Cayman Islands initiated a 600,000-gallons/day reverse osmosis sea-water-to-drinking-water project. Based in part on the success of this project, a similar reverse osmosis process is being put into operation to supply the city of Cambria in California with its drinking water. In commercial operations it is common to remove up to 90 percent of the total dissolved solids by reverse osmosis. This method is also frequently used by electroplaters to **polish** incoming city-supplied water prior to use. Periodic back flushing of the wastes that concentrate on the backside of the membrane is required.

Reverse osmosis and ultrafiltration are applied whenever both recovery and/or reuse of the permeate fluid as well as removal or recovery of the dissolved constituents is attractive; for example, in recovery of water and chemicals from dilute electroplating rinse water. To make the process more economical, ultrafiltration is frequently used as a pretreatment process prior to reverse osmosis, if there is a colloidal or large molecule component in the waste.

There are several problems involved in membrane technology application. One is the economic implication of maintaining the high pressures (100-1,000 psi) required for successful separation. The second has to do with the chemical and/or biological fouling of the membrane and – particularly with reverse osmosis membranes – the deterioration of the membrane. The latter may occur by physical,

chemical, and/or biological mechanisms. For example one of the common membrane substances – cellulose acetate – deteriorates when it comes into contact with disinfectants such as chlorine. The cost of planned and unplanned membrane replacement may be the most significant factor in the economics of membrane separation technologies.

Electrodialysis

Electrodialysis also relies on synthetic membranes, but these are placed in a direct-current electric field to cause the ions from a waste solution to be separated according to charge. Like reverse osmosis, electrodialysis membranes are sensitive to fouling and become coated with waste that must be back washed periodically. Figure 3-24 shows a typical electrodialysis cell.

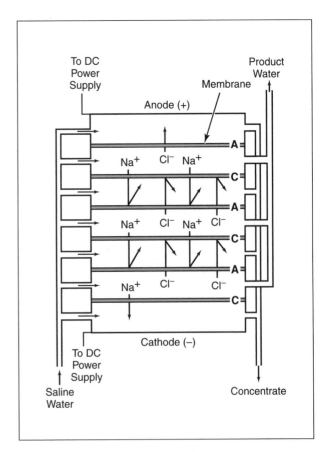

Figure 3-24: Separation of a NaCl solution in an electrodialysis process.

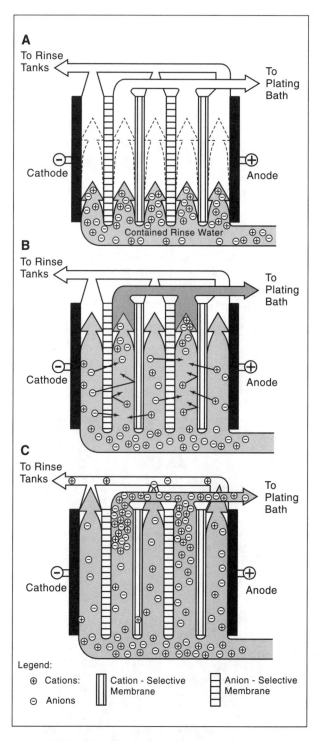

Figure 3-25: Flow schematic for separation of a plating bath using electrodialysis.

A   As contaminated water flows through the chamber, ions are attracted toward cathodes.

B   The membrane selectively causes the ions to collect in alternating chambers.

C   The ion-concentrated stream is returned to the plating bath; the ion depleted stream is used for makeup rinse water.

The cell consists of an anode and a cathode separated by cation-selective and anion-selective membranes. The waste stream to be treated is fed into the areas between the anionic- and cationic-selective membranes. The cationic-selective membrane will pass only the positively charged cation, Na^+, and the anionic-selective membrane will pass only the negatively charged anion, Cl^-. When a direct current is applied to the cell, cations migrate toward the cathode and anions migrate toward the anode. Ions in the waste stream pass through the membranes according to their charge, and are concentrated in one chamber and eliminated from the other.

By appropriately connecting the chambers, recovery of nearly ion-free solvent, and concentrated ion-containing brine is possible. As shown in Figure 3-25, electrodialysis is well suited to the removal or concentration of ions; therefore, the metal-finishing and electroplating industries are the greatest users of this technology.

Checking Your Understanding

1. Microfiltration can remove particles down to what size range?

2. What are two problems associated with the use of ultrafiltration and reverse osmosis?

3. Which of the membrane technologies uses the application of electrical energy to help separate ions?

3-7 Vapor Pressure Dependent Separation Technologies

It is commonly observed that small quantities of liquids evaporate. Equal quantities of different liquids, however, do not evaporate at the same rate. Molecules at the surface of the liquid have a tendency to escape from the remaining liquid and enter the gaseous phase. The pressure exerted by the escaping molecules is known as the vapor pressure of the substance.

Vapor pressures tend to be related to both the molecular mass and the polarity of the substance. Vapor pressure also varies with the temperature. As the temperature goes up, so does the vapor pressure of that substance. For example, the vapor pressure of polar water molecules is quite low (0.38 psi or 19.8 torr) at room temperature (72° F or 22°C). As the temperature of the water is raised to 212° F or 100°C, however, the vapor pressure reaches normal atmospheric pressure (14.7 psi or 760 torr). This point, at which the vapor pressure equals the atmospheric pressure, is by definition the boiling point of the liquid. It is at this point that the liquid starts to change very rapidly from its liquid to its gaseous form.

To illustrate the effect that the nonpolar/polar nature of the substance has on its vapor pressure, consider the following. Methane, CH_4, is a substance that has only a slightly smaller molecular mass than water. However, it is a nonpolar substance that reaches 14.7 psi or 760 torr at –259°F or –161.5°C. Nonpolar molecules have little attraction for each other; therefore, they tend to have very high vapor pressures even at room temperatures.

Figure 3-26: Mineral deposits at Mono Lake.

Evaporating Ponds

The Dead Sea, the Great Salt Lake, Mono Lake, and the Salton Sea are examples of very large naturally occurring **evaporating ponds**. Over the years water running across the land has dissolved and carried minerals into these basins. The sun and wind have supplied the energy to evaporate the water and to carry it away. As a result, the concentration of the remaining minerals has continued to increase; today only a few creatures can survive within such brackish waters.

The typical industrial evaporating pond is generally built – not naturally occurring – and much smaller than the examples described above. Aqueous waste streams or brines containing small amounts of valuable minerals are pumped into the basin and allowed to concentrate or dry completely. Except for the pumping, evaporation ponds are very energy efficient, since the work of the wind and sun are free of charge, but they do require large amounts of land. As a consequence, they need to be located in geographical areas with inexpensive land, little rain, and adequate sun. Depending on their contents and location, regulatory constraints may also affect their construction and use.

Air Stripping

Air stripping takes advantage of the different vapor pressures of compounds found in a liquid mixture. It is performed by pumping huge amounts of air into the bottom of a tank filled with wastewater and aggressively agitating the whole mixture to separate the substance with the higher vapor pressure from the rest. It can also be performed *in situ*, which is called **air sparging**; this entails drilling a series of holes into the ground and forcing air into selected holes, which carries high vapor pressure liquid into the remaining holes where it is collected (see Figure 3-27). In either case, by bringing large amounts of air into contact with the mixture, the substances with higher vapor pressures will evaporate more rapidly, resulting in the separation.

If the mixture involves nonpolar hydrocarbons (gasoline, diesel, etc.) and polar water molecules, the hydrocarbons will evaporate before water, producing very good separation results. When this treatment method is used on contaminated soils or water, the released hydrocarbon vapors are either captured with an activated carbon adsorber or incinerated – which prevents them from entering the atmosphere.

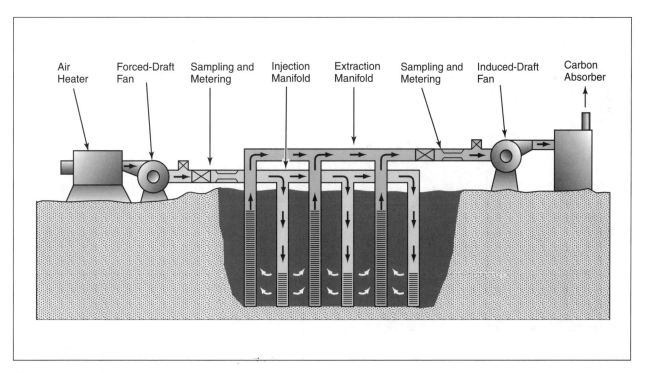

Figure 3-27: An *in situ* air sparging system.

Distillation

Distillation as a physical separation process has been in existence for thousand of years. Alchemists used it in their attempts to find the Elixir of Life and to transmute base metals into gold. They used a container called a retort to heat substances and collected its condensed vapors from its long spout. In the modern laboratory application, a mixture of liquids with different vapor pressures (boiling points) is placed into a distillation apparatus, commonly called a **still**. As the temperature of the liquid in the vessel is raised, the substance with the highest vapor pressure reaches its boiling point first. These vapors pass

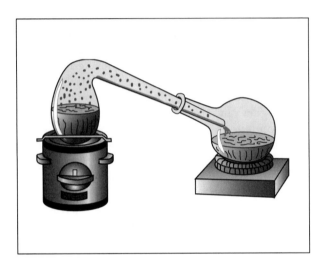

Figure 3-28: An alchemist's retort used for simple distillation.

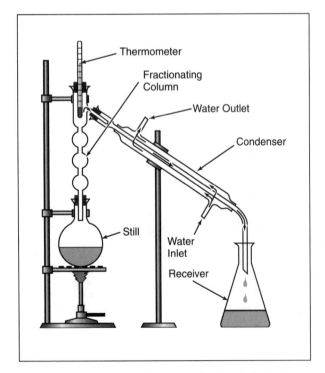

Figure 3-29: A modern laboratory fractional distillation apparatus.

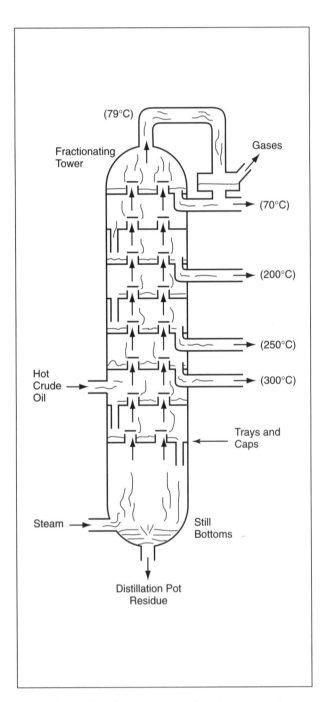

Figure 3-30: Fractionating tower showing trays and bubble caps.

through a water-cooled tube where they are condensed back into a liquid. When all of that substance is removed from the mixture, the temperature again rises and the second substance starts to boil. By changing the receiving container each time the temperature increases, the original mixture can be nearly completely separated into its constituent compounds.

Distillation is the major industrial process that occurs at a petroleum refinery. In this instance, the hot oil vapors are allowed to rise in a **fractionating column** that has a series of collection trays – each at a different temperature. This makes the process continuous and allows for the separation of the crude oil mixture into a series of fractions, each with a different boiling point.

Distillation is also used to concentrate ethyl alcohol from fermented liquids, thereby producing hard liquor among other things. Since ethyl alcohol kills the enzyme-producing yeast when its concentrations reach about 18 percent, it is only through distillation that higher concentrations can be achieved, as for example in the content of grain alcohol, which is as high as 95 percent (190 proof).

Distillation gives good separation of different boiling point substances, but suffers from the disadvantage of being an energy-hungry process. In most industrial applications, heat exchangers are used to cool the emerging vapors while at the same time increasing the temperature of the liquid entering the distillation pot.

Checking Your Understanding

1. What is the effect of temperature on the vapor pressure of a substance?

2. What is the relationship between the vapor pressure and the boiling point of a substance?

3. What are the major sources of energy that cause an evaporating pond to work?

4. Does air stripping tend to remove the substance with the higher or lower vapor pressure?

3-8 Electrostatic Separation Technologies

Did you ever notice that your TV picture tube or computer screen seems to act like a dust magnet? The **electrostatic** or **Cottrell precipitator**, named after its inventor Frederick G. Cottrell, operates on the principle that particulate matter in a gaseous waste stream can be removed by subjecting it to a strong electric field.

The first step in the process involves the generation of a strong electric field by applying a large potential difference (high voltage) across two electrodes. The applied voltage, when large enough, results in a **corona**, which appears as a luminous blue glow. The corona is the result of a process in which the molecules in the gaseous waste stream are ionized by electron collisions. The strong electric field close to the discharge electrode (wire) accelerates the free electrons. These electrons acquire sufficient energy to ionize a molecule in the gas upon collision, producing a positively charged ion and additional free electrons.

The additional free electrons create more positive ions and free electrons as they collide with additional molecules in the gas, in a process called avalanche multiplication. This process occurs in the corona glow region and continues outward from the discharge electrode until the electric field has insufficient energy to perpetuate ionization.

In the second step, the positively charged gas ion migrates toward the negatively charged electrode. If the positively charged gas ion encounters a dust particle in its path, it will attach to the particle. The positively charged particle-gas ion continues to

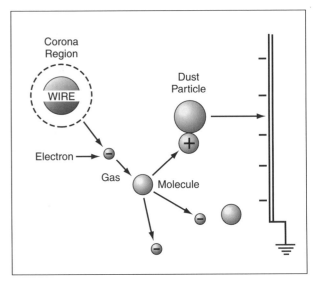

Figure 3-32: An electron from the corona region converts a molecule into a positive ion, which becomes attached to a dust particle. The charged dust particle is removed by being drawn to the negative electrode.

be attracted to the negatively charged electrode, resulting in the deposition of the dust particle on the surface of the electrode.

Periodic cleaning of the electrode is accomplished by rapping mechanically, which frees the particles that are collected in a hopper below. Electrostatic precipitators are expensive and require a large amount of space to install. However, they remain a frequent choice for treating hot dirty air streams, where baghouse fabrics would be unable to withstand the temperatures.

Checking Your Understanding

1. What is a corona?

2. How are the collected dust particles freed from the charged electrode?

3. What are advantages and disadvantages of using an electrostatic precipitator?

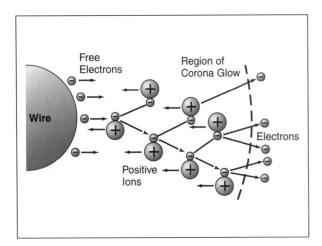

Figure 3-31: Electrostatic precipitator corona generation.

3-9 Surface Phenomena Separation Technologies

Experts differ on whether adsorption and ion exchange should be considered as chemical or physical treatment technologies. One argument is that the particles chemically attach to the surface of the adsorbent or **resin**. The other side of the argument is that the substance can later be released from the surface, without having changed its chemical nature. In this text we will treat ion exchange resins and adsorption as physical treatment technologies.

Ion Exchange

The process of **ion exchange** is probably most familiar to you in the form of a home water softening unit. When its operation is reduced to simplest terms, it involves hard water entering and soft water exiting the unit. While inside the unit, the hard water is brought into contact with a polymeric resin that removes unwanted calcium, magnesium, and iron

ions and exchanges them for a chemically equivalent number of sodium ions. Later, through a back-flushing process using a concentrated sodium chloride solution, the calcium, magnesium, and iron ions are removed from the resin and replaced with more sodium ions. The water rich in the unwanted ions is disposed of to the sewer. The unit is again ready for use.

Ion exchange, therefore, rarely represents an option for the disposal of hazardous wastes, since the process does not actually eliminate the waste. Its role is usually to reduce the magnitude of a problem by reducing its toxicity, reducing its volume, or converting the waste stream into a form where it can be reused.

It is also important to note that ion exchange, unlike many other separation technologies, provides the option to either separate or concentrate the pollutant. Exchange resins are cation or anion selective. Most resins are insoluble solid acids or bases that are capable of holding or releasing either cat-

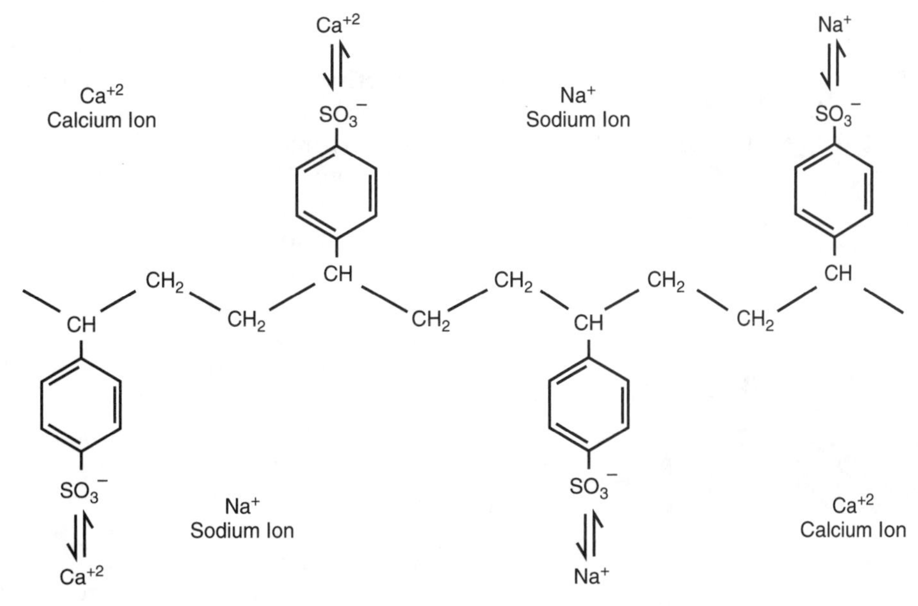

Figure 3-33: Calcium and sodium ion exchange occurring on an ion exchange resin.

ions or anions. The general equation for a typical cation exchanger would be as follows:

$$Ag^+ + R - H \longrightarrow R - Ag + H^+$$

STRONG-ACID CATION RESIN

By the selection of the proper resin, low levels of silver remaining in electroplating and photographic processing waste streams, for example, can be concentrated and recovered.

Sorption

Sorption is a term that is used to describe the surface phenomenon when it is not known whether the process is **absorption**, **adsorption**, or a combination of the two. The distinction between the two terms is that absorption is the penetration of one substance (**sorbate**) into the inner structure of another (sorbent); in the case of adsorption the sorbate is attracted and held on the surface of the sorbent by chemical bonds and/or physical forces. Cotton balls are frequently referred to as being absorbent – that is, they allow the liquid to enter or

Sorbent	Application
Activated carbon	Solvent recovery, elimination of odors, purification of gases.
Alumina	Drying of gases, air, and liquids.
Bauxite	Treatment of petroleum fractions: drying of gases and liquids.
Molecular sieves	Selective removal of contaminants from hydrocarbons.
Silica gel	Drying and purification of gases.

Table 3-4: Common sorbents and their applications.

penetrate their inner structure. The combination of an organic vapor with activated carbon, on the other hand, is generally considered to be a surface phenomenon and therefore correctly referred to as adsorption.

Adsorption can be used to remove substances from either liquid or gaseous waste streams. Although there are several other adsorbents (alumina and silica gel), **granular activated carbon** remains most popular and is used with both types of waste streams.

Granular activated carbon is prepared by the **destructive distillation** (heating in the absence of oxygen) of carbon-containing materials. Although wood is frequently used, animal bones, coconut shells, and other carbonaceous materials can be used. For the resulting carbon to become "activated" it is heated to 800-900°C in an atmosphere of steam or carbon dioxide, which turns it into a porous substance. If magnified, it would appear much like the surface of a piece of popcorn. It has been estimated that the surface area of one gram of granular activated carbon is 10,000 ft² (930 m²) or more than the area of three tennis courts!

A packed tower is often used for the granular activated carbon gas adsorption processes, while large enclosed tanks are used for treatment of liquid waste streams. When the material has accumulated on the activated carbon, it is removed and treated if it is a waste, or recovered if it is valuable. The activated carbon is subsequently regenerated by heating to a high temperature in the absence of oxygen.

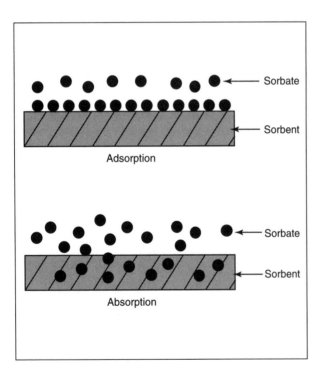

Figure 3-34: Adsorption vs. absorption.

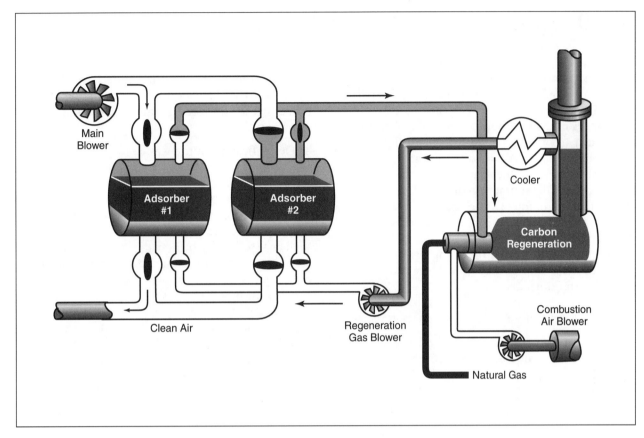

Figure 3-35: Example of an activated carbon sorption-regeneration process.

Checking Your Understanding

1. What is the function of an ion exchange resin?

2. How are calcium ions trapped on a resin removed?

3. Define and explain the relationship among the terms sorption, adsorption, and absorption.

4. What are the steps used to prepare activated granular carbon?

3-10 Immobilization Technologies

The physical waste treatment method discussed in this section is also described by other terms such as **solidification**, **stabilization**, **fixation**, **cementation**, and **immobilization**. In general, all of these terms refer to processes used to convert liquid and semi-solid wastes into a solid that is stable and exhibits physical properties that make it suitable for land disposal. The ultimate goal, therefore, is to retard further migration of its hazardous contaminants. Because of the similarities among the methods, they will all be referred to by the more generalized term – immobilization – in the following discussion.

Immobilization is used to achieve any one of the following objectives:

— Improvement of waste handling characteristics;

— Solidification of liquid phases and immobilization of any highly soluble components;

— Reduction in the potential contact area between the waste and any liquids that may come in contact with the waste to minimize leaching potential; and

— Detoxification of the waste.

Portland cement, the product used to make concrete driveways and sidewalks, is widely used for immobilization because of its ready availability. Other pozzolanic (porous volcanic ash) materials, such as fly ash, may be available at a lower cost, but regulations governing land disposal of hazardous bulk liquids do not permit its use because it does not fully immobilize the waste.

One of the advantages of this method is that the equipment required for treatment is widely available, as for example standard cement mixing and handling equipment. A second advantage is that the

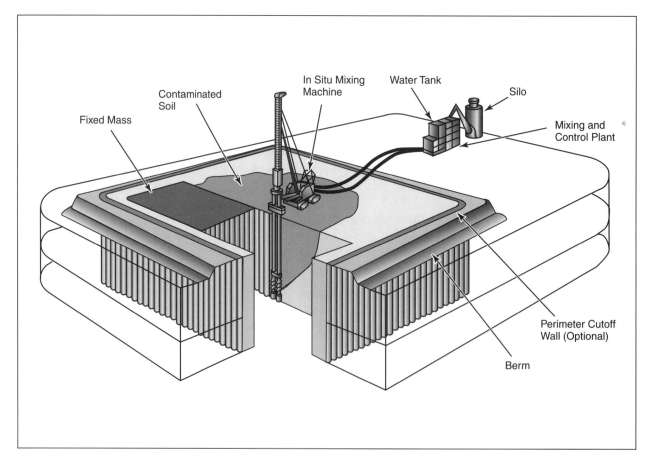

Figure 3-36: *In situ* solidification/stabilization using soil-cement mixing wall method.

techniques of cement mixing and handling are well developed and the process is reasonably tolerant to variations in the waste stream and/or soil matrix. Heavy equipment such as backhoes, specialized hydraulic augers, cement mixers, and dump trucks can be used for both excavation, mixing, and hauling needs, although some companies have developed specialized equipment such as injectors and augers that simultaneously inject cement and mix the matrix.

Immobilization has been found to be well suited for treating sludges and soils that contain heavy metals; organics; asbestos; inorganics such as sulfides; and solidified plastics, resins and various types of latex. Wastes containing fine organic particles, elevated levels of organics, soluble salts of specific metals, and sulfates are not appropriate for immobilization because they interfere with the setting and curing of the cement.

Checking Your Understanding

1. What is the ultimate goal of immobilization?

2. What common material is most commonly used to immobilize the waste?

3. Immobilization is particularly good treatment technology for which types of waste?

Summary

Physical properties include characteristics that can be easily determined using our physical senses, e.g., color, odor, density, etc. Within the context of waste, physical treatment technologies do not alter the characteristics of the waste, but rather make use of them to concentrate or remove waste from the waste stream.

If a density difference exists or can be caused to exist by using one of the air flotation techniques, suspended solids and liquids can be separated by using either sedimentation or flotation techniques. Because of the flexibility and ease of use, a variety of filtering processes have been developed. The many examples of these include bar screens, belt filters, filter presses, dual media filters, basket centrifuges, baghouses, synthetic and reverse osmosis membrane filtering operations. Waste streams that contain colloidally suspended particles may need chemical clarification techniques used to coagulate and flocculate the solids prior to their economical removal. In either case, the size of the unwanted particles to be removed depends solely on the nature of the filtering medium selected.

Vapor pressure differences explain the ability of evaporation ponds, air stripping, and distillation techniques to separate a mixture of components with differing boiling points. The electrical properties of matter can also be used to remove wastes by means of electrostatic precipitation, electrodyalisis, or ion exchange. Sorption, including both absorption and adsorption, rely on the attraction between a substance and the surface of a sorbent like granular activated carbon. For those wastes that are liquid or semi-solid, an immobilization or solidification/stabilization process may be used to encapsulate the waste components in an insoluble cement tomb.

Finally, it is important to remember that physical treatment technologies do nothing to change the toxic nature of the substances they remove. Often they are used to simply concentrate the waste stream so that other treatment technologies can be economically applied. They are seldom used in isolation, but rather just as a step in an overall waste stream management scheme.

Critical Thinking Questions

1. Analyze the benefits of using coagulation and flocculation as a pretreatment process prior to a filtering or clarifying process.

2. Explain why two substances of equal mass (one polar and one nonpolar) can be separated by air stripping.

3. Based on the treatment processes discussed in this chapter, explain three different combinations of treatment processes that could be used to separate a mixture of oil, water, and salt.

PRESERVING
THE LEGACY

4

Chemical Treatment Technologies

by Howard Guyer

Chapter Objectives

Upon completing this chapter, the student will be able to:

1. **Describe** the chemical properties of a substance that can serve as the basis for the various chemical treatment technologies.

2. **Explain** the concept of ionization and the role it plays in the formation of acids and bases.

3. **Apply** the concept of pH to the process of neutralization.

4. **Apply** the concept of insoluble salt formation to the process of chemical precipitation.

5. **Explain** the concept of oxidation-reduction and its application to various chemical treatment processes.

6. **Describe** the difference between electrolysis and cementation.

Chapter Sections

4-1 Introduction

In recent years the public has become increasingly **chemophobic**; that is, it has become increasingly fearful of anything said to contain **chemicals**. This fear, like most fears, is also based on a lack of understanding. In this chapter we will explore the nature of chemical substances and the way their properties can be used to reduce their toxicity and/or remove them from the waste steam.

In the last chapter, some of the physical treatment technologies used to remove and concentrate wastes were considered. In this chapter the focus shifts to the chemical treatment technologies, which – unlike the physical methods – tend to alter the chemical properties of the waste in such a way as to eliminate or reduce its toxicity by converting it into one or more new substances.

In Chapters 5 and 6 thermal and biological treatment methods will be discussed. It is important to understand that although they are treated in separate chapters, both are also chemical reactions. Thermal treatment methods are typically brought about by heating or burning in the presence of oxygen, while biological treatment methods are chemical reactions performed by living organisms.

4-2 The Composition of Matter

Before launching into a discussion about the various chemical treatment methods, it is important to understand that elements and compounds are the chemical building blocks of all the things around us. In this section we will review some basic chemistry to help you understand the principles behind the more common chemical treatment methods. A review of these fundamentals will also help in understanding the changes discussed later in the chapters on thermal and biological treatment methods.

Basic Atomic Theory

By definition, **matter** is anything that occupies space and has mass. If matter is variable in its composition, it is called a **mixture**, but if the composition is always the same, then it is called a **pure substance**. As shown in Figure 4-1, pure substances are again subdivided into two groups – **elements** and **compounds**.

There are nearly 90 naturally occurring elements and a few more that have been man-made. All known elements are listed on the **periodic table** and include, for example, carbon, magnesium, gold, and chlorine. Compounds are composed of two or more elements that have bonded together. Two familiar examples include water (H_2O) and salt (NaCl). Chemical reactions can also be used to break compounds down into their separate elements. Both H_2O and NaCl, for example, can be decomposed by using electricity in a reaction called **electrolysis**.

Pure substances can be mixed in varying amounts forming mixtures such as salt water, air, blood, and milk. Generally, mixtures can be separated back into pure substances on the basis of their physical and/or chemical properties. Simple examples would include the separation of oil and water mixture by skimming, or the evaporation of water from a salt.

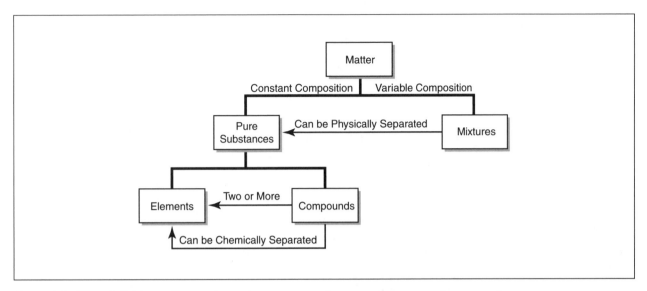

Figure 4-1: The subdivisions of matter into mixtures, pure substances, elements, and compounds.

The **atomic theory** states that all elements are composed of **atoms**. The Greek philosopher Democritus proposed the concept of the atom around 350 BCE (Before Common Era). He stated that all matter was composed of particles of matter so small as to not allow any division without changing the characteristics of the matter; these "uncuttable" components are the atoms. Atoms are very small, with an estimated diameter of about 1×10^{-8} cm; but ways have been found to break them into even smaller parts.

Modern atomic theory states that each element has its own kind of atom, and that at its center there is a very, very small and dense **nucleus**, the diameter of which is 1×10^{-13} cm. Within the nucleus there are **protons** that carry a positive charge and **neutrons** that have no charge. Around the central nucleus there are **energy levels** that contain orbiting **electrons** that have a negative charge. Since atoms do not have an overall electrical charge, the number of positively charged protons and negatively charged electrons within an atom must always be equal.

The number of protons in the nucleus determines the element's atomic number and name. No two elements, therefore, can have the same number of protons. The mass of the proton (1.67×10^{-24} g) is the basis for the relative scale of atomic mass units (a.m.u.). Although neutrons do not contribute to the atom's charge, they weigh about the same amount as protons. Since electrons weigh very little (9.1×10^{-28} g) the sum of the protons and neutrons determines the atom's atomic mass.

Atoms of the same element may vary in the number of neutrons, giving rise to **isotopes**. All carbon atoms have six protons and six electrons. Some carbon atoms, however, have six neutrons and are known as carbon-12, and others have eight neutrons and are called carbon-14. The ratio of carbon-12 to carbon-14 found in archeological remains is the basis for radiocarbon dating.

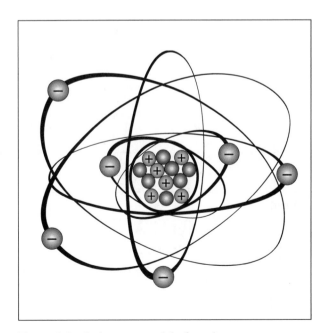

Figure 4-2: A planetary model of a carbon atom.

POSITIVE Charge of the Ion			
(+)	(2+)	(3+)	(4+)
Hydrogen H^+	Calcium Ca^{2+}	Iron(III) (Ferric) Fe^{3+}	Tin(IV) (Stannic) Sn^{4+}
Lithium Li^+	Magnesium Mg^{2+}	Aluminum Al^{3+}	Lead(IV) (Plumbic) Pb^{4+}
Sodium Na^+	Barium Ba^{2+}		
Potassium K^+	Radium Ra^{2+}		
Silver Ag^+	Zinc Zn^{2+}		
Copper(I) (Cuprous) Cu^+	Mercury(II) (Mercuric) Hg^{2+}		
Ammonium NH_4^+	Tin(II) (Stannous) Sn^{2+}		
	Iron(II) (Ferrous) Fe^{2+}		
	Lead(II) (Plumbous) Pb^{2+}		
	Copper(II) (Cupric) Cu^{2+}		

NEGATIVE Charge of the Ion			
(–)	(–)	(2–)	(3–)
Fluoride F^-	Dihydrogen Phosphate $H_2PO_4^-$	Oxide O^{2-}	Nitride N^{3-}
Chloride Cl^-	Acetate $C_2H_3O_2^-$	Sulfide S^{2-}	Phosphide P^{3-}
Bromide Br^-	Chlorate ClO_3^-	Sulfite SO_3^{2-}	Phosphate PO_4^{3-}
Iodide I^-	Permanganate MnO_4^-	Sulfate SO_4^{2-}	Arsenate AsO_4^{3-}
Hydroxide OH^-	Hydrogen Sulfite (Bisulfite) HSO_3^-	Carbonate CO_3^{2-}	
Nitrite NO_2^-	Hypochlorite ClO^-	Hydrogen Phosphate HPO_4^{2-}	
Nitrate NO_3^-		Dichromate $Cr_2O_7^{2-}$	
Hydrogen Carbonate (Bicarbonate) HCO_3^-		Chromate CrO_4^{2-}	
Hydrogen Sulfate (Bisulfate) HSO_4^-		Oxalate $C_2O_4^{2-}$	

Table 4-1: Table of common ions.

In an attempt to gain chemical stability, atoms frequently undergo either a loss of electrons called **oxidation**, or a gain of electrons called **reduction**. The electrons affected are only those in the outermost energy level. The resulting charged particles are known as **ions** of the element; for example, in Figure 4-3 when a sodium atom, Na, loses an electron, it becomes a positively charged sodium ion, Na^+. An ion may be composed of either a single nucleus or a group of nuclei bonded together; the latter is called a polyatomic ion. All positively charged ions are called **cations** and all negatively charged ions are called **anions**, regardless of their composition.

Origin of Chemical and Physical Properties

As noted in the previous chapter, all chemical substances have a unique set of physical and chemical characteristics. Atomic theory attributes the chemical characteristics of a substance to the arrangement of its electrons. In the early 20th century the **octet theory** was proposed, suggesting that if an element did not have eight electrons in its outer energy level, it would react with another elements to obtain eight electrons. (The first few elements are exceptions to the octet rule, since only two electrons are required to stabilize the first energy level.) The inert or **noble gases** at the right-hand end of each row on the periodic table are considered to have ideal configurations or the most stable electron configurations possible and, therefore, do not react with other atoms.

In general, those elements that have less than four electrons in their outer energy levels tend to lose one, two, or three electrons, thereby emptying their outer energy level. The resulting positively charged cation then has the same electron configuration as the inert gas at the end of the previous row. If, on the other hand, the atom has more than four electrons in its outer energy level, then gaining one, two, or three electrons allows it to complete its outer octet of electrons. The resulting negatively charged anion has the same number of electrons as the inert gas at the end of that row.

This concept of losing (oxidation) and gaining (reduction) electrons to achieve greater stability gained rapid acceptance and, in addition, allowed for the prediction of the formulas for compounds resulting from interaction of different elements. It was also noted that the closer an element is positioned to one of the inert gases, the more vigorous its reactions are to achieve the electron configuration of the inert gas. For example, the elements in the **halogen** family (fluorine, chlorine, bromine, and iodine) are only one electron short from having an inert gas configuration and are the most reactive **nonmetal** elements. Their ability to gain the additional electron makes them strong **oxidizing agents** that are highly corrosive and toxic. Once they achieve that inert gas electron arrangement and become fluoride, chloride, bromide, and iodide ions, however, both their chemical and physical properties become different than their parent atoms.

In a similar but opposite manner, the alkali metals that occupy the first left-hand row on the periodic table represent the most reactive metals. By losing their one and only outer electron they can achieve the same electron configuration as the preceding inert gas. These metals are strong **reducing agents** and react vigorously with many other substances – even water – almost explosively. Once achieved, however, the ions become chemically and physically different from their parent atoms.

As an extension to the octet theory, the American chemist G. N. Lewis proposed that some elements do not gain or lose electrons to reach electrical stability, but rather participate in a sharing of electrons (Figure 4-4) with one or more atoms. The resulting partnership of atoms is called a **molecule**, which is the second major way atoms become stable.

These two explanations of how elements achieve their outer octet of electrons give rise to the two major types of chemical compounds: **ionic** and **covalent**. The physical and chemical characteristics of these compounds are quite different. As careful examination of Table 4-2 reveals, a vast majority of the covalently bonded examples are **organic compounds**. An organic compound is defined as any covalently bonded compound that contains the ele-

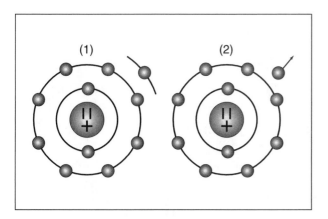

Figure 4-3: A sodium atom (1) becomes a sodium ion (2) by losing one electron.

Inorganic Compounds	Organic Compounds
Aluminum Sulfate, $Al_2(SO_4)_3$	1,1,1-Trichloroethane, CCl_3–CH_3
Ammonia, NH_3	Benzene, C_6H_6
Ammonium Hydroxide, NH_4OH	Butane, CH_3–CH_2–CH_2–CH_3
Calcium Sulfate, $CaSO_4$	Carbon Tetrachloride, CCl_4
Ferric Chloride, $FeCl_3$	Chloroform, $CHCl_3$
Hydrochloric Acid, HCl	Diethyl Ether, CH_3–CH_2–O–CH_2–CH_3
Lead Carbonate, $PbCO_3$	Ethyl Alcohol, CH_3–CH_2–OH
Potassium Nitrate, KNO_3	Methane, CH_4
Sodium Chloride, NaCl	Methyl Alcohol, CH_3–OH
Sodium Hydroxide, NaOH	Octane, CH_3–CH_2–CH_2–CH_2–CH_2–CH_2–CH_2–CH_3
Sulfuric Acid, H_2SO_4	Propane, CH_3–CH_2–CH_3

Table 4-2: Some common inorganic and organic compounds.

ment carbon, with the exception of CO_2, CO, carbonates, CO_3^{2-}; bicarbonates, HCO_3^-; and cyanides, CN^-. If the element carbon is not present, as in the example of hydrogen chloride, HCl, then the compound is considered to be an **inorganic compound**.

In summary, all matter is composed of chemicals that may be either pure elements, compounds, or mixtures. Two or more elements combine into a compound, which is held together by either ionic or covalent bonds. Compounds can be broken down into their elemental components, but elements themselves cannot be further decomposed.

It is the arrangement of electrons in the outer energy level that determines the chemical and physical properties of an element. If the number of electrons in the outer shell is changed, it becomes an ion. Ions have different chemical and physical properties than their parent atoms. The physical and chemical properties of compounds also differ from the elements of which they are composed.

Checking Your Understanding

1. Atoms are composed of what three subatomic particles?

2. What is gained or lost when an atom becomes an ion?

3. What are three physical properties of an element?

4. What is the difference between an ionic and a covalent bond?

5. What element must be present in an organic compound?

Figure 4-4: Covalently bonded HCl molecule.

4-3   Acids, Bases, and the pH Scale

The remaining fundamental principles that must be mastered to fully understand the nature of most chemical treatment methods involve the intertwined concepts of acids, bases, and **pH**. Although only the treatment process described as **neutralization** is solely dependent on balancing acid and base, the success of many other chemical and biological treatment technologies are equally dependent on maintaining the proper acid/base balance, measured as pH.

Ionization of Water and Characteristics of Acids and Bases

As explained in the previous section, when atoms gain or lose one or more electrons they become ions. **Ionization** is the term used to describe the conversion of any uncharged atom or molecule into an ion.

All ionic and some covalently bonded molecules can undergo a process that results in the formation of ions. However, when the term ionization is applied to these compounds, it refers to the molecule breaking apart, resulting in the formation of both a positively and a negatively charged ion. Water is an important example of a covalently bonded molecule that ionizes; not only does it undergo ionization, but also it frequently plays a significant role in causing other molecules to ionize. Here is how it all happens. Water is a **polar covalent molecule**. Polar covalent means that although the electrons within the bond are being shared, they are not being equally held by the two nuclei. Within the water molecule the oxygen atom pulls the electrons closer to its nucleus and away from each of the hydrogen nuclei. As a result, the oxygen portion of the molecule becomes partially negative, $\delta-$, and the hydrogen portion becomes partially positive, $\delta+$.

When two polar covalent molecules are placed together, there is a tendency – in a manner reminiscent of tiny magnets – for the partially negative end of one molecule to be attracted to the partially positively charged end of a neighboring molecule. In addition to this attraction, the bumping and jostling of the molecules can result in one of the molecules breaking apart (ionizing) and forming two ions.

When the ionization of a water molecule occurs (Figure 4-5), the results are one **hydronium ion**, H_3O^+, and one **hydroxide ion**, OH^-. If the substance undergoing ionization results in the formation of a hydronium ion and some other negatively charged ion, then it is called an **acid**. For example, when hydrogen chloride gas, $HCl_{(g)}$ and water are put together (Figure 4-6), the water molecule causes the ionization of the polar covalent $HCl_{(g)}$ to occur, resulting in a solution containing hydronium and chloride ions ($H_2O + HCl_{(g)} \longrightarrow H_3O^+ + Cl^-$). This is why solutions of HCl are referred to as hydrochloric acid.

It is common to consider only the ions formed by the molecule undergoing ionization. This simplifies the previous example to just $HCl \longrightarrow H^+ + Cl^-$. By eliminating the water from the equation, it can be said that, upon ionization, acid molecules always produce **hydrogen ions**. When reviewing the formulas for the acids listed in Table 4-3, it should be noted that the hydrogens at the beginning of each formula are always the ionizable hydrogens.

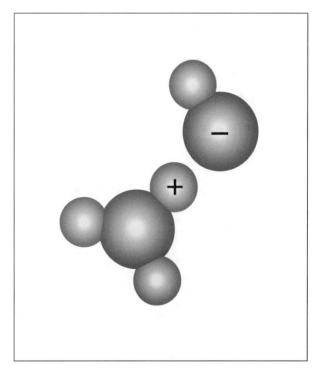

Figure 4-5: The ionization of water results in a hydronium ion, H_3O^+, and a hydroxide ion, OH^-.

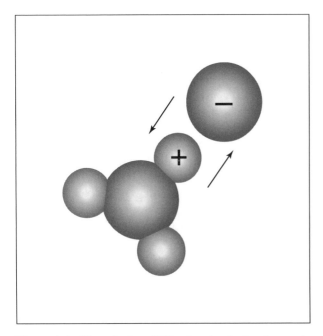

Figure 4-6: Water ionizes $HCl_{(g)}$ forming hydronium and chloride ions.

When a polar covalent substance like ammonia, $NH_{3(g)}$, is bubbled through water, the ammonia molecules are attracted to the polar water molecules. Again, assisted by the constant jostling and molecular collisions, water molecules become ionized, resulting in the formation of positively charged ammonium ions, NH_4^+, and negatively charged hydroxide ions, OH^-. The equation for the reaction would be: $NH_3 + H_2O \longrightarrow NH_4^+ + OH^-$. Any substance that produces additional OH^- ions when placed in water is called a **base**. It should also be noted that the formula for each of the bases listed in Table 4-4 always contains one or more hydroxide ions.

Hydrolysis of Salts

Salts are substances that dissolve in water and produce positive and negative ions other than H^+ and OH^-; however, they may produce the same effect as

Inorganic Acid		Organic Acid	
Name	Formula	Name	Formula
Hydrochloric acid	HCl	Hydrobromic acid	HBr
Sulfuric acid	H_2SO_4	Hydrofluoric acid	HF
Nitric acid	HNO_3	Hydriodic acid	HI
Phosphoric acid	H_3PO_4	Perchloric acid	$HClO_4$
Acetic acid	$HC_2H_3O_2$	Formic acid	$HCHO_2$
Boric acid	H_3BO_3	Carbonic acid	H_2CO_3

Table 4-3: Some common inorganic and organic acids.

Bases			
Name	Formula	Name	Formula
Sodium hydroxide	NaOH	Lithium hydroxide	LiOH
Calcium hydroxide	$Ca(OH)_2$	Magnesium hydroxide	$Mg(OH)_2$
Potassium hydroxide	KOH	Barium hydroxide	$Ba(OH)_2$
Ammonium hydroxide	NH_4OH	Aluminum hydroxide	$Al(OH)_3$

Table 4-4: Some common bases.

adding an acid or base. This is the result of a process called **hydrolysis**, in which one of the salt ions binds and effectively removes either the hydrogen or hydroxide ions. The solution then has an excess of the other ion, resulting in its acidic or basic characteristics. Examples of these salts that are often used in industry include: ferric chloride, $FeCl_3$ that results in an acidic solution; and sodium carbonate (soda ash), Na_2CO_3, or sodium hydrogen carbonate (baking soda), $NaHCO_3$, both of which produce basic solutions. Soda ash is a frequently used and inexpensive choice for neutralizing waste acidic solutions.

The pH Scale

From the preceding section it can be seen that acidic solutions contain an excess of hydrogen ions and that basic solutions contain an excess of hydroxide ions. The need for having a simple way to express the degree of excess was met in 1909 by the Danish biochemist Søren Sørensen. He developed the **pH scale**, which uses logarithms to convert the **concentration** of hydrogen ions into a simple number scale ranging between zero and fourteen.

On this scale pH 7 equals a neutral solution and indicates that the number of H^+ is equal to the number of OH^-. The mathematical application of Sørensen's equation – $pH = -\log [H^+]$ – is beyond the scope of this discussion; however, it can be seen from Figure 4-7 that, as the number of hydrogen ions in a solution increases, the pH numbers decrease. As a result, a solution with a pH = 4 is more acidic than a solution with a pH = 5. It should also be noted that because this is a logarithm scale, a change of one number on the pH scale, pH = 5 to pH = 4, corresponds to a change of 10 times the number of hydrogen ions.

When small amounts of base are added to a solution, they lower the number of hydrogen ions present by converting them back into water molecules. Simply stated, when the number of hydroxide ions is increased, the number of hydrogen ions decreases, and the pH numbers get larger. From Figure 4-7 it can be seen that basic solutions always have pH numbers greater than seven.

Today we are fortunate to have many technologies that allow us to quickly and conveniently monitor the pH of solutions. A variety of pH **indicator papers** are on the market. The best known is called

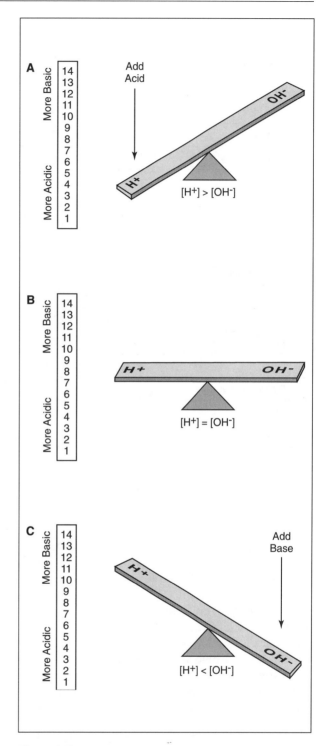

Figure 4-7:
A   When number of $[H^+] > [OH^-]$, the solution is acid, pH < 7.
B   When number of $[H^+] = [OH^-]$, the solution is neutral, pH = 7.
C   When number of $[H^+] < [OH^-]$, the solution is basic, pH > 7.

Substance	pH
Gastric juices	1.0 - 3.0
0.01 M HCl	2.0
Soft drinks	2.0 - 4.0
Vinegar	2.4 - 3.4
Apples	2.9 - 3.3
Grapefruit	3.0 - 3.3
Strawberries	3.0 - 3.5
Tomatoes	4.0 - 4.4
Urine	5.0 - 7.0
Soil, optimal	6.0 - 7.0
Saliva	6.5 - 7.5
Milk	6.5 - 7.6
Bile	6.8 - 7.0
Water, DI	7.0
Blood plasma	7.3 - 7.5
Water, tap	6.6 - 8.0
Sea water	7.8 - 8.2
0.01 M NaOH	12.0

Table 4-5: pH of some common substances.

Figure 4-8: Dr. Arnold Beckman, inventor of the pH meter.

trode to give continuous pH readings to the nearest ± 0.01 pH unit. In fact, today pH has been redefined as the number reading obtained from a calibrated pH meter! Table 4-5 lists the pH of some common substances.

litmus paper, which is often used in elementary science and/or chemistry classes. Like most other indicators, litmus is derived from a plant source and behaves much the same as the pigment that makes some cabbages and onions red. In spite of its low cost and ease of use, litmus suffers from at least two drawbacks. First, its accuracy: litmus is red at pHs below 4.5 and blue at pHs above 8.3; this leaves a critical window near the neutral point in which litmus is useless in determining pH. Second, it cannot be used as a continuous monitor on a liquid waste stream.

Thanks to Dr. Arnold Beckman, both of these problems were solved in 1935 with the development of the electronic **pH meter**. Dr. Beckman devised the original electronic circuitry that could measure small changes in electrical conductivity inside two glass **electrodes** when they were immersed into solutions of differing pHs. The original instrument – designed to test for the acidity of lemon juice – has been refined many times and now uses a single elec-

Checking Your Understanding

1. Which ion is in excess in acid solutions?

2. Name the process in which water breaks a molecule into ions.

3. Solutions that are neutral (pH = 7) have an equal number of which two ions?

4. What are two ways the pH of a solution can be determined?

5. Would a solution with a pH less than seven be an acid or a base?

4-4 Neutralization

Neutralization, which is also called pH adjustment, is one of the most common industrial chemical treatment methods. It is used to treat acidic and **alkaline** (basic) wastes, as well as those defined as corrosive in RCRA.

Title 40 of the Code of Federal Regulations, Section 261.22, defines a waste as a D002 **characteristic hazardous waste**, if it exhibits the characteristics of **corrosivity**. As defined in this section, a waste is considered to be corrosive if it has a pH ≤ 2 or ≥ 12.5, or if it is a liquid that corrodes steel at a rate greater than a designated corrosion rate. Some listed hazardous wastes (K062), such as spent pickling liquors generated by metal finishing operations, are also corrosive wastes that require neutralization.

Depending on the parameters of the discharge permit, if average measurements of a spent acid or alkaline waste stream have a pH that falls outside the 4.3 to 8.3 range, then the waste stream will need to undergo a neutralization process. A pH meter or indicator paper is used to follow the progress of the neutralization process. If the solution contains an excess of H^+, its pH can be moved toward pH = 7 by the addition of an equivalent amount of OH^- or some other substance that reacts with and reduces the number of H^+.

Because of its low cost, lime, CaO, is a frequent industrial choice when neutralizing acids. Lime becomes calcium hydroxide, $Ca(OH)_2$, when it is added to water in a process known as slaking. Sodium hydroxide, NaOH, also known as **caustic** or lye, offers the advantages of easy handling and storage, but is more expensive.

Conversely, solutions that have an excess of OH^- can be brought back toward a pH = 7 by adding equivalent amounts of H^+. This may be done by adding an expensive acid like hydrochloric acid, HCl, or by bubbling a nonmetal oxide, like carbon dioxide, CO_2 through the solution. Carbon dioxide gas dissolves in water forming carbonic acid, H_2CO_3. In industrial applications, neutralization is performed both as a batch and a continuous flow process.

Neutralization reactions can be performed using simple off-the-shelf equipment. As shown in Figure 4-9, the equipment may consist of little more than a reaction tank, a chemical feed system, and a rapid mixing device. By adding a continuous pH monitoring device, the neutralization process can be followed and adjusted according to the needs of

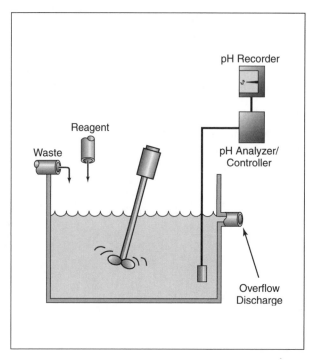

Figure 4-9 Equipment used in a simple neutralization process.

the waste stream. Depending on the requirements and products, this may be followed by other physical or chemical treatment processes to collect and remove undesirable by-products.

Acid/base neutralization is most often used on liquid waste streams, but it can also be used to treat sludges, slurries, and gases. For example, hydrogen sulfide, H_2S, is a common gaseous waste stream in the wastewater treatment industry. When H_2S is brought into contact with water, weak hydrosulfuric acid H_2S is formed. Therefore, the use of a **wet scrubber** to treat the gaseous hydrogen sulfide with a sodium hydroxide solution neutralizes the waste according to the following reaction:

$$H_2S + 2\,NaOH \longrightarrow Na_2S + 2\,H_2O$$

Nearly every aqueous waste method involves some degree of neutralization. On occasion, however, its goal may not be total neutralization, but rather pH adjustment to optimize the performance of another treatment process, such as chemical precipitation.

In 1976 EPA established the hazardous waste management hierarchy. This hierarchy places emphasis on source reduction and recycling; however, for those processes that cannot avoid generating wastes, treatment is still the next best option. One way industry treats its acidic or alkaline waste streams is by using one to neutralize the other. Also, through synergistic arrangements, even neighboring companies can on occasion minimize neutralization treatment costs by using each other's waste streams. Regardless of its implementation, neutralization remains one of the most fundamental and commonly used chemical treatment processes.

Checking Your Understanding

1. Solutions are considered hazardous if their pHs exceed which two values?

2. What are the two advantages of using NaOH to adjust a solution's pH?

3. What are two other names for NaOH?

4. Why will bubbling CO_2 through a basic solution lower its pH?

5. How can companies cooperate to reduce their waste treatment costs?

4-5 Chemical Precipitation

In the last section neutralization was discussed as a treatment method for the purposes of neutralizing or adjusting the pH of a waste stream. This section will explain how adjustment of pH can also result in the **chemical precipitation** of toxic metals from a waste stream.

Chemical precipitation is defined as any treatment process that causes a soluble substance to become less soluble or insoluble in its solvent. Reduction in solubility can result from converting the substance into an insoluble form or from changing the amount or characteristics of the solvent to reduce the solubility of the substance in it. This process is particularly effective in the treatment of hazardous wastewater containing toxic metals such as arsenic, barium, cadmium, chromium, copper, lead, mercury, nickel, selenium, silver, thallium, and zinc.

Ionic compounds containing such ions as sulfide, carbonate, phosphate, sulfate, or chloride may be used to **precipitate** many metal ions, but **hy-**droxide precipitation – using slaked lime, $Ca(OH)_2$ – is by far the most widely used method. After precipitation occurs, a chemical coagulant may need to be added to assist in aggregating the small solid particles. With or without the addition of a coagulant, the resulting particles formed are typically separated out using a clarifier or filter. The filter press is the most commonly used filtration equipment to accomplish the final dewatering and filter cake recovery.

Precipitate Formation

When ionic compounds are put into water, they are either **soluble** or **insoluble**. If they are soluble, all of the positive and negative ions become surrounded by a number of water molecules and are said to be **solvated**. The solution, therefore, consists of a mixture of these solvated ions and water molecules.

When a second solution is added, there is a possibility for two new ion combinations to form.

For example, when common table salt, NaCl, is dissolved into water, the resulting solution contains both solvated sodium, $Na^+_{(aq)}$, and chloride, $Cl^-_{(aq)}$, ions. When silver nitrate dissolves in water, a solution containing solvated silver, $Ag^+_{(aq)}$ and nitrate, $NO_3^-_{(aq)}$, ions is formed. When these two solutions are combined, an insoluble metal salt quickly forms. The precipitate is silver chloride, AgCl. From an energy point of view, it requires less energy for the silver and chloride ions to become a silver chloride precipitate than to remain in solution as solvated ions. The opposite appears to be true for the solvated sodium and nitrate ions.

After years of observing which ions tend to stay in solution and which tend to form precipitates, general guidelines for predicting their solubility have been developed. Table 4-6 lists these general solubility rules. Application of the general metal salt solubility rules to the previous example predicts the precipitation of AgCl, since the combination of chloride ion and silver ion is one of the exceptions stated in Rule 3.

A second, more precise way to make the above prediction would be to consult Table 4-7. In this table, a solubility limit of 1g/100g of water has been arbitrarily set as the cutoff point for soluble (S) and insoluble (I). This table reveals that the combination of silver and chloride ions would be expected to produce an insoluble salt. Further examination of the columns in Table 4-7 reveals that several other metals can be successfully precipitated through formation of insoluble salts.

1.	The salts of all nitrates and acetates tend to be soluble.
2.	The salts of all lithium, sodium, potassium, and ammonium compounds tend to be soluble.
3.	The salts of all chlorides, bromides and iodides tend to be soluble, except when the positive ions are silver, lead or mercury.
4.	The salts of most other metal ions are insoluble and will precipitate from solution.

Table 4-6: General solubility rules for metal salts.

	Br $^-$	CO$_3$ $^{2-}$	Cl $^-$	CrO$_4$ $^{2-}$	OH $^-$	I $^-$	PO$_4$ $^{3-}$	SO$_4$ $^{2-}$	S $^{2-}$
Ba $^{2+}$	S	I	S	I	S	S	I	I	d
Co $^{2+}$	S	I	S	I	I	S	I	S	I
Cu $^{2+}$	S	I	S	I	I	I	I	S	I
Fe $^{3+}$	S	I	S	–	I	–	I	S	I
Pb $^{2+}$	I	I	I	I	I	I	I	I	I
Mg $^{2+}$	S	I	S	S	I	S	I	S	d
Ag $^+$	I	I	I	I	–	I	I	I	I
Zn $^{2+}$	S	S	S	I	I	S	I	S	I

KEY: S = soluble in water
      I = insoluble in water less than 1 g/100g H$_2$O)
      d = decomposes in water

Table 4-7: Solubility table for selected metal salts.

Hydroxide Precipitation

The major sources of metal-containing wastes are metal plating and finishing, steel and other nonferrous metal production, inorganic pigments, mining, and the semiconductor and printed wire board industries. Waste streams containing metal ions are also generated by the hazardous waste site cleanup industry.

As can be seen from Figure 4-10, the solubility of metal hydroxides varies considerably at different pHs. Knowledge of the optimum pH for the metal to be removed improves the overall efficiency of the treatment process. Typically, precipitation of metals is followed by flocculation and clarification; this, in turn, is followed by removal through either a sedimentation or filtration process.

At Southern California's NPL site, known as the **Stringfellow Acid Pits**, the organic-acid-metal laden waste stream is initially treated by hydroxide precipitation, using slaked lime as the precipitant of choice. The generalized chemical formula for the precipitation of a divalent cation (M^{2+}) would be the following:

$$M^{2+} + Ca(OH)_2 \longrightarrow M(OH)_2 \downarrow + Ca^{2+}$$

After flocculation, clarification, and filtering are used to remove the toxic metal hydroxide precipitate, the remaining waste stream is subjected – prior to discharge – to an activated carbon adsorption pro-

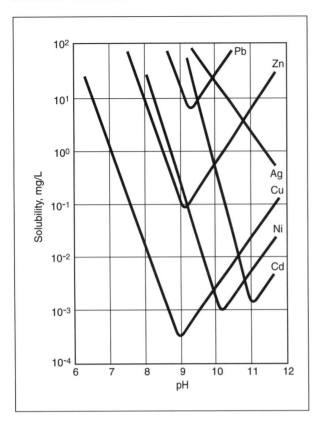

Figure 4-10: Solubilities of metal hydroxides as a function of pH.

cess to remove the remaining organic components. The solubilities of metals vary considerably not only from metal to metal, but also within each metal depending on the oxidation number. An example that will be discussed in more detail in another section is hexavalent chromium, Cr^{6+}, which must be reduced to trivalent chrome, Cr^{3+}, before hydroxide precipitation becomes an effective treatment method.

Sulfide Precipitation

Reference again to Table 4-7 reveals that a number of metal sulfides are insoluble. In fact, sulfide precipitates of several metals are even less soluble than those of the corresponding hydroxides. The generalized equation for the precipitation of a bivalent metal sulfide is the following:

$$M^{2+} + S^{2-} \longrightarrow MS \downarrow$$

As noted above, the oxidation number of chromium has to be changed prior to hydroxide precipi-

tation. Perhaps the one place where **sulfide precipitation** is most useful is the direct precipitation of hexavalent chromium, in the form of dichromate ion, without having to reduce it to its trivalent form. The overall generalized reaction is the following:

$$Cr_2O_7^{2-} + 3\ S^{2-} + 8\ H^+ \longrightarrow 3\ S + 2\ Cr(OH)_3 \downarrow + H_2O$$

The source of sulfide ions, S^{2-}, for sulfide precipitation can be provided by sodium sulfide, Na_2S, ferrous sulfide, FeS, or any other soluble sulfide. However, care must be taken to maintain pH > 8 to prevent evolution of toxic hydrogen sulfide gas.

The use of ferrous sulfide has been found to be the most satisfactory source of sulfide ion industrially. Ferrous sulfide is more soluble than other metal sulfides and will therefore precipitate other metals, but its solubility is low enough that few free sulfide ions are present in the solution. This, therefore, virtually eliminates the problem of hydrogen sulfide gas evolution within the reaction container.

Carbonate Precipitation

Table 4-7 also reveals that several metal carbonates are insoluble. Some metals, e.g., lead and cadmium, can be effectively removed by **carbonate precipitation**. In addition, whereas pH ≥ 10 is required to precipitate them as hydroxides, pHs of only 7.5 to 8.5 are sufficient to form the carbonate precipitate. The generalized equation for the precipitation of a divalent metal is the following:

$$M^{2+} + CO_3^{2-} \longrightarrow MCO_3 \downarrow$$

The salt sodium carbonate, Na_2CO_3, not only proves to be an inexpensive source of carbonate ions for this reaction, but as previously noted, undergoes hydrolysis with water shifting the solution's pH into the appropriate range. The following hydrolysis equation shows the reaction that generates the extra hydroxide ions:

$$H_2O \longrightarrow H^+ + OH^-$$
$$Na_2CO_3 \longrightarrow 2\ Na^+ + CO_3^{2-}$$
$$\overline{H_2O + Na_2CO_3 \longrightarrow HCO_3^- + 2\ Na^+ + OH^-}$$

As can be seen by the above equation, carbonate ions capture the hydrogen ions produced by the ionization of water forming bicarbonate ions. The sodium ions, however, fail to combine with the hydroxide ions, resulting in an excess of OH⁻ in the solution.

In some cases, chemical precipitation reactions can only proceed after the waste has been pretreated. For example, metal-bearing solutions that also contain **cyanide** ions, CN⁻, generally cannot be precipitated without first destroying the cyanide ion; this due to the tendency of cyanide ions to form very strong complex ions with other metals. Therefore, metal finishers often destroy cyanides in their wastewater streams by using an alkaline chlorination pretreatment process.

Sodium Borohydride Precipitation

Sodium borohydride, NaBH$_4$, is a mild reducing agent that will result in the precipitation of certain soluble metal ions by converting them into their insoluble metal form. This process works well for removing lead, mercury, and nickel from waste solutions. **Sodium borohydride precipitation** can also be used to extract several precious metals, e.g. members of the gold and platinum groups, thus providing a good method for their recovery.

When the white free-flowing crystals of sodium borohydride are dissolved into water they produce hydrogen gas and sodium hydroxide. Most metals can be effectively reduced in the resulting high pH solution, according to the following simplified equation for a divalent cation:

$$NaBH_4 + 4\ M^{2+} + 8\ OH^- \longrightarrow 4M + NaBO_2 + 6\ H_2O$$

In summary, chemical precipitation always results in forming two process streams: 1) the filtrate and 2) the precipitate containing most of the contaminants originally present in the waste stream. Precipitation is, therefore, most often simply a volume-reduction process and may or may not be a destruction process.

Upon initial removal, the precipitate may still contain, by weight, up to 99 percent water that is generally removed before final disposal occurs. Dewatered precipitates typically remain characteristic hazardous wastes based on TCLP criteria, or are listed hazardous wastes because of the industrial process from which they were generated. Such precipitates may be further treated prior to metal recovery processes or sent for landfilling in accordance with current land disposal restrictions.

By studying the solubility of metal salts listed in Table 4-7 or other appropriate references, it should be clear that there are many different ion combinations that can lead to the successful precipitation of a given toxic metal from a waste stream. The method selected, however, must always take into consideration a variety of physical, chemical, and economic considerations.

Checking Your Understanding

1. What is the definition of chemical precipitation?

2. What chemical is most often used in the hydroxide precipitation process?

3. List one advantage and one disadvantage of the sulfide precipitation process.

4. What is the name of the process that results in sodium carbonate making a basic solution?

5. What are three metals that can be removed by the use of sodium borohydride?

4-6 Oxidation-Reduction Reactions

As explained in an earlier section, the loss or gain of electrons from an atom is defined as oxidation and reduction, respectively. By definition, metals are substances that tend to lose electrons during chemical events. In a similar way, nonmetals are substances that tend to gain electrons during chemical reactions.

For example, a substance like chlorine will take an electron from a sodium atom. In the process, sodium has become oxidized to the sodium ion, Na^+, and the chlorine atom is the oxidizing agent. At the same time the chlorine atom has gained an electron from the sodium atom and is reduced to the chloride ion, Cl^-. The sodium atom is, therefore, considered to be the reducing agent.

$$Na^o + Cl^o \longrightarrow Na^+ + Cl^-$$

Substance Oxidized Substance Reduced
Reducing Agent   Oxidizing Agent

Oxidation-reduction reactions result in a change in the oxidation state of both the reducing and oxidizing agents. In many instances this process can be used to reduce or destroy the toxicity of the waste. There are various oxidation-reduction reactions – also called **redox** – that play important roles in the treatment of wastes. Table 4-8 lists some of the various types of wastes that can be treated by either oxidation or reduction reactions.

These wastes include solutions containing hexavalent chromium, cyanides, sulfide, and many organic wastes such as amines, mercaptans, pesticides, and phenols. Table 4-9 lists some of the oxidizing and reducing agents more commonly used.

Destruction of Cyanide Wastes by Oxidation

Electroplating, organic chemical manufacturing, and ore extraction solutions frequently contain cyanide or – more accurately – cyanide ions, CN^-. Due to the toxic nature of this ion and its potential for generating lethal hydrogen cyanide gas, HCN, in an acidic environment, it is imperative that any wastewater treatment process first destroy all of the cyanide ions.

As a treatment safeguard, the first step involves raising the solution's pH to between 10 and 11 through the addition of **caustic soda**, NaOH. Next, a strong oxidizing agent – like **sodium hypochlorite** or bleach – is introduced, assisted by rapid stirring to insure adequate contact. For example, when

Oxidizing Treatment	Reducing Treatment
Benzene	Chromium (VI)
Phenols	Mercury
Most organics	Lead
Cyanide	Silver
Arsenic	Chlorinated organics (PCBs)
Iron	Unsaturated hydrocarbons
Manganese	

Table 4-8: Wastes commonly treated by redox reactions

Oxidizing Agents	Reducing Agents
Ozone	Ferrous sulfate
Sodium or Calcium hypochlorite	Sodium sulfate
Hydrogen peroxide	Sulfur dioxide
Chlorine	Iron (ferrous)
Potassium permanganate	Aluminum
UV/ozone	Zinc
	Sodium borohydride

Table 4-9: Commonly used oxidizing and reducing agents.

a solution containing copper(II) cyanide, $Cu(CN)_2$, is treated, the following oxidation-reduction occurs:

$$2\ CN^- + 10\ OH^- \longrightarrow N_2 + 2HCO_3^- + 4\ H_2O + 10\ e^-$$
(Oxidation Half-Equation)
$$5\ (ClO^- + H_2O + 2\ e^- \longrightarrow Cl^- + 2\ OH^-)$$
(Reduction Half-Equation)

$$5\ NaClO + Cu(CN)_2 + H_2O \longrightarrow$$
$$3\ NaCl + CuCl_2 + 2\ NaHCO_3 + N_2$$
(Overall Reaction)

Even though 99 percent or more of the cyanide ions are typically destroyed in the first oxidation step, it is common to follow this by a second oxidation, in which more sodium hypochlorite is added to insure total cyanide ion destruction. With the assurance that all cyanide ions have now been eliminated, the third step is to adjust the pH to near seven, by either adding acid or allowing it to be neutralized by an acidic waste stream.

If a metal ion, like copper(II), Cu^{2+}, remains in the solution, it must be removed before the wastewater is discharged. This may be accomplished by additional pH adjustment, chemical precipitation, flocculation and coagulation, or a combination of these treatment processes.

Dehalogenation Reactions

In an earlier section of this chapter the elements in Group VII on the periodic table were identified as the halogens. The group includes fluorine, chlorine, bromine, iodine, and the radioactive element astatine, which has no known commercial applications.

For more than 100 years organic chemists have been creating new **organohalogen** compounds. Many of these compounds have unique properties and have been found to be ideally suited for particular applications. Examples include many of our pesticides, solvents, refrigerants, fire retardants, paints, polychlorinated biphenyls (PCBs) used in capacitors and electrical transformers, as well as plastics such as polyvinyl chloride (PVC), Teflon®, and Saran Wrap®.

We have since become aware of the persistence of many of these compounds in our environment. The wakeup call came in the form of a book, Silent Spring, by Rachel Carson published in 1962. Within

this book Carson attributed the disruption of the reproductive processes of songbirds to the widespread use of pesticides and accumulation of dichlorodiphenyltrichloroethane, or DDT, in their bodies. She further argued that public health and the environment are inseparable, and that the continued indiscriminate use of pesticides was a public health issue. Frightened by Carson's message, public policy soon demanded severe restrictions on the number and use of pesticides, including a ban on the use of the insecticide DDT.

Since then, several methods to destroy or **dehalogenate** organic compounds have been developed. Although incineration remains the most often used technology for their destruction (see discussion in Chapter 5) other methods have been used or demonstrated. These include base-catalyzed decomposition (BCD), chlorinolysis, wet air oxidation (WAO), supercritical water, ultraviolet and other radiation sources, molten-salt chemistry, photolysis, and microbial dehalogenation techniques.

Since the specific halogen most often involved in these compounds is chlorine, in the following sections we will focus our attention specifically on **dechlorination** reactions. As indicated by the name, dechlorination is any process in which chlorine is chemically removed from a chlorinated organic compound. Two of the most troublesome chloroorganic compounds are **polychlorinated biphenyls** or **PCBs** and the 70 isomers of 2,3,7,8-tetrachlorodibenzo-p-dioxin (TCDD), which are collectively known simply as **dioxin**. At present, most dechlorination cleanup activities center around treatment of capacitors, transformers, and PCB-laced soils.

In an adaptation of a PCB-dechlorination technique originally developed by Goodyear Tire and Rubber Company, a proprietary alkaline-based reagent and polyethylene glycol (APEG) are used. The mechanism for the reaction involves the displacement of the chlorine atoms by the APEG to form a chloride salt and a chlorine-substituted organic polymer. The process can tolerate very small amounts of water, but its presence adversely affects the rate of the reaction. In addition, the reagents used are air sensitive so the process must take place in an inert atmosphere.

The base-catalyzed decomposition, or BCD, process can also be used to dechlorinate soils contaminated with PCBs, dioxins, and furans. The soils are screened and then processed in a crusher and in a **pug mill**, while all along they are being mixed with sodium bicarbonate. This mixture is then heated to above 330°C in a rotary kiln to partially decompose and volatilize the contaminants. Whereas APEG residuals contain chlorine and hydroxyl ions, which make them water soluble and slightly toxic, the BCD process produces less toxic, water insoluble biphenyl and low boiling point hydrocarbons, as well as sodium chloride.

As of this time there have been no commercial applications of this technology, but the BCD process has been demonstrated and received approval by EPA's Office of Toxic Substances under the Toxic Substances Control Act (TSCA) for PCB treatment.

Reduction of Hexavalent Chromium

Although EPA considers all compounds of chromium to be toxic, the compounds that contain chromium in its 6+ oxidation state, in particular, are the ones of major toxicological concern. These compounds are known carcinogens and have been shown to produce kidney damage in humans. Chromium(VI), or **hexavalent chrome** compounds – as they are commonly known – are also both strong oxidizing agents and corrosives.

Although our cars no longer sport chrome-plated bumpers, the use of chromium metal as a surface coating on plumbing fixtures, door knobs, guitar string tuning pegs, golf club heads, etc. remains popular. Once in place, chromium metal is quite resistant to oxidation and other types of chemical reactions, making it an ideal surface coating for more reactive base metals. For these reasons many small and a few large electroplating operations persist.

When a work piece is to be chrome plated, it is immersed in the process bath containing the hexavalent chrome plating solution. The work piece acts as the cathode. While in contact with the solution, hexavalent chrome ions are attracted and gain enough electrons (reduction) to be converted into insoluble chromium atoms. After a layer of chromium atoms of sufficient thickness has accumulated, the piece is removed, rinsed, and polished to a bright finish.

Reduction of the spent hexavalent chromic acid plating solution is typically accomplished by treatment with an economical reducing agent such as **sodium hydrogen sulfite**, $NaHSO_3$, according to the following equations:

$$3 \, (OH^- + HSO_3^- \longrightarrow SO_4^{2-} + 2H^+ + 2e^-)$$
(Oxidation Half-Equation)

$$Cr_2O_7^{2-} + 14 \, H^+ + 6 \, e^- \longrightarrow 2 \, Cr^{3+} + 7 \, H_2O$$
(Reduction Half- Equation)

$$H_2Cr_2O_7 + 3 \, NaHSO_3 \longrightarrow Cr_2(SO_4)_3 + 3 \, NaOH + H_2O$$
(Overall Reaction)

As is apparent in the above reduction half-equation, the reduction of hexavalent chrome forms an equivalent amount of trivalent chromium ions. Trivalent chromium compounds, although still considered toxic, are not believed to be carcinogenic. As previously discussed, trivalent chromium ions can also be easily removed from the solution by hydroxide precipitation. This merely requires adjusting the solution's pH to strongly basic. At high pHs, chromium(III) hydroxide, $Cr(OH)_3$, precipitate forms that can be efficiently flocculated, coagulated, and removed from the waste stream.

Ozonation

The element oxygen combines with itself to create two molecular forms: diatomic oxygen, O_2, and triatomic oxygen, O_3, called **ozone**. At earthly altitudes, molecular oxygen comprises approximately 21 percent of the blanket of air we breathe. Ozone only makes up a tiny part of that blanket and only exists in significant concentrations at distances of 16-30 km above the earth in what is called the ozone layer.

The source of the ozone in this layer is from a series of reactions in which an oxygen molecule first gets broken into its monoatomic form. These highly-reactive, single oxygen atoms then collide with the diatomic oxygen molecules ($O\bullet + O_2 \longrightarrow O_3$) resulting in the formation of the triatomic ozone molecules. These ozone molecules, in turn, serve the beneficial role of absorbing ultraviolet (UV) radiation from the sun when the reaction is reversed ($O_3 + UV \longrightarrow O\bullet + O_2$), reforming both molecular and monoatomic oxygen. In short, Mother Nature has given ozone the job of being the earth's sun screen, protecting us from excessive amounts of UV radiation!

When used for industrial applications, ozone is typically generated onsite by passing oxygen through a sparking gap between two electrodes. This breaks the oxygen molecules in much the same way as described above, forming elemental oxygen, which in turn reacts with oxygen molecules to form ozone.

Ozone is fairly unstable, with a half-life of about 12 hours in the air, and only 20-30 minutes when bubbled into pure water at 20°C. Because of the energy required for its formation, ozone is considered to be an expensive oxidizing agent, and therefore must be used efficiently. It is only slightly soluble (8.9 mg/L at 20°C) in water, and is usually brought into contact with water in the form of small bubbles. This may be accomplished through the use of diffuser plates, bubble-plates, packed beds, spray towers, or sieve-plate towers.

For more than 100 years, **ozonolysis** – also known as **ozonation** – has been used for the treatment of drinking water. Ozone is an effective germicide used to purify drinking water supplies. Unlike chlorine gas, it will not form harmful chlorinated organic compounds, like the trihalomethanes. This has made it an increasingly attractive choice in recent years. Even though it is a powerful oxidizing agent, there are still some substances that ozone oxides very slowly or not at all.

It has been found that ozone's ability to oxidize some substances can be greatly enhanced through the addition of a high-energy source in the form of UV light. Solutions containing PCBs and other halogenated solvents – e.g., methylene chloride, **trichloroethylene (TCE)**, carbon tetrachloride, and vinyl chloride – that are otherwise unaffected by ozone can be quickly and efficiently destroyed through use of the UV-ozone combination.

In these processes, not all the ozone generated is absorbed, and not all absorbed ozone gets used as an oxidizing agent. Therefore, it may be necessary to treat the waste – and the gases escaping from it – to reduce the amount of unused ozone. Ozone has been described as having a pungent, irritating odor. Breathing air with as little as 0.15 ppm of ozone for two hours results in some respiratory difficulties. These findings have caused EPA to establish a standard that limits the concentration of ozone to 0.12 ppm during any hour of the day. When cities like Los Angeles experience concentrations in excess of this limit they must declare a smog alert or health advisory warning to limit the activities and exposures, particularly of young children and elderly people.

Electrolysis

In general, the term electrolysis refers to any process that results in the passage of an electric current through an **electrolyte**. An electrolyte is defined as any solution that contains ions and conducts electricity. In the process, the positively charged ions migrate toward the cathode where they are reduced, and the negatively charged ions migrate toward the anode where they are oxidized.

Perhaps the most often demonstrated example of electrolysis is the conversion of water into elemental hydrogen and oxygen gases. The equation for the reaction is:

$$2\ O^{2-} \longrightarrow O_2 + 4\ e^- \text{ (Oxidation Half-Equation)}$$
$$4\ H^+ + 4\ e^- \longrightarrow 2\ H_2 \text{ (Reduction Half-Equation)}$$
$$\overline{2\ H_2O \longrightarrow O_2 + 2\ H_2 \text{ (Overall Equation)}}$$

An application of the above principle is used to recover significant amounts of silver metal from spent photographic developer solution. By passing an electrical current through this solution, the silver ions are attracted to the cathode. Once the ions come in contact with the negatively charged electrode, they gain an electron and are reduced to metallic silver according to the following reduction half-equation:

$$Ag^+ + e^- \longrightarrow Ag^o \text{ (Reduction Half-Equation)}$$
Silver Ion   Silver Metal

The recovered silver metal can be sold to help defray the treatment process costs.

To insure that the amount of silver contained in the effluent of this process is below NPDES limits, the waste stream is subjected to a second process called **cementation**. Cementation is defined as a process in which metal ions in solution are reduced by atoms of another metal that is higher on the **electromotive series**.

In general, the electromotive series – partially shown in Table 4-10 – is a ranking of the reducing or electron-giving strength of each metal. The metals found at the top of this list are also found on the left-hand side of the periodic table and are good electron donors. The metals found toward the bottom right of the periodic table have a lesser ability to give electrons and are found lower on the elec-

K	Potassium
Ca	Calcium
Na	Sodium
Mg	Magnesium
Al	Aluminum
Zn	Zinc
Fe	Iron
Pb	Lead
H_2	Hydrogen
Cu	Copper
Ag	Silver
Au	Gold

Table 4-10: Partial listing of metals in the electromotive series.

tromotive series. Some of our most prized metals – gold, silver and copper – are the poorest electron donors. When a waste stream containing silver ions is brought into contact with iron, Fe^o, in the form of steel wool, the following redox reaction will occur:

$$Fe^o \longrightarrow Fe^{2+} + 2e \text{ (Oxidation half-reaction)}$$
$$2\,Ag^+ + 2\,e \longrightarrow 2\,Ag^o \text{ (Reduction half-reaction)}$$

$$Fe^o + 2\,Ag^+ \longrightarrow Fe^{2+} + 2\,Ag^o$$
(Overall Equation)
Iron metal   Silver metal

The use of an inexpensive metal that is higher on the electromotive series allows for the recovery of the remaining silver metal, without the input of any additional energy. A process similar to this can be used for removal of copper from wastes generated by the printed wire board industries, and the reduction of hexavalent to trivalent chromium in wastes from the electroplating industry.

Checking Your Understanding

1. When an atom loses one or more electrons, is it defined as oxidation or reduction?

2. Ozone is a form of which element?

3. What are two major drawbacks of using ozone as a treatment technology?

4. What is a precipitate?

5. What does the process of electrolysis use to remove ions from a waste stream?

6. For a metal to be removed by cementation, must it be above or below the one in solution on the electromotive series?

4-7 Photolysis

As mentioned previously, ultraviolet radiation converts ozone into oxygen, which limits the amount of short, harmful, chemical bond-disrupting UVC rays reaching the earth. Because of this screening effect, most of the UV rays reaching the surface of the earth are of the UVA and UVB-type, with lower energies, and longer (> 290 nanometer) wavelengths.

The discovery of a hole in the ozone layer has generated considerable concern and discussion. Some claim the hole will lead to a decreased agricultural production as well as an increased incidence of cataracts and skin cancer. The damage is the result of UV light **photodegrading** organic compounds in organisms, in the atmosphere, on earth, and in bodies of water.

When a substance is intentionally exposed to UV radiation to bring about its destruction, the process is called **photolysis**. Photolysis is defined as the photodegradation of a substance into simpler substances as the result of exposure to nothing but light energy.

The equipment used to conduct photolysis consists of two components: a UV light source and a vessel to contain the material being irradiated. The light source is either low- or medium-pressure mercury vapor lamps. The containment vessel must be made of a substance that is transparent to UV light wavelengths and is typically made from borosilicate glass or fused silica. The lamps may be placed either externally or directly in the waste stream. When external placement is used, a two-compartment vessel is used; one holding the substance, while the second houses the light source.

As suggested in the section on ozonation, a combination of ozone or other strong oxidizing agents and UV light can be used to oxidize many substances

that can not be decomposed through the use of ozone alone. The role of the UV light is to supply the energy needed to cause the displacement of an electron, thereby creating a more reactive free radical. Photodecomposition has been shown to be a successful treatment process whenever aldehydes, ketones, azo, or organometallic compounds are involved.

As noted in a previous section, one of the more important uses of photolysis is in the area of pesticide degradation or dechlorination. Most chlorinated aromatic and polynuclear compounds, e.g. PCBs, DDT, 2,4-D, Aldrin, Toxaphene, Atrazine, Alachlor, and Dieldrin, are particularly resistant to biodegradation and hydrolysis reactions. In short, because they cannot be easily broken down, they tend to be persistent in the environment. As a result of their persistence and mobility, many of the previously banned pesticides have found their way into the groundwater. One of the most important photodegradation reactions is the following generalized equation for the **photodehalogenation** of a chlorinated aromatic compound:

The reaction involves the use of high energy UV light to disrupt the carbon to chlorine bond in the aromatic halide. In the presence of a hydrogen source (R-H), photodechlorination of the aromatic compound results.

Successful photodegradation of dioxins (2,3,7,8-TCDD) as well as of a variety of compounds associated with munitions, like TNT (2,4,6-trinitrotoluene), has been reported. However, the process is expensive and therefore not currently considered an economic way to process toxic wastes.

UV Subcategory	Wavelength nanometers (nm)	Health Effects
UVA	315 - 400	Little damage
UVB	290 - 315	Causes sunburns
UVC	100 - 290	Chemical bond disrupting

Table 4-11: UV-light subcategories.

Checking Your Understanding

1. What type of radiation is used in photolysis?

2. Why do pesticides persist in the environment?

3. In photodechlorination, what element is being removed from an aromatic compound?

Summary

Chemical treatment technologies fall into several broad categories, but all have one thing in common: they reduce or eliminate the toxicity of a waste by changing the chemical nature of the substance being treated.

Neutralization involves the addition of an acid or base to solution. The solution's pH is used to determine the progress of the reaction. Precipitation results from the combination of cations and anions forming an insoluble precipitate.

There are many different types of reactions that depend on the exchange of electrons from atoms to ions or vice versa. Collectively, these reactions are called oxidation-reduction, or redox, reactions. In each of these reactions the oxidizing agent takes electrons from the reducing agent. This exchange of electrons results in the oxidizing agent being reduced and the reducing agent being oxidized. The toxic nature of many substances can be reduced or eliminated by this process.

Photolysis is the use of ultraviolet light to degrade compounds. It has been used primarily for the photodegradation of chlorinated aromatic compounds and in conjunction with ozone to chemically alter particularly persistent chemicals.

Neutralization and pH adjustment reactions remain the most commonly used chemical reactions. As is often the case, pH adjustments are required to improve the efficiencies of many other chemical, physical, and biological reactions.

Critical Thinking Questions

1. Explain the process of neutralization.

2. Explain why oxidation and reduction must happen at the same time.

3. Propose a list of negative ions that would effectively precipitate Ag^+, Pb^{2+}, and Cu^{2+} ions from a waste stream.

4. Using information presented in this chapter, predict if zinc metal ions could be removed from a wastewater stream by treatment with sodium chloride solution. Why or why not?

5. Predict if copper metal could be removed from a solution by cementation using a piece of aluminum metal. Why or why not?

5

Thermal Treatment Technologies

by Howard Guyer

Chapter Objectives

Upon completing this chapter, the student will be able to:

1. **Explain** the differences in products resulting from stoichiometric vs. nonstoichiometric combustion.

2. **Compare** the various types of incinerators that use stoichiometric amounts of oxygen.

3. **Discuss** the principles involved in the pyrolysis (nonstoichiometric) method.

4. **Discuss** the principles involved in the wet air oxidative and supercritical water oxidation process.

5. **Compare** the various chemistries of acid gas controls used in scrubbers.

6. **List** and describe the steps required to permit a hazardous waste incinerator.

Chapter Sections

5-1 Introduction

From the consumer point of view, we enjoy one of the highest standards of living in the world. These standards, however, continue to extract an environmental price – the generation of hazardous waste. Even before the first Earth Day in 1970, Americans were becoming more aware of their environment. In 1976, the Congress passed the Resource Conservation and Recovery Act (RCRA), which radically changed the way we were disposing of our wastes. Not only did it make dumping a crime, but it also required industry to treat waste before its disposal. One of the treatment methods for organic hazardous wastes approved by EPA was incineration.

In 1984, RCRA was amended by the Hazardous and Solid Waste Amendments (HSWA), which established even tougher waste management standards for industry. According to HSWA, incineration was considered to be the **best demonstrated available technology (BDAT)** for many types of industrial wastes. As a result, HSWA continued to require industry to use incineration for these wastes.

Today, hazardous waste incinerators are regulated and permitted by EPA and often must meet individual state environmental regulatory requirements as well. The basic EPA regulations for incineration are set forth in Title 40 of the *Code of Federal Regulations* (CFR), Part 264, and are supplemented by EPA technical and permit guidance. Most states are authorized by EPA to administer these regulations, under EPA oversight. In order to be authorized, a state's regulations must be at least as stringent or more stringent than the federal regulations. The regulations are then enforced by both EPA and the state.

Part 264 has many provisions that apply to all hazardous waste management facilities, plus some that apply only to incinerators. The general provisions require such elements as a worker training program, a facility inspection program, a waste analysis plan, an emergency prevention and response plan, and environmental liability insurance. The specific incineration requirements are contained in Subpart O of Part 264. The key requirements of Subpart O relate to:

— Permits;

— Trial burns;

— Destruction and removal efficiency (DRE);

— Required operating parameters;

— Waste feed cut-offs; and

— Emission controls monitoring.

Incineration is defined as the controlled high-temperature oxidation of primarily organic compounds producing carbon dioxide and water. In addition to carbon dioxide and water, inorganic substances such as acids, salts, and metallic compounds may also be produced depending on the composition of the waste. In general, thermal treatment technologies include incineration and all other processes that use elevated temperatures to destroy the waste.

Pyrolysis differs from the above method because it is designed to prevent the hot organic materials from coming into contact with oxygen. Therefore, burning cannot occur; rather, the material is destroyed by separating it into its volatile and nonvolatile components, a process known as **destructive distillation**. Depending on the economic value of these substances, the volatile components can either be collected and reused, or later brought into contact with stoichiometric amounts of oxygen and burned in a secondary combustion chamber. If the material contains sufficient **BTU value** (heat), it may be used as a fuel source. The nonvolatile components, too, can either be disposed of or recovered depending on their economic value.

The **wet air oxidative (Zimpro)** and **supercritical water oxidation** methods also use heat to destroy organic materials, but without the use of flames. The organic waste material is dissolved or suspended in water and brought into contact with varying amounts of oxygen under conditions of high pressure and temperature. Under these conditions, cell walls rupture and – depending on the contact time – complete oxidation of the various substances occurs. An advantage of these processes is that they produce little, if any, NOx and SOx and no **fly ash**.

Although not every type of thermal treatment technology will be discussed, it is the intent of this chapter to familiarize the reader with the scientific principles involved in destroying unwanted substances through the use of oxidation processes and/or heat. The regulatory framework governing its use will also be examined.

5-2 Incineration and Oxidation Reactions

Chemists typically use the term oxidation in a very broad context. For most of us, however, the term is typically associated with burning or **combustion**, but it can also be applied to any process that results in a substance combining with oxygen. Carbon and a variety of other elements can react with varying amounts of oxygen. For the purposes of this discussion we will use the term **stoichiometric** to refer to those oxidation processes where the substance combines with the maximum chemically allowable amount of oxygen and **non-stoichiometric** for those processes that do not. As an example, the element carbon, C, can combine with oxygen, O_2, in two different ratios:

$$C + O_2 \longrightarrow CO_2 \quad \text{(stoichiometric)}$$

$$2\,C + O_2 \longrightarrow 2\,CO \quad \text{(non-stoichiometric)}$$

In the first equation, oxygen is present in high quantities, and carbon uses its maximum combining capacity to react with one oxygen molecule to form carbon dioxide as the product. In the second equation, due to a limited amount of oxygen, each carbon atom uses only one-half of its combining capacity, resulting in the formation of two molecules of carbon monoxide. We could describe the latter as an oxygen-starved condition. If at a later time more oxygen were to become available, the carbon monoxide could still react with more oxygen and be converted into carbon dioxide. The equation for the second step of this reaction would be:

$$2\,CO + O_2 \longrightarrow 2\,CO_2$$

Because some products of combustion may be further oxidized, the use of a secondary combustion chamber behind or following an incinerator can further complete the oxidation of substances like CO. The use of a catalytic converter on a car's exhaust system performs much the same function.

The goal of industries that use incineration to treat waste is the complete oxidation of all organic matter present to minimize hazardous emissions. For this reason, the original art of combustion has become a carefully engineered process that makes use of what is euphemistically referred to as the **Three Ts** – time, temperature, and turbulence. The combination of a sufficiently long reaction time – coupled with high temperatures– and enough turbulence to promote adequate mixing of the fuel and oxygen will result in the complete destruction of any organic matter contained in the waste.

A variety of furnaces and incinerators, also known as **thermal treatment units (TTUs)**, make use of stoichiometric amounts of oxygen in either the primary, or later in the secondary combustion chamber, to achieve these goals. Depending on the incinerator size and the nature and amount of waste being fed into it, this one-two punch results in nearly complete oxidation of all the organic matter present.

There are numerous everyday observations that demonstrate combustion under oxygen-abundant and oxygen-starved conditions. For example, it is rare today to see exhaust from an automobile; however, remember that black plume of smoke coming from the exhaust of a diesel truck or bus? What are the similarities and differences in the exhaust gases and particulates produced by gasoline and diesel-burning engines?

Within the cylinder of an internal combustion engine, the fuel – which is a mixture of hydrocarbons – is always burned under a **starved oxygen** set of conditions. The tendency for elements to combine with each other favors the hydrogen-oxygen reaction over all others. For this reason, whenever there is an insufficient amount of oxygen present, hydrogen atoms get first choice and combine with the oxygen to form water. Any remaining oxygen then combines, in varying ratios, with carbon atoms forming colorless carbon monoxide, CO, and carbon dioxide, CO_2, gases, and some carbon may not be burned at all. It is this uncombined carbon, C, that is responsible for the black smoke seen in the diesel exhaust.

Today automobiles are required to have a catalytic converter. In the converter, C atoms and incompletely oxidized CO have a second chance to combine with oxygen. This second chance results in the conversion of most carbon and carbon monoxide into colorless carbon dioxide and eliminates the black exhaust plume. Diesel engines, as of yet, have not been required to use catalytic converters, so the carbon atoms remain a visible part of its exhaust.

To illustrate the different degrees of combustion we will use octane – one of the components of gasoline – as an example. Depending on the amount of oxygen available in the cylinder, octane, C_8H_{18}, and oxygen, O_2, can react in a variety of ratios.

a) $C_8H_{18} + 4\frac{1}{2}\,O_2 \longrightarrow 8\,C + 9\,H_2O$ + least energy

b) $C_8H_{18} + 6\frac{1}{2}\,O_2 \longrightarrow 4\,C + 4\,CO + 9\,H_2O$ + more energy

c) $C_8H_{18} + 7\frac{1}{2}\,O_2 \longrightarrow 2\,C + 6\,CO + 9\,H_2O$ + even more energy

d) $C_8H_{18} + 10\frac{1}{2}\,O_2 \longrightarrow 4\,CO + 4\,CO_2 + 9\,H_2O$ + even more energy

e) $C_8H_{18} + 12\frac{1}{2}\,O_2 \longrightarrow 8\,CO_2 + 9\,H_2O$ + most energy

Equations a – d are examples of the result of using non-stoichiometric amounts of oxygen. As can be seen, as the oxygen to octane ratio improves, less and less carbon and carbon monoxide, collectively known as **products of incomplete combustion (PICs)**, are produced. Only in equation e, however, where the combustion is complete, are there only **products of complete combustion (POCs)** formed. It should also be noted that as the ratios of oxygen and octane improve, the amount of energy produced also increases. In cars this would translate into getting better gasoline mileage.

In addition to PICs and POCs, it is not unusual to have varying amounts of other combustion products formed. For example, if the waste contains chlorinated hydrocarbons, then hydrogen chloride gas, HCl, will be a product. Frequently, sulfur oxides, SOx, and nitrogen oxides, NOx, are simultaneously produced as the organic components are burned. The SOx are the result of burning sulfur-containing compounds, such as mercaptans. The NOx may be

> Oxygen ↑   PICs ↓   POCs ↑   Energy ↑
>
> If oxygen increases so do POCs and energy, while PICs decrease.

from either burning nitrogen-containing compounds, such as amines, or the reaction between atmospheric nitrogen and oxygen at the high temperatures present in the incinerator.

Small silica-based particulate matter, called fly ash, is also produced. These unwanted products are typically removed by the **air pollution control system (APCS)** equipment, which may include acid scrubbers, bag houses, and electrostatic precipitators. This equipment, its design and use, was discussed in Chapter 3.

Checking Your Understanding

1. The availability of which element determines if combustion will occur stoichiometrically or non-stoichiometrically?

2. What is the effect on the amount of energy produced as the combustion of a hydrocarbon becomes more complete?

3. What is the function of a catalytic converter on an automobile?

4. Which substance must be absent for pyrolysis to occur?

5. What is an advantage of wet air oxidation?

5-3 Incinerators

Liquid Injection Incinerators

This is the most common incineration method for hazardous waste disposal. **Liquid injection incinerators** may be used to destroy pumpable wastes such as fumes, liquid, sludges, and slurries. Liquid injection incinerators may be of two kinds: either specially designed to burn pumpable wastes or simply the addition of an injection system to an incinerator used to burn other kinds of waste. When specially designed, they are generally composed of 1) a cylindrical, primary combustion chamber lined with refractory material, and 2) a secondary combustion chamber.

Depending on the material to be treated, these units typically operate at temperatures from 1,000-1,700°C. The waste is either burned directly in the burner or injected into the flame zone of the incinerator via an **injector nozzle**. The most critical parts of the incinerator are the injector nozzles that convert the liquid stream into uniform and finely atomized droplets.

Atomization of the liquid increases the surface area exposed to contact with the surrounding com-

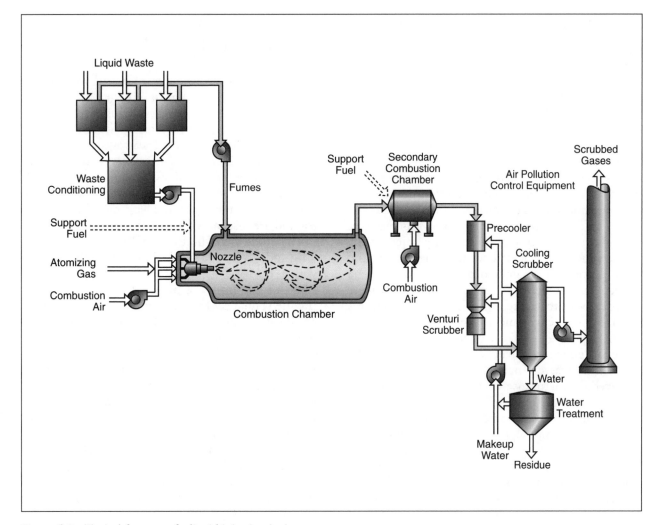

Figure 5-1: Typical features of a liquid injection incinerator.

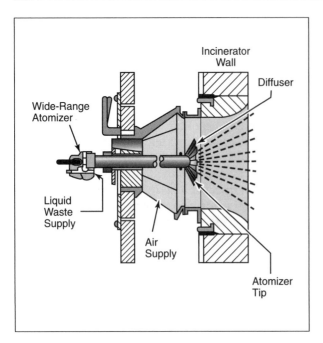

Figure 5-2: Liquid injector nozzle atomizes the liquid.

bustion air. Proper mixing of the combustion air with the liquid droplets is necessary to bring about complete oxidation. As the liquid is vaporized and heated to ignition temperature, oxygen reacts with the hydrocarbon vapors releasing its energy. As this occurs, there is a sudden rise in the surrounding temperature. This increases the velocity of the gases in the zone around the droplet, resulting in increased turbulence and completion of the oxidation reaction.

As the viscosity of liquids being injected increases, the droplet size tends to get larger and the boiling point of larger hydrocarbons tends to be higher. To completely vaporize a droplet of a larger hydrocarbon, therefore, both higher temperatures and more residence time are necessary.

A high degree of turbulence is desirable for achieving effective destruction of the organics in the hazardous waste. Increased turbulence can be achieved, with the use of high-intensity burners, which allows this reaction to happen more rapidly. If the hydrocarbons should contact an area of low oxygen concentration, pyrolysis would result in the formation of hydrogen, carbon, and carbon monoxide. This would increase the amount of time required for complete combustion to occur. Following oxidation of the flue gases in the secondary combustion chamber, the exiting gases are cooled and treated with air pollution source control devices to

remove unwanted particulates and to absorb or neutralize the acid gases.

The advantage of liquid injection incineration is that it can be used on almost any pumpable, atomizable organic substance, including substances containing PCBs, dioxins, phenolic sludges, halogenated and non-halogenated organics, pesticides, and acid sludges. Even wastes with a high moisture content can be destroyed.

The major disadvantage of liquid injection incineration is that the waste stream must be free of (or preprocessed to remove) solids that prevent pumping and satisfactory atomization. It also needs to be free from inorganics that tend to fuse with the refractory material at incineration temperatures or that **sublime**/vaporize to yield hard-to-collect fumes. This would include those wastes that contain soil, high amounts of inorganic salts, or heavy metals.

Rotary Kilns

Rotary kilns are considered to be the most versatile type and the second most often used incinerator in hazardous waste disposal systems. They can accommodate all types of wastes (solids, liquids, and gases). Solid and drummed wastes are usually fed into them by a conveyor or ram system. Liquids, gases, and pumpable sludges are introduced through the use of injection nozzles. Noncombustible metal and other residues travel the length of the incinerator and are discharged as ash at the end of the kiln.

The kiln is a cylindrical shell, 3-9 m (10-30 ft) long and 1.5-4 m (5-12 ft) in diameter, that is lined with refractory material and that is mounted at a slight incline from the horizontal. The rotation of the shell enhances the mixing of the wastes with the combustion air; it also helps transport the solid wastes and ash through the kiln. Residence time for solids fed into a kiln is controlled by the feed rate, rotational speed, angle of inclination, as well as an internal structure (e.g. dams, chains, etc.) that is specifically designed to slow the waste's progress through the kiln, thereby adding to the turbulence experienced by the waste.

The kiln is initially fired by natural gas and – depending on the BTU value of the material(s) being introduced – may or may not need continued natural gas support to maintain its operating temperature of 800-1,200°C. Most rotary kilns are also fitted with a secondary combustion chamber where

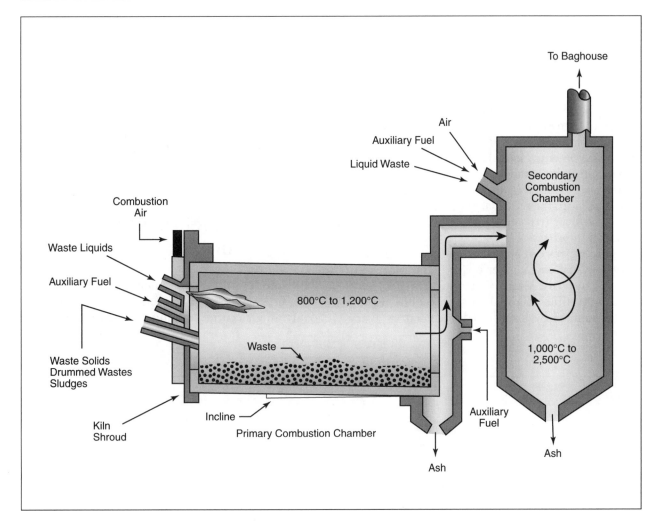

Figure 5-3: Typical features of a rotary kiln incinerator.

final oxidation of any escaping volatile gases occurs. It is operated at temperatures ranging between 1,000 and 2,500°C to ensure complete oxidation. The exiting gases are then introduced into the air pollution control unit to remove acid gases and particulates. The gases – typically HCl, nitrogen, and sulfur oxides (NOx and SOx) – are removed by means of acid scrubbers and either a bag house or an electrostatic precipitator used to capture the fly ash.

As previously indicated, the major advantage of a rotary kiln is its versatility. It can be operated at high temperatures and accommodate the mixing of incoming wastes. Operating conditions can be varied from complete (stoichiometric) to starved oxygen (non-stoichiometric) conditions that result in pyrolysis of the waste. Rotary kilns are effective for treating wastes containing PCBs, dioxins, halogenated and non-halogenated organics, pesticides, and soil contaminated with organics.

The major disadvantages of rotary kilns are that they have high capital and operating costs. Maintenance costs can be high because of the corrosive characteristics of the waste burned and of the exposure of moving parts to high temperatures. Rotary kilns also do not work well for wastes that have a high inorganic salt content, which has a tendency to degrade the refractory lining of the kiln, and which causes **slagging** (fusing) of the ash. Wastes that have a high heavy metal content also do not work well, due to the difficulty of removing these metals with traditional air pollution control equipment.

Fixed Hearth Incinerators

Fixed hearth incinerators are the third most commonly used incinerators for the treatment of hazardous wastes. They usually have small capacities, but can handle both liquid and solid wastes. The incinerator consists of a primary and secondary combustion chamber. The first chamber operates in a non-stoichiometric mode. Liquid wastes are injected into the primary chamber where they are mixed with air, but are incompletely burned. The resulting PICs then move into the secondary combustion chamber where more air (oxygen) is added to complete the combustion process.

Solid wastes can also be introduced to the primary combustion chamber by feed rams pushing it onto the **hearths**. Mixed wastes, including solvents and combustible solids, such as infectious wastes, can be handled by fixed hearth incinerators. They are rarely used, however, for bulky solids or drummed materials.

These incinerators have a limited capacity and can treat only between 1,000 and 2,000 pounds of waste per hour. The primary combustion chamber operates at temperatures ranging from 315-870°C and the secondary chamber at 650-980°C. Because of these low temperatures, fixed hearth incinerators have a limited ability to destroy chlorinated liquid wastes.

Multiple Level Hearth Incinerators

Multiple level hearth incinerators are composed of a vertically mounted, refractory-lined steel shell containing a central rotating shaft, a series of verti-

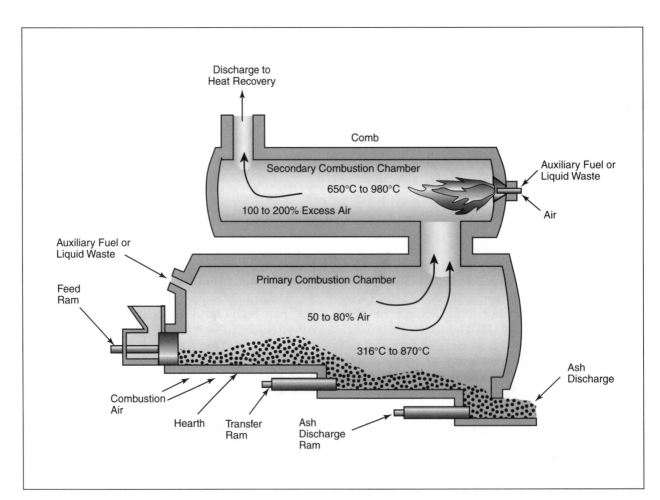

Figure 5-4: Typical features of a fixed hearth incinerator.

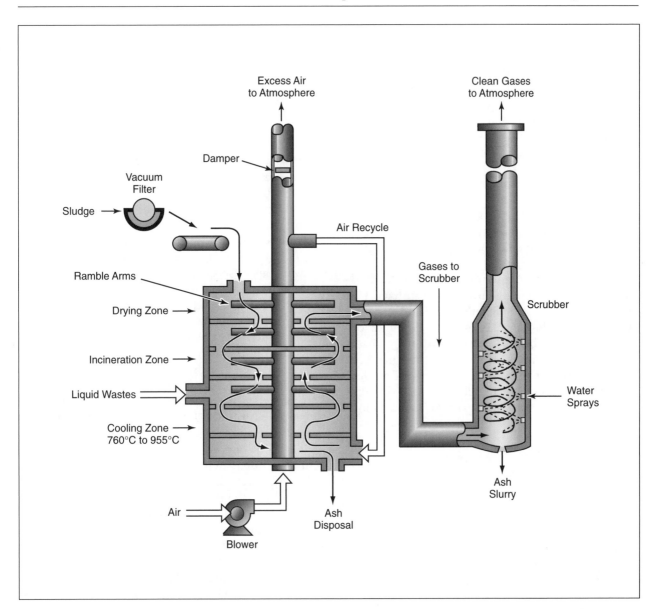

Figure 5-5: Typical features of a multiple level hearth incinerator.

cally stacked flat hearths, and a series of **rabble arms** that plow the burning material successively across the hearths and create some turbulence. The material to be incinerated is introduced at the top of the furnace and the ash is discharged at the bottom. Liquid and gaseous wastes can also be injected through side ports.

Multiple level hearth incinerators are seldom used for the treatment of hazardous wastes, but remain an important tool because of their use in the regeneration of activated carbon and their ability to treat up to 1,000 pounds per hour (dry weight) of municipal wastewater treatment plant sludges.

Depending on the level, the hearths within the incinerator fall into three zones. The upper hearths are used for drying and preheating the incoming material. The central hearth is the oxidizing (incinerating) zone where stoichiometric combustion occurs at temperatures ranging from 760-955°C. The lower hearths are where cooling of the ash occurs, prior to exiting the incinerator. A blower provides stoichiometric amounts of air that enters at the bottom and moves up through the incinerator. Air pollution control equipment – including a scrubber and a particle-removal system – is normally a part of the complete system.

Fluidized-Bed Incinerators

Fluidized-bed incinerators used to treat hazardous waste are the result of an adaptation of industrial technology that has been in use for many years. Initially, the technology was used for coal gasification and later extended to petroleum refineries where the application is known as fluidized-bed catalytic cracking.

Fluidized-bed incinerators are ideally suited for the thermal destruction of solids, liquids, and sludges, either separately or mixed. The versatility of the technology is due to its excellent gas-to-solids contacting, its stable temperatures, and its ability to control residence times.

There are two common variations of the fluidized-bed technologies: the **bubbling fluidized-bed** (Figure 5-6), and the **circulating fluidized-bed** (Figure 5-7) **incinerator**. In the bubbling fluidized-bed incinerator, there is a vertical oxidizing chamber that is lined with insulating refractory material. In the bottom of the reactor is a layer of graded, inert granular material – usually silica sand. Fluidizing-air fans force large volumes of preheated air into the bottom of the sand layer. This results in it becoming a swirling gas-solids mixture with physical properties resembling that of a fluid. This vigorous agitation results in uniform temperature distribution throughout the fluidized bed and distributes stoichiometric amounts of oxygen. The material to be treated is either injected by spray nozzles directly into or just above this swirling mass. The upper portion of the vessel serves as a secondary combustion chamber, which provides additional time for complete combustion.

Depending on the application, fluidized-bed temperatures typically range from 650-750°C. To attain **destruction and removal efficiencies (DREs)** greater than 99.99 percent requires – if hazardous wastes containing chlorinated hydrocarbons are to be injected – the bed to be operated at or above 775°C and the secondary portion of the chamber at 1,200°C or more. Off-gas cleanup is accomplished with the same type of wet or dry scrubber, bag house, or electrostatic precipitator equipment utilized by other incinerators.

The circulating fluidized-bed incinerator uses even higher air velocities and slightly higher temperatures (760-870°C). The higher air velocities and recirculation of solids create a highly turbulent incineration zone, which allows for sufficient residence time to adequately destroy the undesirable waste. Any solids remaining in the circulating bed are separated from the flue gases using a cyclone filter and are reintroduced into the incinerator.

Although the temperatures achieved by the bubbling and circulating fluidized-bed incinerators are typically lower than rotary kilns, they are still capable – when incinerating hazardous wastes – of achieving the DREs within the regulatory limitations of RCRA by providing high heat transfer efficiency and extended contact time with the waste.

Wastes containing halogenated and non-halogenated organics, contaminated soil, PCBs, phenolics, and pharmaceuticals have been satisfactorily treated by the fluidized-bed technologies. There are several wastes that are not suitable for fluidized-bed units: 1) oversized wastes that cannot be shredded to less than one inch in size for circulating fluidized-bed or three inches for bubbling fluidized-bed incinerators; 2) inorganic salts with high sodium content that cause the ash to be fused and degrade the kiln's refractory lining; 3) wastes with a high heavy metal content; and 4) wastes with a low melting point constituent (< 850°C).

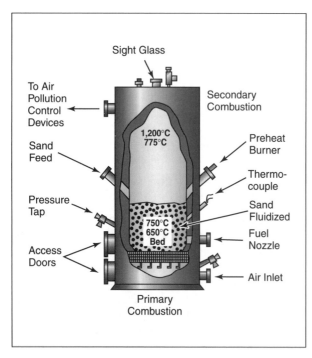

Figure 5-6: Typical features of a bubbling fluidized-bed incinerator.

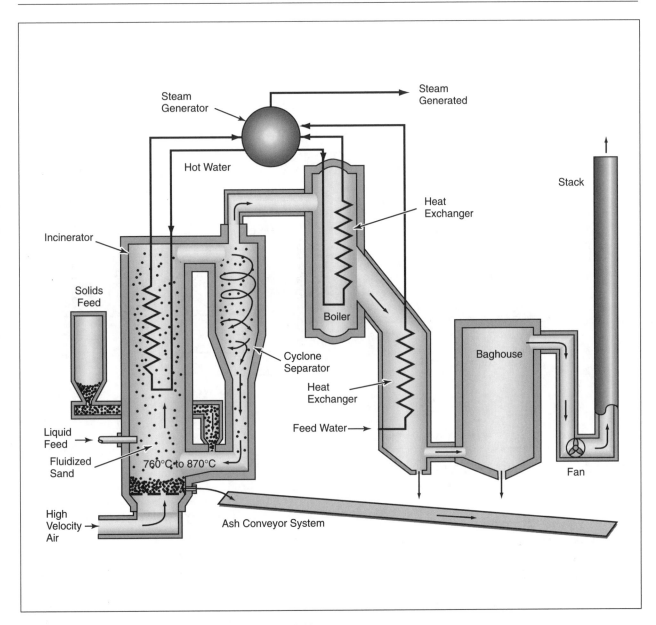

Figure 5-7: Typical features of a circulating fluidized-bed incinerator.

Reciprocating Grate Incinerators

A **reciprocating grate incinerator** (Figure 5-8) is composed of a long chamber lined with **refractory** materials. The floor of the chamber consists of a stair-step arrangement composed of moving grates. Waste to be incinerated is fed onto the grates at the upper end of the staircase. To ensure the Three Ts of combustion, the feeding rate, the speed of grate movement, and the airflow rate are constantly monitored and modified according to needs. Large fans provide a constant stream of air that moves upward through grates as they slowly reciprocate forward and back, causing the burning waste to tumble down the fiery staircase. At the end of the staircase, the ashes and any unburned material are quenched by being dropped into a container of water.

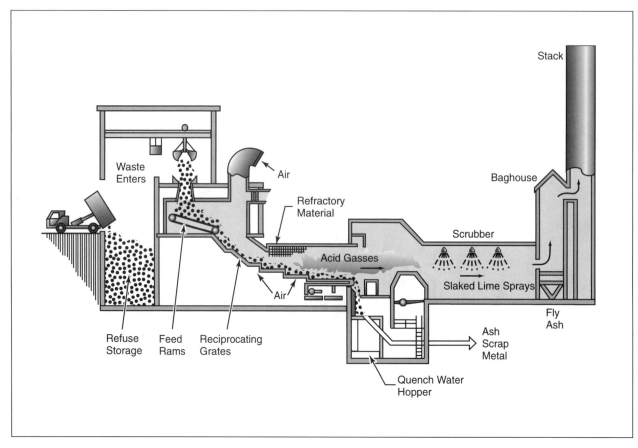

Figure 5-8: Typical features of a reciprocating grate incinerator.

This type of incineration is typically used for waste streams such as municipal trash where the waste stream components are nonhazardous and vary greatly in size and composition. Because of the size and composition variations of the materials entering, those that fail to burn completely are separated from the ash and sent through the incinerator again. The ash waste stream can also be separated into metal and nonmetal waste. The metals can be sent to a salvage operation for recycling, while the nonmetal wastes are mixed with cement, made into a concrete product that is crushed, and used for road-bed construction material.

The hot gases produced by the burning process are typically treated by a series of acid scrubbers to remove the acid gases (nitrogen oxides, NOx, and sulfur oxides, SOx) and a bag house or electrostatic precipitator is used to capture the fly ash. The heat generated by this process is also a useful product. Due to their ability to accept nonhazardous waste in a wide range of types and sizes, several waste-to-energy co-generation plants use them to also produce electricity.

Plasma Thermal Treatment Units

One of the more exotic types of incinerators is Retech's **plasma thermal treatment unit**. At its core is a copper-lined steel centrifugal crucible where the tip of the plasma arc torch (natural gas burner) serves as one electrode and the copper cladding serves as the other. Boosted by the electrical discharge, the core of the burning gas **plasma** may reach temperatures ranging up to 15,000°C. These high temperatures result in the pyrolysis of all organic wastes and the fusing or vitrification of the noncombustible metals and minerals. Although this treatment method may not detoxify the metals and minerals, it does result in their permanent immobilization by fusing them into a non-leachable **glassy slag**.

The gaseous organic wastes driven off by pyrolysis move on to a secondary combustion chamber where they are stoichiometrically burned. The flue gases exiting the secondary chamber then enter the typical air pollution control equipment described in each of the previous sections.

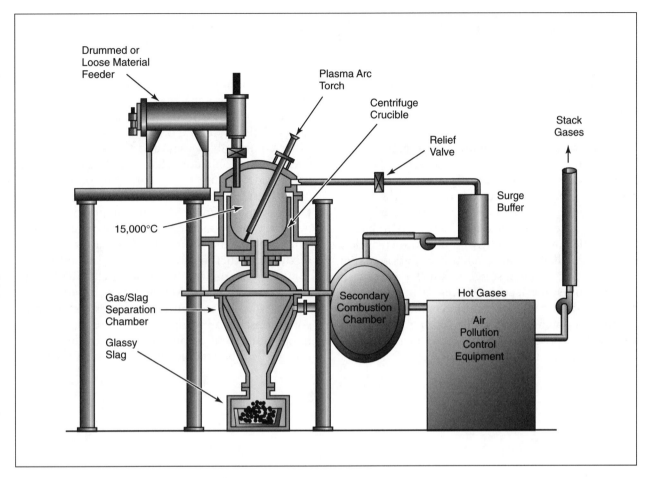

Figure 5-9: Features of Retech's plasma centrifugal furnace.

The use of this technology to treat hazardous wastes is limited by production cost, lack of operational experience, and lack of incinerators. The process also suffers from both limited treatment capacity and high treatment costs. However, due to the extremely high temperatures possible, this method may be effectively used to treat metal, PCB, and dioxin-contaminated soils or other mixed wastes.

Checking Your Understanding

1. What is the function of a secondary combustion chamber? *to add O₂ to assure complete Combustion*

2. Which types of gases are typically removed by acid scrubbers? *HCl, N, SOx, NOx*

3. What types of substance are removed by the use of a bag house? *Fly Ash*

4. Which type of incinerator allows for sludges to dry before they are incinerated? *multiple level hearths*

5. Which type of incinerator uses an inert substance to transfer heat uniformly to the waste being burned? *Fluidized-bed*

6. Which type of incinerator is used for nonhazardous substances that vary greatly in both size and shape? *Reciprocating grate*

7. Which type of thermal treatment process produces the highest temperatures? *Plasma*

5-4 Other Elevated Temperature Methods

Pyrolytic Thermal Treatment Units

Pyrolytic or starved oxygen treatment refers to the heating of waste materials under non-stoichiometric conditions. The amount of oxygen present can range from only a few percent down to the complete absence of oxygen. Under these conditions, the waste is generally separated into its volatile and nonvolatile components by a process known as destructive distillation. Although the separation of the waste in this manner suggests smaller particles, through an **isomerization process** it may, in fact, result in the formation of even larger molecules. This is the phenomenon that occurs during the petroleum coking process. If any of these fractions have economic value, they can be recovered.

The typical **pyrolytic thermal treatment unit** consists of two compartments. The pyrolysis chamber where the waste is heated causing it to separate into volatile water vapor, carbon monoxide, and other combustible gases, and the non-volatile ash components containing **char**, metals, and salts. The heat used to cause this pyrolysis to occur is provided by using the heat produced by the combustion of the unwanted volatile components and, if necessary, supplemented by an auxiliary fuel source. The auxiliary fuel source may be natural gas or a mixture of other injected liquid or gaseous wastes.

In the secondary combustion chamber, or afterburner, the volatile components driven off are mixed with air and burned under stoichiometric conditions. By allowing for ample residence time, temperature, and turbulence, the hazardous components can be completely destroyed, if they have

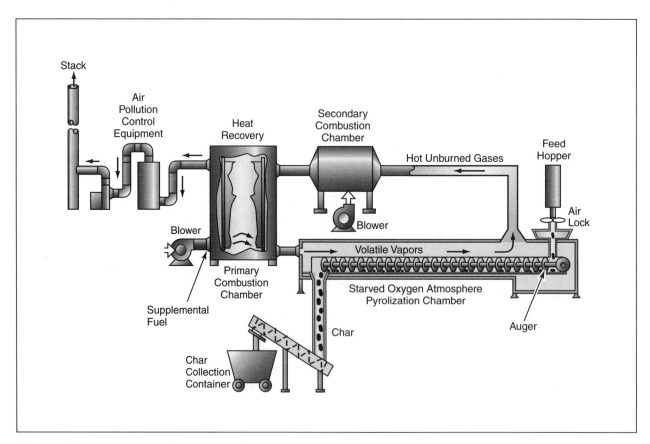

Figure 5-10: Pyrolytic thermal treatment unit.

no recovery value. Pyrolysis may be expensive and is generally limited in its input capacity. Since the separated components do not necessarily undergo combustion, however, those of economic value may be recovered and sold to offset operating costs.

Wet Air Oxidative Methods

Zimpro Wet Air Oxidative Method

The wet air oxidative or Zimpro method is a thermal treatment technology that does not include burning in a flame, but is frequently used to dewater and oxidize sludges composed of organic materials. Through the use of elevated temperatures and pressures, water that is tightly held by various chemical groups – known as **bound water** – is released from the organic material, and oxidation of the finely divided organic material occurs.

In this process, aqueous organic wastes are mixed with compressed air. The waste-air mixture is then preheated in a **heat exchanger** before entering the corrosion-resistant reactor. In the reaction chamber, pressures of up to 2,000 psi and temperatures of 300°C are reached. Under these conditions, the cell walls of the bacteria and other organisms in the sludge are ruptured and their organic contents are mixed with the water. In addition, the oxygen portion of the compressed air, under these pressures, dissolves into the water making it available to oxidize the released organic materials. This reaction is **exothermic** and produces a great amount of heat.

The heat from the stream exiting the reactor is used in the heat exchanger to warm the incoming sludge before it enters a separator. In the separator, the vapors (i.e. non-condensable gases consisting primarily of water vapor, nitrogen, and carbon dioxide gases) are separated from the liquid phase (water and remaining unoxidized organic materi-

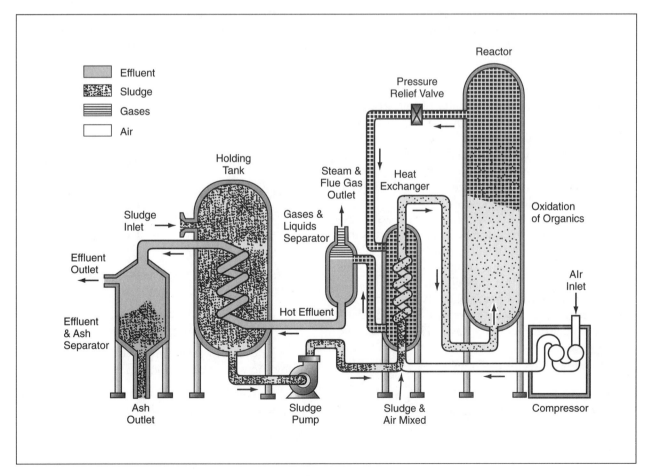

Figure 5-11: Typical features of the Zimpro wet air oxidative method.

als). By varying the contact time, the pressure, and the temperature, oxidation can range from incomplete to complete.

There are several advantages to wet air oxidation. Due to the heat released during the oxidation phase, the process is generally self-sustaining. It also works well for a waste stream that has a **chemical oxygen demand (COD)** between 10,000 and 200,000 mg/liter. This makes it ideally suited for those waste streams that are too concentrated for biological treatment, but too diluted for incineration. The oxidation products typically remain dissolved or suspended in the water, resulting in very low concentrations of nitrogen and sulfur oxides in the off-gas. Off-gas treatment, however, may still be required to control releases of unburned hydrocarbons. The process is typically used to treat waste streams with less than five percent organics, which may include small amounts of pesticides, phenolics, organic sulfur, and cyanide; its use on aromatics and highly halogenated inorganics and organics is restricted.

Supercritical Water Oxidation

Recently a new technology that closely resembles the wet air oxidative process and known as the supercritical water oxidation process has been demonstrated. It is based on the ability of water to perform as an excellent solvent for organic materials, when it is put under supercritical conditions. For water the **critical temperature and pressure** is at 374°C and 3,200 psi. At or above this temperature, water exists as a dense gas which cannot be liquefied by exerting additional pressure.

When oxygen in the air is mixed with aqueous wastes above the critical temperature and pressure of water, organic substances dissolve in the dense vapor and are rapidly and completely oxidized to carbon dioxide and water. In addition, inorganic salts precipitate out of the supercritical mixture. The oxidation reaction produces energy in excess of that needed to heat the incoming waste. This excess heat can be used to produce high pressure steam to be used for some other purpose, such as electricity generation.

The process involves the use of a high pressure pump to bring an aqueous solution or slurry of haz-ardous waste up to system pressure, before being heated in a heat exchanger to supercritical conditions. Under these conditions, large organic molecules are broken into smaller ones. High pressure air is then injected into the reactor, rapidly oxidizing these smaller molecules. A base, such as sodium hydroxide, may need to be added to neutralize any inorganic acids formed during the oxidation.

This technology can be applied to aqueous solutions or slurries with organic concentrations from one to nearly 100 percent. The concentration of organic material in the waste fed into the process will determine the heat produced per pound of waste material. The heat content of waste fed into the process is maintained at 1,800 BTU/lb. Therefore, wastes with a heat content below this value will require the addition of auxiliary fuel. Waste materials with a heat content above this require either dilution with water or blending with another waste that has a lower heat content.

Wastes that have been successfully demonstrated include PCBs, dioxins, solvents, pesticides, and still and tank bottoms. It is believed that the supercritical water oxidation process can be adapted to a wide range of feed mixtures and scales of operation, potentially making it a feasible mobile technology. At this time, however, supercritical water oxidation has only been demonstrated in pilot plant operations and is still considered a potential new member of the thermal treatment technology group.

Checking Your Understanding

1. What is another name for pyrolysis?

2. What are two things that can happen to volatile organic substances driven off in the pyrolysis process?

3. What is the source of oxygen in the wet air oxidation process?

4. What is at least one advantage to the wet air oxidation process?

5. What type of wastes cannot effectively be treated by the wet air oxidation process?

6. Why has the supercritical water oxidation process not been more widely applied?

5-5 Scrubber Chemistry

In the previous discussion on hydrocarbon POCs and PICs it was noted that combustion of wastes containing chlorinated hydrocarbons produces hydrogen chloride gas. In general, chlorinated hydrocarbons tend to be slower-burning and soot-forming. As a result, systems are often operated with excess air to promote better mixing of the oxygen with the hydrocarbons and to minimize the soot formation. It is also desirable to convert the chlorine to HCl, rather than let it form free chlorine, Cl_2, which is much more corrosive to metals. For example, when vinyl chloride, $CH_2=CHCl$, is burned in excess oxygen, HCl gas is produced, as shown in the following equation:

$$2\ CH_2=CHCl + 5\ O_2 \longrightarrow 4\ CO_2 + 2\ H_2O + 2\ HCl$$

Hydrogen chloride gas is an acid gas that is converted into hydrochloric acid HCl in the presence of water. Other incineration products formed by burning chlorinated hydrocarbons such as pentachlorophenol, chlorinated phenols, or from non-

chlorinated compounds reacting with phenol and chloride ions are unwanted dioxins (dibenzo-p-dioxins) and **furans** (dibenzofuran). These unwanted substances can be formed in the cooler (250-350°C) regions of an incinerator or air pollution control equipment if the temperatures are not carefully controlled.

The combustion of sulfur-containing compounds results primarily in the formation of sulfur dioxide, SO_2, although some sulfur trioxide, SO_3, may also be produced. The reaction for their formation, for example, can be shown by the combustion of ethyl mercaptan, C_2H_5SH, and the further oxidation of the sulfur dioxide to sulfur trioxide:

$$2\ C_2H_5SH + 9\ O_2 \longrightarrow 4\ CO_2 + 6\ H_2O + 2\ SO_2$$

$$2\ SO_2 + O_2 \longrightarrow 2\ SO_3$$

In a similar way, the formation of nitrogen dioxide, NO_2, or nitrogen monoxide, NO, can be shown by either the interaction of oxygen and ni-

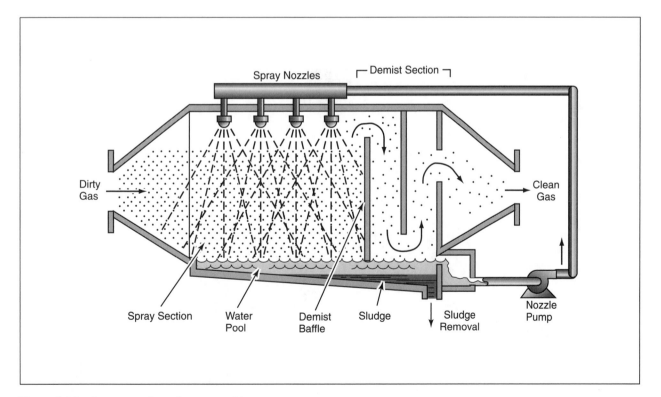

Figure 5-12: A cross-section of a wet scrubber.

trogen gases at incinerator temperatures or by the combustion of a nitrogen-containing organic compound such as methyl amine, CH_3-NH_2.

$$N_2 + 2\ O_2 \longrightarrow 2\ NO_2$$

$$N_2 + O_2 \longrightarrow 2\ NO$$

$$4\ CH_3\text{-}NH_2 + 11\ O_2 \longrightarrow 4\ CO_2 + 10\ H_2O + 4\ NO$$

The resulting acid gases – HCl, NOx, and SOx – from these reactions can be removed in a wet or dry scrubber by the use of several different substances. Aqueous ammonia is frequently used for the removal of HCl and NOx. For example, HCl or nitrogen monoxide, NO, can be removed in a wet scrubber according to the following equations:

$$HCl + NH_3 \longrightarrow NH_4Cl$$

$$6\ NO + 4\ NH_3 \longrightarrow 5\ N_2 + 6\ H_2O.$$

Both HCl and sulfur dioxide, SO_2, may be removed, and a useful product produced by treatment with inexpensive slaked lime. In the first step, hydrogen chloride gas reacts with the slaked lime according to the following equation:

$$Ca(OH)_2 + 2\ HCl \longrightarrow CaCl_2 + 2\ H_2O$$

The calcium chloride solution resulting from this reaction then comes in contact with the SO_2, where the following reaction occurs.

$$2\ H_2O + CaCl_2 + 2\ SO_2 \longrightarrow Ca(HSO_3)_2 + 2\ HCl$$

The HCl produced in this step can be recycled to the previous step and the calcium bisulfite, $Ca(HSO_3)_2$, can then be oxidized, producing usable gypsum, $CaSO_4$, according to the equation:

$$Ca(HSO_3)_2 + Ca(OH)_2 + O_2 \longrightarrow 2\ CaSO_4 + 2\ H_2O$$

The gypsum, $CaSO_4$, produced as a result of the acid gas neutralization is a valuable product that can be used, for example, in making the building material known as drywall.

Checking Your Understanding

1. The combustion of what kinds of compounds results in the formation of acid gases?

2. What substance can be used to remove either HCl or NOx compounds from a gaseous waste stream?

5-6 Hazardous Waste Incinerator Regulations

As was noted earlier in the chapter, CFR, Title 40, Part 264, Subpart O specifies the regulations for hazardous waste incineration. Each of the key requirements under Subpart O is discussed below:

Permits

All hazardous waste incinerators must obtain operating permits from EPA or a state agency authorized by EPA. To obtain a permit, the facility must submit a detailed, multi-volume permit application, providing detailed engineering and other data that describe the incinerator design and how it will operate.

After an extensive review of the application by the regulatory agency and the public, the incinerator must conduct a trial burn (discussed below), which demonstrates the incinerator's ability to operate within regulatory limits. A draft permit, detailing operating and other requirements that will govern the incinerator's performance, is then issued for public review and a hearing.

Finally, after all regulatory requirements have been met to assure protection of public health and the environment, a final operating permit is issued. The incinerator must operate at all times in accordance with its permit. This is assured, in part, by extensive and sophisticated control equipment, by continuously recorded monitoring, and by automatic waste feed cut-offs (discussed below). Further assurance is provided by frequent inspections by the regulatory agencies.

Further clarification of the public's participatory role in the rule-making process is now incorporated into the *RCRA Expanded Public Participation Rule*. This new rule applies to hazardous waste facilities that are seeking an initial permit or renewal of a permit under Subtitle C of RCRA. The intent of the rule is to make it easier for citizens to become involved earlier and more often in the process of permitting hazardous waste facilities. It also expands public access to information about facilities. It is believed that the rule will eliminate many of the confusions and delays in future permitting processes by involving the public in a more timely manner.

Trial Burns

A **trial burn** is a test of an incinerator's ability to meet all performance standards when burning hazardous waste under the worst possible set of conditions. The conditions demonstrated in the trial burn then become the conditions specified in the permit and those that must be used for all future operation.

During a trial burn, which is conducted under direct, onsite regulatory agency supervision, measurements are taken of 1) the waste feed characteristics and volumes; 2) the combustion temperatures; 3) the combustion gas velocity (which is a key element in determining combustion efficiency); 4) the levels of carbon monoxide, hydrogen chloride, heavy metals, and particulates in the stack emissions; and 5) other important parameters, particularly the emissions of **principal organic hazardous constituents (POHCs)**.

POHCs are difficult-to-burn compounds such as highly halogenated compounds like freon-113 or pentachlorophenol, which are easily detected. Although the incinerator operator recommends selection of the POHCs to be used in the trial burn, the regulatory agency reviewer has the final authority for the selection. Successful combustion of the selected POHCs assures that the incinerator, in its daily operations, will properly destroy the hazardous constituents in the waste.

Furthermore the trial burn is conducted under worst-case conditions. First, POHCs are selected from constituents judged to be present in the waste and that are the most difficult to burn. Secondly, waste feeds are selected to contain maximum expected levels of ash and chlorine. These two characteristics directly relate to the capability of the air pollution control system (APCS) to achieve the particulate and hydrogen chloride (HCl) emission limits. Thirdly, the incinerator is operated, during the trial burn, at the projected worst case limit of the expected range for each critical operating condition.

Destruction and Removal Efficiency (DRE)

A hazardous waste incinerator, in order to be permitted as such, must also demonstrate during the trial burn a DRE of at least 99.99 percent (four nines). In fact, most incinerators demonstrate a DRE of 99.999 percent (five nines), ten times greater than required.

A DRE of 99.99 percent means that 99.99 percent of the POHCs – the hazardous constituents in the waste – have been destroyed by the incineration process or removed into the ash or the air pollution residues, for further treatment and disposal. In fact, EPA reports that destruction and removal efficiencies reported for incinerators are almost entirely the result of destruction in the incinerator rather than removal into the ash or other residues.

Required Operating Parameters

Once an incinerator has achieved the required DRE, based on the operating conditions during the trial burn, it receives a draft permit that specifies 1) wastes acceptable for incineration, 2) waste feed rates, and 3) a minimum combustion temperature, combustion gas velocity, and carbon monoxide levels in the stack gas – all of which are determinants or indicate the quality of the combustion process. In addition to these required operating conditions, the regulatory agency may add other conditions, depending on the cirumstances.

Waste Feed Cut-offs (WFCOs)

All hazardous waste incinerators must be designed and operated with a system that will cut off waste feed automatically, if key operating conditions vary from the limits established in the permit. The parameters include at least: 1) carbon monoxide levels in the stack, 2) combustion temperature, 3) combustion gas velocity, 4) critical APCS control con-

ditions, and 5) any others determined by the permitting agency to be necessary to ensure that the required performance standards are met.

Emissions

The incinerator is required, as a minimum, to limit emissions to not more than four pounds per hour of hydrogen chloride (HCl) and not more than 180 milligrams of particulate matter per cubic meter of stack gas. Almost invariably, the permit will include other emission limits, such as for carbon monoxide and various metals, and may also specify other, more stringent, limits.

A risk analysis is conducted, based on the projected emissions of metals, HCl, etc., as demonstrated during the trial burn. The resultant risk must be significantly less than the future risk levels set by the Clean Air Act (CAA) for manufacturing and other emission sources.

In March 1996, EPA proposed a new rule that would revise the technical standards for hazardous waste combustion facilities under the joint authority of the CAA and RCRA. The proposal fulfills EPA's commitment to upgrade the emission standards for hazardous waste-burning facilities, as stated in its 1993 Hazardous Waste Minimization and Combustion Strategy. In addition to advancing EPA's mission to promote waste minimization, it also establishes a common-sense approach that provides for coordinated CAA and RCRA permitting of hazardous waste-burning facilities.

If approved, the proposal will limit emissions of dioxins and furans, mercury, semi-volatile metals (cadmium and lead), low volatile metals (arsenic, beryllium, chromium and antimony), particulate matter, acid gas emissions (hydrogen chloride gas and chlorine), hydrocarbons, and carbon monoxide. The standards will be based on **Maximum Achievable Control Technologies (MACTs)**, an approach required by the CAA. The rule will also require continuous emissions monitors for particulate matter and mercury. Currently they are required for only carbon monoxide, total hydrocarbons, and oxygen.

The proposal will also change the current small quantity burner exemption. In the past, facilities that burned up to 2,000 gallons of hazardous waste per month were exempt. Under the proposal, that number will be dropped to less than 27 gallons of hazardous waste per month. In particular, the proposal

Figure 5-13: Monitoring the control panel of an oven incinerator.

ensures that facilities will be able to avoid two potentially different regulatory compliance schemes by integrating the monitoring, compliance testing, and recordkeeping requirements of CAA and RCRA. It also significantly promotes regional, state, and local agency flexibility by enabling the various agencies to coordinate their resources for permitting, compliance, and enforcement efforts.

Monitoring

There must be continuous monitoring and recording of 1) combustion temperature, 2) waste feed rate, 3) combustion gas velocity, and 4) carbon monoxide after the combustion and before the release to the atmosphere. These are the key parameters that determine whether the incinerator is operating properly in accordance with the permit. Normally these parameters are recorded on strip charts or computer data logs, which keep a continuous and permanent record of the incinerator's operation. These records are periodically reviewed by the environmental regulatory inspectors, to assure ongoing compliance with the incinerator's permit.

On November 18, 1994, EPA released its Strategy for Hazardous Waste Minimization and Combustion. The strategy was a major milestone in the agency's effort to determine how best to integrate source reduction and recycling into the national hazardous waste program and to ensure safe operation of hazardous waste combustion facilities. In summary, the strategy's goals are to:

—Ensure that combustion technology is safely regulated;

—Continue to aggressively enforce compliance of waste boilers, industrial furnaces, and incinerators to safeguard public health and safety;

—Grant highest permitting priority to those existing facilities that will have the greatest potential for realizing environmental benefits or reducing risk to the public;

—Provide opportunities for public involvement in the combustion facility permitting and enforcement processes; and

—Ensure that permits protect against unacceptable risk using up-to-date, comprehensive methodologies.

After spending millions of dollars on technology upgrades, current incinerators are designed to meet an eventual particulate matter limit of 0.015 g/m^3, which is more stringent than the current limit of 0.030 g/m^3. The anticipated MACT PM limit of 0.030 g/m^3, therefore, will not require any new capital investment for most commercial incinerators.

For those incinerators designed and used for the destruction of PCBs, the standards are found under the Toxic Substances Control Act (TSCA) and are considerably more stringent than those of RCRA. In general, a combination of time, temperature, and turbulence must result in a DRE ≥ 99.9999 percent (six nines) or no more than 0.001 g of PCB per kilogram of the PCB waste introduced. If, for any reason, the incinerator should fall below these conditions, the flow of PCBs into the incinerator must be automatically stopped. The rate and quantity of PCBs fed to the incinerator must be measured and recorded at regular intervals of no more than 15 minutes. A scrubber must be used to control the HCl produced during incineration, and the stack gases must be monitored for all other exhaust emissions.

Checking Your Understanding

1. In what portion of the CFRs are the regulations for governing incineration found?

2. What is the meaning of the acronym DRE?

3. What are three things that a permitted incinerator must be monitored for continually?

4. What is the meaning of the term "six nines"?

Summary

Depending on the amount of oxygen present, thermal treatment processes can result in a wide variety of products. When stoichiometric amounts of oxygen are present, water, carbon dioxide, and lesser amounts of other substances are produced. When an inadequate amount of oxygen is present, there is less complete combustion and a greater variety of residue products, collectively known as products of incomplete combustion or PICs are formed.

Complete incineration results in the destruction of the chemical characteristics of a hazardous waste. It also frees the generator from the liability risks associated with other methods of disposal. Volume reductions often exceeding 90 percent leave only a small amount of easily managed residue. If the waste has heat value, additional benefits can be realized from its recovery.

Of all thermal destruction technologies, incineration is the most often used for the treatment of hazardous wastes. Incineration has become a carefully engineered process that makes maximum use of the operating conditions of time, temperature, and turbulence to maximize hazardous waste destruction and minimize hazardous emissions. Secondary combustion chambers and air pollution control equipment are typically used to meet local, state, and federal air regulations.

Pyrolysis is the process of heating in an oxygen-deficient atmosphere. It results in the separation of the material into its nonvolatile and volatile components. Both of these can be recovered, if they have any economic value; or they can be burned to recover their energy value.

Wet air and supercritical water oxidation are processes used for destroying organic materials that are contained in sludges or the liquid phase. The processes rely on oxygen that is applied under conditions of high or critical temperature and pressure. The reaction is exothermic, which assists in maintaining the necessary temperatures. Complete oxidation of most inorganic and organic compounds is possible, although halogenated organics are usually too stable for complete oxidation. Typically, 80 percent of the organic substances will be completely oxidized to carbon dioxide and water.

Hazardous waste incinerators are regulated and permitted by EPA regulations set forth in Title 40 of the *Code of Federal Regulations* (CFR), Part 264. The specific incineration requirements are contained in Subpart O of Part 264. Most states are authorized by EPA to administer these regulations, under EPA oversight. In order to be authorized, state regulations must be at least as stringent or more stringent than the federal regulations. State regulations are then enforced by both EPA and the state itself. Incinerators that are used to destroy PCBs fall under TSCA and are held to a much more stringent set of standards. For example, these incinerators must have a DRE $\geq$ 99.9999 percent (six nines).

Even with the considerable efforts being directed toward the minimization of waste, our society is projected to continue to produce large volumes of hazardous waste well into the next century. In spite of some public disagreement on the subject, a wide range of governmental, scientific, and environmental experts have concluded that modern, regulated, high temperature hazardous waste incinerators are clearly the best available technology for the destruction of organic wastes.

Critical Thinking Questions

1. What differentiates stoichiometric and non-stoichiometric amounts of oxygen used in incineration?

2. Discuss the Three Ts and explain the role of each with respect to causing complete combustion to occur.

3. Discuss the differences between combustion and pyrolysis.

4. What are the advantages of wet air oxidation processes when compared to combustion?

5. Compare and explain the DRE standards for the combustion of hazardous wastes vs. those that contain PCBs.

6

Biological Treatment Technologies

by Howard Guyer

Chapter Objectives

Upon completing this chapter, the student will be able to:

1. **Describe** the nature and classification of microorganisms.

2. **Identify** the microorganisms that are naturally present in the soil.

3. **Differentiate** the products of aerobic, anaerobic, and facultative metabolism.

4. **Describe** the role of microorganisms in suspended growth and fixed-film treatment methods.

5. **Evaluate** the use of hyperaccumulating plants in phytoremediation technologies.

Chapter Sections

6-1 Introduction

Nature's transformation of plant and animal matter into useful nutrients for the growth of the next generation is undoubtedly the oldest application of biological treatment principles. Were it not for the relentless action of the microbial world, today we would certainly be buried beneath mountains of plant and animal debris. Unknowingly, our ancestors used **microorganisms** for thousands of years during fermentation to make wine and beer as well as to make such foods as sauerkraut, cheese, and yogurt. Yet the **microbes** responsible for these processes remained undetected until the early seventeenth century when microscopic studies revealed their teeming presence.

By the latter half of the eighteenth century the science of microbiology emerged, labeling microbes as the germs responsible for the spread of disease. Understandably, the public's perception of them quickly became a negative one. For many years this association overshadowed the facts we know today – that microbes are energetic and tireless workers, and the vast majority is harmless to humans. Today, the use of microbes enjoys widespread approval, which appears to be based on the perception that

their use is natural, and – in the public's mind – if it is natural it must be good.

In this chapter we will see that some biological treatment methods could be placed in other treatment categories. As noted in the last chapter, several chemical processes are actually performed by plants. Some plants are capable of chemically changing very toxic Cr^{6+} to less toxic Cr^{3+}, thereby greatly reducing its threat. In other applications, biological treatment processes are used to concentrate wastes, a role more traditionally served by physical treatment processes. This is exemplified by the use of plants to economically concentrate low levels of toxic metals found over large areas.

While microbial activity has long been used by nature and man for the transformation and reduction of organic matter, its deliberate application to the removal of specific organic wastes is relatively recent. The choice of biological treatment methods is typically the result of a balance between its economic benefits and time constraints: biological treatment methods are generally quite economical, but may require days, months, or even years to accomplish their goals.

6-2 Classification of Microorganisms

Based on interpretation of fossil records, it is believed that **bacteria** – the most primitive organisms known – have been present on the earth for more than three billion years. Since their appearance, countless numbers of increasingly complex organisms have evolved and adapted to varying environments. At the core of this diversity is the evolution of two very different kinds of cellular structures: those with nuclei and those without. Depending on the criteria used, several systems for classifying organisms have also developed. One system, based on cell type, places them in five broad kingdoms: **Monera** (sometimes listed separately as Eubacteria and Archaebacteria), **Protista**, Plantae, **Fungi**, and Animalia.

All bacteria are members of the kingdom Monera and have **prokaryotic** cells; this means that they do not have a nucleus, a cellular structure that is present in all other types of organisms. Bacteria are the most varied and abundant organisms in the world. They live in every possible environment from Antarctic glaciers to hydrothermal vents on the bottom of the oceans. All are single-celled, but as a group, they exhibit more nutritional diversity than all other organisms combined. Many are decomposers obtaining their nutrients from consuming the dead bodies and bodily wastes of other organisms. This is an important function since it returns elements and compounds to the environment transforming them into nutrients for other living things.

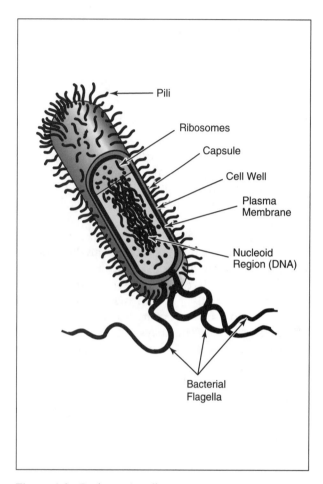

Figure 6-2: Prokaryotic cell.

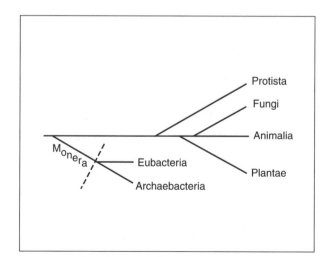

Figure 6-1: Classification of bacteria kingdoms.

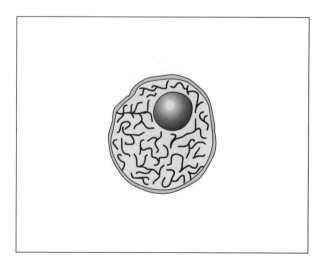

Figure 6-3: Eukaryotic cells.

The cells of the members of the other four kingdoms are **eukaryotic**, which means they have a true nucleus. Eukaryotic microorganisms that are primarily one-celled are placed in the kingdom Protista. Those that are multicellular are put in the Plantae, Animalia, or Fungi kingdoms, depending on their primary mode of nutrition.

Nutritional Characteristics

For bacteria and other microorganisms to grow and convert organic wastes into simpler substances, a balance of certain nutrients must be available. Table 6-1 lists the major, minor, and trace nutrients required.

The growth of microorganisms also involves a balancing of two opposing processes: **anabolism** and **catabolism**. Anabolism, or biosynthesis, is the process by which new cellular materials such as proteins, nucleic acids, etc. are made from simpler materials. For this to occur, however, an energy source is required. This may come from a chemical process, such as an oxidation-reduction reaction, or from sunlight captured by photosynthesis. Catabolism is the process of breaking down more complex chemical substances into simpler ones, coupled with the release of energy in a form that is useful for biosynthesis.

Another way of categorizing organisms is based on how they obtain food. For example:

—**Autotrophs** are self-reliant organisms – e.g., cyanobacteria, algae, and plants – that make their own organic compounds from an inorganic carbon source such as CO_2, by photosynthesis or from redox reactions involving inorganic chemicals. They catabolize organic molecules to obtain energy for growth, movement, and other energy-demanding processes.

—**Heterotrophs**, on the other hand, are organisms that must rely on carbon from organic compounds as their food source. Most heterotrophs also get their energy from catabolizing organic compounds, but some – e.g., purple nonsulfur and green nonsulfur bacteria – can also obtain it by photosynthesis.

Whereas the autotrophic bacteria are more closely related to plants in their general pattern of metabolism, the heterotrophs are more animal-like. Although it is convenient to divide bacteria into different groups on the basis of their nutritional diversity, there is no clear-cut boundary of separation between these groups, but rather a gradual transition from one to another. In theory, autotrophic organisms are the more primitive ancestor from which the heterotropes have evolved.

Depending on the carbon and energy sources used for the anabolic and catabolic processes, microorganisms can also be broadly described as follows:

—Lithotrophs are organisms whose biosynthesis relies on only inorganic carbon sources like carbon dioxide, CO_2, carbonate ions, CO_3^{2-}, and bicarbonate ions, HCO_3^-.

—Organotrophs are organisms whose biosynthesis requires an external source of organic carbon, such as other dead organisms.

—Phototrophs are organisms that use sunlight to synthesize organic molecules, which they later metabolize to obtain energy.

—Chemotrophs are organisms that use redox reactions to catabolize chemicals from an external source to obtain energy.

As shown in Table 6-2, combinations of the various nutritional characteristics lead to four types of organisms. By far the largest single group of bacteria is made up of chemoheterotrophs. They are nutritionally similar to animals in that they obtain both their energy and carbon from organic molecules. They are so diverse that almost any organic molecule can be broken down by one of their species.

Major Nutrients	Minor Nutrients	Trace Nutrients
Carbon, C	Phosphorous, P	Calcium, Ca
Hydrogen, H	Potassium, K	Manganese, Mn
Oxygen, O	Sulfur, S	Iron, Fe
Nitrogen, N	Magnesium, Mg	Copper, Cu
		Zinc, Zn
		Vitamins

Table 6-1: Required microorganism nutrients.

Nutritional Type	Energy Source	Carbon Source	Examples
Photoautotroph	Sunlight	Carbon dioxide, CO_2	Captures energy from sunlight and does not require organic carbon. Includes algae and higher plants.
Photoheterotroph	Sunlight	Organic compounds	Captures energy from sunlight, but requires organic carbon. Includes purple and green nonsulfur bacteria, none of which are important for waste treatment.
Chemautotroph	Inorganic compounds	Carbon dioxide, CO_2	Requires external chemical source, but does not require organic carbon source. Includes the nitrifying bacteria.
Chemoheterotroph	Organic compounds	Organic compounds	Requires external organic carbon for both biosynthesis and energy. Includes most bacteria, all fungi, and higher organisms. Most important category for waste treatment.

Table 6-2: Organisms by nutritional characteristics.

Chemoorganotrophs and chemolithotrophs both obtain their energy by redox reactions on external chemicals. If the oxidizing agent (electron acceptor) is a distinct external chemical, such as oxygen, the process is called **aerobic respiration**, but if there is no distinct external oxidizing agent then the process is called **anaerobic respiration** or **fermentation**.

Microorganism Population Growth Curve

The rate at which a population of microorganisms grows is directly related to the amount of nutrients available. This concept is important in biological treatment methods since it determines the length

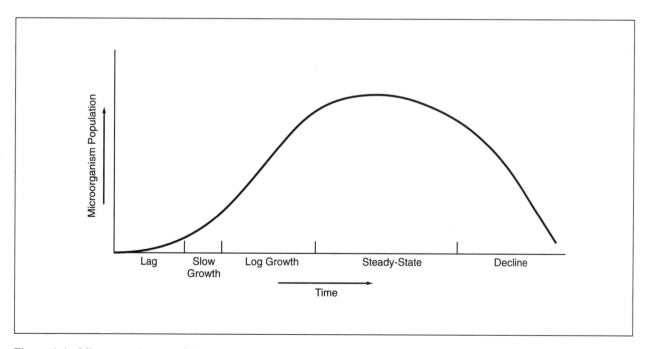

Figure 6-4: Microorganism population growth curve.

of treatment time, or – in a continuous system – the flow rate at which the unit can treat.

At the beginning of a treatment process the number of microorganisms is typically low and their population grows slowly. This phase is known as the lag phase. During this time the microbial population is adapting to its environs and preparing to grow. As is shown in Figure 6-4, this is followed by the log growth phase in which the microorganism population grows exponentially. This is followed by a period of time during which the population is stable, because the number of microorganisms dying is equal to the number being produced. This equilibrium period is called the steady-state phase. If the population of microorganisms is relatively low compared to the amount of food or **substrate** present, this phase may be extended considerably. As the supply of nutrients declines, or if toxic by-products accumulate, the population growth rate rapidly decreases leading to the decline phase.

The shape of the life-cycle curve is also affected by pH and temperature fluctuations or by the availability of nutrients like nitrogen and phosphorus.

For example, there is a direct correlation between increasing temperatures and microbial activity until some maximum is reached that is dependent on the thermal sensitivity of the microbe. Then there is a drastic drop in the population size. In short, the effectiveness and speed of a biological treatment process will be greatly affected by ambient temperatures.

Checking Your Understanding

1. What is the basic difference between a prokaryotic and a eukaryotic cell?

2. How do autotrophs and heterotrophs differ?

3. What is the energy source for a phototroph's growth, movement, etc.?

4. List two ways in which microorganisms obtain energy.

5. What is the name of the phase during which the microorganism population soars exponentially?

6-3 Soil Microorganisms

The soil is an environment that most of us have never examined closely. To the casual observer it appears to be a dense solid, but a closer analysis reveals that only about one-half of its volume is actually solid material. The remaining one-half is actually composed of microscopic spaces that are occupied by air, water, and a dizzying array of microorganisms. Depending on the moisture content, the spaces may contain nearly the same percentage of oxygen as the air we breathe. But due to the metabolic activity of the microorganisms, the soil often contains a higher percentage of carbon dioxide than the air.

The smallest and most abundant microorganisms in the soil are the bacteria. They range in length from 1 to 3 μm (micrometers) and average 0.5 μm in diameter. Under high magnification most soil bacteria appear to be either spherical or rod shaped. Because of their small size, it is not surprising to find estimates of 1×10^5 to as high as 1×10^8 bacteria per gram of dry soil.

Although they appear quite simple, as a group, bacteria have the ability to reproduce at a prolific rate and perform complex chemical reactions. Furthermore, while animals must depend on their food sources for essential nutrients, most bacteria can manufacture everything they need from a combination of organic molecules, trace metals, phosphates, and nitrogen in various chemical forms.

Soil Nitrogen Compounds

All organisms require nitrogen to make proteins and genetically important nucleic acids, such as DNA. The atmosphere is the earth's vast storehouse of elemental nitrogen, N_2, but because of its relative inertness it can only be converted into other chemically useful forms by specialized **nitrogen-fixing bacteria**. These bacteria can generally be found in the soil, in water, or in nodules on the roots of legumes like peas, peanuts, alfalfa, clover, and beans.

In a process known as **nitrogen fixation**, these bacteria transform atmospheric nitrogen into ammonia, NH_3. Depending on the pH of the surrounding watery environment, the ammonia may quickly dissolve to form solutions containing ammonium ions, NH_4^+. Other specialized bacteria convert ammonium ions into nitrate, NO_3^-, and nitrite, NO_2^-, ions in a process known as **nitrification**. Nitrate and ammonium ions are the main source of nitrogen used by plants to synthesize amino acids and other nitrogen-containing compounds.

Other microorganisms in the soil, known as **decomposers**, can also make nitrogen available to plants and other organisms. The decomposition process releases nitrogen from metabolic wastes and proteins of dead organisms. Once decomposition is started the process is assisted by the **nitrifying bacteria** that chemically reduce the nitrogen-containing waste products into ammonia. The ammonia formed is, in turn, converted into ammonium ions, nitrite ions, and nitrate ions in subsequent steps in the cycle.

At the same time these processes are occurring, **denitrifying bacteria** operating in an anaerobic environment are converting a part of the nitrite ions back into nitrogen gas, which is released to the at-

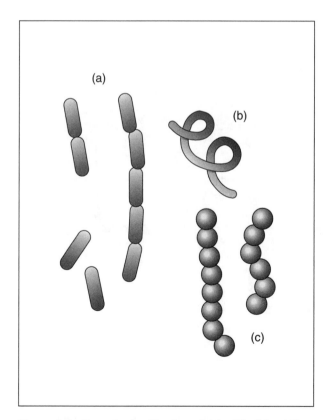

Figure 6-5: Bacteria classifications: (a) bacilli, (b) spirilla, and (c) cocci.

mosphere. This process is called **denitrification**. As will be discussed later in this chapter, decomposer bacteria are the backbone of the modern wastewater treatment plant.

As shown in Figure 6-6, the processes described above are only a part of the much more complex nitrogen cycle. After carefully comparing the roles of nitrogen fixing and decomposing bacteria, it becomes apparent that all other heterotrophic forms of life – including man – are dependent on these bacteria to transform wastes as well as the more stable forms of inorganic nitrogen into ammonia and organic nitrogen compounds. The ammonium and nitrate ions present in the soil are the principal source of nitrogen for all higher plants and animals. Clearly, without the activities of both the nitrogen fixing and the nitrifying bacteria, life as we know it would not exist.

Soil Fungi

Fungi of various kinds are also vitally important members of the soil community. Some fungi, such as yeast, are unicellular, but most others – such as bread molds, mushrooms, and puffballs – are multicellular. They lack chlorophyll and cannot manufacture their own food. Their main role is as a decomposer of the larger, more resistant organic substances that are common in plant material. Substances like cellulose, lignin, starch, and gum, which are resistant to the action of soil bacteria, are readily decomposed by the fungi. As suggested in the last section, they also break down protein, which releases essential nitrogen compounds to the soil to nourish future plant life. It is true that bacteria are more numerous than the fungi, but in many ways fungi are equally important in maintaining the soil's ecosystem.

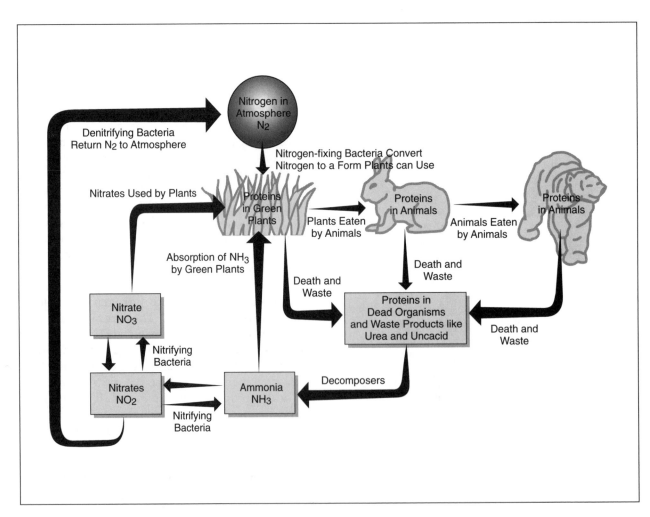

Figure 6-6: The nitrogen cycle.

Figure 6-7: Soil fungi.
Basidiomycetes – Mushroom
Ascomycetes – Chaetomium

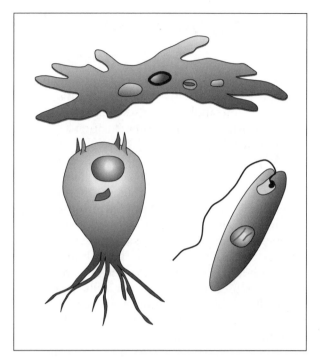

Figure 6-8: Protista: amoebae, eglypha, and euglena.

Soil Protozoa

Finally, there are the single-celled organisms in the Protista kingdom, generally referred to as protozoa. They are often highly specialized and function similarly to small multicellular animals. They are generally much larger than bacteria, which they use as food source. Most protozoa are very mobile and swim rapidly in water by using a whip-like motion of their flagella or cilia (hair-like structures).

Of the 17,000 species known, more than 250 of them have been identified in the soil. Most species are aquatic and live in bodies of water. The species living in the soil also depend on water; however, they are limited to the thin film of water surrounding soil particles and the tiny channels of water trapped in the spaces between the soil particles. Some have diets that consist mostly of one species of bacteria. The most important function the protozoa fulfill within the soil ecology, therefore, is to control the various bacterial populations.

Checking Your Understanding

1. How do the percentages of oxygen in the soil and in the air compare?

2. Which is the smallest microorganism found in the soil?

3. In what kingdom have bacteria been placed?

4. In the nitrogen cycle, what are the various roles played by bacteria?

5. What types of materials are decomposed by the fungi?

6. In what media do protozoa usually live?

7. What is the most important soil ecology function provided by protozoa?

6-4 Aerobes, Anaerobes, and Facultative Bacteria

Microbial metabolism is an extremely complex subject and an area of active research. In this section only those metabolic processes that have application to wastewater and bioremediation treatment methods will be considered.

As previously discussed, when chemoorganotrophs and chemolithotrophs obtain energy by using oxygen as the oxidizing agent, the process is called aerobic respiration. Most bacteria use oxygen and are, therefore, known as **aerobes**. Like animals, aerobes consume food that is composed mostly of organic compounds. Some of the carbon from the compounds is converted into carbon dioxide and the rest is used to produce new cellular material. Generally, aerobic respiration is regarded to be the most complete and efficient way to degrade organic substances. Bacteria for which oxygen is an absolute necessity are known as strict or **obligate aerobes**.

Aerobic respiration works much the same for bacteria as it does for other organisms. In the presence of oxygen and the bacteria's enzymes, complex organic substances are broken down into water and carbon dioxide, a process accompanied by a release of energy. The reaction can be summarized by the following generalized equation:

Aerobic Respiration:

$$C_6H_{12}O_6 + 6\ O_2 \longrightarrow 6\ CO_2 + 6\ H_2O + energy$$

Anaerobic bacteria obtain their energy in the absence of oxygen by performing both oxidation and reduction on the food. Strictly speaking, **obligate anaerobes** are killed by exposure to oxygen and can survive only in the absence of oxygen. Anaerobic respiration is also called fermentation. In fermentation, complex organic compounds such as sugars, organic acids, and dead cellular material act as both the electron donor and acceptor producing by-products such as carbon dioxide, methane, ammonia, alcohols, and simple organic acids.

Anaerobic respiration is a sequential, biologically destructive process in which more complex organic compounds are converted to ones that are less complex. In most cases in which anaerobic respiration (fermentation) occurs, a mixture of several kinds of bacteria is present. The two kinds of bacteria commonly involved are acetogens and methanogens. The acetogens attack the complex organic substances and convert them into simpler organic compounds. These compounds typically include alcohols and small fatty acid molecules, such as acetic and butyric acids. The methanogens are then capable of consuming the alcohols and smaller acids as food and – in turn – produce carbon dioxide and methane. This combination of bacteria is pH sensitive and can only slowly proliferate in a pH 6.5-8.0 range.

The following is a typical anaerobic reaction sequence, performed by acetogenous and methanogenous bacteria:

Fermentation:

(Acetogens)

$$C_6H_{12}O_6 \xrightarrow{enzymes} \boxed{2\ CH_3CH_2OH} + 2\ CO_2 + energy$$

(Methanogens)

$$\boxed{2\ CH_3CH_2OH} + CO_2 \xrightarrow{enzymes} \boxed{2\ CH_3 \overset{\overset{\displaystyle O}{\|}}{C} - OH} + CH_4 + energy$$

(Methanogens)

$$\boxed{2\ CH_3 \overset{\overset{\displaystyle O}{\|}}{C} - OH} \xrightarrow{enzymes} 2\ CH_4 + 2\ CO_2 + energy$$

Summary: $C_6H_{12}O_6 \xrightarrow{enzymes} 3\ CH_4 + 3\ CO_2 + energy$

There is less total energy released in this anaerobic respiration sequential than if it had occurred aerobically, since the product methane, CH_4, still contains recoverable energy.

A third group of bacteria, called **facultative** (technically, facultative aerobic), combine the best of both worlds. They can survive both in the presence and absence of oxygen, by possessing enzymes that can use either molecular oxygen or some other oxygen-containing material as the oxidizing agent. Other substances that may be used as the electron acceptor include sulfates, SO_4^{2-}, nitrates, NO_3^-, and carbonates, CO_3^{2-}. The following is an example of facultative oxidation reaction, where sulfur – in the form of sulfate ion – is acting as the oxidizing agent:

Facultative Reaction:

$$2 \ CH_3CH_2 \overset{\overset{\displaystyle O}{\|}}{C} - OH + SO_4^{2-} \xrightarrow{\text{enzymes}} 2 \ CH_3 \overset{\overset{\displaystyle O}{\|}}{C} - OH$$
$$+ CO_2 + S^{2-}$$

In conclusion, all living organisms must have adequate supplies of both energy and elements, especially carbon, in order to survive. Most bacteria obtain both of these from one source – organic compounds found in dead plant and animal matter. The source of energy may be from sunlight, from a reducing agent such as ferrous ion, Fe^{2+}, or from an organic compound. If the energy source is other than an organic compound, then the organism must have a supply of carbon dioxide to be able to perform biosynthesis and make cellular materials like proteins, carbohydrates, fats, and nucleic acids. Due to enzyme/coenzyme activity many organisms also require traces of metals and other nutrients such as phosphate and nitrate ions.

Although the chemistry of aerobic, anaerobic, and facultative bacteria has been discussed separately, it is critical to understand that pure populations of these bacteria rarely exist in nature. Such a situation would not be beneficial, since many microorganisms have very complex and mutually dependent relationships with one another. Ruminants such as cows contain mixtures of microorganisms that are involved in a complex set of interactions. Most plant matter is composed of complex carbohydrates such as cellulose – materials that are typically resistant to degradation by normal digestive enzymes. Cows overcome this problem by use of a second, blind stomach, called the rumen. Within the rumen there is a bacterium that can break down the cellulose fibers and another type of bacteria that can degrade the breakdown products into simple sugars and organic acids. The sugars can now be acted upon by the digestive enzymes, and the simple organic acids are either absorbed and used for energy or further broken down by anaerobic bacteria yielding – among other things – methane gas. The release of large volumes of methane from feedlots with thousands of cows has prompted EPA to examine the issue, given the fact that methane is one of the gases that contributes to the greenhouse effect.

It is important to bear in mind that – individually or collaboratively – aerobic, anaerobic, and facultative bacteria resemble miniature chemical factories that are capable of breaking down a wide variety of organic materials. In the following sections we will discuss several treatment technologies that make use of this ability.

Checking Your Understanding

1. Aerobic respiration uses what substance for the oxidizing agent?

2. What type of bacteria can use either oxygen or another substance as an electron acceptor?

3. Which form of bacteria is the most efficient in producing energy?

4. Which type of bacteria results in cows producing methane gas as a digestive by-product?

6-5 Wastewater Treatment Methods

Although microorganisms were not discovered until modern times, it was known in early Roman times (600 BCE) that pools of sewage go through a type of self-purification process. Today, municipal wastewater usually includes storm water, commercial and industrial discharges as well as domestic sewage. Although domestic sewage is usually the major component, the other sources add substances that often have significant effects on the total treatment process.

Environmental laws and regulations require that wastewater treatment plants remove solids, toxic materials, and soluble inorganics, such as nitrates and phosphates prior to discharge. Phosphates and nitrates pose a threat by stimulating algae and plant growth. These growths provide food for animals, but increase turbidity and rob the rivers and lakes of dissolved oxygen when they die during periods of low water or in the winter.

Primary Treatment Processes

Upon entering a municipal wastewater treatment plant, the effluent is subjected to two brief primary treatment processes. The first involves bar screening to remove large objects that could damage treatment plant equipment; the second removes the large insoluble particles called grit. The wastewater is then ready to enter large settling basins where the smaller suspended solids settle out due to gravity, while the less dense oils and greases are removed by a skimmer. After this, the effluent from this process is ready to enter the next treatment phase where secondary treatment methods are used to remove the remaining organic molecules.

Secondary Treatment Processes

There are many types of secondary biological treatment processes, the most common being variations on either the **suspended growth** or the **fixed-film** processes. In both of these, microorganisms are allowed to come into contact with the suspended and dissolved organic wastes. **Activated sludge**, **surface impoundment (lagoon)**, **sequencing batch reactor (SBR)**, and **anaerobic suspended growth** are the names of some of the more commonly used processes discussed below.

Fixed-film processes rely on the microorganisms to grow on a supporting medium and form a biological slime layer. Nutrients – in the form of organic molecules – in the wastewater diffuse into the slime layer. As the nutrients are consumed by the microorganisms, smaller organic molecules and CO_2 are produced. Excessive growth of microorganisms on the media are sloughed off, maintaining a rela-

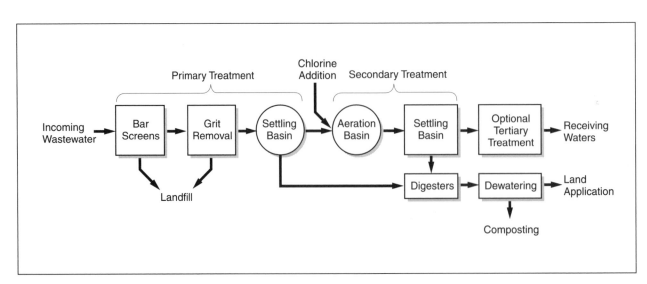

Figure 6-9: Diagram of a wastewater treatment plant.

tively constant level of biological population. **Trickling filters, rotating biological contactors (RBC), anaerobic filters, anoxic denitrification** (in some applications), and bioreactors (discussed later) are the more common applications of these methods. Regardless of the design, all of these treatments rely on the same aerobic, anaerobic, and facultative bacteria discussed in the earlier sections.

Suspended Growth Methods

Activated Sludge Process –In the activated sludge process, nutrient-laden wastewater from the primary treatment stage enters an aeration tank, basin, or bioreactor, where it is mixed thoroughly with microorganisms and oxygen. The oxygen is provided by either surface aeration or diffusers. In the less expensive surface aeration method, air is introduced by lifting a shallow layer of the liquid on the surface of large impellers. By contrast, diffusers can also be

used to introduce compressed air or pure oxygen into the solution. This greatly improves the availability of oxygen, but also increases the treatment costs.

The microorganisms in the aeration tanks tend to grow in brownish clumps. These clumps – known as floc – contain a wide variety of microbes that feed on each other and the nutrients present in the wastewater. The floc consists mainly of aerobic bacteria, but it often includes fungi, animal-like protozoa that eat both bacteria and fungi, a few insect larvae, and worms that feed primarily on the protozoa. This mixture of organisms is known as **zooglea**.

The oxygen is necessary to support this amazing cast of characters. For six or more hours the floc growth – stimulated by an abundant oxygen supply and dissolved nutrients – becomes the center of biological activity; this is why the sludge is defined as activated sludge.

The effluent entering the secondary treatment can be supplied either continuously or in batches.

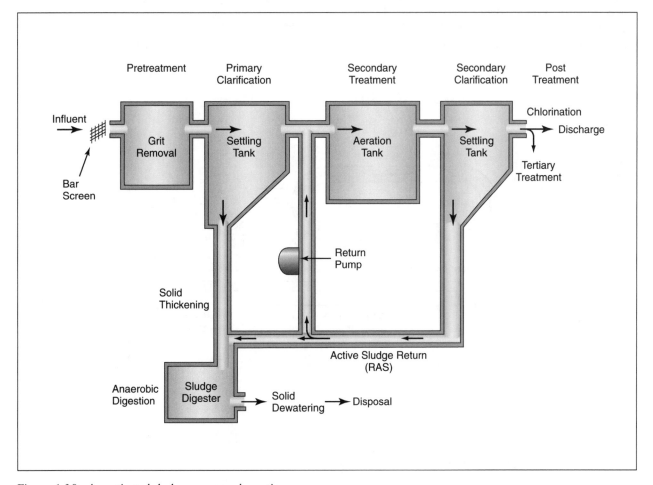

Figure 6-10: An activated sludge process schematic.

As the nutrients are consumed, new cell materials are produced, so the effluent of this process contains both sludge-solids and water. This mixture is separated by settling; the effluent water then continues to either a tertiary treatment process or it is discharged.

The sludge solids are typically separated into two streams: one continuing on to sludge treatment and disposal, and the other recycled to the processing tank to serve as the seed for the next feeding frenzy. Some of the other types of activated sludge systems are described in Table 6-3.

The activated sludge process is a demonstrated technology for the treatment of municipal wastewaters and aqueous hazardous wastes having less than one percent suspended solids. It is frequently the choice for canneries, paper and pulp mills, refineries, breweries, petrochemical plants, and pharmaceutical plants. It involves a continuous recycling of suspended microorganisms through an aeration basin, bioreactor, or tank. The recycling process allows for microorganisms to become acclimated to the wastewater composition. By ensuring complete mixing conditions and high dissolved oxygen levels, waste streams containing large amounts of organic waste can be successfully treated. The separation of sludge and water takes place in a settling basin or clarifier. A portion of the sludge is recycled to the aeration basin or tank to serve as seed for continuation of the process.

Plug Flow	Long, narrow aeration basins, with floc recycled to the head of the aeration basin. Produces good settling sludge, but process is very sensitive to toxic or inhibiting contaminants.
Complete Mixed	Tank conformation, feeding arrangements, and aeration method produce uniform condition throughout the aeration basin. More resistant to shock loading, but sludge settling not as good as in the plug flow system.
Extended Aeration	Can be of the plug flow or complete mixed system, but with minimum sludge wasting. Growth rates and sludge yields are therefore low, but they have a high oxygen demand.
Oxidation Ditch	A loop-reactor system, where a single unit provides both the aeration and clarification functions.
Pure Oxygen	Aeration is provided by introduction of oxygen into a series of well mixed reactors. Onsite generation of oxygen is typically required.

Table 6-3: Types of activated sludge systems.

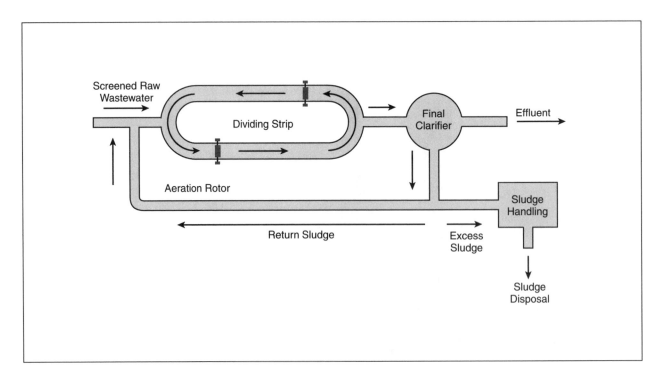

Figure 6-11: Diagram for an oxidation ditch activated sludge system.

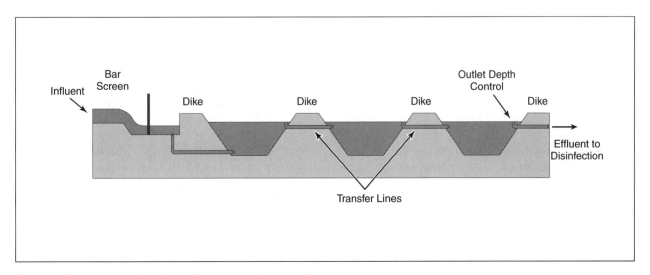

Figure 6-12: Cross section of a surface impoundment or lagoon.

The major advantage of the activated sludge process is that it can – in a reasonable length of time – produce a cleaner effluent than most other biological processes. It is also fairly insensitive to external temperature conditions. In general, it has proved to be an effective method for the treatment of organic matter found in domestic sewage and some industrial wastewaters, but suffers from the disadvantage of requiring considerable amounts of energy.

Surface Impoundments or Lagoons – The activated sludge methods discussed in the last section are useful for producing treated water of good quality in a reasonable length of time and on a limited area. When processing time and land are not limiting factors, the most economical biological wastewater treatment method is a variation on surface impoundments or lagoons.

Lagoons tend to be rectangular containments with a length-to-width ratio of two to one. They are typically 8-12 feet deep, fitted with a liner to prevent seepage, and may be equipped with aerators to supply sufficient air to maintain dissolved oxygen at all depths.

There are several different classifications of lagoons, based on oxygen availability. The types are listed in Table 6-4. In the **facultative lagoon**, wastewater is continually added to the lagoon providing some oxygen; as the wastewater enters the lagoon, some of the larger solids settle to the bottom where they are acted upon by anaerobes. In the intermediate zone, facultative anaerobes capable of decomposing wastes by either aerobic or anaerobic methods, thrive. In addition to the typical microbes seen in wastewater treatment systems, lagoons also have **algae**. The role algae play is to help stabilize the lagoon by utilizing the CO_2, sulfates, nitrates, and phosphates during photosynthesis and to give off oxygen to the upper waters of the lagoon. Thus, a well-balanced lagoon provides an environment in which the processes of microbial respiration and photosynthesis combine to break down organic compounds. They can be operated in a manner similar to activated sludge units, but without sludge recycling; they also do not need settling basins or clarifiers. Periodically, the solids that settle to the bottom must be removed.

Aerobic	Liquid depths of 1-5 feet, and although it is not force aerated, it is kept aerobic through natural or mechanical mixing.
Aerated	Liquid depths of 5-20 feet, and uses mechanical or diffused aeration to keep the mixture aerobic.
Facultative	Liquid depths of 3-9 feet, and uses no forced aeration, which results in the upper layers operating aerobically and the lower levels anaerobically. In the middle level, facultative bacteria thrive.
Anaerobic	Liquid depths of 8-16 feet deep, with some having a low surface-to-volume ratio. They use no type of forced aeration and will accept wastes with a high organic content, yet produce little sludge.

Table 6-4: Lagoon types based on oxygen availability.

The advantages of a lagoon are that 1) they are the cheapest of all methods for treating organic waste and 2) due to their large volume they can tolerate the introduction of potentially toxic materials because these are diluted as they enter. The main disadvantages are: 1) lagoons fall under RCRA and are required to have double liners and leachate collection systems similar to landfills, therefore coming under stricter regulatory scrutiny than other secondary treatment processes; 2) if lagoons are not properly aerated, anaerobic digestion may result in odor production; and 3) lagoons are sensitive to temperature fluctuations and become less effective during the colder months.

Sequencing Batch Reactor (SBR) – The sequencing batch reactor is of relatively recent origin and one of the simplest suspended-growth processes. The technology is a modification of the activated sludge process just discussed and is carried out in a single tank or basin. Each cycle of the reactor involves several phases.

In the first step, called the fill phase, the reactor containing a small amount of reserve or seed sludge is filled with the untreated wastewater. Mixing is started during the fill phase and continues until the desired volume is reached. This step is called the react phase. During the react phase, a portion of the microbial suspension is withdrawn to provide the reserve sludge population for the next batch. At a predetermined time, mixing is stopped and settling occurs. During the third step, called the settling phase, the sludge sinks to the bottom. The decant phase follows, in which the supernatant liquid is removed, treated, and released to the environment. The final idle phase results in accumulation of the reserve sludge, which will be used to seed the next cycle.

There are several benefits to the use of SBR over the activated sludge process. For example, larger variations in the wastewater content are less harmful to SBR treatment performance; solids separation is more efficient due to the lack of turbulence from solids removal; periodic input flow and operation is possible; and SBR requires less space. Finally, the process may be kept totally aerobic or anaerobic or it may be cycled through various degrees of oxygenation.

Anaerobic Suspended Growths – Anaerobic suspended growth processes are similar to activated sludge or lagoon processes, except that a sealed container is used to exclude oxygen. The following specific designs have been developed for this purpose:

—Anaerobic Digesters – They are widely used for sludge stabilization and volume reduction in wastewater treatment plants. (This process will be discussed in more detail later in the chapter.)

—Anaerobic Contact Process – This process is similar to the complete mixed activated sludge process, except that the reactor is not aerated and the methane is removed between the reactor and clarifier.

—Upflow Anaerobic Sludge Blanket – Here wastewater flows upward through a blanket of biologically formed granules, which consume the waste and produce methane. Effluent solids are separated and returned to the reactor.

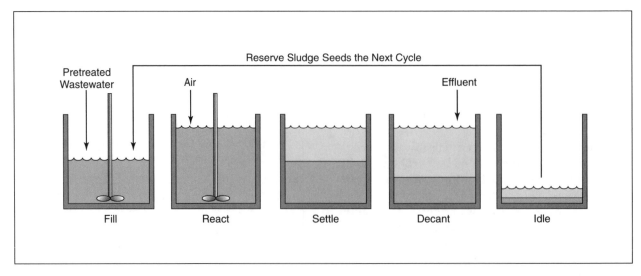

Figure 6-13: Operational steps in a sequencing batch reactor (SBR).

— Fluidized Bed Reactor – Anaerobic organisms are developed on a sand bed, and wastewater is pumped upward through the bed at a rate sufficient to fluidize the bed. A portion of the effluent is recycled through the reactor.

Advantages of anaerobic processes include the production of methane gas that can be used for fuel and for better removal of certain halogenated hydrocarbons. On the other hand, anaerobic processes typically require long startup times, ranging from two to four months. Anaerobic processes also work more efficiently at higher temperatures, therefore requiring supplemental heat, particularly during colder months.

Fixed-film Processes

Trickling Filters – The trickling filter is the most common fixed-film secondary treatment process used in wastewater treatment plants. A trickling filter is in fact not a filter, but rather a process that provides an opportunity for microorganisms to consume organic wastes. It is an attempt to concentrate and duplicate the microbial layer nature has provided on the surface of rocks in flowing streams since the beginning of time, but in a smaller space and with higher concentrations of organic wastes.

In the modern version, 100-200 ft. diameter tanks are filled to depths of 6-8 ft. with rocks or with a synthetic medium. Through the use of rotat-

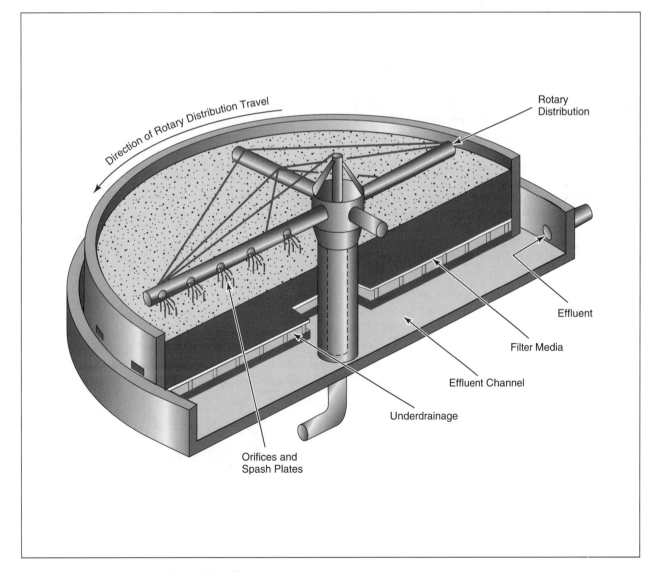

Figure 6-14: Cross-section of a trickling filter.

ing distributor arms, primary treated wastewater is sprayed onto the fixed media, absorbing oxygen as it falls. Fed by this oxygen- and nutrient-rich wastewater, a thick, fixed-film gelatinous zoogleal slime grows. On the outer layer of the slime, where food and oxygen concentrations are high, the microorganisms are aerobic. At depths beyond 0.3 mm, the oxygen and food concentrations are reduced, creating an environment where anaerobic bacteria reside. Overall, the zoogleal slime contains bacteria, protozoa, fungi, and a few higher organisms like worms, insect larvae, rotifers, and algae.

Periodically, chunks of the slime slough off, and a clarifier, or settling basin, is used to capture them as well as the sludge-solids that are separated from the solid-free effluent. The sludge-solids are removed and eliminated in a process called **sludge wasting**, and the clear effluent is either sent back through the process or released to the environment.

Two advantages of trickling filters over the activated sludge processes are: shorter residence times and a greater ability to recover from sudden large changes in the waste stream's nutrient load. On the other hand, the ability to remove organic matter is less efficient, making it necessary to recycle the effluent through the trickling filter several times in order to remove enough organic waste to achieve acceptable **biochemical oxygen demand (BOD)** levels.

As would be expected from living organisms, the zoogleal slime is sensitive to temperature fluctuations. Trickling filters work best at warmer temperatures. The effluent emerging after up to 18 hours in this process is subjected to a final, tertiary treatment process or released directly to the environment.

Rotating Biological Contactor (RBC) – A rotating biological contactor is another type of attached growth or fixed-film system. As shown in Figure 6-16, the contactor consists of a series of porous discs that are coated with a fixed-film microbial growth. The disks are positioned over a tank and rotate so that at any given time 40-50 percent of their surface is immersed in the wastewater, while the remaining 50-60 percent is exposed to the air. As the disks rotate, they are repeatedly brought into contact with the effluent and then again exposed to the air. As the film grows, the excess is sloughed off and removed from the effluent in a clarifier.

The advantages of a rotating biological contactor include having low energy requirements and the ability to handle large surges in waste concentrations. Disadvantages include high installation costs and moderately high operational costs. If temperatures drop so does the microbial activity; in colder seasons the disks must, therefore, be covered for temperature stabilization.

Anaerobic Filters – The anaerobic filter is an additional modification of a fixed-film process. It is a relatively recent development in the field of domestic and industrial wastewater treatment. Like trickling filters, anaerobic filters are filled with solid

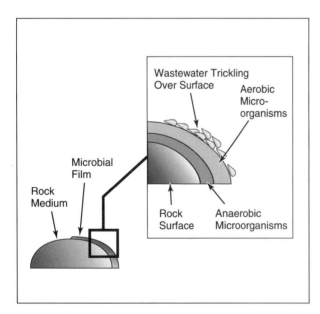

Figure 6-15: Cross-section of microbial film on rock (trickling filter medium).

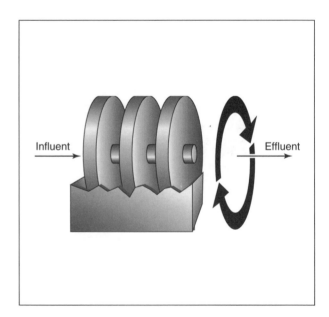

Figure 6-16: The basics of a rotating biological contactor.

media of various types located within a vertical column. The wastewater may flow gently either upward or downward through the column, bringing the organic wastes into contact with the anaerobic bacteria growing on the media. Because the fixed-film retains the organic waste and the remaining waste stream continues, anaerobic filters are best suited for the treatment of waste streams that have dilute organic concentrations at ambient temperatures.

Anoxic Denitrification – Denitrification is the removal of nitrogen, in the form of nitrate, by converting it to nitrogen gas. This can be accomplished by use of microorganisms known as denitrifying bacteria under **anoxic** (without oxygen) conditions. Both suspended-growth and fixed-film denitrification applications exist. Suspended-growth denitrification is usually carried out in a plug-flow type of activated-sludge system following nitrification to convert the ammonia and organic nitrogen into nitrates. The denitrifying bacteria obtain energy for growth from the conversion of nitrate to nitrogen gas, but since most nitrified effluents are low in carbon-containing matter, an external source of carbon may be required for cell synthesis.

Fixed-film denitrification is carried out in a column reactor containing stone or synthetic media upon which the bacteria grow. The action of the denitrifying bacteria is the same as in the suspended-growth method, resulting in the conversion of nitrates to nitrogen gas. Also, similarly to the suspended-growth denitrification process, an external carbon source is usually necessary for cell synthesis to occur.

An increasing number of compounds is being found that can be degraded by denitrifying bacteria operating under anoxic conditions. Such troublesome contaminants as naphthalene and benzene can be completely destroyed. This is especially good news, since it is often difficult to provide adequate supplies of oxygen to perform *in situ* (onsite) remediation of groundwater contaminants. In this application the bacteria use the organic molecules as their carbon source, completely converting them into carbon dioxide and cell mass. The nitrate ion, NO_3^-, is the electron acceptor and is converted into nitrogen gas, N_2, and nitric oxide, NO.

Sludge Treatment

Sludge Processing – As a part of all secondary treatment processes the resulting semi-liquid sludges are collected through either a sedimentation/clarifier or air flotation process. Sludges from activated sludge processes typically contain about 0.5-1 percent solids, while those from the trickling filter contain 2.5-5 percent solids. Once separated, the solids can either be dewatered (dried) and incinerated (see multiple hearth incineration in Chapter 5) or enter a stabilization process. One of the more popular stabilization processes is anaerobic digestion because it facilitates water removal later and reduces the volume of organics by as much as 50 percent.

Anaerobic Digestion – The final treatment of most sewage sludge takes place in large (20-45 ft. deep by 115 ft. in diameter), cylindrical, air-tight tanks called **anaerobic digesters**. In the digester, settled solids and sludges containing 5-7 percent solids are mixed with anaerobic microbes that metabolize the organic matter and use it as food; this results in a 40-60 percent reduction in volume.

Fermentation of complex organic matter by acetogens results in the production of short-chained carboxylic acids and alcohols. Maintaining a healthy environment within the digester is very important and requires constant monitoring. The carboxylic acids tend to reduce the pH within the digester, and lime must be periodically added to neutralize the excess acids. The anaerobic methanogens are capable of utilizing part of the smaller carboxylic acids and alcohols for their food, producing additional carbon dioxide and methane.

There are three advantages of anaerobic digestion: 1) it produces about 20 times less solids than aerobic digestion, which in turn lowers the amount of sludge-solids that must be handled, dried, and transported; 2) it does not require expensive mechanical agitation and aeration, while at the same time achieving a very high organic destruction rate; and 3) the methane gas produced by this process can be collected and used to supplement the power requirements of the treatment plant. After a few weeks in the digester, the stabilized sludge can be removed and either incinerated, landfilled, or used as a soil supplement.

Tertiary Effluent Treatment

As noted in previous sections, the solid-free effluent resulting from secondary treatment methods may still require some final or **tertiary treatment** before it can be released to the environment. Most often these tertiary treatment processes are used by industries that have their own wastewater treatment facilities, rather than by publicly owned treatment works. Industries that use large volumes of process water have their own primary and secondary treatment processes that remove nearly 85 percent of the BOD and suspended solids. By following these processes with tertiary treatment of the effluent to remove the remaining nitrates, phosphates, and heavy metals, they can recycle much of the water for in-plant purposes.

The tertiary treatment processes used to remove these unwanted substances are primarily chemical in nature, but they also rely on some physical treatment processes. Typically, they involve addition of a coagulating agent, followed by precipitation, sedimentation, and filtering. Depending on the substances to be removed, activated carbon, ion exchange, reverse osmosis, electrodialysis, ultrafiltration, molecular sieves, and other processes may be used. Many of these were described in detail in Chapters 3 and 4.

Checking Your Understanding

1. What types of materials are removed in a pretreatment process?

2. Most secondary wastewater treatment methods are of which treatment type?

3. A sequencing batch reactor is a modification of which process?

4. What is the major advantage and disadvantage of using a lagoon process?

5. What are two advantages of trickling filters over the activated sludge process?

6. What does the term anoxic mean?

7. What substance has to be added to an anaerobic digester to maintain its pH?

8. Tertiary treatment processes are primarily of what two treatment types?

6-6 Bioremediation Methods

In the previous section, the focus was on the ability of bacteria to remove the organic components of wastewater, as measured by reduction in biochemical oxygen demand (BOD). Although it was noted that the effluent entering a wastewater treatment facility is primarily domestic sewage, it usually contains additional components as the result of storm water as well as commercial and industrial discharges.

Bioremediation is the term used to describe the use of biological processes to concentrate waste or transform it into a less toxic or nontoxic substance. To accomplish this goal, bioremediation makes use of the metabolic processes of either naturally occurring bacteria – called **intrinsic bacteria** – or genetically altered bacteria designed to metabolize specific waste. Since toxic waste components are often of manmade origin, and therefore considered foreign to the environment, they are referred to as **xenobiotics**. The ability of microorganisms to remove xenobiotic substances is varied and sometimes complex.

In the simplest case, the xenobiotic substances become absorbed in the biomass and get removed from the waste stream when physical separation methods are employed. In other situations the microbes may use the xenobiotic substance in a more traditional sense – as a primary food and energy source. In yet other situations, the microbes may use some other substance as their primary food, while simultaneously degrading the xenobiotic substance. Some familiar organic compounds that occur naturally or that are xenobiotic are presented in Table 6-5 according to their degree of biodegradability. Table 6-6 shows several general observations regarding the biodegradability of the various groups of compounds presented in Table 6-5.

Over the last 100 years, underground storage tanks and their associated pipes have leaked vast quantities of petroleum products into the soil. Degreasing operations using trichloroethylene (TCE), dry cleaning operations using perchloroet-

Degradable	Moderately Degradable	Least Degradable
Benzene	Gasoline	Carbon tetrachloride
Toluene	Styrene	Chloroform
p-Xylene	Methyl iosbutyl ketone	Methane
Acetone	Acrylonitrile	Perchloroethylene (PCE)
Methyl ethyl ketone (MEK)	Ethylene glycol	Trichloroethene (TCE)
Organic acids		PCBs
Alcohols		o - Xylene
Phenols		m - Xylene
Jet fuels		Ethylbenzene
Kerosene		

Table 6-5: Degree of biodegradability of common substances.

1. Highly branched compounds are more resistant to biodegradation.

2. Short side chains are less rapidly degraded than long side chains.

3. Highly oxidized materials, such as halogenated or oxygen-rich materials, may resist further oxidation under aerobic conditions, but may be more rapidly degraded under anaerobic conditions.

4. More highly polar compounds tend to be more biodegradable than less polar compounds.

5. Unsaturated aliphatic compounds (alkenes and alkynes) are more readily biodegraded than alkanes compounds.

6. Alcohols, aldehydes, acids, esters, amides, and amino acids are more susceptible to biological decomposition than their analogous alkanes, olefins, ketones, dicarboxylic acids, nitriles, and chloroalkanes.

7. Increased substitution on straight chained hydrocarbons slows their biodegradation.

8. Increased halogenation slows biodegradation.

9. Methyl, chloro, nitro, and amino groups on aromatic rings inhibit biodegradation of the ring.

Table 6-6: Biodegradability generalizations based on compound characteristics.

hylene (PERC), and the indiscriminate use of pesticides have further polluted the environment with a wide array of xenobiotic non-chlorinated and chlorinated hydrocarbons. These materials have been migrating through the soil for years and are now being detected in the soil and groundwater in alarming amounts and at thousands of different sites across the nation.

Once a water-soluble contaminant reaches the groundwater, it enters the natural flow and continues to spread forming large underground plumes. One of the problems associated with the nationwide move to using cleaner burning gasoline is that it contains the octane booster methyl tertiary butyl ether (MTBE), which is a water-soluble substance. By 1996, a plume of MTBE had contaminated several drinking water wells in Santa Monica, California; this resulted in their use being discontinued.

Most hydrocarbon products are not water-soluble and less dense than water. Highly chlorinated hydrocarbons are also not water-soluble but are denser than water. Contaminants of the first category tend to stay in the water-unsaturated soil above the aquifer, known as the vadose zone, while those belonging to the second category descend through the vadose and the saturated zone forming a layer of dense, non-aqueous phase liquid at the bottom of the aquifer. In the past, the favored method for treating either type of subsurface contamination was **pump-and-treat technology**. This involved the drilling of several wells in the contaminated area, or just ahead of the contaminant plume, followed by the pumping of the groundwater to the surface where the contaminant could be removed by air stripping or treatment with activated carbon. Although the technology works, the consensus among experts is that it is energy-intensive and does not work well enough. Even though reductions in the level of contamination occur, they are usually insufficient to regard the application as a complete success. In addition, pump-and-treat technology still leaves a disposal problem: that of the original contaminant!

As an alternative, *in situ* **bioremediation** shows great promise as an alternative treatment technology. It is cheaper and more effective for a wider range of contaminants, and it also results in the contaminant being completely destroyed. Experience has shown that the critical factor in successful engineered bioremediation is the careful characterization of the site. The information needed includes such elements as the nature of the contaminant, the underlying

geological formation, and a biotreatability assessment. Because each site is unique, laboratory and pilot testing needs to be done to optimize and verify the process and progress at a given site.

If the results of the characterization reveal that the intrinsic bacteria are capable of accomplishing the cleanup, then their growth can be encouraged by providing them with the necessary amounts of nutrients and oxygen. The nutrients may be supplied along with the water that is pumped into the formation. Oxygen can be supplied by either forcing compressed air into the formation or by introducing hydrogen peroxide into the water supply. The progress of the bioremediation can be followed by decreases in the level of contaminant and by changes in the numbers of bacteria and protozoa present in samples withdrawn from monitoring wells.

It is always preferable to use the native bacteria found at a site. By their mere presence, they are demonstrating adaptation to the surrounding conditions and some ability to degrade the particular contaminant. If, on the other hand, the contaminant present is not susceptible to degradation by the intrinsic bacteria, the area may need to be inoculated with an appropriate designer strain of bacteria.

One novel approach that has been used at some sites is to further contaminate the site with yet another substance. The introduced substance must be similar in chemical structure to the already present one, as for example the introduced toluene would be similar to benzene already contaminating the soil. The introduced substance is a known food source to the bacteria, which inadvertently consume the contaminant at the same time they are consuming the introduced substance. Since the same enzymes degrade both substances, the bioremediation goal is accomplished. This process is known as **co-metabolism**.

In the following sections, a variety of engineered bioremediation applications are discussed in more detail. It is important to remember, however, that regardless of the application, bioremediation depends on living organisms, nutrients, water, oxygen – if aerobic bacteria are involved – as well as an organic contaminant that it can use for food.

Landfarming

Landfarming – also called land cultivation, land application, and land treatment – involves mixing ex-

Figure 6-17: Aerating the soil in a landfarming operation.

such as acids, bases, cyanides, and ammonia are not considered candidates for landfarming operations. Wastes contaminated with toxic metals that are likely to leach into the groundwater – known as EP toxicity metals – also require more specialized treatment techniques and cannot be landfarmed.

As with other land disposal facilities, assessments of the land treatment site should include evaluation of regional and site geology, hydrology, topography, soil, climate, and land use to determine if they will allow for effective bioremediation application. Careful management of landfarms, including testing and monitoring, is necessary to ensure that hazardous waste constituents are not introduced into crops that will enter the human food chain. Of particular concern is the control of run-on and runoff waters from the site.

It is crucial to protect the natural waters by designing a system of collection channels and a retention pond. A permit for the pond must also be obtained. In some areas it must meet the 100-year flood design or, more commonly, it must be designed to capture the runoff from the 25-year, 24-hour return period storm. Once captured, the runoff can either be stored or treated and released under the conditions stated in the NPDES permit.

Overall, this treatment method is a simple and effective surface process that requires no special engineering skills or exotic tools. It degrades, transforms, or immobilizes the waste, thereby reducing the long-term liability. Although it is management-intensive, its costs are lower than landfill and incineration options. When completed, the treated soil can be converted to beneficial uses, such as parks, playgrounds, or wildlife habitats.

Biological Treatment Cells

Biological treatment cells, or **biocells**, provide another way to allow hydrocarbon-contaminated soil to come into contact with aerobes. Biocells is an *ex situ* (off site) bioremediation process, in which the excavated soil is placed in a pile or cell, and the necessary oxygen, nutrients, and water are provided. Typically, a cell contains 6-8 feet of contaminated soil placed over a series of slotted PVC pipes that have been covered by a layer of gravel. The pile is then covered with a slotted plastic canopy to control air emissions and dust and to maintain the moisture content of the soil.

cavated contaminated soil to be bioremediated with clean topsoil. The name landfarming comes from the fact that the excavated soil is laid out in thin layers that resemble a farmer's freshly plowed field. By occasionally tilling the soil to add oxygen, fertilizing it with other required nutrients, and keeping it moist, intrinsic bacteria accomplish the remediation process. Landfarming is used extensively by the petroleum industry for the treatment of hydrocarbons on what they call oily wastes.

Xenobiotic substances, such as halogenated hydrocarbons, are more difficult to biodegrade and are therefore a potential problem for a landfarming operation. In addition, radioactive, volatile, reactive, or flammable liquid wastes as well as inorganic wastes

PVC pipes are then placed vertically through the plastic to provide for air entry and diffusion into the cell. A vacuum pump is used to create a negative pressure in the bottom slotted pipes. This draws the air down through the pile, where it becomes available to the aerobic bacteria consuming the hydrocarbons. The air-vapor mixture exiting the pile is passed through activated carbon canisters to remove any volatile organic substances prior to re-entering the atmosphere.

The major difference between landfarming and biological treatment cells lies in how the required oxygen is provided. In landfarming a fresh supply of oxygen is periodically tilled into the soil, while in the biocell the air is drawn through the cell by operating a vacuum pump. All of the previously discussed considerations, including runoff, site geology, hy-drology, topography, soil, climate, and land use also apply to the siting of a biological treatment cell. The types of contaminants that can and cannot be effectively treated are also the same. Typically, both methods are performed onsite, thereby eliminating transportation liability; also, both methods allow for the treated soil to be used for beneficial purposes or returned to its excavation site.

Air Sparging

The term **air sparging** literally means air sprinkling. Like the scientific principles discussed and applied to air stripping in Chapter 3, it relies on the principle of differential vapor pressure to separate liq-

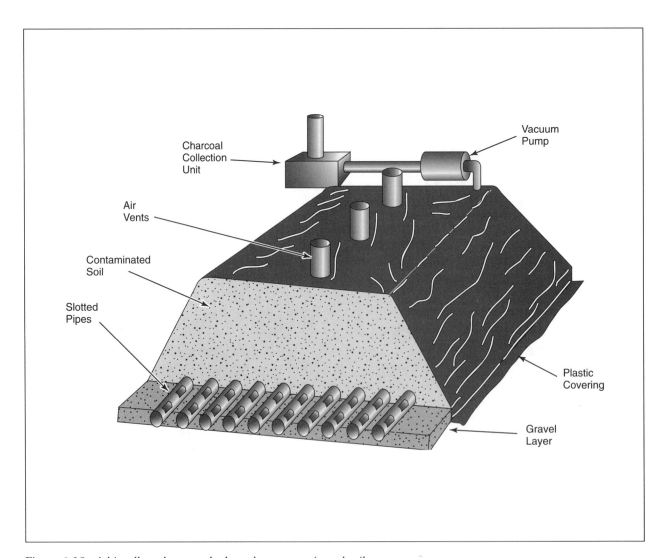

Figure 6-18: A biocell used to treat hydrocarbon-contaminated soil.

uids that have become commingled. In general, when air is bubbled through a mixture, the substance with the higher vapor pressure evaporates more rapidly.

To establish an *in situ* air sparging system, a series of openings must be made into the soil and a vapor collection system installed. By means of a compressor, ambient air is forced into the soil or groundwater, where it travels both vertically and horizontally from the slotted pipes to form an oxygen-rich zone. Volatile organic substances with high vapor pressures are transferred into the air stream and are captured by an aboveground vapor extraction system. The volatile organic component collected by the vapor extraction system must then be used, treated, or destroyed by some other treatment technology.

Were it not for the fact that this process also re-oxygenates the surrounding soil, it would likely be considered as a physical treatment method. Large amounts of organic substrate present in the soil or groundwater cause increased microbial activity, which can quickly deplete the oxygen supply. This results in a slowing of microbial activity, until the oxygen supply is renewed. Reinvigorated by the oxygen, intrinsic bacteria once again assist in the cleanup action. In this respect, air sparging becomes a process with many similarities to biocells discussed above.

This method has the advantage of being an *in situ* method thereby reducing the expense and liability of excavating and transporting the contaminated soil. It is not, however, well suited for all sites.

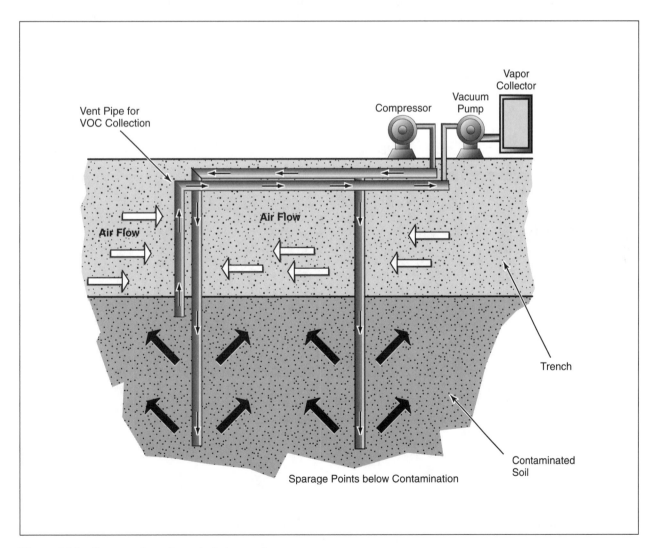

Figure 6-19: Cross-section of a typical air sparging system.

It is difficult, for example, to force air through soils with high clay content or past barriers that prevent efficient distribution of oxygen to the surrounding soil.

Fixed-film Bioreactors

Contaminated groundwater or other wastewater streams that contain an organic component may be treated by a **fixed-film bioreactor**. The bioreactor unit may be an enclosed tank, a series of enclosed tanks, or a chambered tank that allows for the controlled exposure of a contaminated effluent to a biological fixed-film while at the same time limiting the loss of volatile organic vapors to the atmosphere. Within the unit a series of metal honeycombs provide the surface on which the biological slime grows. As the wastewater slowly passes from tank to tank or chamber to chamber it comes into contact with the fixed-film that absorbs and consumes the colloidally suspended and dissolved organic matter.

To maintain a healthy bioreactor, it is important 1) to maintain the pH and the amount of dissolved oxygen (supplied by compressed air), and 2) to have an adequate food supply for the microbes. It is also very important to insure that the air and food supplies for the biomass are never interrupted. If microbes die off, it can take up to a month to restart the system.

From the above description, it is apparent that there are several similarities between the previously discussed rotating biological contactor (RBC) and a bioreactor. Both make use of a fixed-film microbial growth, but in a bioreactor it is attached to a submerged metal honeycomb rather than to a rotating disk. In both processes, oxygen must be supplied to the microbial growth. In a bioreactor air is bubbled through the solution rather than periodically bringing the solution into contact with the air.

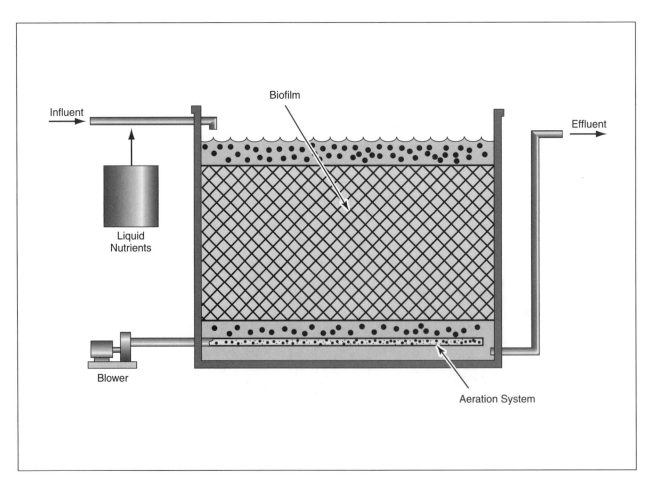

Figure 6-20: Honeycomb support lattice in a fixed-film bioreactor.

In a field previously dominated by air stripping and carbon adsorption, bioreactors are becoming more popular. The primary reason for this trend is that both air stripping and carbon adsorption are mass transfer technologies, rather than destruction technologies. In the current legal framework, many companies are concerned about the long-term responsibility of waste products. As a result, any process that destroys the contaminant will be favored over a process that may expose the company to future liability.

Checking Your Understanding

1. What is a xenobiotic material?

2. What are intrinsic bacteria?

3. What types of organic compounds are typically denser than water and not soluble in it?

4. What does the term *in situ* bioremediation mean?

5. Compressed air and what other substance can be used to supply oxygen for *in situ* bioremediation?

6. What are some of the advantages of the bioremediation method known as landfarming?

7. How is the air supplied to a biological treatment cell?

8. What are the important factors that need to be controlled to maintain a healthy fixed film bioreactor?

9. What is the reason for selecting a bioreactor over carbon adsorption or air stripping?

6-7 Phytoremediation

Until recently the most common method of eliminating soil contamination was to remove and isolate the soil in a hazardous waste landfill. Contaminated soils may also be incinerated to destroy or burn off contaminants, with the residue returned to the site as a sterile material. Reclamation of the site may require replacement of the soil. The 1997 estimates for this type of cleanup range from $200 to $600 per ton of soil.

Today, **phytoremediation** is a promising new treatment method that utilizes the ability of certain plant species to absorb and concentrate metals from the soil and/or water. Euphemistically known as the green clean, it is considered to be one of the most environmentally-friendly methods for removing toxic metals – with the added bonus of providing the possible opportunity to harvest the metals for reuse. Phytoremediation is, perhaps, best suited for treating large areas of soil that are contaminated with relatively low concentrations of toxic metals.

Successful phytoremediation depends on plants known as **hyperaccumulators**, or metal scavengers, that are able to absorb and concentrate up to 100 times normal amounts of metals in their roots and aboveground shoots. Most hyperaccumulators readily absorb nickel, but may also concentrate cobalt, copper, zinc, manganese, lead, and calcium. One possible explanation for this behavior is that these plants use the metal toxins to arm themselves against predators that might otherwise use them for food.

Such plants as the members of the Brassica family have been shown to be good accumulators of Pb and Cr^{6+}. One species of Indian Mustard (*Brassica juncea*), when fertilized with sulfates and phosphates, has been shown to be particularly adept at concentrating Cr^{6+}, Cd, Ni, Zn, and Cu in its shoots.

Use of Indian Mustard as a remediation tool was demonstrated on a piece of property that had served as a lead battery-manufacturing site for 30 years. The abandoned, lead-polluted site had lead contaminant levels of 200-1,800 mg/kg of soil. Prior to treatment, about 60 percent of the area was considered to be clean, meaning that the soil contained less than the regulatory limit of 400 mg of lead/kg of soil. After three crops of Indian Mustard were grown and harvested, an additional 15 percent of the area fell below the lead regulatory limit. Analysis of the plant shoots showed that Indian Mustard accumulates as much as 3,900 mg of lead/kg of dry plant weight – enough to be classified as a hazardous material. Once the plants are harvested, the lead metal can either be recovered, or disposed of in an appropriate disposal facility.

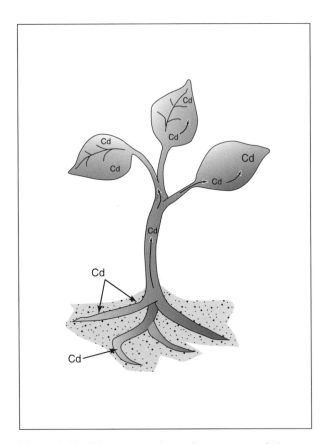

Figure 6-21: Hyperaccumulator plant store metals in their roots, shoots, and leaves.

Traditional Method	Phytoremediation Method
Removal of soil	*In situ* process
Isolation, incineration, chemical treatment	Biological treatment
Soil replacement	Land returned to production
High cost	Low cost
Immediate solution	Long-term solution

Table 6-7: Comparison of traditional and phytoremediation soil cleanup methods.

Box 6-1 ■ Mutant Alfalfa Strain Used to Remediate Fertilizer Spill

In a bit of a switch, phytoremediation work is also being done on the development of an unusual alfalfa mutant. As a member of the nitrogen-fixing legumes, alfalfa – with the assistance of its symbiotic bacterium (*Rhizobium meliloti*) – normally converts atmospheric nitrogen into ammonia. However, in the mutant alfalfa the plant performs the opposite role, extracting nitrogen from the soil.

There are two major forms of nitrogen fertilizer in the soil: ammonium, NH_4^+, and nitrate, NO_3^-, ions. Ammonium ions do not migrate in the soil because their positive charge is attracted to negatively charged soil clay particles. Nitrate ions, on the other hand, are negatively charged and are therefore not attracted to the soil clay particles. As a result, nitrate ions tend to migrate through the soil, often reaching groundwater.

It was quickly realized that this naturally occurring mutant alfalfa could be a useful tool for removing unwanted nitrate ions from soils and groundwater contaminated by such things as fertilizer spills, livestock manure, municipal sludge, or food-processing wastes. In 1989, a train derailment spilled thousands of pounds of anhydrous ammonia and granular urea onto the land and provided an opportunity to test the hypothesis. The hypothesis is still being tested, as nitrates that have reached the groundwater from the spill site are being pumped out and used to irrigate a plot of mutant alfalfa along with a plot of other control plants. The goals of the mutant alfalfa project are not only to test how the alfalfa performs as a remediation agent and if it can reduce cleanup costs, but also to determine if it produces large enough alfalfa yields to be profitably cultivated.

McLaughlin, Lloyd, editor, *Novel Alfalfa Cleans Fertilizer Spill*, Agricultural Research, January, Agricultural Research Service, U.S. Department of Agriculture, Washington, DC (January 1997).

Most plants can accumulate up to 400-500 mg of zinc/kg of dry plant weight. The hyperaccumulator Alpine pennycress (*Thlaspi caerulescens*) has been shown to be able to accumulate up to 25,000 mg of zinc/kg of dry plant weight. Unlike most plants that can accumulate 5 to 100 mg/kg of cadmium, it can also accumulate up to 5,000 mg of cadmium/kg of dry plant weight without causing damage to the plant.

Phytoremediation is best suited for large volumes of soil with low amounts of contaminants. Among the many advantages of phytoremediation is the fact that it uses the same equipment and materials common to agricultural practice. In some cases it is possible to equate the costs of phytoremediation to local costs of planting crops. Often, however, sites designated for remediation have been used for industrial operations that have left the ground compacted and devoid of vegetation. These sites require intense preparation before they will sustain vigorous plant growth. The major drawback of this treatment method is the amount of time required to accomplish the cleanup. Several growing seasons are often required to reach acceptable contaminant levels.

Although funding for phytoremediation research has been limited in the past, some researchers are proposing that it be reconsidered from a mining ore point-of-view. For example, when alpine pennycress is pyrolyzed, its ash can contain as much as 40 percent zinc. That puts it into the same category as a high-grade ore. It also has an advantage over naturally occurring ore in that it doesn't contain other metals that often interfere with the recovery of the zinc. Researchers have calculated that growing hyperaccumulators on metal-rich, but agriculturally poor soils could double farm income, if metal accumulators were substituted for the more traditional crops of alfalfa or corn.

Checking Your Understanding

1. Under what conditions is phytoremediation likely to be the best treatment choice?

2. What is the major drawback to phytoremediation as a treatment choice?

3. What is a hyperaccumulator?

Summary

Long before humans left their first footprints on earth, nature was using an unseen army of microbial servants to keep the streams clear and the land from becoming cluttered with a collection of dead animals and plants. Humans learned of and applied these silent servants to produce food and drink and to treat sewage wastes. Today their use has become the backbone of the wastewater treatment industry.

The servants of biological treatment technology remain the aerobic, anaerobic, and facultative bacteria that have been put to work in a series of processes subdivided into two groups: suspended growth and fixed-film methods. When coupled with the aerobic and anaerobic bacteria, the results are several processes such as activated sludge treatment, RBCs, aerated lagoons, trickling filters, fixed-film rotating biological contactors, SBRs, anaerobic filters, anoxic denitrification, bioreactors, and anaerobic digesters.

In recent years a new technology emerged utilizing the servants of old, plus a few new genetically altered bacteria, collectively known as bioremediation. Bioremediation is defined as any process that uses living organisms – either naturally occurring or genetically altered – to concentrate or decompose contaminants. This process is being applied both *in situ* and *ex situ* depending on the nature and location of the contaminant. Processes utilizing these technologies have names like landfarming, air sparging, biological treatment cells, co-metabolism, and biological reactors.

The newest servants being brought into the remediation arsenal are the higher plants. In an area known as phytoremediation, plants known to be hyperaccumulators are being used to collect low concentrations of toxic metals from large areas of soil by accumulating them in their roots, stems, and leaves. It has been demonstrated that the harvesting and treating of these plants can effectively and economically remove unwanted metals and perhaps provide a new method for mining these metals.

Critical Thinking Questions

1. Compare and contrast the metabolic products derived from aerobic and anaerobic metabolism. Why is aerobic metabolism considered to be the most efficient metabolic process?

2. Define a fixed-film and explain how it differs from a suspended growth method.

3. Propose other uses for hyperaccumulating plants. What would be the advantages and disadvantages of their use?

4. Visit your local wastewater treatment plant; obtain information about the secondary waste treatment processes used and whether tertiary treatment is required before release.

PRESERVING
THE LEGACY

7

Minimization of Waste and Pollution Prevention

by David Y. Boon

Chapter Objectives

Upon completing this chapter, the student will be able to:

1. **Describe** the waste management hierarchy (source reduction, recycling, treatment, and disposal).

2. **Compare** the regulatory requirements for pollution prevention and minimization of waste.

3. **List** the key elements in a pollution prevention and minimization of waste assessment.

4. **Discuss** the benefits of pollution prevention.

5. **List** the key pollution prevention and minimization of waste opportunities.

6. **Analyze** EPA's and industry's voluntary waste reduction program goals.

7. **List** sources of pollution prevention and minimization of waste information.

Chapter Sections

7-1 Introduction

The management of hazardous wastes became a national issue in the early 1970s. Federal and state regulations were enacted requiring businesses to record the amount of waste produced and how it was being treated. In the years that followed, regulators and industry began to realize the problems inherent in generating, treating, and burying waste. By September 1988, the EPA Science Advisory Board announced, "The EPA should shift the focus of its environmental strategy from end-of-pipe control to preventing the generation of pollution." This came at a time when more than 99 percent of all environmental funds were being spent on controlling pollution after it was created rather than preventing its production.

To most people this shift in focus made good sense. Certainly the only permanent and practical solution to the many environmental problems caused by the generation of waste was to reduce its production. Thus, the cornerstone was laid for developing minimization of waste and **pollution prevention (P2)** strategies. Starting in the late 1980s and continuing through the 1990s, the procedures and processes required to implement these waste elimination strategies has shifted the attention from end-of-pipe treatment to front-of-pipe prevention.

EPA continues to promote P2 using several strategies. One of the more successful efforts is the **33/50 Program** that started in 1991. Its accomplishments will be the topic of discussion later in this chapter. In the fall of 1996, President Clinton signed an executive order establishing the **Federal Government Environmental Challenge Program** within EPA. The focus of the program is to foster waste reduction efforts by federal agencies and contractors. As a result, agencies like the Department of Defense (DOD) and the Department of Energy (DOE) are required to include P2 provisions and environmental compliance in their agreements with outside contractors.

EPA has plans to launch yet another initiative called **Innovations in Building Sustainable Industries (IBSIN)**. It is designed to provide incentives for industry, government, and communities to work together to promote sustainable development goals such as reducing the release of toxic substances and implementing product stewardship programs.

It is important to understand that until 1997 the focus of all waste minimization was only on RCRA-defined hazardous wastes. In developing a pollution prevention program, however, it is important that all pollutants emitted into air, water, and land be considered. Later in this chapter, information is presented on the basic principles used to develop minimization of waste and pollution prevention programs.

7-2 Governmental Action

Hazardous and Solid Waste Amendments (HSWA)

Over the years waste reduction has been the target of a patchwork of laws and a number of executive orders. In 1984, for example, Congress passed the Hazardous and Solid Waste Amendments (HSWA) to the Resource Conservation and Recovery Act (RCRA), which states:

> The Congress hereby declares it to be the national policy of the United States, whenever feasible, that generation of hazardous waste is to be reduced or eliminated as expeditiously as possible. Waste that is nevertheless generated should be treated, stored, or disposed of so as to minimize the present and future threat to human health and the environment.

HSWA requires hazardous waste generators to identify, in their biennial reports, the efforts they are undertaking to reduce the volume and toxicity of waste generated. It further asks them to report on the reductions they have actually achieved. In addition, generators must certify that they have a program in place to minimize waste generation. The certification statement, found on the Uniform Hazardous Waste Manifest, reads as follows:

> If I am a large-quantity generator, I certify that I have a program in place to reduce the volume and toxicity of waste generated to the degree I have determined to be economically practicable and that I have selected the practicable method of treatment, storage, or disposal currently available to me which minimizes the present and future threat to human health and the environment, or, if I am a small-quantity generator, I have made a good-faith effort to minimize my waste generation and select the best waste management.

Pollution Prevention Act of 1990 (PPA)

The **Pollution Prevention Act of 1990 (PPA)** establishes – as a national policy – that whenever feasible, pollution should be prevented at its source. The PPA then identifies an environmental management hierarchy (Figure 7-1) and further states that pollution

> ...should be prevented or reduced whenever feasible; pollution that cannot be prevented should be recycled in an environmentally safe manner, whenever feasible; pollution that cannot be prevented or recycled should be treated in an environmentally safe manner whenever feasible; and disposal or release into the environment should be employed only as a last resort.

In short, preventing pollution before it is created is preferable to trying to manage, treat, or dispose of it after its creation.

Minimization of Waste = Source Reduction
Attempts to prevent waste from being produced in the first place and includes internal recycling.
Waste Management Activities
External Recycling, **Treatment**, and **Disposal**. Focus is on reducing the amount and toxicity off waste generated.

Figure 7-1: EPA waste priorities summary.

According to the Pollution Prevention Act, source reduction is meant to reduce the generation and release of hazardous substances, pollutants, wastes, contaminants, or residuals at the source. It also states that source reduction

> ...includes equipment or technology modifications, process or procedure modifications, reformulation or redesign of products, substitution of raw materials, and improvements in housekeeping, maintenance, training, or inventory control. The term 'source reduction' does not include any practice which alters the physical, chemical, or biological characteristics or the volume of a hazardous substance, pollutant, or contaminant through a process or activity which itself is not integral to or necessary for the production of a product or the providing of a service.

In effect, source reduction means reducing the amount of a pollutant that enters a waste stream or that is otherwise released into the environment prior to **out-of-process recycling**, treatment, or disposal. To implement the Pollution Prevention Act of 1990, EPA developed a formal definition of pollution prevention and a strategy for making pollution prevention a central goal. According to the Act's definition, pollution prevention means source reduction, but also "...includes other practices that reduce or eliminate the creation of pollutants through such things as (1) increased efficiency in the use of raw materials, energy, water, or other resources, or (2) protection of natural resources by conservation."

The PPA further defines source reduction as

> ... any practice which reduces the amount of any hazardous substance, pollutant, or contaminant entering any waste stream or otherwise released into the environment (including fugitive emissions) prior to recycling, treatment, or disposal; and reduces the hazards to public health and the environment associated with the release of such substances, pollutants, or contaminants.

Unfortunately, neither HSWA nor the regulations EPA established to implement HSWA defined what actually constitutes a waste minimization program. The law only requires that a program be established, and that waste minimization be evaluated.

Executive Orders

Several Executive Orders that specifically require federal agencies to comply with pollution prevention, waste minimization, and recycling policies have also been issued. **Executive Order 12780** – the **Federal Agency Recycling and the Council on Federal Recycling and Procurement Policy** signed on October 31, 1991 – and **Executive Order 12873** – the **Federal Acquisition, Recycling, and Waste Prevention** order signed on October 20, 1993 – require federal agencies to promote cost-effective waste reduction and recycling of reusable materials from waste generated by federal government activities.

Executive Order 12856 – the **Federal Compliance with Right-to-Know Laws and Pollution Prevention Requirements** signed on August 3, 1993 – establishes federal policy and requires federal agencies to incorporate the activities required by the Emergency Planning and Community Right-to-Know and the Pollution Prevention Act of 1990 into all aspects of agency programs.

These executive orders challenge the federal government to lead by example and apply source reduction strategies in the management of its facilities and acquisition practices. By preventing pollution, the federal government not only protects the environment, it also saves taxpayers money by reducing waste management costs and long-term liability for expensive cleanups.

Executive Order 12856 specifically requires federal agencies to exercise pollution prevention by:

— Incorporating the Pollution Prevention Act of 1990 in planning, management, and acquisition actions.

— Complying with Emergency Planning and Community Right-To-Know Act (EPCRA), including the **Toxic Release Inventory (TRI)** reporting.

> "We must spur the development of a new generation of technologies that prevent pollution..."
>
> **President Bill Clinton**
> (NSTC, 1994)

—Developing voluntary goals to reduce total releases and off-site transfers of TRI toxic chemicals or toxic pollutants by 50 percent by 1999.

—Establishing a plan and goals for eliminating or reducing acquisition, manufacture, processing, or use of products containing extremely hazardous substances or toxic chemicals.

—Developing a pollution prevention policy statement and designating a responsible principal official and program coordinator.

—Committing to source reduction as the primary method of environmental protection and compliance.

—Submitting a written plan by August 3, 1994 and including a written list of federal facilities that meet TRI thresholds.

This list of requirements illustrates that the federal government's attempt to initiate and promote minimization of waste and pollution prevention practices has been and continues to be multifaceted. The successful implementation, however, requires changes in attitudes as well as commitments and significant capital investments.

Checking your Understanding

1. Explain the difference between pollution prevention and minimization of waste.

2. Explain the function served by out-of-process recycling.

3. What is the main thrust of the Executive Orders presented in this section?

4. What is the key point of the Pollution Prevention Act of 1990?

7-3 Pollution Prevention and Minimization of Waste Assessments

After careful analysis of many successful minimization of waste programs, there are several key elements that appear to be common to most programs.

Table 7-1 lists those elements and gives a brief explanation for each.

Top Management Support	Support for a minimization of waste program must be provided at a sufficiently high management level to influence production and environmental policy decisions. This support can be provided in the form of written policies, designation of persons responsible for pollution prevention, and allocation of resources to develop a program.
Implement Motivational Factors	A variety of motivating factors have been successfully used to accomplish minimization of waste objectives. In addition to management support mentioned above, motivating factors can include: —reward and recognize successful minimization of waste efforts; —involve operations personnel in planning and implementation; —select a program leader that enthusiastically supports minimization of waste; —integrate minimization of waste with quality improvement; —use the simplest means available to accomplish minimization objectives; —process modification should be simple to operate, reliable, and easy to maintain; —provide employee training to help them recognize waste generation and identify possible corrective measures; and —provide adequate funding to implement a minimization of waste program.
Waste Characterization	Before waste can be minimized, it must be identified and characterized in terms of types, amounts, and constituents. Waste production should be measured and listed by individual production units. Priority should be placed on minimization of wastes that pose the greatest risk to human health or the environment.
Waste Tracking	Periodic assessments that track materials from initial receipt through usage and waste generation should be implemented. This assessment can identify ways to reduce or eliminate wastes.
Cost Allocation	Successful minimization of waste requires that production unit personnel know the true cost of hazardous waste disposal and consider these costs when making management decisions. A cost accounting system that charges individual operating units for the cost of regulatory compliance, administrative oversight, and insurance attributed to waste management has proven to be highly effective in minimizing wastes.
Program Evaluation	Periodic examination of minimization of waste programs must be conducted if their effectiveness is to be adequately measured and objectively evaluated.

Table 7-1: Elements of a successful minimization of waste program.

Pollution Prevention Opportunity Assessment (PPOA)

The **Pollution Prevention Opportunity Assessment (PPOA)** is an EPA program that encourages ongoing, facility-wide efforts to identify opportunities to eliminate or reduce the generation of waste. A Pollution Prevention Opportunity Assessment accomplishes four important tasks. It:

1. assesses the amount of material disposed of as waste during a particular process or at a particular workplace;

2. provides a summary of hazardous materials used and waste produced;

3. targets those operations needing improvement or replacement to accomplish pollution prevention; and

4. establishes a mechanism for prioritizing options developed during the assessment.

EPA has developed two general manuals for conducting P2 assessments: the *Waste Minimization Opportunity Assessment Manual* (EPA 1988) and the revised *Pollution Prevention Facility Manual* (EPA 1992). These manuals explain the management strategies needed to incorporate minimization of waste into company policy, to establish a company-wide P2 and minimization of waste program, to conduct assessments, to implement options, as well as suggestions on how to make the program ongoing.

Through the use of this information, Pollution Prevention Opportunity Assessments can also be used as a tool to provide two major outcomes: a baseline of waste and energy usage from which pollution prevention and energy savings can be measured, and identification and evaluation of opportunities that prevent pollution and promote energy conservation. In brief, the baseline assessment contains a careful review of facility operation records (purchase and disposal) and waste stream characterization; it also selects specific areas to assess. After a particular waste stream or area is chosen as the Pollution Prevention Opportunity Assessment focus, several options with good waste

minimization potential are developed. The technical and economic feasibility of the various options is then evaluated. Finally, the most promising options are selected for implementation.

The four phases involved in conducting a minimization of waste assessment are: 1) planning and organization, 2) assessment, 3) feasibility analysis, and 4) implementation. Each of these will be briefly described below.

Planning and Organization

The first step in formulating a pollution prevention or minimization of waste program is obtaining management commitment. This requires a commitment to provide the resources necessary to develop and implement the program as well as a commitment to support the efforts of all employees through a formal policy statement. The management of a company will typically support a minimization of waste program if it is convinced that the benefits will outweigh the costs. These benefits may include economic advantages, easier compliance with regulations, reduction in waste disposal liabilities, improved public image, and reduced environmental impact.

The organization of a pollution prevention task force is the next step. This task force should include an advocate, management personnel, production managers, process engineers, and line employees. Individuals that work in waste-generating areas are excellent candidates for the task force because of their knowledge of processes and how they relate to waste generation. They are also likely to be able to identify ways to reduce waste.

Pollution prevention and minimization of waste goals that are consistent with management policy must be set. Goals can be general statements, such as: the company will work toward a significant reduction of hazardous substance releases into the environment. However, it is better to establish measurable, quantifiable goals that establish clear guidelines for what the program is expected to accomplish. Program goals should be reviewed periodically, particularly as the pollution prevention program becomes more defined. This often leads to a modification in the goals. Pollution prevention assessments are not intended to be one-time projects; rather, they require modifications as technology, economics, and regulations change.

Assessment

The purpose of the assessment phase is to develop a comprehensive listing of minimization of waste options. This will allow the more feasible options to be identified so that they can receive further analysis in the next phase. The assessment phase involves a number of steps: 1) collect process and facility data, 2) prioritize and select assessment targets, 3) select assessment teams, 4) review data and inspect site, 5) generate options, and 6) screen and select options for the feasibility study. Each of these steps is discussed in the following sections.

Collect Process and Facility Data – Developing a basic understanding of the processes that generate waste at a facility is essential to the Pollution Prevention Opportunity Assessment process. One of the first tasks of the assessment phase is to identify and characterize the facility waste streams. Some of this information can be obtained from hazardous waste manifests, reports to regulatory agencies, permits, production composition and batch sheets, product inventory logs, operator data logs, etc. Flow diagrams and material balances (See Chapter 1) should be prepared to identify important steps in the process and to identify sources of waste. Material balances are particularly important because they quantify losses or emissions that were previously unknown. Material balances should be prepared for all components that enter and leave the process.

Prioritize and Select Assessment Targets – Ideally, all waste streams in a facility should be evaluated for potential minimization of waste opportunities. With limited resources, however, managers may need to concentrate minimization of waste efforts in a specific area. The assessment process should concentrate on the most important waste problems first, and then move on to the lower priority problems as time and resources permit. Items commonly addressed in prioritizing waste streams include regulatory compliance, cost of waste treatment and disposal, environmental and safety liability, quantity, hazardous properties, and potential for minimization.

Select Assessment Team – Assessment teams should include people with direct responsibility and knowledge of the particular waste stream or area of the facility to be modified.

Review Data and Inspect Site – The assessment team evaluates data in advance of the inspection.

The facility inspection should begin with the point where raw materials enter the facility and end at the points where products and waste leave. The team should identify the suspected sources of waste. These may include maintenance operations, storage areas for raw materials and finished products, as well as manufacturing process operations. The inspection should follow an agenda and must include actual observations of the particular operation that is of interest. The operation should be monitored at times when waste generation is highly dependent on human involvement, e.g., when dumping and replacing a process bath. During the inspection, operators, shift supervisors, and foremen involved in the operation under inspection should be interviewed. Housekeeping aspects of the operation should be observed. Finally, organization structure and administrative controls should be investigated. The inspection may result in the formation of preliminary conclusions about minimization of waste opportunities. Additional data collection, analysis, and/or site visits may be required to confirm preliminary conclusions.

Generate Options – Once the origins and causes of waste generation are understood, the assessment process enters the creative phase. The objective of this step is to generate a comprehensive set of minimization of waste options for further consideration. Identifying potential options relies on the expertise and creativity of the team members. The process for identifying options should follow a hierarchy in which source reduction options are explored first, followed by recycling options. Source reduction may be accomplished through improved operating practices, changes in technology and in the materials used in the manufacturing process. Recycling includes in-process use or reuse of a waste stream and out-of-process reclamation.

No options should be ruled out at this time, since technical and economic concerns will be considered later in the feasibility step. Information from the site inspection, as well as from trade associations, government agencies, technical and trade reports, equipment vendors, consultants, plant engineers and operators may serve as sources of ideas for minimization of waste options.

Screen and Select Options for Feasibility Study – The last step in the assessment phase is to screen the options and select those that warrant full technical and economic analysis of their feasibility. Options that appear marginal, impractical, or inferior are eliminated from consideration. The remaining options are selected for the next steps: technical and economic feasibility analysis.

Feasibility Analysis

A pollution prevention or minimization of waste option must be shown to be both technically and economically feasible in order to merit serious consideration for implementation. A technical evaluation determines whether a proposed option will work in a specific application. Typical criteria used in technical evaluation include: worker health and safety, maintenance of product quality, space availability, compatibility with existing operations, labor requirements, installation procedures, and system maintenance. In addition, both process and equipment changes need to be assessed for their overall effects on waste generation and product quality.

Economic evaluation is carried out using standard measures such as payback period, return on investment, and net present value. Capital investment criteria include such costs as site development, permitting costs, contractor's fees, start-up time, and training. Operating costs and savings must also be analyzed and may include reduction in waste management and disposal costs, material costs savings, insurance and liability savings, changes in utility costs, and changes in operation and maintenance.

While degree of profitability is important in deciding whether or not to implement an option, compliance with existing and future environmental regulations may be even more important. An organization operating in violation of environmental regulations can face fines, lawsuits, civil and criminal penalties, and even closure. Therefore, decisions should not be based on short-term profitability alone.

Implementation

Assessment and feasibility analyses provide the basis for implementation. Because projects are not always sold on their technical merits, a clear description of intangible as well as tangible benefits can help edge a proposed minimization of waste project past other projects in the competition for funding. Obtaining funding for a minimization of waste project may be difficult, especially when companies have typically prioritized capital resources toward enhancing future revenues rather than cutting current costs through minimization of waste. Having well-documented technical facts and knowing the level within the organization that has approval authority for capital projects will assist in obtaining funds.

After the minimization of waste option has been implemented, its effectiveness must be measured. One measure of effectiveness for a minimization of waste project is the effect on the organization's cash flow. The project should pay for itself through reduced waste management and raw materials costs. The easiest way to measure waste reduction is by recording the quantities of waste generated before and after a minimization of waste project has been implemented. However, this simple measurement ignores changes in amount of product manufactured. Therefore, a waste generation to production ratio may be a more accurate way to assess effectiveness of minimization of waste efforts.

Finally, minimization of waste is not a one-time effort, but a continuing management system. Once the highest priority waste streams have been assessed and minimization projects implemented, the assessment program should look to areas and waste streams with lower priorities. To be truly effective, minimization of waste must be an integral part of the company's operations.

Pollution Prevention Benefits

The payoffs for a commitment to pollution prevention can be significant and include: reduced liability, more efficient use of natural resources, reduced treatment and disposal costs, lower environmental impacts, reduced regulatory costs, and improved public relations. Short-term waste disposal costs have increased dramatically in recent years and will continue to increase. Potential long-term costs of waste disposal associated with the liability for environmental damage under CERCLA cannot be estimated and have no financial ceiling. Therefore, these short-term disposal costs and the potential for long-term liability have combined to make minimization of waste economically attractive.

Businesses have strong incentives to reduce the toxicity and volume of the waste they generate. The cost per unit produced will decrease as pollution prevention measures lower liability risk and operating costs. A company with an effective, ongoing pollution prevention plan may be the lowest-cost producer and have a significant competitive edge. The company's public image will also be enhanced.

Reduced Risk of Liability

A company will decrease both civil and criminal liability by reducing the volume and toxicity of the waste streams generated in the production process. Waste handling affects public health and property values in the communities surrounding production and disposal sites. Worker's compensation claims are directly related to the hazards of the chemicals used and the volume of waste materials generated. Even materials not currently covered by hazardous waste regulations may present a risk of civil litigation in the future. Environmental regulations at the federal and state levels require that facilities document their pollution prevention and recycling measures and report the percentage or volume reduction achieved. Companies that produce excessive non-hazardous waste and hazardous waste are subject to heavy fines. Mismanagement of wastes may subject managers to fines and imprisonment.

Reduced Costs for Raw Materials, Waste Treatment, and Disposal

An effective pollution prevention program can generate cost savings that will more than offset program development and implementation costs. Reduced operating cost show up immediately and are easy to identify. Reduced cost of future liability is more difficult to quantify. Savings are particularly noticeable when the costs of treatment, storage, and disposal of waste are allocated to the production unit that produces the waste. Costs associated with the purchase of raw materials can be reduced by adopting production procedures that consume fewer resources and create less waste. As wastes are reduced, the percentage of raw materials converted to finished products increases, with a proportional decrease in materials costs.

Pollution prevention directly reduces the cost of waste management and disposal. Increased production efficiencies are likely to result in decreasing energy costs as pollution prevention measures are implemented in the various production areas. Waste handling and management procedures, on the other hand, are generally associated with increased costs. Federal and state regulations require specific training, recordkeeping, and reporting; the costs of complying with these regulations represent a direct cost to business. Also, costs associated with facility cleanup may occur because regulations require the upgrading of underground storage tanks or remediation of soil and groundwater contamination.

Public Health and Environmental Benefits

Reducing wastes and emissions lessens the amount of raw material consumed, which, in turn, reduces ecological damage due to raw material extraction, production, and refining operations. The public health and environmental benefits fall into three categories: 1) reduced cost of dealing with the waste, 2) reduced cost of raw materials because they are used more efficiently, and 3) reduced cost of complying with regulations.

Improved Company Image

As environmental equity becomes an issue of greater importance to society, the company's pollution prevention policy and practices become an important statement to employees and community members. Employees are likely to feel more positive toward their company when they believe that management is committed to providing a safe work environment and is acting as a responsible member of the community by taking a proactive role in waste reduction. By participating in pollution prevention

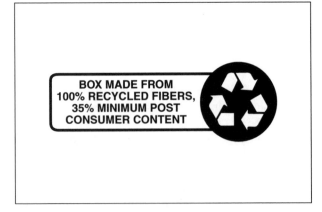

Figure 7-2: A product recycling label.

activities, employees can interact positively with each other and with management. Helping to implement and maintain a pollution prevention program should increase their sense of identity with company goals.

Community attitudes toward companies that operate and publicize a thorough pollution prevention program will be more positive. Creating environmentally compatible products and avoiding excessive consumption of energy resources and discharge of materials will greatly enhance a company's image with customers and within the community.

Checking Your Understanding

1. List the six elements that must be present for a successful minimization of waste program.

2. List the four phases involved in conducting a minimization of waste assessment.

3. Describe the six key elements of the assessment phase.

4. List four benefits of a pollution prevention program.

5. Explain how pollution prevention can reduce risk of liability.

6. Explain how pollution prevention can improve a company's image.

7-4 Pollution Prevention and Minimization of Waste Opportunities

The primary goal of pollution prevention is source reduction – reducing the amount of wastes or contaminants generated at the source. Several of the practices that reduce waste at its source are discussed below.

Purchasing and Inventory Controls

In the past, the basic consideration involved in the purchase of chemicals was cost. Little or no attention was given to the expenses and liability associated with disposal. As a result, chemicals were often purchased before they were needed and in large quantities. In addition, various brands were purchased to take advantage of sales. These practices often result in supplies of chemicals that exceed their shelf life date and are inferior in quality. This creates hazardous waste disposal problems that can be rather easily solved by the following suggestions:

1. Require supervisor approval prior to purchasing hazardous chemicals. This forces workers to justify the purchases and makes management aware of when and how often hazardous chemicals are being added to the inventory.

2. Keep the inventory of hazardous chemicals to a minimum. This makes it easier to rotate the stock and less likely that shelf life will be exceeded. In

---

Box 7-1 ■ EPA Lithography Project

Custom Print Corporation in Arlington, Virginia is a sheetfeed, offset lithographic printer of commercial color products including brochures, folders, and booklets. Inventory and purchasing records showed that the company has more than 80 different chemicals onsite. Often, the less frequently used products would need to be disposed of because they had exceeded their expiration date. The money spent on them was wasted, and by law they had to be properly disposed of – another expense. Many more were product samples, often used once and left to clutter the stockroom until they exceeded their expiration dates. In addition, the large inventory created extra labor costs. Employees had to order and track each chemical, and ensure compliance with government regulations.

To address these problems, Custom Print assembled a team of press operators, purchasing staff, and maintenance personnel. This team not only looked at the causes of the large inventory, but also recommended several ways to reduce it. The solutions they found included:

1. Use multi-task chemicals. Working with the suppliers, the team identified chemicals that can be used for more than one task. Using these products reduced the stock of infrequently used chemicals and resulted in fewer expired chemicals.

2. Eliminate duplication. The team found that in some cases two or three different chemicals were being bought for the same task. To eliminate this duplication, employees who used similar chemicals got together and reviewed all products in use. As a team, they selected only one chemical for each task.

3. Give unused samples back to the vendors. Custom Print asked vendors to pick up their unused or partly used samples each time they dropped off new ones. Custom Print continued testing new and promising products while getting rid of half-used bottles and cans.

These changes reduced the number of chemicals onsite at Custom Print from more than 80 to just 24 – a 70 percent decrease. This reduction operation has cut pollution and waste; it has also minimized potential liability and reduced inventory and related costs resulting in an estimated $5,000 savings per year.

From **EPA's Office of Pollution Prevention Technology Design for the Environment**

addition, when less material is available, workers generally use less.

3. Reduce the number of product brands used for the same purpose. Different brands of the same product may introduce quality or processing variations.

4. Purchase only what will be used within a short period of time. Manufacturing facilities call this **just-in-time (JIT)** inventory management. This will reduce storage problems and the likelihood of contamination before use.

5. Use simple labeling codes, such as an orange sticker, to identify chemicals that contain hazardous constituents. Train individuals who are using these chemicals to recognize the code and to use proper handling and disposal techniques.

Figure 7-3: Example of bad housekeeping practices.

Improved Housekeeping

Improved housekeeping can solve a variety of hazardous waste generation problems. In addition, housekeeping changes can be implemented quickly and with little cost. Sloppy housekeeping, which includes leaks from tanks, pumps, valves, and release of product onto the floor, can dramatically increase the volume of hazardous waste generated. Other practices such as tank overfills, lack of drip boards, and chronic spills and leaks add to the hazardous waste stream. Not only is valuable product lost, but the volume of waste generated is increased due to the materials (rags, floor-dry, and water) used during cleanup. Other housekeeping problems involve improper storage practices, inefficient production startup and shutdown, scheduling problems, and poorly calibrated control devices.

Figure 7-4: Example of good housekeeping practices.

Box 7-2 ■ EPA Good Housekeeping

A remote power generation facility owned and operated by a federal agency had a transformer containing PCBs. The facility was not operating, and the transformer was leaking. The area was fenced and locked, a PCB warning label was in place, secondary containment was installed around the transformer, and the site was inspected on a monthly basis. To remove the transformer and replace it with one that did not contain PCBs was estimated to cost $1,500. During a monthly inspection, it was observed that the lock had been broken off the gate and the transformer had been vandalized resulting in a total release of the PCBs from the transformer onto the surrounding soil. The subsequent soil remediation and cleanup costs totaled more than $30,000.

Production/Process Modifications

Outdated equipment and inefficient traditional production methods can generate large volumes of hazardous waste, especially when production is the goal, and waste generation is largely ignored. Although the capital investment to purchase new equipment or to modify existing equipment can be high, the investment payback is usually significant when waste disposal and liability costs are included in the analysis.

Product Substitution and Reformulation

Substituting a nonhazardous chemical for a hazardous one has obvious benefits for minimization of waste, environmental protection, and worker health and safety. Many products are being reformulated by chemical manufacturers due to increased pressure being placed on them by industry.

Waste Segregation

Many wastes are actually mixtures of hazardous and nonhazardous waste, such as chlorinated solvents in waste oil. When this happens, regulatory definitions place the entire waste stream in the hazardous waste category. By segregating key constituents, generators can realize substantial cost savings on waste disposal.

Waste segregation can also assist in recycling. An unsegregated waste stream may be too costly to recycle because of the large component of nonrecyclable waste. Waste separation may also not be possible because of similar boiling points of two or more solvents, which makes it difficult to separate them by distillation or other limitations that interfere with recycling operations.

Box 7-3 ■ EPA Screen Printing

Action Graphics, a commercial printer in Louisville, Kentucky produces point-of-purchase display products such as shelving signs, banners, and window signs used in retail stores and fast food restaurants. Traditional solvent-based inks are used in about 60 percent of the company's printing, and ultraviolet (UV) curable inks are used in 40 percent of the printing operations. Although a small shop with limited resources, it has greatly reduced solvent use through many innovative changes.

The primary change was the introduction of a high-pressure water system, which eliminated the need for ink remover solvent. The ink remover consisted of diacetone alcohol (75 percent), methyl chloroform (10 percent), and 2-butoxyethanol (15 percent). Exposure to these chemicals can cause adverse health effects in the workers.

To reduce these health risks, Action Graphics completely overhauled its screen reclamation process at the end of 1991. The key component of the new process was a high-pressure water system. Operators apply an emulsion remover with a brush, then rinse the screen by shooting water through them at a pressure of 3,000 psi.

The combination of not allowing the ink to dry and using the high-pressure water stream allows workers to flush out both ink and emulsion without the use of ink remover. Eliminating the need for ink remover decreased the company's annual solvent use by approximately 770 gallons, reduced worker exposure to potentially harmful chemicals, and saved more than $13,000 per year in purchasing costs.

In addition, the company substituted safer screen reclamation chemicals; purchased a distiller to recover used solvents; switched to a slower-evaporating solvent for screen cleaning during press runs to increase ink setting time; changed to a safer ink thinner; and developed a rag-reuse policy to reduce the need to purchase as many rags.

From **EPA's Office of Pollution Prevention Technology Design for the Environment**

Box 7- 4 ■ EPA Produce Reformulation

Large mining operations in the Wyoming Powder River coal basin often utilize draglines to remove the overburden to get to the coal. Traditionally, the gear greases contained the carrier chemical, trichloroethylene (TCE), to facilitate the movement of the grease to the gears.

The contaminated grease was periodically removed, recovered, and drummed. In a mining operation, this would result in well over 90,000 pounds of waste from a single dragline. Since the waste grease failed the toxic characteristic leaching (TCLP) procedure for TCE, it was classified as a hazardous waste. The mining industry approached the grease vendors about the problem. After

approximately one year, the grease manufacturers were able to reformulate the gear greases with the complete elimination of TCE. The waste grease is now classified as a nonhazardous waste.

The mining companies quickly switched to the non-TCE containing greases; it took approximately one year to eliminate all of the older TCE-containing grease. This example of product reformulation not only eliminated the production of a hazardous waste, but also created an easily marketable product for the grease manufacturers and vendors.

New Use, Reuse, and In-process Recycling

When a waste material can be reused – as in recycling – or when a new use can be found for the material, several advantages result. First, disposal costs are reduced or eliminated, and second, raw material purchase costs are reduced. Companies are being encouraged to seek out new and environmentally sound uses for waste materials that were previously treated and/or disposed of. Assistance – in the form of **waste exchange** programs – is being established across the United States. The goal of waste exchanges is to minimize waste disposal expenses and to maximize the value of reusable manufacturing by-products. In the ideal situation, waste exchanges serve as a matchmaker between companies, resulting in one company's waste becoming another company's feedstock.

There are two basic types of waste exchange: information exchanges and material exchanges. Information exchanges simply act as clearinghouses to match waste availability to possible users. Most information exchanges are nonprofit organizations that receive funds from governmental agencies. In contrast, material exchanges take actual physical possession of the waste and may initiate or actively participate in the transfer of the waste from generator to user. These waste exchanges are usually run by private companies.

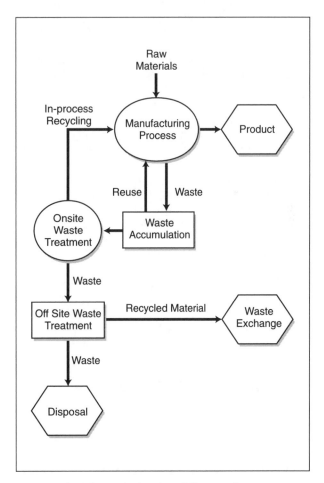

Figure 7-5: Schematic showing differences between reuse and in-process recycling.

Box 7-5 ■ EPA Waste Stream Separation

An electronics facility in New York purchased a still to recycle waste heptane. However, product specification could not be achieved because other solvents in the waste stream were carried over during the distillation process.

The problem was easily solved by separating the heptane waste streams from other solvent waste streams prior to distillation.

Box 7-6 ■ Printing Wiring Board – Case Study 3

Printed Circuit Corporation is located in Woburn, Massachusetts and is a manufacturer of double-sided and multilayer printed wiring boards for the electronics industry. Printed Circuit Corporation employs 300 people and produces 1.8 million surface square feet of board per year. As a surface finish, Printed Circuit Corporation uses solder-mask-over-bare-copper with hot air-solder-leveling. The solder mask – composed of a mixture of tin and lead – prevents copper oxidation and facilitates soldering during the later assembly process.

Before panels can undergo nickel/gold tab plating (also called finger plating, connector plating, or micro-plating) to improve the tabs' electrical conductivity and resistance to environmental degradation, the solder mask must be removed from the panel. The solder mask removal process is accomplished by using a reverse electroplating process, in which the panel is the anode in a methanesulfonic acid solution.

In the past, it was necessary to change the corrosive methanesulfonic acid solution every 30,000 tabs or approximately every six weeks. Methanesulfonic acid is a very expensive acid (~$21/gallons) and costs approxi-

mately $17,000/year to replace. Spent solution was sent off site for disposal at a cost of approximately $5,600/year. Therefore the annual cost of using methanesulfonic acid was about $22,600/year.

Printed Circuit Corporation installed diffusion dialysis technology that allowed for the recycling of the methanesulfonic acid. This process has resulted in an annual savings of $14,500 for process chemicals and $5,600 for waste disposal, for a total annual savings of $20,100. The capital equipment costs for the acid recycling system was $10,800. The investment paid for itself in a little more than six months.

Additional benefits included reducing long-term liability associated with hazardous waste shipping and disposal as well as the reduced employee exposure to the methanesulfonic acid associated with bath dumping.

From **EPA's Office of Pollution Prevention Technology Design for the Environment**

Out-of-process Recycling

Like source reduction, pollution prevention as defined by the Pollution Prevention Act does not include out-of-process recycling, waste treatment, or combustion of waste for energy recovery. The exclusion of recycling from the official definition of pollution prevention activities has been a source of considerable controversy. Strictly speaking, recycling is not a form of prevention. However, sometimes in-process recycling in which the materials are directly reincorporated back into the same process, is considered a form of pollution prevention. Nevertheless, EPA has held fast to the more strict interpretation of pollution prevention, which excludes recycling because wastes that are recycled – even though effectively – have not been prevented. Although it is not recognized as P2, the position of recycling as the second highest option both in Congress and in the EPA pollution prevention/waste management hierarchy attests to its desirability in those instances where the waste cannot be feasibly prevented. Both onsite and off site recycling result in substantial environmental improvements and can contribute to conserving valuable resources.

The most commonly recycled hazardous wastes are probably organic solvents. These wastes can often be recycled by distillation. Small, commercially available recovery units, which may have payback times of less than one year, can handle up to 55 gallons of waste at a time. Off site recycling is also commercially available, especially for organic solvents. The recycler picks up the waste from the generator, treats it, and returns it to the generator. This process, called **tolling**, is available in most metropolitan areas. In some cases, portable recycling units are brought to the generator and the recycling takes place onsite.

In other cases, simple filtration will allow the waste stream to be recycled. This method is useful in machining operations where cutting oils are contaminated with metal shavings that can be easily removed by filtration. Ultrafiltration and distillation units are also readily available for the recycling of such things as automotive antifreeze.

Checking Your Understanding

1. Explain how purchasing and inventory controls can reduce waste generation.

2. Describe a production or process modification that reduces waste.

3. Describe a product reformulation that reduces waste.

4. Describe the difference between the two types of waste exchanges: information exchange and material exchange.

7-5 EPA Voluntary Reduction Programs

There are several voluntary programs that have been developed to encourage minimization of waste and pollution prevention. Some of the programs managed by EPA include the 33/50 Program, **Green Lights Program**, **Energy Star Program**, **Design for the Environment**, and the **Green Chemistry Program**. Each of these will be discussed in more detail below.

EPA 33/50 Program

In the late 1980s senior EPA officials began a series of meetings with executives of various business and industry associations and major corporations to explore ways in which releases of seventeen high-priority pollutants that are reported on the Toxic Release Inventory (TRI) might be reduced. These toxic chemicals were being produced in high volumes and offered great potential for successful pollution prevention activities. An agreement was reached to use the 1.48 billion pounds on the 1988 TRI report as the baseline for the new program. The program's name, 33/50, was selected because of its target goals: a 33 percent reduction in the seventeen targeted TRI chemicals by 1992 and a 50 percent reduction by 1995. After addressing each of the participant's concerns, the final structure of the program was established. Its formal announcement was made in February 1991, and the management responsibilities placed with the EPA's Office of Pollution Prevention and Toxics (OPPT).

About 1,300 companies – operating more than 6,000 facilities nationwide – responded by volunteering to participate and pledging to reduce their releases and off site transfers of the targeted TRI chemicals. The results of the program were quickly realized. Industry exceeded the 33/50 Program's national interim 33 percent reduction by more than 100 million pounds in 1992. National emission of the targeted chemicals were reduced by an additional 94 million pounds in 1993 and 62 million pounds

Seventeen TRI Targeted Pollutants
Benzene
Cadmium and its Compounds
Carbon Tetrachloride
Chloroform
Chromium and its Compounds
Cyanide Compounds
Dichloromethane
Lead and its Compounds
Mercury and its Compounds
Methyl Ethyl Ketone
Methyl Isobutyl Ketone
Nickel and its Compounds
Tetrachloroethylene
Toluene
1,1,1-Trichloroethane
Trichloroethylene
o, m, and p-Xylene

Table 7-2: The 17 targeted TRI pollutants.

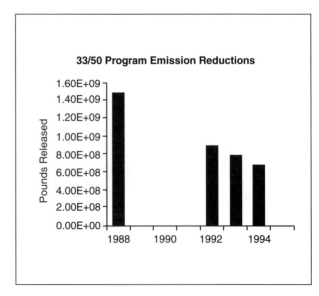

Figure 7-6: 33/50 Program emission reductions 1988-1994.

more in 1994, bringing the total reductions since 1988 to more than 757 million pounds or 50.7 percent of the baseline amount. The program's ultimate 50 percent reduction goal was exceeded – a full year ahead of schedule. The fact that companies participating in the 33/50 Program succeeded in reducing toxic emissions at a rate faster than non-participating companies, suggests that voluntary action on the part of industry can be as effective as the EPA's command and control methods used in the past. In addition, many companies reported that their reduction efforts actually saved them money!

While the 33/50 Program's ultimate 50 percent reduction goal was targeted to end in 1995, EPA encouraged companies to set their own reduction goals and not be constrained by the Program's national time frame. Some 33/50 Program companies responded by setting pollution reduction goals that extended well past 1995. While many company reduction projects might have been completed by then, EPA still had to wait for the 1995 TRI data to be reported and assembled in 1997 in order to be able to analyze and report on the full extent of the 33/50 Program's achievements. At this time, however, the known achievements of the 33/50 Program include:

— The 1,300 participating companies have projected reductions to continue beyond the end of the program.

— The seventeen chemicals targeted by 33/50 have been reduced at nearly three times the rate of other TRI chemicals, since 1991 when the program began.

Box 7-7 ■ Carol Browner Statement on TRI Releases

Releases of hazardous chemicals into the environment by manufacturing companies declined for the eighth straight year in 1995, according to the latest data from the EPA's Toxics Release Inventory. The data released in May 1997 cover emissions of 286 chemicals new to the inventory, which now encompasses a total of 643 chemicals.

"From 1994 to 1995, the amount of chemicals released into the environment was down nearly 5 percent," said EPA Administrator Carol Browner. This continues "...a trend we've seen since 1988, when this Community Right-to-know Initiative began. Since that time, industrial facilities required to report their toxic releases have reduced their emissions by 46 percent."

Box 7-8 ■ EPA's 33/50 Program Receives Hammer Award

On June 18, 1996 Vice President Al Gore presented the Hammer Award to EPA's 33/50 Program. This award symbolically displaying a $6 hammer mounted in an aluminum frame – is presented to governmental programs that are in marked contrast with past "performances" of government efficiency, based on the well-publicized instance of hammers being purchased for $600 each.

—33/50 participants have gone well beyond their commitments, achieving 50 percent more than the amount of reductions originally pledged to the program.

—33/50 participants are achieving reductions at a much faster rate than non-participating companies: 50 percent versus 30 percent (for non-participating companies) from 1991 through 1994, and 60 percent versus 35 percent since 1988.

—Overall generation of production-related waste for the seventeen targeted TRI chemicals has declined slightly since 1991 and is projected to continue declining, even as waste produced for all other TRI reported chemicals is increasing.

—The seventeen chemicals targeted by the 33/50 Program are more frequently selected for source reduction activities than other TRI-listed wastes.

Green Lights Program

Most of the electricity in this country is produced by burning fossil fuels. This results in the release of a large amount of harmful pollutants into the air. It has been estimated that of all U.S. air emissions, electricity generation accounts for 35 percent of the carbon dioxide, 75 percent of the sulfur dioxide, and 38 percent of the nitrogen oxides. Of the electricity sold, approximately 20 percent of the total is for lighting.

The EPA's Green Lights Program was designed to convince companies that lighting should be viewed as an investment opportunity, rather than a cost of doing business, therefore promoting energy efficiency through investment in energy-saving lighting. The program hopes to demonstrate that profitability and environmental protection can go hand in hand.

Participation in this program is voluntary and initiated by signing a Green Lights Memorandum Of Understanding (MOU). In signing the MOU the participant agrees that energy-efficient lighting is one of its high priorities. It also obligates the participant to conduct a total facility lighting survey and, within five years, upgrade at least 90 percent of the square footage that can be profitably upgraded, without compromising lighting quality. (Federal agency partners must complete all upgrades by 2005.)

The participant further agrees to appoint an implementation manager to oversee the progress and to report at least annually to EPA on their progress. EPA agrees that the commitment to survey buildings and complete lighting upgrades is contingent upon the availability of appropriate funds or third party financing.

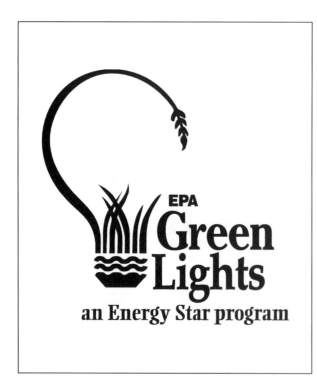

Figure 7-7: Green Lights logo.

Figure 7-8: Energy Star program logo.

Box 7- 9 ■ Walt Disney World Green Lights Program

On August 27, 1996 the Walt Disney World Company became one of the Green Lights Program participants. When all of the lighting retrofit projects at Walt Disney World are completed, it will be equivalent to removing six thousand cars from America's roads, or planting twelve thousand acres of trees.

Walt Disney World Resort President Al Weiss stated, "Through this incredible program we will save enough electricity to power our new 500-acre theme park–Disney's Animal Kingdom – through its entire first year when it opens in 1998." That's a really bright idea!

Box 7- 10 ■ Energy Star Compliant Exit Signs

There are more than 100 million exit signs in buildings throughout the country that operate continuously. Each sign consumes between 44 and 350 kwh of electricity per year, which amounts to about $1 billion annually.

Energy Star compliant exit signs require less than five watts of electricity per side; they exceed NFPA's Life Safety Code for luminance; they have a five-year warranty against defective parts; and they have the potential to save U.S. businesses more than 800 million kilowatts of electricity, which corresponds to about $70 million per year.

In less than five years, the Green Lights Project attracted more than 2,000 members, covering the spectrum of the working world, including corporations of any size, nonprofit organizations, and all levels of government agencies.

It is estimated that if the project were fully implemented across the nation, it could: 1) save more than 150 billion kwh of electricity annually; 2) result in an annual electric bill savings of $12 billion; and 3) result in an average return rate of 30 percent. These figures translate into a reduction in air pollution, solid waste, and other environmental impacts of electricity generation, equivalent to five percent of all U.S. utility emissions.

Earth Day Address – 1993

President Clinton announced that the federal government is committed to purchasing only computers and printers that meet EPA Energy Star guidelines for energy efficiency. This is projected to save the government $40 million per year in terms of reduced electricity use.

Encouraged by the early success of the Green Lights Project, a variety of other programs have been started. EPA contacted computer, printer, fax machine and copier manufactures throughout the world and asked them to join the **Energy Star Office Equipment** program. As a result, for example, partnership agreements have been signed with the manufacturers who sell 70 percent of all desktop computers and 90 percent of all laser printers sold in the United States. These computers, monitors, and printers can automatically power down to save energy when they are not being used. It is estimated that this "sleep" feature will cut a product's annual electricity use by about 50 percent. Office equipment containing this internal energy-saving system is easily recognized by its Energy Star logo.

Through the **Energy Star Loan** program EPA has made arrangements for special financing rates and loans to help homeowners afford homes, heating, cooling, and air handling equipment purchased under the comprehensive **Energy Star Buildings** program. Homeowners who purchase Energy Star-labeled heating and cooling products will not only experience reduced energy bills, but total monthly savings – even with the loan payments taken into account. Finally, because home electric use accounts for 20 percent of the U.S. carbon dioxide emissions, EPA created the **Energy Star Homes** program, which assists new homes builders in developing energy-efficient homes and identifies financing opportunities with participating lending institutions for both the builder and the new home buyer.

If you already own your home, you can still save money by joining the **Energy Star Residential HVAC** program and installing Energy Star heating ventilation and air conditioning devices. More information on these expanding programs can be obtained by contacting the Green Lights/Energy Star Programs, U.S. EPA (6202 J) 401 M Street SW, Washington, D.C. 20460.

Design for the Environment (DfE)

The EPA's **Design for the Environment (DfE) Program** is another program that seeks voluntary participation of business and industry. Its goals are to develop and distribute pollution prevention and risk assessment information on alternative products, processes, and technologies that reduce environmental impact. It targets small- and medium-sized businesses such as dry cleaning, printed wiring board manufacturing, and printing (screen and lithography) industries.

Figure 7-9: DfE logo.

The DfE program provides **Cleaner Technologies Substitutes Assessments (CTSA)** for targeted industries. At the present time CTSAs have been developed for the printed wiring board (PWB), dry cleaning, graphic art screen printers, flexography, lithography, and metal finishing industries. CTSAs provide detailed information about traditional and alternative manufacturing methods and technologies, including information about human and environmental health, risks associated with exposure, economic information, and performance characteristics needed by businesses to determine if changes in manufacturing processes are practical. The ultimate goal of the CTSA is to offer the most complete picture of the issues associated with traditional and alternative products, processes and technologies, so businesses can make more informed choices. To help industry implement some of the new technologies identified during CTSA development, DfE provides a variety of outreach tools, including fact sheets, bulletins, pollution prevention case studies, software, videos, and training materials.

A new approach is being used to compare the risk, performance, and cost tradeoffs of the alternatives. In this approach, a **use cluster** – that is, a set of chemicals, processes, and technologies that can substitute for one another in performing a particular function – is evaluated. This method differs from traditional pollution prevention approaches in that it does not focus strictly on minimization of waste at each step in a manufacturing process. Instead, the cluster approach explicitly arrays alternatives along with performance and cost in a systematic way. During the process of identifying alternatives, attention is focused on finding newer, cleaner substitutes and comparing them with traditional ones.

Green Chemistry and the Green Chemistry Challenge

Following the passage of the Pollution Prevention Act of 1990, EPA – through its Office of Pollution Prevention and Toxics (OPPT) – began to explore ways of improving existing chemical products and processes while eliminating the use of toxic substances. In 1992, OPPT launched a research grant program called **Alternative Synthetic Pathways for Pollution Prevention**. The major objective of the program was to insure the inclusion of pollution prevention thinking as a part of chemical manufac-

Figure 7-10: Green Chemistry logo.

turing processes. This new approach to pollution prevention, through the environmentally conscious design of chemical products and processes, is the central focus of the EPA's Green Chemistry Program.

In short, Green Chemistry promotes changes that reduce or eliminate the use or generation of toxic substances at all stages of manufacture and use. The program further supports research into environmentally benign chemistry, promotes voluntary partnerships with industry in developing green chemistry technologies, and works with other federal agencies in building green chemistry principles into their operations. As part of the Reinventing Environmental Regulations Initiative, on March 16, 1995 President Clinton announced the **Green Chemistry Challenge** to promote pollution prevention and sustainability through a new Design for the Environment partnership with the chemical industry.

The program has two thrusts. First is the recognition of accomplishments in chemistry that have been used to achieve pollution prevention goals. To accomplish this, the voluntary program offers awards, grants, presidential recognition, and publicity about the participants. Second, the program promotes basic research through EPA grants and

encourages industrial and university collaboration to develop innovative approaches to achieve pollution prevention.

EPA will partner with industry to provide technical assistance in designing safer chemicals and processes, and track the reductions achieved in the manufacture, use, and release of harmful chemicals. The real challenge is for industry to find cleaner, cheaper, and smarter ways to produce the materials we need. For more information about how to participate in the Green Chemistry Challenge, contact the Toxic Substance Control Act Assistance Information Service, the Pollution Prevention Information Clearinghouse in Washington, D.C., or e-mail ppic@epamail.epa.gov.

Industry Reduction Programs

Industry has also developed a number of internal programs that reduce waste. For example, Chevron's **Save Money and Reduce Toxics (SMART)** program was initiated in 1987. It focuses on source reduction, it substitutes toxic chemicals with less toxic ones, and it recycles hazardous and nonhazardous

Figure 7-11: Responsible Care® logo.

Box 7-11 ■ 1997 Winners – The Green Chemistry Challenge

Alternative Synthetic Pathways Award – This award was given to BHC Dallas, which – in a joint venture with BASF and Hoechst Celanese – developed a process that eliminates the large quantities of wastes and solvents generated during the bulk synthesis of the anti-inflammatory drug ibuprofen. This process reduces the traditional six steps to only three, while at the same time improving the atom utilization efficiency from 40 to 80 percent. Two of the three steps are solventless and all three are catalyzed, therefore significantly improving efficiency and pollution prevention.

Alternative Solvents/Reaction Conditions Award – Imation Corp., Oakdale, MN received this award for the DryView film development technology used in medical imaging. This process eliminates the use of silver halides, hydroquinone, and various acids that are used to develop the traditional photographic films. Through the use of Imation's photothermography process, the exposed film is developed by heat, thereby eliminating the series of chemical baths. It is estimated that this process will eliminate more than three million gallons/year of developer, more than five million gallons/year of fixer, and more than 900 million gallons/year of wash water, in the medical imaging industry alone.

Designing Safer Chemicals Award – Albright & Wilson of Richmond, VA received this award for developing tetrakis(hydroxy-methyl) phosphonium sulfate (THPS) biocides that have a relatively benign toxicology.

THPS is considered to be a "magic bullet" for controlling bacteria in water, because it is more toxic to the bacteria than it is to humans or other organisms. Once it is discharged it loses its antimicrobial activity and degrades to a virtually nontoxic compound. It is estimated that more than 42 million lbs./year of this nonoxidizing biocide will be used for water treatment within the U.S. alone.

Small Business Award – Legacy Systems of Fremont, CA received the award for developing a chilled ozone process that removes photoresists in the flat-panel display and the micromachining industries. In Legacy's award-winning process only water and oxygen are used to remove organic material on silicon wafers to expose the circuitry etched on the wafer. This eliminates the use of concentrated solutions containing sulfuric acid, peroxides, and other chemicals. The alternative method reduces water consumption by more than 3.4 million gallons/year/silicone wafer wet station and more than five million gallons/year/flat-panel display station. There are thousands of these stations worldwide.

Academic Award – Joseph M. De Simone, a chemistry professor at the University of North Carolina in Chapel Hill received this award for developing surfactants that can be used with a carbon dioxide solvent. These surfactants will permit carbon dioxide to be used as a cleaning agent in applications where 30 billion lbs. of volatile organic compounds (VOCs) and chlorofluorocarbons (CFCs) are currently being used.

---

solid wastes. From 1987 to 1990, Chevron reduced its hazardous waste generation by 60 percent and saved more than $10 million in disposal costs.

Dow Chemicals **Waste Reduction Always Pays (WRAP)** program was started in 1986 with two primary goals: source reduction and onsite recycling. The overall release of SARA Title 313 reportable compounds was down from 12,252 tons in 1987 to 9,659 tons in 1989, a 21 percent reduction. Off site transfers of wastes were reduced from 2,855 tons in 1987 to 2,422 tons in 1989, a reduction of 15 percent. Air emissions for 1989 showed a 54 percent decrease over the 1984 figures.

In 1988 the Chemical Manufacturers Association (CMA) initiated **Responsible Care®**. The intent of Responsible Care® is for the chemical industry to promote continuous improvement in health, safety, and environmental quality. It was not intended to be a public relations program, a quick fix, or an overnight cure, but rather an ongoing process. The program's stated primary goals are to improve chemical processes; enhance practices and procedures; reduce every kind of waste, accident, incident, and emission; provide reliable communication and dialogue; and heighten public scrutiny and input.

Box 7-12 ■ ISO 14000 – Environmental Management Standard

In September 1996 the first **International Organization for Standardization (ISO)** set of environmental management standards (EMS) – collectively known as **ISO 14000** – was passed. The primary goal of the ISO 14000 standard is to help any organization address environmental issues in a systematic way, thereby improving its environmental performance. ISO 14000 is intended to provide organizations with a common and systematic environmental management system. It is important to understand that ISO's standard is a process and not a performance standard. This means that ISO 14000 does not tell companies what environmental performance they must achieve, but rather offers companies a system that will help them achieve their own goals.

A basic assumption in ISO 14000 is that an effective environmental management system can help a company manage, measure, and improve the environmental aspects of its operations. This, in turn, can lead to more effective compliance with mandatory and voluntary environmental requirements. It can also help companies infuse effective environmental management practices into their overall business operations. The ISO 14000 standard is based on a simple equation:

Better environmental management = Better environmental performance, increased efficiency and a greater return on investment.

The work of the ISO Technical Committee 207 (ISO 14001, 14004, 14010, 14011 and 14012) includes standards in the following areas: 1) environmental management systems (EMS), 2) environmental auditing (EA), 3) environmental performance evaluation (EPA), 4) environmental labeling life cycle assessment (LCA), 5) environmental aspects in product standards (EAPS), and 6) terms and definitions. Three of the standards (EMS, EA, and EPE) are used to evaluate the organization. The EMS standards provide the basic framework for the management system. Environmental auditing and environmental performance evaluation are management tools that play a critical role in the successful implementation of the environmental management system. The other two standards (LCA and EAPS) address those aspects in product standards that play important roles in the environmental arena. The emphasis, however, remains on the evaluation and analysis of product and process characteristics.

The process of obtaining ISO 14000 certification is not a fast or inexpensive one. It has been estimated that the registration process could take as long as one year and cost several thousand dollars.

Green products and companies, however, are taking a closer look at all environmental aspects of product design, production, packaging, distribution, and disposal. Foreign customers may soon require U.S. suppliers to be ISO 14000 registered. ISO 14000 may, indeed, become a company's passport to the global marketplace.

Governments worldwide are looking at what role ISO 14000 can play in their regulatory systems. EPA is considering ways to incorporate ISO 14000 into its Common Sense Initiative, which aims at developing regulations that address multi-media permitting for specific industry sectors. ISO 14000 may also become a benchmark for EPA's environmental leadership program where it will waive routine inspections if companies can prove they have a program in place to meet environmental standards.

To the extent that non-compliance with regulations is caused by system deficiencies, implementing ISO 14000 standards can reduce the number of non-compliance fines and increase overall operating efficiency. It is important to remember that pollution prevention cuts costs by reducing the use of materials and energy, while end-of-pipe controls save money only by avoiding fines and penalties due to non-compliance.

There are about 185 CMA member companies that support Responsible Care®. This group of companies represents 90 percent of the U.S. industrial chemical productive capacity. Responsible Care® is built around a set of Guiding Principles and six Codes of Management Practices to put it into action. One of the six Codes is pollution prevention, which is designed to improve the chemical industry's ability to protect people and the environment by generating less waste and minimizing emissions. The Code seeks voluntary, ongoing, long-term reductions in all pollutants released to the environment; progress is measured annually. Their efforts have been very effective. As an example, the eleven members of the Maryland Chemical Industry Council achieved – between 1988 and 1995 – a 74 percent reduction in pollutants released.

Checking Your Understanding

1. Explain the 33/50 Program and its accomplishments.

2. Describe the goals of each of the Green Star programs.

3. Describe the goals of the Green Chemistry Challenge.

7-6 P2 and Minimization of Waste Information and Resources

Successful pollution prevention programs make use of ideas and techniques developed through interactions with other companies, federal agencies, trade associations, technical assistance programs, or professional consultants. Sharing successful pollution prevention techniques and case studies greatly increases the acceptance of ideas that reduce waste and improve financial statements. A variety of pollution prevention assistance and information dissemination mechanisms provide resources on pollution prevention techniques and strategies.

U.S. Environmental Protection Agency, Office of Pollution Prevention and Toxics, Design for the Environmental (DfE) Program

Businesses operating in the 1990s face a variety of competing demands – keeping costs low and quality high, staying competitive in a global marketplace, and meeting consumer preferences for more environmentally friendly products. DfE is a program that helps companies develop an effective strategy for meeting these demands. Building on the DfE concept pioneered by industry, the EPA's DfE program helps business incorporate environmental considerations into the design and redesign of products, processes, and technical and management systems.

Through the DfE program, EPA develops strategies and provides businesses with information to help them make environmentally informed choices. The DfE program forms voluntary partnerships of industry, public interest groups, universities, community colleges, research institutions, and other government agencies to develop environmentally friendly alternatives to existing products and processes.

DfE is working with several industries to identify cost-effective pollution prevention strategies that reduce risks to workers and the environment. DfE helps businesses compare and evaluate the performance, costs, pollution prevention benefits, and human health and environmental risks associated with existing and alternative technologies. The goal of these projects is to encourage businesses to use cleaner processes and technologies and to produce cleaner products. DfE initiatives currently include: printed wiring board, dry cleaning, screen printing, flexography, lithography, and metal finishing projects. DfE is also working with the accounting, insurance, and finance industries to identify and quantify the economic and environmental savings that can be achieved by implementing innovative pollution prevention methodologies in their businesses.

EPA is also collaborating with the U.S. General Services Administration on a long-term project to promote the use of environmentally preferred cleaning products. This effort involves developing standards for cleaning products, performing integrated risk assessments, and evaluating product performance. The project coincides with a federal Executive Order mandating that government agencies use environmentally preferred cleaners. For more information contact Pollution Prevention Information Clearinghouse (PPIC) U.S. EPA, 401 M Street, SW (PM-211A), Washington, D.C. 20460 (202) 260-1023 or e-mail ppic@epamail.epa.gov.

The **Pollution Prevention Information Clearinghouse (PPIC)** is a multimedia clearinghouse of technical, policy, programmatic, legislative, and financial information dedicated to promoting pollution prevention through efficient information transfer. PPIC is composed of four different elements: 1) a repository of technical information, 2) the Pollution Prevention Information Exchange System (PIES), which is an electronic information exchange, 3) a hotline for additional information, and 4) an outreach program. For more information on PPIC contact: Pollution Prevention Division, U.S. EPA, 401 M. Street SW, Washington, D.C. 20460, (202) 382-4418 or e-mail ppic@epamail.epa.gov.

The **National Pollution Prevention Center for Higher Education** – located at the University of Michigan – was established in 1991. This center has developed pollution prevention resource compendiums for undergraduate and graduate engineering, business, and natural resources classes. These materials are available for distribution to colleges and universities nationwide. The Center has focused, to date, on six key areas of pollution prevention: accounting, business law, chemical engineering, environmental studies, industrial ecology, and industrial operations and management.

Plans are underway to develop resource materials for courses in chemistry and architecture. For further information contact: The National Pollution Prevention Center for Higher Education, University of Michigan, Dana Building, 430 E. University, Ann Arbor, Michigan 48109-1115.

The **National Roundtable of Pollution Prevention Programs** holds annual conferences and is the most current and comprehensive source of information on state newsletters and fact sheets. The National Roundtable of Pollution Prevention Programs can be reached at the following address: The National Roundtable of State Pollution Prevention Programs, 218 D Street S.E., Washington, D.C. 20003.

The **Department of Energy (DOE)** produces a newsletter on advances in pollution prevention. The DOE Pollution Prevention Advisor Newsletter can be obtained by contacting the Pollution Prevention Advisor Newsletter, 109 South Riverside Drive, Elizabethton, TN 37643.

The **Pollution Prevention Publications, Fact Sheets, and Guidance Documents** is a large and rapidly growing body of case materials, fact sheets, resources materials, and other industry-specific information that has been published and summarized. Since most state environmental regulatory programs produce fact sheets and newsletters, this is a valuable and rapidly growing resource.

EPA's Pollution Prevention Research Branch publishes a list of documents available from the Center of Environmental Research Information. These documents include: Guidance Documents, Waste Minimization Opportunity Assessment Project Summaries, Case Studies, Project Summaries, and Environmental Research Briefs.

The **United Nations Environmental Program (UNEP)** – in cooperation with the United Nations Industrial Development Office (UNIDO) – has developed an audit and waste reduction manual to assist in the identification of industrial emission and waste sources. The *Audit and Reduction Manual for Industrial Emissions and Wastes* can be obtained

from UNEP, 1889 F Street, NW, Washington, D.C. 20006. In addition, the UNEP has published information on case studies on pollution prevention related to clean production from around the world.

The **Pollution Prevention Review** is a quarterly publication that focuses entirely on pollution prevention articles that include case studies and regulatory and compliance reviews. These can be ordered by contacting; Pollution Prevention Review, Executive Enterprises Publications Co., Inc., 22 West 21st Street, New York, NY 10010-6990.

Computer Software, Databases, and the Internet

Computer software and databases have become increasingly popular for accessing pollution prevention information and resources. Several of the larger and publicly available databases are described below.

The **Pollution Prevention Information Exchange System (PIES)** is a computerized information network for the Pollution Prevention Information Clearinghouse (PPIC). PIES helps organizations 1) identify grant programs and technical pollution prevention documents, 2) contact experts in the field, and 3) identify other pollution prevention resources. For more information on PIES contact: U.S. EPA, Office of Environmental Engineering and Technology Demonstration, 401 M. Street SW, Washington, D.C. 20460, (202) 475-7161.

The **Solvent Alternative Guide (SAGE)** was designed to provide recommendations for solvent replacements in cleaning and degreasing operations. The system leads the user through a question-and-answer session and, based on the responses, provides a brief description of the solvent alternatives. For more information on SAGE contact: Air and Energy Engineering Laboratory, U.S. EPA, Research Triangle Park, NC 27711 (919) 541-7633.

The **Solid Waste Information Clearinghouse (SWICH)** was established by EPA to increase the availability of information on solid waste resource reduction, recycling, composting, and other education and training issues. For more information on

SWICH contact: The Solid Waste Information Clearinghouse, Box 7219, Silver Spring, MD 20910, (800) 67-SWICH. Twenty-four hour modem access is also available to SWICH EBB at (301) 585-0204, set software to 0 parity, 8 data bits and 1 stop bit.

The **Hazardous Solvent Substitution Data System (HSSDS)** was developed by the Idaho National Engineering Laboratory (INEL). The system ranks more environment-friendly solvent alternatives according to specific industry needs (i.e., a degreaser for stainless steel that is not a chlorinated solvent or aerosol). The advantage of this data system is that it has substantial analytical data backing its development and it is accessible on the Internet. For more information on the HSSDS Contact: Idaho National Engineering Laboratory, P.O. Box 1624, Mailstop 1604, Idaho Falls, ID 83415-1604, (208) 526-6956, or http://es.inel.gov/sses/hssdstel.html.

Pollution Prevention World Wide Web Sites (P2 WWW)

The Internet is rapidly becoming littered with a profusion of P-related Web sites. A simple key word search on the Web for "pollution prevention" will net hundreds of hits. In addition to many state and regional plans, case studies, and reports, EPA also maintains a vast number of P2-related Web sites. To narrow the search, key words such as "pollution prevention case studies" or "pollution prevention printed circuits" may need to be used. For EPA Gopher, use gopher.epa.gov or for University of Virginia's Division of Recoverable and Disposable Resources EcoGopher, use http://ecosys.drdr.virginia.edu.

Other information sources such as **U.S. EPA Online Library System (OLS)** may prove to be helpful. This site is a computerized list of bibliographic citations compiled by the EPA library network. Since the citations come from many different sources, they may not be available in all EPA libraries. For more information, contact U.S. EPA (3405R), 401 M Street SW, Washington, DC 20460. A complete directory of pollution prevention databases can be obtained through http://www.medaccess.com/pollution/ppd_15.htm or http://www.gnet.org/gnet/gov/usgov/epa/ppdir.htm.

Checking Your Understanding

1. What is the major function of the DfE program?

2. List at least five sources of information on minimization of waste.

Summary

The Federal Pollution Prevention Act was signed into law in November 1990 and established pollution prevention as a national objective. The Act notes that there are significant opportunities for industry to reduce or prevent pollution at the source through cost-effective changes in production, operation, and raw material use. This legislation established a hierarchy of environmental protection measures, declaring that the most desirable means of pollution abatement are prevention or reduction. Pollution that cannot be prevented should be recycled in an environmentally safe manner. When this is not possible, pollution should be treated in the most environmentally safe manner. Disposal or other releases to the environment should be used only as a last resort.

The Act defines pollution prevention as source reduction. The list of practices that may be employed includes the following: equipment modernization and modifications; improved maintenance; improved operator practices; inventory control; process and/or product modifications; substitution; and in-process recycling. The key elements of a P2 program require the following: top management support, implementation of motivational factors, waste characterization, waste tracking, cost allocation, program evaluation, and pollution prevention benefits.

EPA has developed several manuals to assist in conducting P2 assessments. The Pollution Prevention Opportunity Assessment is a systematic procedure for identifying ways to reduce or eliminate waste. The four phases in the assessment are: 1) planning and organization, 2) assessment, 3) feasibility analysis, and 4) implementation. During the assessment phase, an attempt must be made to identify the minimization of waste options. The assessment phase should involve the following steps: collect process and facility data, prioritize and select assessment targets, select assessment team, review data and site inspect, generate options, and screen and select options for feasibility study. The assessment and feasibility analysis then provides the basis for implementation of the plan.

The payoff for implementing a P2 program can be significant, including reduced liability; more efficient use of resources; reduced treatment and disposal costs; lower environmental impacts; reduced regulatory involvement, including reduced fines and penalties; money savings; and increased public relations.

Pollution prevention and minimization of waste opportunities can be broken into two major categories: source reduction and recycling. A variety of ways have been developed to accomplish P2 and minimization of waste goals; among these are: purchasing and inventory controls, improved housekeeping, production/process modifications, product substitution and reformulation, waste segregation, new use and reuse, and onsite and off site recycling. Source reduction and recycling should always be considered before treatment and/or disposal.

A variety of EPA sponsored volunteer programs have been implemented to encourage pollution prevention and minimization of waste. Some of the EPA-managed programs include: 33/50 Program, Green Lights Program, Design for the Environment, Green Chemistry, Green Chemistry Challenge, and the Energy Star Programs. In addition, a number of businesses and industries have developed their own P2 programs. Perhaps, the most widespread of these is the Chemical Manufacturers Association's Responsible Care®.

A wide array of P2 and minimization of waste information resources are now available. For example, EPA's Office of Pollution Prevention and Toxics, Design for the Environment (DfE) is working with a number of industries to identify cost-effective P2 strategies. DfE provides a variety of outreach tools, including fact sheets, bulletins, case studies, software, videos, and training materials. Searches of the World Wide Web using key words like "pollution prevention" or "pollution prevention case studies" will typically result in hundreds of responses.

Critical Thinking Questions

1. Analyze and put into your own words the elements in a pollution prevention program.

2. As defined by the PPA, contrast the terms pollution prevention and waste minimization.

3. Compare the cost of implementing a P2 program with the possible monetary and liability benefits.

Checking Your Knowledge

1. Waste minimization is specifically mandated by
   a) Hazardous Materials Transportation Act
   b) Federal Insecticide, Fungicide and Rodenticide Act
   c) Hazardous and Solid Waste Amendment
   d) Superfund Amendments and Reauthorization Act

2. Which of the following is the correct sequence in the Hazardous Waste Management Hierarchy? (Assume that left-to-right indicates decreasing desirability from an environmental protection point of view.)
   a) Recycling > Treatment > Disposal > Source Reduction
   b) Source Reduction > Recycling > Disposal > Treatment
   c) Source Reduction > Treatment > Recycling > Disposal
   d) Source Reduction > Recycling > Treatment > Disposal

3. Which two terms constitute minimization of waste?
   a) Source reduction and treatment
   b) Source reduction and in-process recycling
   c) Treatment and disposal
   d) In-process recycling and treatment

4. EPA defines pollution prevention as
   a) Source reduction and recycling
   b) Source reduction, recycling, and treatment
   c) Source reduction
   d) Any activity that reduces the volume and/or toxicity of waste streams

5. Source reduction strategies using changes in technology would include all of the following except
   a) Process changes
   b) Additional automation
   c) Off site recycling
   d) Equipment, piping, and layout modifications

6. Where must generators certify that they have a waste minimization program in place?
   a) On the Uniform Hazardous Waste Manifest
   b) On SARA Title III, Section 313 forms
   c) In corporate policy statements
   d) In biennial reports

8

PRESERVING THE LEGACY

Life Cycle Design for General Manufacturing

by Jack Cavanaugh

Chapter Objectives

Upon completing this chapter, the student will be able to:

1. **Describe** the concept of life cycle design and its importance in environmentally conscious manufacturing.

2. **Differentiate** between life cycle design and design for the environment.

3. **Identify** the critical elements of a product's life cycle and the potential impact of each element on the environment.

4. **Explain** how life cycle design can minimize the total risks and impacts to the environment over the life of the product.

5. **Explain** how environmentally conscious manufacturing fits within the context of total product system requirements.

6. **Describe** the elements of a program for monitoring progress toward reaching goals associated with environmentally conscious manufacturing.

Chapter Sections

8-1 Introduction

Life cycle design is an approach to manufacturing that incorporates into a product's design all decisions related to its manufacture, distribution, use, maintenance, recovery, and disposal. The underlying principle of life cycle design is that the way a product is manufactured affects its quality, cost, usefulness, and its impact on the environment.

Life cycle design recognizes that most environmental consequences result from design decisions that are made long before manufacture. Its key goal is to reduce the total environmental burden of product manufacture, use, and disposal, including resource depletion and ecological and human health effects. It seeks to achieve **sustainable development** by promoting sustainable resource use as well as to promote 0 rather than end-of-pipe control. Figure 8-1 highlights key considerations.

Life cycle design is the most comprehensive approach available for incorporating environmental considerations into product development. It places heavy emphasis on the total impact of a product through its life cycle, but is not limited to environmental considerations. It also evaluates all processes from raw materials acquisition through distribution, use, service, and **end-of-life management**. In analyzing a **product life cycle**, this design approach considers requirements related to the environment, product cost and performance, as well as cultural and legal issues.

This chapter discusses the general concept of life cycle design and emphasizes the importance of decisions that relate to environmentally conscious

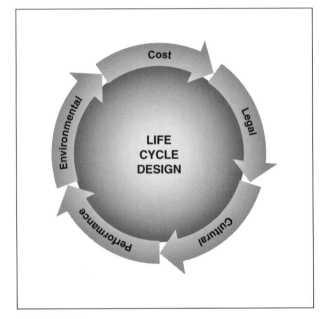

Figure 8-1: Key life cycle design considerations.

manufacturing. These decisions involve looking at the entire product life cycle and include: 1) the processes and materials used in manufacturing; 2) the environmental impact of using the product; 3) identifying how the product – or its components – can be reused, recycled, repaired, or refurbished; and 4) the impact of disposal of the product or components when they are no longer useful, also known as end-of-life management.

8-2 Life Cycle Material Processes

Selecting the material from which a product will be manufactured requires many decisions. Examples will be provided for how environmentally conscious design decisions can both impact product usefulness and provide flexibility in end-of-life management. Figure 8-2 illustrates the processes that must be considered in doing a life cycle analysis of a product. The processes include production, the use of the product during its **useful life**, and its end-of-life management.

Material Selection and Refinement

In a life cycle design, there are cost, performance, cultural, legal, and environmental trade-offs when determining the nature of materials to be used. The use of recycled materials may offer environmental and cost advantages over the use of virgin materials or newly produced components. Decisions about what materials will be used can determine the extent to which the product can be recycled, remanufactured, or reused.

Refinement

The removal of unwanted components of the material as it is adapted for use is called **refinement**. For example, aluminum is produced through a refinement process that involves removing the oxygen from bauxite. In refining a product, both the waste generated and the cost associated with managing the waste must be considered. Recycling the waste is better for the environment than having to dispose of it. In most cases, a design that includes recycling of wastes also results in lower costs.

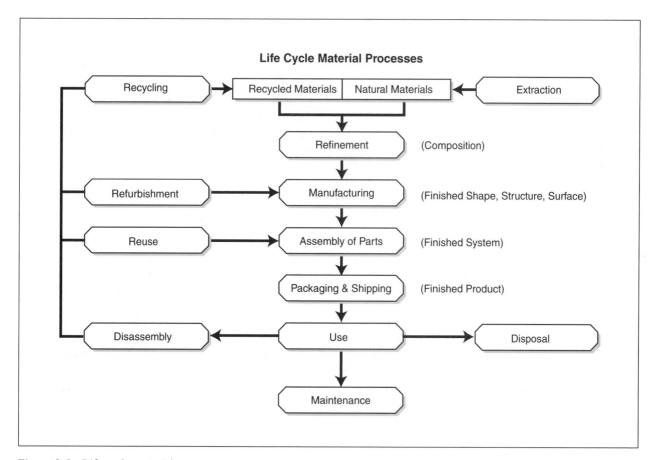

Figure 8-2: Life cycle material processes.

Manufacturing Processes

Manufacturing can include processes such as **shapemaking**, **structural treatment**, and **surfacing**. Each of these processes has environmental implications. Shapemaking processes are those that form materials into the shape of a product. Examples of shapemaking processes are machining, grinding, and extruding. These processes often result in scrap materials, which can either be disposed of or recycled. Ideally, the shapemaking processes result in the manufacture of a part without any waste. While this ideal condition is difficult to meet, the closer the manufacturing process comes to it, the lower the material costs will be because of reduced waste. In addition to waste minimization advantages, the costs of waste disposal and labor are reduced.

Structural treatment is used to enhance one or more of a product's characteristics. Heat treatment of steel is an example of structural treatment that increases the strength – and therefore the life – of the steel. Structural treatment processes can also affect the potential for refurbishment or reuse of the product. Semi-solid metalworking (SSM) is a technology that combines shapemaking and structural treatment to achieve life cycle advantages. With SSM technology, the metals are heated to a semi-solid state, then processed through a die casting machine. The parts produced by SSM have higher structural integrity, quality, and soundness than cast parts. They can be produced at a lower cost since there are fewer manufacturing steps and – since using SSM results in a near **net shape** part – there is little waste produced.

Surfacing is used to protect the product from corrosion or wear, or to enhance the appearance of the product. Some surfacing processes such as painting and plating add layers of material to the product. Other surfacing operations, like chemical milling, remove small amounts of surface material. Surfacing processes have the potential for significant environmental impact. Surface treatment can also impact the opportunity for disassembly or remanufacture of the product following its original useful life.

Surfacing processes, such as painting and plating, typically use large amounts of hazardous materials that impact air and water quality, generating large quantities of hazardous waste. Painting processes can release high levels of volatile organic compounds (VOCs) into the air. Painting processes with poor transfer efficiency generate large amounts of hazardous waste. Traditional spray paint operations have **transfer efficiencies** in the 20 to 40 percent range. This means that only a small amount of the paint actually adheres to the surface of the part being painted. The other part is lost as overspray – a hazardous waste that must be disposed of in compliance with hazardous waste regulations – or is dispersed into the atmosphere as a VOC that contributes to smog formation. Advanced technology to improve transfer efficiency and reduce VOC emissions can help the manufacturer meet cost and quality requirements, while providing significant environmental, health, and safety advantages.

An alternative to spray painting is powder coating. The powder used in a powder coating process contains no solvents, and therefore emits no VOCs. Some powder coating systems can achieve transfer efficiencies that approach 100 percent, which greatly reduces the generation of hazardous waste. Powder coating has some limitations in its application to certain large parts and those with complex geometry, but it does offer significant life cycle advantages in the areas it can be applied.

Many plating processes pose serious health and safety risks. The metals used in plating operations – such as chromium and cadmium – are **carcinogenic**, but they are still used because of particular characteristics they impart to surfaces that are hard to duplicate. The plating process itself may be redesigned to make it more efficient and to extend the life of the plating bath. Recycling or recovering metals in the bath to minimize the amount of hazardous waste generated is another alternative.

Assembly of Parts

Assembly results in the joining of all mechanical parts to produce a final assembly. Some examples of assembly are mechanical fastening, adhesive bonding, and welding. Efficient assembly will minimize the use of hazardous materials. The assembly method should also take into account the ease of disassembly when the product is to be discarded. This enhances the opportunity for recycling or reuse. Design for disassembly is an end-of-life management issue and will be covered in more detail in a later section of this chapter.

Packaging and Shipping

Protection and preservation of the product during distribution, storage, and marketing are key design elements related to packaging. Life cycle design will consider functional requirements of packaging, but will also weigh the cost of managing the waste generated from inefficient packaging against short-term functionality of the packaging. It will consider the potential to recycle or reuse packaging materials to reduce costs and minimize environmental effects.

Use and Maintenance

The environmental impact of the use of a product often relates to energy sources and efficiency. This is translated into the pollution generated during the operation of a piece of equipment. Automobiles, lawn mowers, power tools, and many types of industrial equipment are examples of products that can have a significant negative environmental effect.

The importance of maintaining or overhauling a product must also be considered during the design phase. The type of paint or the metals used in the surface finish of a product could affect the environment when stripping, cleaning, and resurfacing the product is required to extend its useful life. The removal and disposal of chromium metal, for example, can pose a serious environmental problem. Cleaning agents used in repair and maintenance operations can also be hazardous.

End-of-life Management

When a product is used up or is no longer usable, something must be done with it and with its components. For most products, disposal is thought to be the easiest, most cost effective, and preferred end-of-life alternative. Disposal costs continue to rise for a number of reasons, many of which relate to environmental, health, and safety issues. As these costs increase, alternatives to disposal become more attractive.

Design for disassembly and selecting materials conducive to recycling, reuse, or **remanufacturing**, are all integral elements of life cycle design. Through life cycle design, we can consider end of useful life alternatives that are more environmentally attractive. Disassembling the product into individual parts for further use is one of these alternatives. Careful design can make reuse or recycling easier by constructing parts from one kind of material. Easy disassembly can allow some parts to be reused in new units or as repair parts for older units. Others can be refurbished for further use by being taken back through one or more of the manufacturing phases. Still others can be used as recycled feedstock for making new parts.

Individual companies – and in some cases entire industries – are taking a proactive approach to dealing with life cycle product design. AT&T Bell Labs, for instance, is using life cycle design to achieve greater efficiency in the use of materials, improved recyclability, and reduced toxic constituents in its business telephone terminals. The United States automotive industry is actively exploring earth-friendly products and processes through life cycle design approaches to manufacturing. Many governments are taking more aggressive action to force industry to incorporate more environmentally conscious approaches to production. In Germany, for example, manufacturers are required to produce reusable or recyclable cars and components and to accept their products back for recycling.

Checking Your Understanding

1. What is the key goal of life cycle design?

2. What are five trade-offs to be considered in the development of life cycle design?

3. What are the three processes involved in manufacturing?

4. Describe semi-solid metalworking and explain its advantages.

8-3 Making Manufacturing Process Changes

In this section we examine two approaches for infusing life cycle design into the manufacturing processes. The first is a multi-criteria matrix used by AT&T in the design of its new business telephone terminals. It provides a comprehensive, three-dimensional way to evaluate the influence of different design elements. The second is a risk-based approach that can focus on key risk factors from a life cycle perspective.

Multi-Criteria Matrices for Decision Making

Figure 8-3 shows the criteria used to make life cycle design decisions. These criteria take into account 1) product system components, 2) the three stages of a product life cycle, and 3) the societal requirements for each stage of the life cycle.

1. The product system components include 1) the product; 2) the processes used in all stages of the product life cycle, such as the equipment used to paint the product during manufacture or overhaul; and 3) the distribution components used in handling, transporting, storing, and marketing the product.

2. The three stages of a product's life cycle are 1) manufacture, 2) use/service, and 3) end-of-life management. The manufacture stage includes the time involved in the production processes. The use/service stage is the period when the product is either performing its function or being maintained so it can continue to operate properly. The end-of-life management stage starts when the product is no longer functional and something must be done with the product or its components.

3. Societal requirements that must be met by the product system components for each stage of the life cycle fit into the following five categories: environmental, performance, legal, cost, and cultural.

Each design decision would affect at least one subcategory of each of the three major criteria mentioned. A life cycle design decision, for example, could affect any one, or all of the system components. The type of materials used would affect the product quality, the process options available, and the distribution methodology used when handling and storing the product. For example, a decision to use titanium or graphite when making a set of golf clubs affects the product quality, the process options available in manufacturing and, potentially, the packaging used to distribute the clubs to retail outlets.

Some design decisions are only applicable to the manufacturing stage. Others come into play during the product's life, and may include its serviceability. Until now, the most frequently ignored factor was the end-of-life management stage. The implications of treating and disposing of products at the end of their useful lives makes the end-of-life management stage a crucial design consideration.

Traditional design models only address product cost and performance requirements. However, a life cycle design model places comparable weight on the environmental, legal, and cultural requirements in addition to cost and performance. Using the multi-criteria matrix, the implication of each of these requirements for each of the system components and during each of the life cycle stages helps to improve the decision-making process.

The cost requirements, for example, include the cost of the product packaging materials (Product), the cost of the process used to package the product

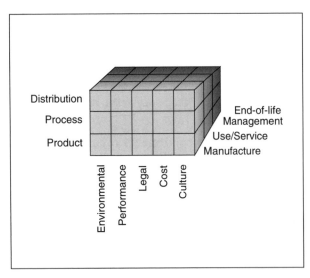

Figure 8-3: Three-dimensional matrices for making life cycle design decisions.

during manufacture (Process), and the cost of distributing the product to its ultimate user (Distribution). These requirements would fit in the cube extracted from the matrix illustrated in Figure 8-4. Each cube in the matrix has potential for several decisions that could be important in the life cycle of a product line.

Performance requirements are affected when making decisions concerning, for example, the materials used in the manufacturing stage. They determine the product utility during its useful life stage and, ultimately, come into play at the end of the product's life. Environmental requirements continue to become more important with increasing consciousness of the impact the products, processes, and distribution systems have on our health and safety. Decisions related to product, process, and distribution during each stage of the product life cycle impact environmental requirements.

Legal requirements are often related to environmental laws such as the Clean Air Act or the Clean Water Act. The many laws enacted in recent years may limit product, process, and distribution options. These laws normally impact manufacturing stage options; they sometimes affect the useful life stage, and almost always influence the end-of-life stage alternatives.

Cultural requirements are not as encompassing as the other requirements. They are important in scenarios where perceptions and traditions have important influence over purchase, use, and disposition of products. The processes used in manufacture and the distribution components may sometimes affect cultural requirements, but the useful life stage is most frequently impacted by cultural requirements.

The multi-criteria matrix decision-making approach to design is a complex, but very comprehensive life cycle design methodology. The use of this methodology can play a major role in reducing risk and improving quality.

Risk Considerations in Decision Making

Most risks encountered by manufacturers fall into four categories: technical, financial, production, and compliance. Each of these risk categories are defined and examined in the following sections.

Once the manufacturers have identified which factors to consider in the life cycle design, they must decide how to address those factors. The manufacturer's goal is to produce the best product while meeting the requirements related to cost to the environment and its own production goals. Improving performance using a life cycle design approach will often require changing something in the operation. The manufacturer takes risks when it changes something related to materials, processes, or manufacturing logistics. But from a business perspective, the company cannot afford to take unreasonable risks, regardless of the environmental benefits derived from making changes. For this reason, life cycle risk reduction must be a key element of the overall strategy.

Technical Risk

Technical risks are those that relate to the way technology is being applied to solve a problem. Life cycle design often results in altering technologies and manufacturing processes. Changing a material or process might make complete sense from an environmental perspective, but if making this change will impact product performance, then there is a risk it will not be accepted by the customer. The manufacturer must evaluate the technical risk and minimize the effect on customer acceptance of the product. In situations where there is more than one technological alternative to meet the technical requirement, the risk of not selecting the best alternative can be costly. Technology verification using

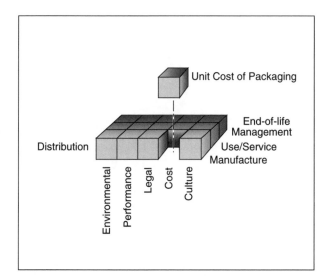

Figure 8-4: Unit cost of packaging cube from the distribution matrix.

modeling, simulation, bench-scale demonstration, or full-scale demonstration requires an investment, but it could lower risk and result in lower costs in the long run.

Financial Risk

Financial risks are those that relate to whether or not the costs of making a change will result in an appropriate return on investment. The return on investment can be based on a specified payback period, cash flow, or an impact on the long-term profitability of the company. By reducing the overall impact on the environment, life cycle design can have a positive financial impact. The time period for realizing the economic benefits must be acceptable to those with financial interest in the manufacturing operation.

Production Risk

Production risks are those that relate to the ability to maintain production rates or minimize production losses from making a change. When making a change in operations, there is a transition period where production stops or slows. The impact of that stoppage, or delay, poses a risk to the manufacturer. In making a change, the manufacturer encounters the risk of not being able to maintain the production that is required to meet customer demand. The ability to validate new methods of production in a non-threatening environment can help minimize risks related to production capacity. A life cycle design approach considers production risks affected by making process changes and evaluates alternatives to minimize the impact of these changes.

Compliance Risk

The manufacturer's ability to comply with laws and regulations become **compliance risks**. In a life cycle design environment, evaluating compliance risks takes into consideration future regulatory direction as well as the current regulations. There is no crystal ball to predict with 100 percent accuracy what the regulatory changes will be, but the use of a multi-criteria matrix will help minimize the impact of risks associated with regulatory changes in both the short- and the long-term.

Checking Your Understanding

1. Describe two approaches that can be used to infuse life cycle design changes into manufacturing processes.

2. Name the three stages in a product's life cycle.

3. What are the five requirements that must be met by the product system components for each stage of the life cycle?

4. Name four risk categories encountered by manufacturers and briefly describe each.

8-4 Environmentally Conscious Manufacturing (ECM) Strategies

A manufacturer must develop a strategy for addressing environmental, performance, legal, cost, and cultural requirements. This strategy must focus on minimizing the risks associated with making technical, financial, production, and compliance decisions. In this section we will identify the elements of an **environmentally conscious manufacturing (ECM)** strategy, give examples of approaches for achieving ECM goals, and discuss ways to measure the impact of life cycle design.

Elements of an Environmentally Conscious Manufacturing Strategy

One way to organize an ECM strategy is by developing environmentally focused objectives that include reduction, remanufacture, and reuse.

Reduction

In selecting materials and processing them during the manufacturing stage, there are opportunities to reduce both the quantity of materials used and the amount of waste generated. The distribution stage offers an additional opportunity to reduce materials used and waste generated. Manufacturers can reduce waste streams through a number of product, process, and technology changes. Source reduction activities focus on:

Material Input Changes – For example, using aqueous cleaners rather than ones with high levels of chlorofluorocarbons (CFCs).

Operational Improvements that Lead to Loss Prevention – For example, improving pretreatment processes to require less stripping, cleaning, and resurfacing of parts.

Production Process Changes – For example, changing from conventional spray painting to a powder coating paint system.

Product Reformulation – For example, increasing the ratio of solids to solvents when spray painting to reduce the amount of volatile organic compounds released into the atmosphere.

Inventory Control – For example, using a first-in, first-out method of managing hazardous materials to prevent them from exceeding their normal shelf life and becoming hazardous waste.

Administrative and Organizational Activities – For example, training technicians on painting systems to improve transfer efficiency.

Remanufacturing

Remanufacturing refers to repair, rework, or refurbishment of equipment. It has obvious advantages that extend the life of a product, ultimately reducing the consumption of raw materials and pollutants generated in manufacturing. Life cycle design is critical to making remanufacturing a viable option. Concepts such as design for disassembly can help ensure that remanufacturing is cost effective. Surface finishing decisions in the original manufacturing processes, for instance, can have environmental consequences in the remanufacturing stages. When a life cycle design approach is taken, the surface coating and the impact of the coating during remanufacture are evaluated.

Life cycle design evaluates the feasibility of remanufacturing. Remanufacturing is a viable alternative when: 1) product and process technology does not change; 2) a product fails functionally, but is not consumed or has a restorable core; 3) the core is capable of being disassembled and restored to its original condition; 4) the product is factory built rather than field assembled; and 5) the recoverable value is high relative to its market value and original cost. Items like an aircraft propulsion system, an automobile engine, and complex pieces of industrial equipment are good examples of products that would be excellent candidates for remanufacturing.

Recycle/Reuse

In an ideal world, all components of all products would be recycled or reused. In the real world, however, only a small fraction of product components can be recycled or reused cost effectively. Life cycle design cannot create a demand for recycled materials and components, but it can have a major impact on the cost effectiveness of recycling and reuse. The choice of raw materials in initial manufacture, the surface finishing materials, and the structural treatment process used all influence whether a component can be recycled or reused. A design for disassembly can also have an effect on the feasibility of recycle or reuse.

Approach to Achieving ECM Goals

Concurrent design, life cycle design, and **Total Quality Management (TQM)** are related practices that will help the manufacturer achieve ECM goals. Concurrent design is a technique that integrates product and process design by having everyone responsible for various stages of a product's life cycle participate in the project from the beginning. This includes the engineers, the procurement and production personnel as well as personnel responsible for packaging, storing, and distributing products. Life cycle design brings a new dimension to traditional design by including into the manufacture process individuals with knowledge of the costs and maintenance of the product through its useful life span as well as individuals with end-of-life management expertise such as regulators, landfill operators, and treatment, storage, and disposal facility operators. Participants from the following constituencies are included in either a concurrent design or life cycle design:

—Accounting

—Community

—Distribution/Packaging

—Environment, health and safety

—Government regulators

—Industrial designing

—Legal

—Management

—Outreach and communication

—Marketing and Sales

—Process Engineering

—Procurement

—Production

—Research and Development

—Service

—Suppliers

—Waste Management

TQM is a broad approach to ensuring quality in an operation. Concurrent design is often a key element of a TQM process. Likewise, life cycle design and TQM are closely linked with their focus on long-term goals. Key elements that TQM shares with concurrent design and life cycle design are: focus on customers, multi-disciplinary teamwork, and cooperation with suppliers. The framework of life cycle design also incorporates societal and environmental requirements.

Measuring the Impact of Life Cycle Design

The ability to measure the impact life cycle design has on the environmentally conscious manufacturing (ECM) posture of an organization requires:

1. Specific criteria that help the organization focus on areas where it should improve in order to become more environmentally conscious.

2. An accounting process to help the organization quantify the changes related to environmental consciousness.

3. An environmental management system to help the organization perform auditing functions required to ensure long term success.

The National Center for Manufacturing Sciences (NCMS) has developed a tool that spells out specific criteria to focus the efforts of organizations to assure continuous improvement. The NCMS tool is called Greenscore™, which is a self-assessment tool that evaluates twelve areas that are important to the organization's environmentally conscious manufacturing posture. These include:

1. Policy
2. Management
3. Planning
4. Cost
5. Stakeholders
6. Human Resources
7. Operations and Facilities
8. Supplies
9. Compliance
10. Technology and Research
11. Auditing
12. Measurement of Continuous Improvement

Through a series of questions about each of these twelve categories the manufacturer can score itself to see where it stands. Areas that need improvement are identified to provide the foundation for a continuous improvement program.

The second element of the measurement process is an appropriate accounting system. An accounting system that tracks life cycle costs, not just those associated with direct production costs, is essential in determining the real impact of a life cycle design approach. Reducing costs through life cycle accounting methodology also makes sense from a business perspective. The costs can be effectively evaluated when using a life cycle accounting methodology. The measures we hope to maximize from an environmental perspective in a life cycle accounting methodology include:

— Recycled content
— Recyclability
— Reusability
— Remanufacturability
— Degradability

The goal is to minimize factors such as hazardous/toxic waste content, water pollution, soil pollution, air pollution, and resource/energy usage.

Appropriate life cycle design takes into account not only the environmental consequences of a product's life, but also the costs incurred throughout its life cycle. Changing products, processes, and technology may cost more in the initial manufacture, but can show an overall reduction in costs when viewed from a life cycle cost perspective. Costs related to compliance with government regulations, solid and hazardous waste management as well as occupational health and safety are examples of life cycle costs that may not be included in the cost of manufacturing and distributing a product. Appropriate life cycle design focuses on minimizing these costs while meeting all performance, environmental, legal, and cultural requirements.

The third element of a measurement process consistent with life cycle design is having an **environmental management system (EMS)**. The EMS specified by the International Organization for Standardization (ISO) is a comprehensive means by which an organization can implement an EMS. The ISO 14000 series focuses specifically on environmental management and can be the basis of a process to ensure ECM practices are an integral part of business and operational processes. The ISO 14000 series is a proactive approach to environmental management. It includes many of the same concepts specified in Greenscore™. It has specific sections that focus on life cycle assessment.

Effective evaluation, accounting, and auditing processes are all important to measuring the impact that life cycle design can have on a manufacturing business. A proactive life cycle design approach will almost always have a positive impact on product quality, on the environment and – subsequently – on the competitiveness of the manufacturer.

Checking Your Understanding

1. Name the focus of three environmentally conscious manufacturing strategies.

2. What are six source reduction activities?

3. What are three key elements that TQM shares with concurrent design and life cycle design?

4. What new dimension does life cycle design bring to traditional design considerations?

5. What are the twelve areas assessed by Greenscore™?

Summary

The underlying principle of life cycle design is that the way a product is designed and manufactured has an impact on its quality, cost, usefulness, and environmental effects. One of the key goals of life cycle design is to reduce the total environmental burden of the manufacture, use, and disposal of a product. Life cycle design is an approach to manufacturing that incorporates this goal in all the decisions regarding material selection and refinement, assembly methods, packaging and shipping, use and maintenance, and, most importantly, end-of-life management. In addition to performance, it also considers cultural and legal issues when analyzing a product life cycle.

Two methods have been developed to infuse life cycle design into a manufacturing process: 1) a multi-criteria matrix and 2) a risk-based approach, which places the focus on key risk factors from a life cycle perspective. In the multi-criteria matrix, three stages in a product's life cycle are considered: manufacture, use/service, and end-of-life processes. Also, five requirements must be met by the product system components for each of the stages of the life cycle: environmental, performance, legal, cost, and cultural. The use of the multi-criteria matrix decision-making approach is complex, but provides a very comprehensive life cycle design methodology. In the risk decision-making model, risks fall into four categories: technical, financial, production, and compliance. As decisions are made, each of these risk factors must be considered. One of the ways to approach the posed risks is to adopt an environmentally conscious manufacturing strategy. The environmentally focused objectives in an environmentally conscious manufacturing strategy are reduction, remanufacture, and recycle/reuse. By adding Total Quality Management (TQM) and concurrent design to the mix, the manufacturer is most likely to achieve the environmentally conscious manufacturing goals. TQM is a broad approach to insure quality in an operation. Concurrent design advocates having all individuals responsible for the various stages of a product's life participate in its design from the beginning.

To measure the success that life cycle design has on the environmentally conscious manufacturing posture of an organization requires 1) the identification of specific criteria to focus on areas that need improvement; 2) an accounting process that quantifies the changes; and 3) an environmental management system to help organize audits. A tool called Greenscore" has been developed as a self-assessment tool to evaluate the twelve areas found to be important to an organization's environmentally conscious manufacturing posture.

Critical Thinking Questions

1. Discuss two approaches you could take to improve the transfer efficiency of a painting operation to reduce negative impact on the environment?

2. How could selection of surface coating materials used in the manufacture of a product impact the environment when the product is refurbished?

3. Give an example of a life cycle design decision. How would it impact each dimension of the multi-criteria matrix?

4. Describe an approach that a manufacturer could use to reduce technical risks of making a process change.

5. Give an example of a product that would be a good candidate for remanufacture at the end of its normal life. Why would it be a good candidate?

Hazardous Waste Regulations for the Cleanup Industries

by Sylvia M. Medina and Richard C. Marty

Chapter Objectives

Upon completing this chapter, the student will be able to:

1. **Describe** the types of waste that may be generated during a site cleanup.

2. **Discuss** the management and applicable or relevant and appropriate requirements of a Superfund cleanup action.

3. **Identify** and describe the onsite and off site waste disposition options.

Chapter Sections

9-1 Introduction

The preservation of pristine areas and the remediation of contaminated sites are two goals shared by many industrialized nations. Several regulations have been enacted in the United States to accomplish these goals; however, because they sometimes overlap and are therefore confusing they have resulted in continued and – in some instances – additional environmental harm.

The Comprehensive Environmental Response, Compensation, and Liability Act (CERCLA) and the subsequent Superfund Amendment and Reauthorization Act (SARA) are two examples of legislation with shortcomings. These Acts were intended to fund – through a tax placed on chemical feedstocks – the identification, evaluation, and cleanup of environmentally contaminated sites across the nation. However, many companies perceived the liability provisions of CERCLA to be so burdensome that they elected to develop new sites rather than deal with the complex cleanup requirements of the polluted ones. This lead to a preponderance of heavily polluted abandoned sites across the country, commonly called **Brownfield sites**.

The intent of this chapter is to introduce the complex and often conflicting regulations that are involved in the characterization, cleanup, and management of wastes found on these sites. The discussion will focus on **Superfund** or **National Priority List (NPL)** sites, but many of the underlying problems also affect cleanups of other types of sites, i.e., underground storage tanks regulated under RCRA.

9-2 Wastes

Remediation Wastes

Small remediation projects, such as those involving underground storage tanks, generally do not require a lengthy investigation phase and can often make use of routine cleanup strategies. The phases involved in any site remediation project are as follows: 1) site characterization, 2) technology selection, and 3) cleanup. Each phase generates some wastes, but both the waste type and its volume vary. Site characterization generates **Investigation-derived Waste (IDW)**. This type of waste may include:

— Drill cuttings;

— Soil/mud;

— Decontamination fluids;

— Waste from well installation and monitoring (purge water and well development fluids);

— Spill response materials;

— Samples and analytical residues (which may or may not be in their original state); and

— Personal protective equipment (gloves, tyvek outer suits, respirator cartridges, shoe covers, and related materials).

Wastes generated during the technology selection phase may include:

— Unaltered samples of waste, soils, or water used in treatability studies;

— Secondary wastes of altered waste, soil, or water generated during the treatability studies;

— Contaminated bench-scale equipment;

— Personal protective equipment (similar to that generated during an investigation); and

— Contaminated pilot scale facilities.

Wastes generated during cleanup commonly include:

— Unaltered contaminated soil or groundwater;

— Construction debris generated to build treatment facilities;

— Secondary wastes generated by treatment (sludges, spent granular activated carbon, spent ion exchange resins, contaminant concentrates, etc.);

— Demolition debris (underground storage tanks, contaminated structures, and similar materials);

— Personal protective equipment; and

— Contaminated treatment facilities and equipment.

Throughout the remainder of this chapter, these wastes will be collectively referred to as **remediation wastes**.

RCRA Wastes

A **hazardous waste**, as defined in 40 CFR 260 and 261, is a waste that poses substantial or potential threats to public health or the environment and that meets any of the following criteria:

— It exhibits one or more of the characteristics of a hazardous waste (see characteristic waste definition below);

— It is specifically listed as a hazardous waste by EPA; and/or

— It is generated by the treatment of a hazardous waste, or is contained in a hazardous waste.

Characteristic Wastes

A characteristic waste is defined by RCRA as one that exhibits 1) corrosive, 2) reactive, 3) ignitable, or 4) toxic characteristics. A characteristic waste, therefore, has a physical or chemical property that makes it a hazardous waste. If the characteristic is removed and the underlying hazardous constituents are treated to standards, the waste is no longer considered a hazardous waste.

Listed Wastes

RCRA defines a listed waste in four separate tables in the 40 CFRs. Each type of waste has a code that identifies its source. F-code wastes, for example, are from nonspecific sources; K-code wastes from specific sources; P-code wastes are from commercial chemical products, are intended for discarding, and considered to be acutely hazardous; and U-code wastes are commercial chemical wastes that are simply hazardous.

If it is known by process knowledge or past records that media at a cleanup site contain a listed waste, the waste generated during remediation must also be managed according to RCRA **substantive requirements** (actions that must be performed). If the site investigator does not have process knowledge or prior knowledge regarding its presence, a CERCLA action does not have to assume it is listed waste. If the waste remains onsite, only the substantive RCRA requirements are applicable in **delisting** the waste [40 CFR 260.22(a)(1) and (2)]. If the waste is removed off site, then RCRA **administrative requirements** (paperwork) – including submitting a petition and undergoing the rulemaking process – would be required.

All listed wastes must meet the **Land Disposal Restrictions (LDR)** [40 CFR 268] treatment standards prior to land disposal. Even after the waste meets the treatment standards, it must still be disposed of at a RCRA-regulated facility. A listed waste continues to be regulated as such unless it is delisted in a **Record of Decision (ROD)** or some other delisting process.

Land Disposal Restrictions – The 1984 amendments to RCRA significantly broadened its scope and specifically banned land disposal of listed wastes, unless they met certain treatment standards. As a result, EPA issued the Land Disposal Restriction (LDR). The standards have subsequently been modified and reconciled, but a **restricted waste** may not be disposed of to the land without first meeting the current treatment standards. Wastes exhibiting RCRA characteristics must be treated so they no longer meet the definition of a characteristic hazardous waste in accordance with 40 CFR 261, Subpart C.

Delisting Requirements – A listed waste that has been treated must continue to be regulated as a listed waste unless it has been delisted. A listed waste that meets risk-based contamination levels and no longer presents a threat to the public and the environment may be delisted in the following ways: 1) by treating the waste with the LDR's **Best Demonstrated Available Technology (BDAT)** to reduce the concentration levels as specified in this regulation; 2) by meeting the substantive requirements of delisting specified in 40 CFR 260.22, and 3) by following instructions in A *Guide to Delisting of RCRA Wastes for Superfund Remedial Response* (OSWER Superfund Publication, 9347.3-09FS).

CERCLA waste residuals and secondary wastes containing RCRA hazardous waste contaminants that require delisting must meet requirements defined in the Record of Decision (ROD). This means that the calculation of delisting levels must be met through use of 1) the EPA Maximum Contaminant Level (MCL) model (see 56 FR July 19, 1991; 58 FR December 30, 1991); 2) the *Docket Report on Health-Based Levels and Solubilities Used in the Evaluation of Delisting Petitions Submitted under 40 CFR 260.20*; and 3) *Use of EPA MCL for Delisting*.

Delisting may be granted during the signing of the ROD. During the delisting process, all remediation waste must be stored within the **Area of Contamination (AOC)** or within a RCRA-permitted or interim status facility and must be handled according to the RCRA substantive requirements. Once the waste is delisted, it is no longer a RCRA hazardous waste and can be land disposed of as solid waste.

Newly Identified Wastes

Waste materials identified or listed as hazardous waste by EPA after November 8, 1984 are not subject to the LDR until new rules for each of these wastes are promulgated. These are commonly referred to as **newly identified wastes**. Within six months of identifying a new waste, treatment standards for these wastes must be issued. This triggers LDR requirements unless a variance is established by EPA. This means that a newly identified waste cannot be land disposed of until treatment standards have been established and are met.

Toxicity Characteristic Leaching Procedure – Hazardous waste contaminants with a **Toxicity Characteristic Leaching Procedure (TCLP)** waste code of D018-D043 are included in the list of haz-

Box 9-1 ■ Characteristic or Listed Hazardous Waste?

The regulatory status of a site may depend only on how it became contaminated. For example, if a soil became contaminated with lindane – a pesticide – as the result of its intended use, it would not be subject to regulation as a RCRA hazardous waste, unless extracts from it exceeded the 0.4 mg/liter lindane TCLP limit. Should that limit be exceeded, the soil would be regulated as a RCRA characteristic hazardous waste (D013). Such wastes, which are either not a characteristic hazardous waste or that can be treated to below the characteristic waste threshold, can be disposed of in a RCRA Class D landfill.

If the lindane contaminated soil were the result of a spill during mixing, or if a drum containing lindane were found in a landfill, the soil and other wastes generated by its cleanup would be treated as a U129-listed hazardous waste, without regard to its concentration in the soil or in any other environmental media. This category of waste would have to meet LDR limits prior to disposal and be disposed of in a RCRA Class C landfill, unless the waste could be delisted.

ardous wastes. Prior to September 25, 1990, waste generators conducted an **EP toxicity test** to determine if a waste was hazardous, but after this date EPA replaced the EP toxicity test with the TCLP.

If a concentration-based treatment has been identified for a particular contaminant, the technology may be used as long as the treatment residues meet the identified concentrations. When the toxicity characteristic or hazardous waste characteristic has been removed and the underlying hazardous constituents have been dealt with, the waste is no longer considered to be hazardous and can be disposed of in accordance with the 40 CFR 241 regulations for management of solid waste.

Waste Disposition

Remediation Waste

One problem commonly encountered in the disposition of remediation wastes is the lack of data necessary to insure its proper disposal. The data collected to support cleanup commonly focus on determining human health and ecological risk as well as on selecting and designing the remedy or the cleanup technology. The data necessary to manage the remediation waste are often overlooked until

later in the cleanup. This problem is exacerbated if the remediation waste is stored until the end of the project, and if changes in the regulations or the regulatory personnel have been introduced at some point during the storage time.

A solid contaminated with a simple hydrocarbon presents few waste disposal problems. A waste that falls under multiple regulatory jurisdictions, however, may pose complex disposal problems. For example, a waste disposal is virtually impossible to manage if the waste contains 1) a RCRA characteristic waste; 2) a RCRA listed waste at levels subject to the LDR; 3) PCBs that are subject to TSCA regulations; and 4) radionuclides that are subject to either Department of Energy or Nuclear Regulatory Commission regulations. As demonstrated in Box 9-2, even if the waste contains only the radionuclides and the PCBs, there are currently few viable disposition paths available.

Industry Waste Types

The most important factors influencing the disposition of remediation wastes are: physical form, media type, type of contaminant, level of contamination, and regulatory authorities applicable to the waste. As shown in detail in Table 9-1, the type of contamination tends to vary according to

Industry	Waste Types		Regulated By	Clean-Up Criteria	Examples of Sites	Treatment Viability	Regulatory Treatment Standard or BDAT**	Potential Secondary Waste***
	Types*	Contaminants						
Agricultural	Drums of Waste	Pesticides, Herbicides	TSCA	5 ppm* or 10 ppm per RCRA	Numerous Sites	If available it is defined in the ROD, otherwise BDAT is under development.	Incineration	Ash Treatment Process Wastes Carbon filters
Mining	Drums of Waste, Runoff, Leachate	Acids	CERCLA RCRA	Characteristic Levels, Delisting Levels, LDR BDAT, etc.	Bunker Hill, ID	If available it is defined in the ROD, otherwise BDAT is under development.	Characteristic Waste – Deactivation Radiological – Vitrification Lead - Thermal recovery Organics - Carbon Adsorption Macroencapsulation Metals – Precipitation Organics – Biodegradation Organics – Stabilization	Geological Media, Leachate
DOD	Soil, Surface and Ground-water, Drums of Waste, Spent Fuel, D & D Waste	Solvents, Metals, Explosives, Radiological, Mixed Waste	EPA NRC DOD	Characteristic Levels, Delisting Levels, LDR BDAT, etc.	Tooele, Utah	If available it is defined in the ROD, otherwise BDAT is under development.	See Mining Column	Treatment Process Residue Part Replacements
DOE	Soil, Surface and Ground-water, Drums of Waste, Spent Fuel, D & D Waste	Solvents, Metals, Explosives, Radiological, Mixed Waste	EPA NRC DOE	Characteristic Levels, Delisting Levels, LDR BDAT, etc.	INEL Rocky Flats	If available it is defined in the ROD, otherwise BDAT is under development.	See Mining Column	Treatment Process Residue Part Replacements
Oil & Gas Production	Oils	Petroleum By-products	EPCRA	Characteristic Levels	Exxon *Valdez*	As determined by clean-up specifications.	Liquids – Stabilization; Metals – Incineration; Organics – Biodegradation; Ignitability – Deactivation	Oil Residue

Table 9-1: Waste types and contaminants generated by various industries.

Industry	Waste Types		Regulated By	Clean-Up Criteria	Examples of Sites	Treatment Viability	Regulatory Treatment Standard or BDAT**	Potential Secondary Waste***
	Types*	Contaminants						
Construction	Oil, Gas By-products	Wood, Metal, etc.	EPA	Characteristic Levels, Delisting Levels, LDR BDAT, etc.	Any Construction Site	If available it is defined in the ROD, otherwise BDAT is under development.	Petroleum Byproducts – Stabilization Organics – Biodegradation Note: Much of construction waste is regulated by the Universal Waste Rule	Metal scrap Fluorescent Tubes Used oil
Painting	Paint Residues Thinners Solvents	Solvents	RCRA CERCLA	Characteristic Levels, Delisting Levels, LDR BDAT, etc.	Any Construction Site	See DOE/DOD Column	See Mining Column	Solvents Organics Characteristic Wastes Heavy metals
Dairy	Animal Waste	Nitrates	State Regulated CWA	Meet MCL Levels, etc.	Feedlots	BDAT	Deactivation	Cattle residue
Medical	Sharps	Biohazards	State Specific	State Specific	Hospitals and Physicians Office	Incineration	Incineration	
Household Waste	Miscellaneous	Solvents, Cleaners, etc.	Non-regulated	N/A	Landfills	Stabilization - liquid	Organics, metals, and other household types of wastes	If used

*Assumes that all types of industrial activities or wastes presented can potentially contaminate groundwater, surface waters, and immediate land area with contaminants.
** Treatment standards for HW are found in Table 40 CFR 268.40
*** Are only some examples

Table 9-1: Waste types and contaminants generated by various industries (continued).

Box 9-2 ■ Disposition of Samples Falling Under Multiple Regulatory Agencies

During the investigation phase of burn pits located on a DOE site, six 125 ml jars of soil were collected. The later disposition of these samples will serve as an example of the difficulty in managing waste that falls under multiple regulatory agencies.

It is known that the burn pits were used to dispose of a variety of materials during the 1950s and 1960s, but there is little reliable information concerning their exact operation. The pits are known to have received oil; PCB contamination was therefore suspected because of its ubiquitous use during that time. However, since PCBs were not identified as a constituent of concern in the CERCLA investigation of that site, there was no data concerning its presence or concentration. Also, the samples did contain low but detectable levels of cobalt-60 and had to be handled as a radioactive waste due to the Department of Energy's "no-radiation-added" policy. In addition, the samples were contaminated with RCRA listed hazardous wastes at levels slightly above the Land Disposal Restriction (LDR) limits.

Considering the regulations applicable, the disposal options would appear to be the following:

1. The samples may be returned to the area of contamination, if they do not significantly increase the risk at the site and have not "incurred placement," which would make them subject to the LDRs.

2. At the discretion of the EPA Region, they may be treated to below the LDR limits and disposed of at a TSCA chemical waste landfill, which is permitted to receive low-level radioactive waste. Unfortunately, no such facilities are currently permitted, making this an **orphan waste stream**, for which there is no legal disposal. Even if a legal disposal becomes possible at some point in the future, management of this waste until that time is likely to be extremely expensive.

industry and site, but the following generalizations can be made:

— Department of Defense (DOD) – fuel and chlorinated solvent contamination;

— Department of Energy (DOE) – chlorinated solvents, radionuclides, fuel, PCBs;

— Private Industry – fuel, chlorinated solvents, and a wide variety of other contaminants;

— Agricultural – fuel, fertilizers, pesticides, herbicides;

— Mining – mine spoils, acidic metal-rich water; and

— Oil and gas production – refined products, crude oil, and drilling wastes.

Contaminant Categories

As previously noted, the source of the contaminant may be equally or even more important than its concentration in determining its management. In the following sections, several of the contaminant categories are given.

Hydrocarbons – Hydrocarbon-contaminated soil, water, and sludge are the most commonly encountered wastes during cleanup operations involving such items as leaking underground storage tanks. Wastes contaminated by crude oil and refined petroleum products are specifically exempted from many regulatory requirements, but their disposal may still be difficult. In most cases hydrocarbon-contaminated media are classified as RCRA solid

Box 9-3 ■ Treatment of Hydrocarbon Wastes

Short-chained, unbranched, aliphatic hydrocarbons and small aromatic molecules tend to biodegrade rapidly. Wastes contaminated with gasoline and diesel, therefore, can be effectively treated by landfarming or other biodegradation treatment strategies. Wastes contaminated with complex aromatic hydrocarbons and/or long-chained, highly branched, aliphatics hydrocarbons tend to biodegrade much less rapidly. Therefore, soils contaminated with fuels that are rich in these compounds, such as bunker-C oil, biodegrade slowly, if at all.

Oil-water emulsions – known as "chocolate mousse" – that form following marine oil spills, also tend to resist biodegradation. The reason appears to be that oxygen availability to the aerobic microbes – that would otherwise consume the oil – is greatly restricted by the relatively thick coating of emulsified oil. Consequently, emulsions may have to be broken before mousse-contaminated wastewater can be effectively biodegraded.

Under certain circumstances, hydrocarbon-contaminated materials may be reused, thereby eliminating the need for disposal. Hot-batch asphalt facilities, for example, can incorporate a small percent of hydrocarbon-contaminated sediment/soil into their mix. The exact percentages depend on the grain size and the concentration of the impurities in the contaminated material. Such beneficial uses, however, are highly dependent on regulatory permissions, which are often difficult to obtain. Most of the hydrocarbon material recovered during the Exxon *Valdez* spill response, for example, was hauled by barge and disposed of at landfills in the lower 48 states, because the regulatory agencies involved were in opposition to any other disposal scheme.

---

wastes; the hydrocarbon concentration beyond which the waste is not accepted is determined by the landfill operator. Since the acceptable limit of hydrocarbon contamination differs among facilities, it may be necessary to blend or pretreat the waste prior to landfill disposal.

PCBs and Dioxin Wastes – As will be discussed later in this chapter, media containing PCBs, pesticides, dioxins, and some other contaminants are subject to additional regulations besides RCRA. TSCA, for example, regulates disposal of PCB-contaminated waste and – as is the case with RCRA – the date and source of contamination are important factors. The different EPA regions have broad discretion in the application of TSCA regulations, so the regional EPA regulators should be contacted for directions whenever PCB contamination is known or suspected.

Radioactive Waste

Many government sites, mining facilities, and civilian nuclear facilities are contaminated with radioactive substances. These substances may have contaminated buildings, soils, groundwater, and other media, and because they are often metals, they cannot be easily destroyed. Normal hazardous waste treatment and disposal methods tend only to move them from one site to another.

Radioactive contamination is regulated under the **Atomic Energy Act**, which established separate regulatory pathways for civilian and government facilities. The **Nuclear Regulatory Commission (NRC)** regulates civilian activities, while the DOE regulates most government activities. NRC requirements do not apply to DOE waste, except when the waste is disposed of in a civilian disposal facility.

Radioactive contamination is legally divided into **low-level radioactive waste (LLW)**, uranium mill tailings, **high-level radioactive waste (HLW)**, **transuranic waste (TRU)**, spent nuclear fuel, and special nuclear material, on the basis of its source and concentration of radionuclides (See chapter 27). LLW is any radioactive waste not classified as HLW, TRU, or uranium mill tailings, and may contain just barely detectable amounts of radioactivity dispersed in a large amount of material. It is the most abundant type of radioactive waste and is generated by uranium enrichment processes, reactor operations, isotope production, medical procedures, and R&D activities. Uranium mill tailings and some LLW have clear paths for disposition, but all other radioactive wastes lack such clarity.

The most common strategy for minimizing radioactive waste is to make the **Radioactive Management Areas (RMAs)** as small as possible and bring material into the RMA only if it is necessary for the job at hand. Free-release criteria are used in civilian RMAs, but DOE has not adopted the free-release criteria and, therefore, any addition of radiation – no matter how small – is grounds for treating the waste as an LLW.

Mixed Wastes

Remediation waste containing both radionuclides (LLW, HLW, TRU, etc.) and RCRA-regulated constituents is considered to be a **mixed waste**. This type of waste is difficult to manage because: 1) treatment standards are lacking in some cases; 2) there are no disposal sites for many possible combinations, and 3) there are conflicting regulatory requirements. NRC-regulated facilities, for example, specify no landfill liners for facilities receiving LLW to avoid the "bathtub effect"(liquid collection) and the enhanced leaching of radionuclides. RCRA facilities, on the other hand, require liners to limit potential migration of contaminants to groundwater. EPA and other federal agencies are currently working to establish methods to manage this waste, but the political pressures against effective management of mixed waste may be insurmountable, at least in the near future.

Checking Your Understanding

1. What are seven types of waste generated during site characterization processes?

2. Which are the characteristics of organic wastes that typically undergo rapid biodegradation?

3. Which federal agency is responsible for the regulation of PCB-contaminated wastes?

4. What is a mixed waste?

5. What must a waste exhibit to be classified as a RCRA characteristic waste?

9-3 Superfund Cleanup

Superfund sites represent only a portion of all the U.S. sites that require cleanup. Hazardous waste generated during the cleanup activities at Superfund sites are regulated by either CERCLA or RCRA. Waste management requirements under CERCLA may be less rigorous than under RCRA, because sites cleaned up under CERCLA authority must meet only the substantive requirements of RCRA and are exempted from administrative requirements (paper work). Administrative and substantive requirements are discussed in more detail in the following sections.

Applicable or Relevant and Appropriate Requirements

Compliance with the **applicable or relevant and appropriate requirements (ARARs)** is a threshold criterion for a CERCLA cleanup. Substantive requirements from other areas of the law are borrowed to define the cleanup requirements. Typically the substantive portion of the ARARs comes from the:

— Safe Drinking Water Act (SDWA);

— Clean Water Act (CWA);

— Clean Air Act (CAA);

— National Standards for Hazardous Air Pollutants (NESHAPs);

— Occupational Safety and Health Administration (OSHA) regulations;

— Resource Conservation and Recovery Act (RCRA);

— Toxic Substance Control Act (TSCA);

— Federal Insecticide, Fungicide, and Rodenticide Act (FIFRA); and

— Other federal, state, and local standards.

CERCLA's "applicable" regulations apply directly to the cleanup situation under consideration. Such requirements may include: 1) cleanup standards, 2) standards of control, and 3) other substantive requirements, criteria, or limitations promulgated under federal or state environmental or facility siting laws. These are specific to the hazardous substance, pollutant, contaminant, remedial action, location, or to other circumstances found at a CERCLA site.

CERCLA's "relevant and appropriate" requirements are less strongly linked to the situation under consideration and may include cleanup and control standards and other substantive requirements, criteria, or limitations promulgated under federal or state environmental laws. While these ARARs are not legally applicable to a CERCLA response action, they have been determined to be relevant and appropriate and address problems or situations sufficiently similar to those encountered at the Superfund site; hence their use is well suited. Box 9-4 describes an example in which the ARAR concept is applied.

To be considered (TBC) requirements are neither "applicable" nor "relevant and appropriate" to a situation; nonetheless, they are requirements that should be considered during cleanup. DOE orders and other non-promulgated requirements are examples of TBC requirements that are still binding on the cleanup operation.

Substantive and Administrative Requirements

Substantive requirements specify actions that must be performed, while administrative requirements specify how the action is to be documented and approved. The exact boundary between substantive and administrative requirements is commonly blurred, but it is important to identify the regulatory agency responsible when determining the waste management requirements that must be followed.

Box 9-4 ■ Determining ARARs

Let's assume that a pond had been contaminated with acids and trichloroethylene, both of which are regulated under the CWA and SDWA. The site also had been abandoned and placed on the NPL, which means CERCLA jurisdiction applied. The site manager, therefore, had to determine the actions necessary in order to legally clean up the site to a "safe" level, while also managing the wastes generated during the cleanup.

During characterization, the site manager would collect liquid samples from the pond. If the samples are to be poured back into the pond, the liquid must meet the risk-based level established in the risk assessment or in the preliminary site investigation. If these values have not been identified, then the "applicable" Maximum Contaminant Level (MCL) SDWA standards may apply; however, in general these requirements are more stringent than CWA's discharge requirements. The CWA may be considered "relevant and appropriate."

In general, the requirements under RCRA are explicitly stated. The most important substantive requirements for onsite waste management are the following:

1. Containers must be in good condition;

2. Wastes must be compatible with the container;

3. Containers must be closed during storage;

4. Container storage areas must have a containment system that will hold 10 percent of the volume of all the containers or 10 percent of the largest container; spilled or leaked waste must be removed from the collection area as necessary to prevent overflow [40 CFR 264, Subpart I and 265 Subpart I].

In a later section, a description of onsite and off site disposal is given, as it applies to the administrative and substantive requirements. The RCRA substantive requirements are used to ensure proper and safe management of waste without the use of permits or other administrative documentation. If full RCRA requirements apply, the following requirements also apply:

1. Use of hazardous waste manifests;

2. Shipment of waste to a RCRA treatment, storage, or disposal (TSD) facility would trigger additional requirements, such as the off site CERCLA policy, which pertains to TSD facility requirements for handling of CERCLA-generated waste;

3. Imposition of the 90-day time clock on **Temporary Accumulation Areas (TAAs)**; and

4. All other administrative requirements.

Applying ARARs

CERCLA also incorporates statutes other than RCRA under the ARAR process. Sites containing PCBs, for example, may be regulated under TSCA, and the PCB waste required management in accordance with the TSCA substantive requirements (see Box 9-4).

Waste Disposal Options

CERCLA sites often involve widespread and dispersed contamination; as a consequence, EPA developed the concept of an Area of Contamination

Box 9-5 ■ Concept of AOC and Applicability to Onsite/Off Site Management

Example: An AOC includes a fenced-in area and extends 25 feet beyond the perimeter of the fence to include the area where the groundwater plume from a pond is located. The site investigator can move waste from within the AOC without:

1. Triggering the Land Disposal Restrictions (LDRs) so long as the waste does not incur "placement";

2. Using hazardous waste manifests;

3. Meeting specific RCRA administrative TAAs storage time limitations (90 days); or

4. Meeting off site requirements for handling of hazardous wastes generated during CERCLA activities.

Box 9-6 ■ Management of Purge Water within the AOC

Example: A CERCLA site has generated 15 drums of purge water as a result of well monitoring activities. The AOC includes the area of the groundwater plume from the waste pond. Since the groundwater plume is within the AOC, the site investigator can take the drums of wastewater from this location in the AOC and pour it into the pond, without having to meet RCRA treatment standards.

(AOC), which is delineated by the extent of continuous contamination and may contain varying types and concentrations of waste. The government imposes different requirements for wastes managed within (onsite) and outside (off site) the AOC. The requirements for waste moved to other locations are more stringent than for waste that is simply consolidated or treated and consolidated onsite.

Onsite Disposal

Under CERCLA regulations onsite disposal requirements apply, which facilitates the management of remedial waste at the AOC. Waste to be managed within the designated AOC does not have to meet the administrative requirements and is exempt from Land Disposal Restrictions. Boxes 9-5 and 9-6 illustrate the onsite AOC disposal options.

Off Site Disposal

If waste is moved out of the AOC or if it incurs placement, the administrative and substantive RCRA requirements are triggered. If the waste is removed from the AOC to another AOC, this is also considered to be an off site activity. In general, management of waste outside the AOC is difficult, and waste should be maintained onsite whenever possible.

Waste Treatment

Waste treated to predefined cleanup levels may be replaced in geologic strata once it is treated to **clean levels**. Clean levels are established using a variety of factors including: 1) chemical specific ARARs (such as BDATs, MACs and LDRs); 2) acceptable risk-based levels; 3) meeting delisting criteria; or 4) meeting predetermined disposal levels. The treatment methods used may generate secondary waste. As shown in Figure 9-1, the treated primary waste may not meet the defined treatment standard, and it may be necessary to develop a new treatment standard or BDAT. This may not be practical in all cases.

Secondary waste streams generated by a treatment method are also subject to regulation and must meet the substantive requirements of RCRA. Soil contaminated with organic chemicals, for instance, may be processed through a charcoal system to adsorb the chemicals onto the activated carbon. The activated carbon will have to be replaced periodically as it becomes saturated with organic chemicals; it can subsequently either be processed for reuse (which minimizes waste) or treated to LDRs and disposed of directly.

Waste Disposal Alternatives

As mentioned earlier, treatment and disposal options are not available for all waste streams generated during cleanup. Cleanups, therefore, should be designed to minimize the generation of orphaned waste streams that may require storage in perpetuity. Disposal of many other possible wastes is contingent upon meeting the applicable treatment levels. If waste treatment cannot or does not meet predefined levels, management requirements for the waste may trigger additional regulatory requirements. Table 9-2 presents general disposal information pertaining to specific contaminants.

Management of Hazardous Waste from Onsite Remedial Processes

As described earlier, RCRA hazardous waste generated as part of treatment can be managed within the AOC in accordance the substantive requirements of RCRA. If the waste is taken off site, however, it must meet both the substantive and the administra-

tive requirements of RCRA, including LDRs, in most cases.

Off Site Management of Hazardous Waste

If processing generates RCRA hazardous waste that is shipped off site, both substantive and administrative requirements of RCRA must be met. The waste must be shipped to a permitted or an interim status RCRA Treatment, Storage, and Disposal (TSD) facility, which can accept hazardous waste (40 CFR 264). All shipments must meet appropriate Department of Transportation (DOT) requirements; facilities accepting waste generated from a CERCLA response action must comply with the CERCLA off site policy.

RCRA Permits Required for Facilities Accepting Hazardous Waste – A facility that treats or disposes of remediation-generated hazardous waste must have a current operating permit – established by EPA or by the authorized state agency – for the acceptance, treatment, and disposal of the specific waste types. The facility must meet the following three requirements: 1) 40 CFR 264: *Standards for Owners and Operators of Hazardous Waste Treatment, Storage, and Disposal Facilities;* 2) 40 CFR 265: *Interim Status Standards for Owners and Operators of Hazardous Waste Treatment, Storage, and Disposal Facilities;* or 3) applicable state standards in states having authorized hazardous waste programs. Waste may not be shipped to any facility with major violations that are currently unresolved.

CERCLA Off Site Requirements for a TSD Facility – Hazardous waste generated within an AOC as a result of a response action under CERCLA is designated as CERCLA waste. Any CERCLA wastes shipped off site to a receiving TSD must comply with the **CERCLA Off Site Rule** [58 FR 49204, Sept. 22, 1993]. The receiving TSD also must notify EPA that it plans to manage CERCLA wastes. This allows EPA to ensure that the TSD meets the CERCLA off site requirements. After confirming that the TSD is not in violation of any of the off site policy requirements, EPA may add the facility to an updated list of acceptable facilities.

Facility-Specific Waste Acceptance Criteria

Any hazardous waste generated by cleanup must be fully characterized and/or profiled so that the waste

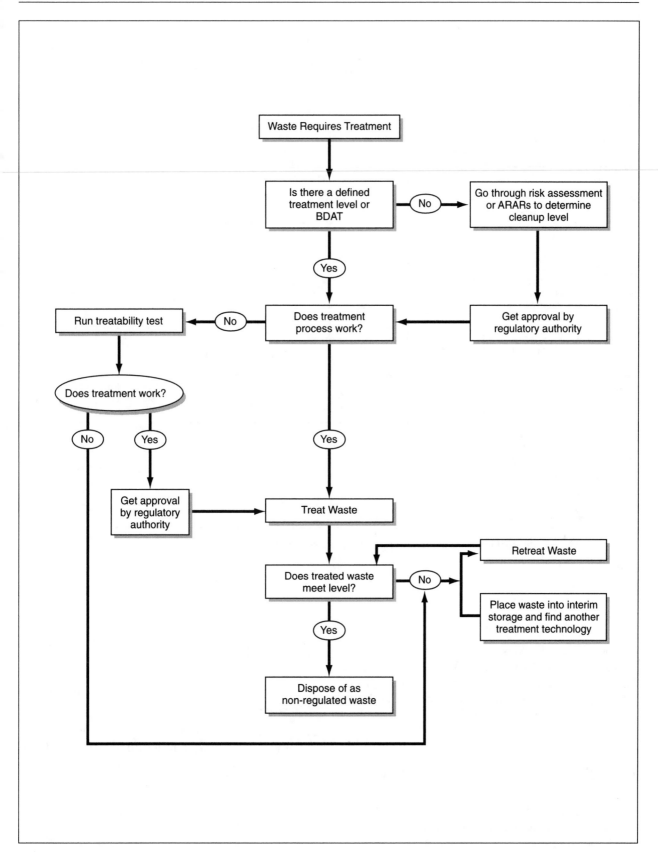

Figure 9-1: Waste Treatment Flow Diagram

Contaminant Type	Regulatory Level Met	RCRA Permitted TSD	LLW Disposal Site	WIPP or Other TRU Disposal/ Storage Facility	Other		
					Return to Origins	Interim Storage Onsite	Determine Optional Disposal & Treatment
Radioactive Low-level Waste	Meets Level	NA	NA	NA	✓		
	Still "Hot"	NA	✓	NA	NA	✓	✓
Mixed Low-level Waste	Meets Level	NA	NA	NA	✓		
	Still "Hot"	✓	NA	NA	NA	✓	✓
TRU (Transuranic Waste)	Meets Level	NA	NA	✓	✓		
	Still "Hot"	NA	NA	✓*	✓	✓	✓
TRU Metal Waste	Meets Level	NA	NA	✓*	✓		
	Still "Hot"	NA	NA	✓	NA	✓	✓
Hazardous Waste	Meets BDAT or LDRs	✓	NA	NA	✓		
	Still Hazardous Waste	✓	NA	NA	NA	✓	✓
Municipal Solid Waste	Solid	NA	NA	NA	NA	Dispose of at Subtitle D Landfill	✓
	Liquid	NA	NA	NA	NA		✓
Other Media, etc. (Groundwater, Soil, Sediment, etc.)	Solid	See above for various waste forms and contaminants as it relates to geologically derived material and water.					
	Liquid						

*Specific to each site and regulatory agency.

Table 9-2: General disposal information pertaining to specific contaminants.

can be evaluated against the off site facility-specific **Waste Acceptance Criteria (WAC)**. Facility-specific WACs ensure that waste sent to the facility meets RCRA permit requirements for treatment and disposal as well as applicable DOT requirements.

Checking Your Understanding

1. Which two federal agencies may regulate hazardous wastes generated during cleanup of Superfund sites?

2. What do substantive requirements specify?

3. What are eight possible sources for substantive requirements that may be borrowed during a CERCLA cleanup?

4. Under what conditions may the federal government treat two or more noncontiguous facilities as one for the purposes of response?

5. What do WASs insure?

Summary

In an attempt to identify, evaluate, and clean up environmentally contaminated sites across the nation, Congress passed the Comprehensive Environmental Response, Compensation, and Liability Act (CERCLA) in 1980 as well as the Superfund Amendment and Reauthorization Act (SARA) in 1986. However, the web of conflicting regulations and liability provisions of these Acts have prompted some to assert that delays have been caused in cleanup efforts leading to unintended consequences.

All phases of the cleanup process generate wastes. Investigation-derived wastes as well as remedy and cleanup wastes are collectively known as remediation wastes. Wastes generated during a Superfund (CERCLA) cleanup fall under the substantive requirements of RCRA. Two types of hazardous wastes are defined by RCRA, those that exhibit one or more of the characteristics of a hazardous waste (corrosive, reactive ignitable, or toxic) and those that are specifically listed as a hazardous waste by EPA. A characteristic waste has physical and/or chemical properties that make it hazardous. If the characteristic is removed through treatment, it is no longer considered hazardous. Listed wastes appear in four separate tables in the 40 CFRs, which use a code to identify their source. Listed wastes must meet the Land Disposal Restrictions treatment standards prior to land disposal in a RCRA-regulated facility. Listed wastes continue to be regulated until they are delisted in a Record of Decision (ROD) or by some other delisting process.

Superfund sites are a subset of all sites that require cleanup. The wastes generated during a cleanup may be regulated through either CERCLA or RCRA requirements. All CERCLA cleanup actions must conform to the Applicable or Relevant and Appropriate Requirements (ARARs). The substantive requirements (actions that must be performed) of the ARARs come from any other federal, state, and local standard. The administrative requirements ensure that the appropriate documentation and other paperwork are completed.

To facilitate the management of waste on a CERCLA site, EPA developed the concept of an Area of Contamination (AOC), which is delineated by the extent of continuous contamination and may contain varying types and concentrations of waste. There are different requirements for waste management within and outside the AOC. EPA has also authorized treating two or more non-contiguous AOCs as one area for the purpose of response. Wastes to be managed within the designated AOC do not have to meet the administrative requirements and are exempt from the Land Disposal Restrictions. Waste moved out of the AOC triggers the administrative and substantive RCRA requirements.

Disposal of wastes is contingent upon meeting the applicable treatment levels. If the waste treatment cannot or does not meet predefined levels, management requirements for the waste may trigger other regulatory requirements. Any hazardous waste generated by cleanup must be fully characterized in order to be evaluated against the off site facility-specific Waste Acceptance Criteria (WAC). Solid wastes that are not a RCRA waste and do not contain free liquids can be disposed of at a municipal Subtitle D landfill.

Critical Thinking Questions

1. Identify the flaw in CERCLA regulations as illustrated in Box 9-1 and propose an amendment that would correct this flaw.

2. Carefully study the differences between RCRA listed and characteristic hazardous wastes. Propose an argument for why listed wastes should remain regulated, based solely on their source.

3. Prepare a table comparing the EP toxicity test and the TCLP. Analyze their similarities and differences, and write a paragraph justifying the change to the TCLP to determine the toxicity characteristic.

10

The Petroleum Industry

by Dave Olsen

Chapter Objectives

Upon completing this chapter, the student will be able to:

1. **Explain** the origin of fossil fuels and early petroleum production attempts.

2. **Identify** and explain the characteristics of petroleum and methods for evaluating its market value.

3. **Explain** the drilling process and well evaluation procedures.

4. **Identify** and explain the major refinery processes and the function served by each process.

5. **Identify** the various methods used to transport petroleum and petroleum products.

6. **Identify** and explain the portions of the Clean Air Act Amendments of 1990 that had a major impact on the petroleum industry.

10-1 Introduction

The fossil fuel reserves on earth are believed to be the product of vegetation and animal life that existed millions of years ago. The early atmosphere surrounding the earth is believed to have consisted mostly of carbon dioxide, CO_2, and only a little oxygen. The carbon dioxide – a greenhouse gas – effectively trapped the warmth of the infrared radiation (IR) from the sun; consequently, the entire earth had a tropical climate. For millions of years, the warm oceans teemed with tiny marine organisms, and plants grew forming lush forests. These processes eventually captured and converted most of the CO_2 into either plant or animal (organic) matter. As shown in the following equation, photosynthesis was not only responsible for removing vast amounts of CO_2 from the atmosphere, but it also replaced it with vast amounts of oxygen.

$$6 \, CO_2 + 6 \, H_2O \longrightarrow C_6H_{12}O_6 + 6 \, O_2$$

Glucose, $C_6H_{12}O_6$ – the other product of photosynthesis – provided the chemical energy for the accumulation of thick layers of decaying vegetation on the forest floors and the formation of an organic ooze on the ocean floors. Through cataclysmic events, including volcanic eruptions and earthquakes, these deposits of organic debris became buried under the layers of the earth and under the oceans and were subjected to tremendous heat and pressure for millions of years. Under these anaerobic conditions, the plant material was converted into deposits of coal, and the organic ooze was transformed into **crude oil**.

Crude oil is a complex mixture of **hydrocarbons**, composed of molecules ranging from a single carbon and four hydrogen atoms in methane, CH_4, to very long carbon chains containing dozens and dozens of carbon and hydrogen atoms. Crude oil is predominantly composed of saturated hydrocarbons, which have only single bonds between the carbons. However, unsaturated hydrocarbons, which have carbon-to-carbon double bonds and aromatic (benzene ring) compounds, are also present in low concentrations. In addition to carbon and hydrogen, small amounts of other elements such as sulfur, nitrogen, and traces of various metals are also present in crude oil.

Petroleum has been seeping to the surface of the earth for centuries and has been found to be useful for a variety of applications. There is evidence, for example, that the Egyptians used petroleum as an embalming agent, while native Americans used it as a medicine and as a heating fuel before the Europeans arrived. Early boat builders found that the thick black material could be effectively used for waterproofing their ships. The potential for an oil industry was first identified by the Canadian geologist Abraham Gesner, who discovered that he could distill a useful fuel – **kerosene** – from either oil or coal.

The modern petroleum industry was born in America in the early 1800s. Entrepreneurs built dams on Oil Creek in western Pennsylvania to trap oil floating on the surface of the water. Blankets were used to soak up the oil, which was then retrieved by wringing the blankets. By the 1840s whale oil, which had been the principal fuel source for indoor lighting, was becoming scarce and expensive. A druggist, Samuel Kier, was selling refined Pennsylvania crude oil as a miracle remedy for the treatment of wounds, rashes, pimples, scratches, and other skin problems. Eventually, someone sent Kier's remedy to chemists at Yale University, who found little evidence that it cured anything; they did, however, find it could be further refined into kerosene. It was the development of the kerosene lamp in 1854 that lead to a demand for kerosene and the formation of the Pennsylvania Rock Oil Company, America's first oil company. Five years later, Edwin Drake, a retired railroad worker, drilled a well in Titusville, Pennsyl-

Crude oil is the elixir that has transformed modern life, powering industry, creating goods, and even fueling politics. On its journey from the depths of the earth to the consumers who thirst for it, this valuable resource must be treated with care in order to preserve and protect the environment.

From: *Petroleum Production and Transportation,*
PRESERVING THE LEGACY video series.

vania. He had made a commitment to drill to the depth of 1,000 feet, if needed, to find oil, but five months after the start of the project, driller Billy Smith returned from church one Sunday to examine the drilling progress. Much to his surprise, the hole was full of oil; the well had struck oil at a depth of only 21 meters (69.5 feet). Kerosene remained the main product refined from crude oil until the development of the internal combustion engine in the 1890s.

In 1859 the total U.S. production of oil was only 2,000 **barrels** (1 barrel = 42.00 gallons). However, within three years the market was saturated and its price was only 10 cents a barrel. The problem faced by the fledging industry was the transportation of the crude oil to the laboratories where it was refined. At first, the oil was transported by horse-drawn wagons and barges, but by 1865, a pipeline was constructed from the oil fields around Titusville to the local railroad station. Delivery of the crude oil to East Coast refining locations became somewhat more efficient. The construction of a 60-mile pipeline from Titusville to the industries of Pittsburgh in 1875, however, firmly established the modern petroleum industry.

10-2 Petroleum Characteristics and Production

Since the mid-nineteenth century petroleum production has exploded, and refined petroleum products have become the major energy resource for the planet. With the exception of Antarctica, oil is being produced on every continent. The economics and political alignments of nations are largely dependent on the availability of cheap crude oil. Increases in petroleum usage over the past 100 years have tended to parallel population growth and a general improvement in living standards. In recent decades, the fear of running out of oil has been temporarily eased by the discovery of significant new reserves. However, as Table 10-1 suggests, petroleum is not a renewable energy source.

As shown in Table 10-2, petroleum is a mixture of various compounds that can be classified according to their fundamental molecular structures.

The products from petroleum refining are distinguished by their uses and fall into three broad categories: fuels and lubricants, chemical feedstocks, and residue.

Fuels and **Lubricants** – Fuels are used in combustion processes for energy. The largest single user

Year	Proven Reserves (Billions of Barrels)	Annual Production (Billions of Barrels)	Remaining Oil Supply (Years)
1950	76	3.8	20
1955	154	5.6	27
1960	256	7.7	33
1965	342	11.0	31
1970	517	16.7	31
1975	569	19.3	29
1980	574	22.0	26
1985	696	19.6	36
1990*	933	21.8	43
*Estimate			

Table 10-1: Relationship between known oil reserves and annual production.

Component of Crude Distillate	Description of Molecular Structure	Example of Structure
Paraffins	Saturated Hydrocarbons	CH_3-CH_2-CH_2-CH_3
Olefins	Unsaturated Hydrocarbons	CH_3-CH=CH-CH_3
Naphthalenes	Polynuclear Aromatics	
Aromatics	Benzene Compounds	

Table 10-2: Components and molecular structure of crude oil.

Fraction	Percent by Volume	Boiling Point °F	°API (Density)	Uses
Gas	–	Up to 50	–	Fuel gas, LPG
Light Naphtha	8	50 to 200	79	Motor fuel
Heavy Naphtha	21	200 to 400	55	Jet fuel
Kerosene	11	400 to 500	42	Jet fuel Kerosene
Gas Oil	15	500 to 650	35	Diesel, Lube Home heat
Residue	45	Over 650	18	Bunker oil Asphalt, Coke

Table 10-4: Fractions, characteristics, and uses of petroleum products.

Number of Carbons	Formula (C_nH_{2n+2})	Name
1	CH_4	Methane
2	CH_3CH_3	Ethane
3	$CH_3CH_2CH_3$	Propane
4	$CH_3CH_2CH_2CH_3$	Butane
5	$CH_3CH_2CH_2CH_2CH_3$	Pentane
6	$CH_3CH_2CH_2CH_2CH_2CH_3$	Hexane
7	$CH_3CH_2CH_2CH_2CH_2CH_2CH_3$	Heptane
8	$CH_3CH_2CH_2CH_2CH_2CH_2CH_2CH_3$	Octane
9	$CH_3CH_2CH_2CH_2CH_2CH_2CH_2CH_2CH_3$	Nonane
10	$CH_3CH_2CH_2CH_2CH_2CH_2CH_2CH_2CH_2CH_3$	Decane

Table 10-3: First ten members of the paraffins series.

of fuels and lubricants is the transportation industry, in which the fuels power internal combustion engines, while the lubricants are used to reduce friction. Residential and industrial heating represents the second largest use of fuels. As shown in Table 10-3, the molecular structures of the gaseous fuels range from the C_1 to the C_4 molecules. Liquid mixtures of longer carbon chains (C_5 to C_8) are blended for gasoline to fuel **internal combustion engines**. Diesel engines and heating oils have even longer carbon chains, ranging from ten to eighteen carbon atoms per molecule.

Chemical feedstocks are the raw materials used for the manufacture of petroleum-based consumer products, including plastics, synthetic fibers, and many household and industrial chemicals. These are composed of small molecules that are easily separated from crude oil in the refining processes. To be useful in product manufacture, however, these raw materials must be processed to a higher degree of purity than is required for fuel blending.

The **residue** from the refining process includes large molecules that are used for tar, asphalt road surfacing, and in the production of petroleum coke.

Table 10-4 shows the percentages of fractions, by volume, and some of their uses.

The Nature of Crude Oil

The mixture of hydrocarbons in crude oil may vary and determines its chemical and physical properties. Two factors determine the value of crude oil: consumer demand of its end products and crude oil composition. The most valued product – determined by consumer demand – is gasoline, followed by diesel fuel/heating oil. The less refining steps are required, and the fewer the impurities to be removed, the lower the production costs and therefore the higher the profits.

Therefore, crude oil that is rich in the gasoline fraction and low in impurities will have the highest market value.

Table 10-5 provides a breakdown of the products that can be obtained by the refining of Arabian crude oil in relation to the U.S. consumer demand. These data reveal the need for extensive molecular reformation to reduce the excessive waste of unneeded products, while at the same time responding to consumer demand. The industry looks at three properties of crude oil to establish a commercial value: density, chemical characterization, and sulfur content. Each of these will be discussed in the following sections.

Density

The density of petroleum and petroleum products is measured in units of **API** (American Petroleum Institute) **gravity** (°API). On this scale, an API gravity of 10 is equal to the density of water, or 1.0 gram/ml. Therefore, API gravity is actually a den-

Percent by Volume of Crude	Fraction Name	Boiling Point °F	Percent by Volume of Product	Product Name	Boiling Point °F
8	Light Naphtha	50-200	5	Fuel Gases, LPG	-40-100
21	Heavy Naphtha	200-400	45	Gasoline	100-400
11	Kerosene	400-500			
15	Gas Oil	500-650			
			5	Jet Fuel/ Kerosene	300-500
45	Residue	>650	25	Fuel Oil	400-600
			5	Lube, Diesel Fuel	>600
			15	Bunker Oil, Asphalt, & Coke	>600

Table 10-5: Product distribution from Arabian crude oil vs. U.S. product deliveries.

sity measurement and gives an indication of the size of hydrocarbon molecules present in the sample. When the sample has a low specific gravity, it is composed of lighter molecules with short carbon chains. Through the application of the equation below, it can be seen that the smaller the specific gravity, the larger the °API value becomes. Reference to Table 10-4 shows that crude oil having high °API values is considered more desirable, because it is composed of hydrocarbons with shorter carbon chains. The American Petroleum Institute has proposed using values of 20°API or less to identify the less desirable heavy crude oils.

$$\text{API Gravity} = \frac{141.5}{\text{specific gravity}} - 131.5$$

Chemical Characterization

The **Watson K factor** provides additional information about the chemical nature of a crude oil sample. The boiling point of crude petroleum varies over a wide range, constantly increasing during the distillation process as the lower boiling fractions are removed from the mixture. The **Watson boiling point** (T_B) is the average of five boiling points of a sample taken when 10 percent, 30 percent, 50 percent, 70 percent, and 90 percent of the sample has been removed. To convert the resulting average temperature to the **Rankine temperature scale** – the Fahrenheit equivalent of Kelvin temperature –459.7 degrees (°R = °F + 459.7) must be added to the average. The Watson K factor is calculated by the following equation.

$$\text{Watson K} = \frac{(T_B)^{1/3}}{\text{specific gravity}}$$

Higher value is placed on samples that give lower Watson K factors since those samples will have higher proportions of the lower-boiling, gasoline-producing molecules.

Sweetness – The sweetness (or sourness) of crude oil relates to the amount of sulfur present in the oil. While all petroleum contains small amounts of sulfur or sulfur-containing compounds, samples having more than 0.5 percent sulfur by mass are considered to be **sour crude** and have a lower market value. It is estimated that 58 percent of the U.S. oil reserves and more than 80 percent of the world reserves are sour crude. Today, modern fuels require the removal of sulfur, necessitating additional processing steps that add to production expense. However, the removed represents an additional marketable product.

Sample Calculation for Watson K:

The boiling point of a 1,000g sample of crude oil that has a specific gravity of 0.91 was monitored as the distillation occurred. The following temperatures were obtained:

Mass distilled:	100 g	300 g	500 g	700 g	900 g
Boiling Pt.	220°F	400°F	600°F	810°F	1,050°F

Average Boiling Pt. =
3,080/5 = 616 °F = 1,075 °Rankine

$$\text{Watson K} = \frac{(T_B)^{1/3}}{\text{specific gravity}} = \frac{(1075)^{1/3}}{0.91} = 11.26$$

Checking Your Understanding

1. What are chemical feedstocks?

2. Are crude oils with a high °API density more or less valuable than the ones with a low °API and why?

3. Are crude oils with a high Watson K factor more or less valuable than the ones with a low Watson K factor and why?

4. What is a sour crude oil and why does it bring a lower market value?

10-3 Petroleum Production and Refining

Petroleum Exploration

Petroleum production and refining encompasses a wide range of processes beginning with the exploration for oil and ending with the marketing of products to individuals and industry. The discovery of oil deposits is as much an art as it is a science. Trained petroleum geologists use **seismographs** to study seismic data (shock waves) generated by above and below ground explosions or by vibrator systems. Sound waves reflected from the interfaces between geologic strata provide useful information about the subsurface rock strata. **Magnetometers** and **gravimeters** measure small variations in the local magnetic and gravitational characteristics; these variations help reveal potential oil reserves situated in the lower strata.

Ground surface data, obtained by using a **gas chromatograph**, are used to identify the presence of petroleum products that have seeped to the surface. The careful recording of data about the geological strata from previously drilled wells provides information that is helpful in locating potential well sites in the immediate area. When these studies reveal a location with a high potential for oil production, a test well may be drilled to investigate specific geologic structures. The operating company must, however, 1) arrange the finances for drilling the well, 2) hire the drilling contractor as well as the service and supply companies, and 3) obtain the necessary permits. The permitting process assures the landowners and state and local agencies that the drilling operation will be performed in an environmentally sound manner. These permits are designed to assure that there will be no contamination of surface waters, groundwater, or soil in the area and that when the operation is completed the surface will be returned to its original pre-drilling condition. All of these conditions are submitted as a part of the work plan.

Drilling

The drilling of Drake's well in the 1850s was done using the **cable tool technique**. In this process, a heavy bit is repeatedly raised and dropped, with each impact punching the hole a little deeper, until the desired depth is reached. As shown in Figure 10-1, today a **rotary drilling rig** is used, with a **rotary bit** (Figure 10-2) on the end of a hollow drill stem that turns and grinds its way into the soil and rock. The hollow drill stem allows **drilling mud** to be pumped into the hole to lubricate and cool the drill bit. The mud flowing back to the surface is used to transport the rock cuttings out of the well hole; its weight also reduces the risk of a **blowout**. The potential for a blowout occurs whenever the drill bit enters a stratum containing gases under high pressure. A second measure that helps control the dangers of a well blowout is to cement steel **casings** in the upper portion of the well. The steel casing also prevents the bore hole walls from caving in, while eliminating the danger of contaminating groundwater aquifers, and keeping the oil from seeping into other strata.

As the well is drilled, the rock cuttings are separated from the drilling mud, cleaned, tested to determine if petroleum is present, and reused. When an oil-containing layer is reached, additional tests are used to estimate the amount of petroleum present.

Well Testing for Production Potential

The typical tests used to evaluate the well production capabilities are: coring, logging, and drill stem testing.

Coring – When a bore hole core sample is needed, the rotary drilling bit is replaced with a coring bit that cuts and brings a section of rock to the surface for further examination and testing for evidence of petroleum.

Logging – In this procedure, the drill pipe and bit are removed from the well and a set of logging instruments is lowered into the well. These instruments analyze the well walls with electrical measuring devices as well as sounding and radiation

Figure 10-1: Rotary drill rig.

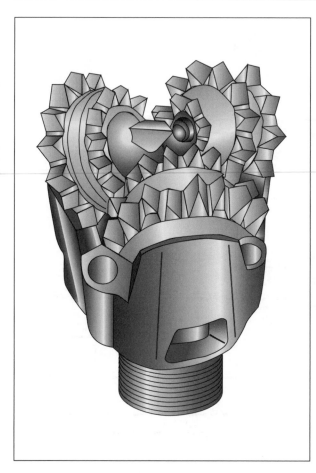

Figure 10-2: Rotary drill bit.

measuring devices. The data are sent to the surface where they are analyzed and compared to data from producing wells. These data provide information about the rock formation's porosity, conductivity, and composition, as well as about the potential presence of petroleum.

Drill Stem Testing – In this procedure, an instrument is lowered to collect samples of the fluids present and to record the fluid pressure in the hole.

Separation and Refining Processes

Petroleum exits the well as a complex mixture of hydrocarbons, water, salts, soil, rocks and other inorganic materials. The first step in producing com-

mercial product requires the separation of the organic hydrocarbons from inorganic materials. Field separation of useful petroleum from inorganic material involves the use of a large tank in which the mixture is allowed to separate. Gravity draws the denser inorganic solids, such as sand and soil particles, to the bottom of the tank. The water, often containing dissolved salts, forms a second layer, while the petroleum portion floats on the top. Some of the smallest hydrocarbons volatilize and are collected and removed from the headspace of the tank. The floating petroleum is drawn out of the tank, washed with water to remove salts, and pumped through pipelines to storage tanks or to the refinery. The inorganic wastes, the water, and the solids are withdrawn from the bottom of the tank and disposed of according to the work plan.

The resulting wastewater may be reinjected in an unused well, or undergo a second separation followed by the use of activated carbon to remove the final traces of hydrocarbons. The clean water may then be reused in the crude oil washing processes or in crop irrigation. The cleaning of the solid waste and oily soil is typically accomplished by bioremediation in a landfarming operation. In this process, the contaminated soils are mixed with an organic substrate on an impermeable surface. Fertilizers and atmospheric oxygen are regularly mixed into the moistened soil by cultivating and turning it. Bioremediation projects sometimes require years of soil agitation and mixing to fully remove all the petroleum residues.

The Refining Process

The simplest separation process is **fractional distillation** carried out at normal atmospheric pressures. A fractionating column (see Figure 30 in Chapter 3) is used to separate the mixture of hydrocarbons into various **boiling point fractions**. In a **fractional distillation tower**, the temperatures are highest at the bottom of the column and decrease with height. The lowest boiling substances – the ones with the shortest carbon chains – move upward through the column to the cooler top, while the heavier molecules recondense to liquid status near the bottom of the column. Fractions with different boiling points are drawn off at specific locations along the column. As shown by the diagram in Figure 10-3, of the

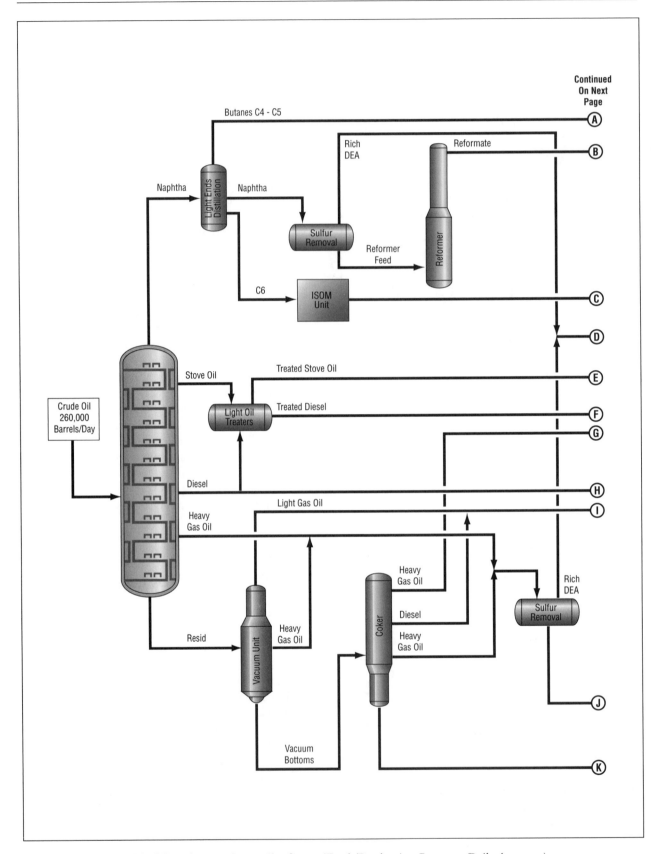

Figure 10-3/1: Simplified flow diagram for an oil refinery. (Feed/Production Rates are Daily Averages)

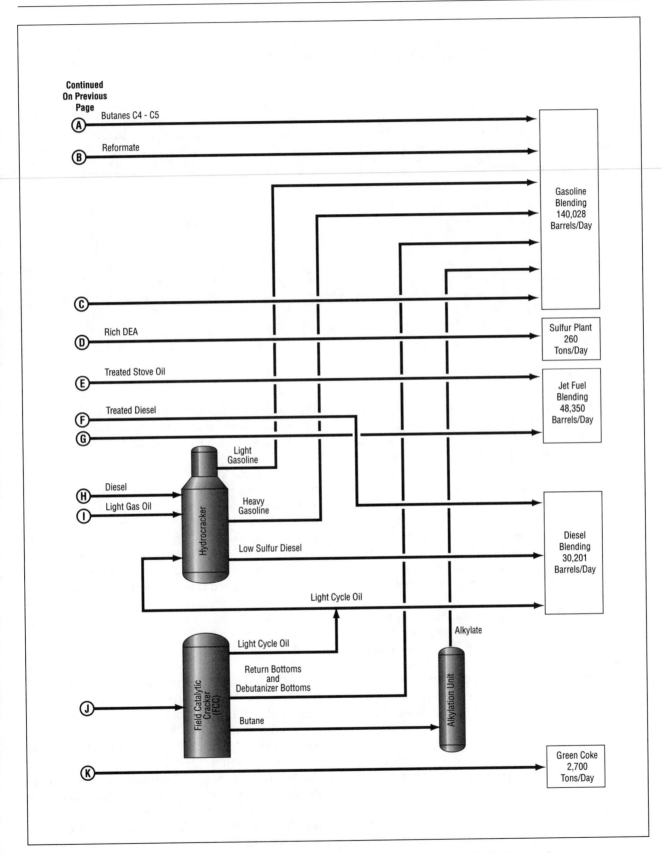

Figure 10-3/2: Simplified flow diagram for an oil refinery. (Feed/Production Rates are Daily Averages)

Box 10-1 ■ What is an Octane Rating?

The evolution of the automobile engine has required a similar evolution in fuel composition. Early automobiles simply needed a fuel that would explode when an oxygen/fuel mixture was provided with a spark. With the advent of high compression and high RPM engines, tailoring of the fuel molecules became a necessity. Fuel octane numbers became the measure of fuel quality. Additives were used to enhance the quality of fuel. Carbon chain branching was found to improve the smooth combustion of fuel molecules. For decades, highly branched tetraethyl lead was used as an effective additive. The Clean Air Act of 1970 responded to the problem of automobile engines spewing lead into the air by banning the additive, and the search for new additives and fuel mixes began.

The octane rating scale is based on two fuel molecules: the highly branched isooctane (2,2,4-trimethyl pentane) and the straight-chained n-heptane. In a standardized internal combustion engine, a pint of each fuel is used to run the engine while measuring several operating parameters. Based on these measurements isooctane is given a 100 octane rating and n-heptane is assigned a 0 octane rating. A 75/25 (isooctane/n-heptane) blend of the two fuels, therefore, would be assigned an octane rating of 75. Gasoline samples are run in the standard engine, and each mixture's performance is measured against the standardized scale. A gasoline sample performing in a similar way to a fuel mixture blend of 87 percent isooctane/13 percent n-heptane is given an octane rating of 87.

hydrocarbons arriving at the refinery only about 25-30 percent falls into the boiling point range to become gasoline. In order to meet consumer demand, the percentage of the gasoline fraction needs to be increased.

To increase the amount of gasoline and other valuable petrochemical feedstocks, various molecular restructuring processes have been incorporated into the modern refining operations. These processes – listed below – are designed not only to convert molecules in the crude oil into more valuable products, but also to reduce the amount of residue.

Catalytic Reforming – Catalytic reforming allows the **naphtha** fraction of crude oil to be reformed into aromatic molecules. These materials are purified for use as chemical feedstocks or used as gasoline additives to enhance the octane rating of the fuel. The hydrogen, formed as a byproduct, can be compressed and utilized in other phases of the refining process.

Hydrotreating – To produce cleaner burning fuels, sulfur and nitrogen contained in organic molecules are removed in the refining process. This is accomplished by bubbling hydrogen gas through the crude and passing the mixture over a catalyst bed that converts the sulfur into hydrogen sulfide,

H_2S, and the nitrogen into ammonia, NH_3. The hydrogen sulfide and ammonia are collected from the process and sold as byproducts.

Hydrocracking – Hydrocracking is similar to hydrotreating in that hydrogen gas is again being added to the reactor feed and the mixture brought into contact with a catalyst. However, the catalyst is in a reactor, where higher pressures of 2,000-3,000 psi are maintained. The raw materials for this process are the heavier residues and gas oil fractions; they are converted into the more desirable hydrocarbons in the naphtha ranges. By increasing the retention time, increased amounts of the lighter fractions can be produced, providing additional control over the product mix.

Alkylation – Alkylation is a process that combines short (C_2 and C_5) carbon chain hydrocarbons – one alkane and one olefin – to form branched molecules that fall into the light naphtha fraction. The reaction has a high potential to form highly desirable products like 2,2,4-trimethylpentane (**isooctane**). In the past, both **sulfuric acid** and **hydrofluoric acid** were in wide use as the catalyst for the reaction. Today, refineries are rapidly phasing out the use of hydrofluoric acid due to its extremely hazardous nature. To prevent the olefins from undergoing polymerization and forming larger residue-type molecules, the reaction is carried out at low temperatures.

Thermal Cracking (Coking) – Large hydrocarbon molecules can be broken into two or more small molecules by maintaining them at temperatures above 650°F for an extended time. This process was initially developed and used to increase the naphtha fraction, but has been mostly replaced by the more efficient catalytic cracking processes. However, in recent years coking of the residue – containing the very large hydrocarbon molecules – and of refinery wastes has been widely used to produce high-grade **petroleum coke**, which is used extensively in metal production and in the electronics industries. The coking process is an excellent example of converting a waste from an industrial process into a valuable product, further reducing disposable waste.

Catalytic Cracking – The complete chemistry of the catalytic breakdown of large hydrocarbon molecules is not fully understood. Several catalysts are used, and the products are a mixture of smaller molecules, but the catalytic cracking process results in the production of fewer petroleum gas molecules (C_1-C_4) than the coking processes. The use of a catalyst allows this reaction to take place at a lower temperature than in thermal cracking offering, therefore, greater control over the kind of products it produces.

Polymerization – Polymerization is another catalyzed reaction that results in the joining of some of the two or three short carbon chain olefins (unsaturated) to form larger hydrocarbon molecules that fall into the gasoline fraction. This process, therefore, reduces the amount of **Liquid Petroleum Gas (LPG)** and increases the amount of gasoline that can be derived from a barrel of crude oil.

Ether Process – Requirements of the 1990 Clean Air Act Amendments for reduced carbon monoxide emissions from automobile engines has led to the introduction of reformulated gasoline or **oxygenated fuels** containing either an alcohol or ether. **Methyl tertiary butyl ether (MTBE)** has become a popular ether for blending and is produced by the reaction between methanol and isobutylene. While MTBE is the current ether of choice for producing oxygenated gasoline, concerns have been expressed about groundwater contamination from leaking MTBE-containing storage tanks.

The products derived from a refinery tend to fall into two broad categories – **petrochemicals** and **refined products**. There is no clear distinction between the two; however, petrochemicals tend to include those substances that are obtained in a high degree of purity and have unique properties. Examples of petrochemicals would include the olefins: ethylene, propylene, and isobutylene and the aromatics: benzene, toluene, and styrene. The refined products are substances that are less pure, having been collected over a wider range of boiling points. These substances are typically marketed as blended products – most often for fuel blending – and a variety of other applications.

Checking Your Understanding

1. What are three functions served by the drilling mud?

2. What are three tests used to evaluate well production capabilities?

3. What are six methods used in an oil refinery to reduce the amounts of less valuable fractions and to increase the amount of more valuable fractions?

4. What are the two functions served by the thermal cracking (coking) process?

5. What is an oxygenated fuel?

10-4 Petroleum Transportation and Environmental Issues

Transportation Methods

Transporting petroleum and its products from the oil field to the refinery and then on to consumers is a major and costly undertaking for the industry. Since large volumes are transported, the potential for major environmental disasters is always a concern. Some of the more important transportation methods are listed below.

Pipelines

Pipelines provide the most economical means of transportation and account for most interstate transportation of large volumes of domestic crude oil and its products. About one third of all petroleum products in the United States are transported from refineries to consumers through a national network of more than 220,000 miles (352,000 km) of pipelines. To minimize the possibility of spills, pipelines are provided with constant automated surveillance for leaks and have spill catch provisions designed to minimize environmental contamination.

Super Tankers

The potential for great environmental harm is easily understandable in light of the fact that today's super tankers can carry more than four million barrels of crude oil per trip. The **Oil Spill Prevention and Liability Act of 1990**, passed in response to the Exxon *Valdez* spill in Alaska, has established rigorous standards for petroleum transport on the high seas. Provisions include requirements for double hull construction, increased hiring and training standards for vessel operators, and establishing emergency response teams. The Act also includes provisions to assure that if a spill occurs there is enough emergency equipment available to assure the response can be immediate and effective. The first effort in any response is to stop the leak and get the product corralled. Finally, the water surface is skimmed to remove as much of the product as possible before it reaches the shoreline. Written procedures and training are also required to insure the safety of the workers.

Tractor-Trailers

The transporting of petroleum products on the nation's highway system presents special risks to citizen safety and the possibility of environmental damage. To provide for safety, DOT regulations require drivers to have special training as well as licensing to transport hazardous materials. Each region of the country has response teams trained in up-to-date techniques to deal with highway emergencies.

Environmental Issues

RCRA Subtitle C Exemption

Waste generation from the drilling process consists of the drilling mud and the water produced from the groundwater aquifers that are intersected during the drilling process. Mixed in with the drilling mud is the rock and soil that was removed to make the well hole. The mud is screened to remove the drill bit cuttings and reused in the drilling process. When the drilling project is completed, the mud is dried with absorbents and disposed of by being used as a component in making concrete. Wastewater is often injected into other wells in the field to force oil production in nearby wells.

In 1978, EPA proposed hazardous waste management standards for several types of large volume wastes. Among the wastes specifically covered were "...gas and oil drilling muds and oil production brines." Generally, EPA believed that these "special wastes" were lower in toxicity than other wastes being regulated as hazardous waste under RCRA. Subsequently, Congress exempted these wastes from RCRA Subtitle C hazardous waste regulations until a study could be done. The oil and gas exemption was later expanded in the 1980 legislative amendments to RCRA to include "...drilling fluids, produced water, and other wastes associated with exploration, development, or production (E&P) of crude oil or natural gas." Because of the study, EPA issued a regulatory determination in 1988 stating that control of E&P wastes under RCRA Subtitle C regulations is not warranted. Although these wastes are exempt from Subtitle C regulations, the federal exemption was not meant to imply that these wastes could not present a hazard to human health and the environment, if improperly managed. As a result, E&P wastes are still subject to RCRA Subtitle D solid waste regulations and other federal regulations, such as the Clean Air Act (CAA), Clean Water Act (CWA), Safe Drinking Water Act (SDWA), Oil Pollution Act of 1990 (OPA), and state regulations.

According to legislative history, the terms "other wastes associated" and "intrinsically derived from the primary field operations" were intended to distinguish exploration, development, and production operations from transportation and manufacturing operations. In general, the exempt status of E&P waste depends on how the material was used or generated, not necessarily whether the material is hazardous or toxic. An easy way to determine if a waste qualifies under the E&P waste exemption, is to answer the following two questions:

— Has the waste come from down-hole, that is, was it brought to the surface during oil and gas E&P operations?

— Has the waste otherwise been generated by contact with the oil and gas production stream during the removal of produced water or other contaminants from the product?

If the answer to either of the above questions is affirmative, then the waste is most likely considered exempt from RCRA Subtitle C regulations!

Clean Air Act Amendments of 1990

Industrial sources of air pollution have been largely controlled. The transportation industry – a mobile source of pollution – has become the next environmental cleanup target. The Clean Air Act Amendments of 1990 (CAAA) and the controversy surrounding the Global Warming Treaty are directly related to energy use and long-term concerns with the use of petroleum. The CAAA focus is on cleaning up automobile emissions by mandating cleaner

burning fuels and requiring the marketing of zero emission cars in areas with the most severe air pollution.

Global warming concerns present yet another problem. The long-term threat from the continued use of fossil fuels is the return of our atmosphere to its early carbon dioxide-rich composition. Monitoring of atmospheric composition over the past 35 years reveals that there has been a 15 percent increase in atmospheric carbon dioxide. That increase presents the potential for continued greenhouse effect and global warming. Indeed, the petroleum industry has brought many changes to life on earth. As a cheap and portable energy source, it has expanded the economic base of the world and provided a higher standard of living for many. Nevertheless, with the blessings of cheap energy come many technological challenges. Technology will have to produce solutions for these challenges in the twenty-first century, if our existence on earth is to continue.

Checking Your Understanding

1. What is the most economical means of interstate transportation for petroleum and petroleum products?

2. What are four of the provisions of the Oil Spill Prevention and Liability Act of 1990?

3. By what percent have atmospheric carbon dioxide levels increased in the last 35 years?

4. What are two factors that determine if a waste falls under RCRA Subtitle C exemption?

Summary

The fossil fuels on earth today – petroleum and coal – are believed to be the product of vegetation and animal life that existed millions of years ago. Petroleum has been seeping to the surface for centuries and has been used for a variety of purposes, including as a waterproofing material and as a medicine. The U.S. petroleum industry was developed during the nineteenth century: the first oil company was founded in 1854 and the first oil well was drilled five years later.

The products from a modern petroleum refinery are distinguished by their uses and fall into the three categories of fuels and lubricants, chemical feedstocks, and residue. The economic value of crude oil is dependent upon three properties: density, chemical characterization, and sulfur content. The density is determined by API gravity (°API); a high value is desirable. Its chemical characterization is determined by the Watson K factor, which is based on the Watson boiling point and specific gravity of the oil. The lower the Watson K factor, the more valuable the oil. The sweetness of the crude oil is also an important factor. Those oils having more than 0.5 percent by mass of sulfur are considered sour and require additional steps and expense to process.

Petroleum exploration combines much scientific data with a bit of luck. Once a site has been selected, the well is drilled using a rotary drilling rig. As the hole gets deeper, tests – such as coring, logging, and drill stem testing – are periodically performed to examine for evidence of petroleum. The oil brought to the surface is first placed in a large tank to allow the various components to separate. The oil is then removed from the top, washed with water, again separated and sent to storage tanks or to a refinery. The water is cleaned and used in the oil washing operation or reinjected into a nearby well to assist in further oil production. The bottom sediments are removed and sent to a landfarming operation, where bioremediation is used to remove the hydrocarbon wastes.

Once the oil reaches the refinery, fractional distillation is used to separate its components into different boiling point fractions. Through a series of processes, the smaller molecules undergo polymerization to increase their size, the larger molecules undergo thermal cracking or catalytic cracking to reduce their size, and some molecules are subjected to reforming processes designed to increase their degree of carbon chain branching.

Petroleum and petroleum products must be moved from the oil field to the refinery and on to the consumers. This process is both costly and environmentally risky. The most economical and environmentally safe interstate transportation of petroleum and petroleum products occurs by pipelines. Super tankers, holding more than four million barrels of oil, have been developed to move oil over great distances, while tractor-trailer rigs are commonly used to move small amounts of product, such as gasoline, to local service stations.

Most of the wastes generated during petroleum exploration, development, and production (E&P), are exempt by the RCRA Subtitle C Exemption. Drilling mud is screened and dried with absorbents and used as a component in making cement. Most refinery wastes are subjected to the thermal cracking (coking) process and converted into useful products. The largest waste product from the petroleum industry is the production of carbon dioxide due to the combustion of all its products. The Clean Air Act Amendments of 1990 target automobile exhaust and require the use of cleaner burning fuels, which contain oxygen. Concerns about the greenhouse effect, which has seen a 15 percent increase in atmospheric carbon dioxide levels in the past 35 years, is providing technological challenges that will continue well into the twenty-first century.

Critical Thinking Questions

1. Is the RCRA Subtitle C Exemption for oil and gas E&P wastes justified? Prepare a defense for the position selected, based on the nature of the wastes generated.

2. A large percentage of the known oil reserves contain sour crude. What are the likely environmental and economic consequences of this fact?

3. If Sample A has a specific gravity of 0.89 and Sample B has a specific gravity of 0.91, what is the °API gravity for each sample? Which of the samples will bring a higher price and why?

4. What is the specific gravity of an oil sample, if it has a Watson K factor of 10.500 and a T_B of 1080°R?

11

The Chemical Production Industries

by Melinda Trizinsky

Chapter Objectives

Upon completing this chapter, the student will be able to:

1. **Describe** the basic processes currently used in the chemical production industries.

2. **Identify** common waste streams generated by the chemical production industries.

3. **Describe** opportunities for waste minimization and pollution prevention that are available to this industry.

4. **Evaluate** the economics of using various source reduction, recycling and treatment technologies.

5. **Cite** specific regulations and permits that pertain to these industries.

11-1 Introduction

The chemical production industries (SIC code 28) supply essential raw materials to other industries. This chapter will discuss industries that produce both inorganic (SIC code 281) and organic (SIC code 286) chemical feedstocks. Chemical production industries use chemical reactions and physical treatment processes to convert raw materials into products. They may be in a solid, liquid, or gaseous phase.

The partial listing of chemical products shown on the left hints at the scope of the chemical industry. It would be impossible to describe each chemical production process in detail. However, the chapter will provide general descriptions of a few major industrial processes, which will contribute to a general understanding of the industry and allow the reader to identify applicable federal regulations and pollution prevention opportunities.

Much of the activity in the chemical production industries involves converting one chemical substance into another. These types of processes rely on the reactive nature of the chemicals involved. Since "reactivity" is one of the four criteria used to define a chemical as being a hazardous material (reactivity, toxicity, flammability, and corrosivity), many of the feedstocks, intermediates, products, and wastes from this industry are hazardous. Therefore, as a class, chemical production facilities are among the heaviest sources of pollution in the United States. Consequently, chemical production facilities are among the most heavily regulated manufacturing industries.

The output from chemical production industries is used mainly by other industries and may end up in products such as:

— Cleaning Products

— Cosmetics and Fragrances

— Explosives

— Fertilizers and Soil Conditioners

— Fire Retardants

— Fuels

— Glues and Adhesives

— Inks and Dyes

— Lubricants and Coolants

— Paints and Finishes

— Pesticides

— Pharmaceuticals

— Plastics

— Solvents

— Surfactants

— Textiles

— Wood Preservatives

11-2 General Production Practices

Because of the diversity of the raw materials, production practices, and products, it is difficult to make generalizations about the chemical production industries. However, in the following section several aspects that are of general concern throughout the industry will be discussed.

Material Handling and Housekeeping

During the manufacturing process, the raw materials, intermediates, products, and wastes associated with chemical production must move to the plant, within the plant, and from the plant. These activities are collectively referred to as **material handling**. Much of the activity within a chemical production facility is associated with material handling, and the most likely time for a spill to occur is during material transfer operations. Liquids and gases are typically transferred via pipelines and with pumps. Dry solids may be transferred via bulk containers, conveyor belts, or screw augers. Alternatively, solid reactants may be suspended in a liquid to form a slurry; which is usually transported via pipelines.

Since many of the materials handled in a chemical plant are hazardous, there are strict OSHA rules that specify handling procedures and set exposure limits for the workers. Workplace exposure is generally controlled by:

1. Requiring workers to wear appropriate protective gear;

2. Controlling material transfers to minimize spills;

3. Providing adequate ventilation to keep atmospheric concentrations within acceptable limits;

4. Using proper storage procedures;

5. Conducting frequent safety audits;

6. Educating workers about the risks and training them to properly handle hazardous materials; and

7. Providing access to Material Safety Data Sheets.

In addition to controlling the worker exposure to hazardous materials, the chemical plant must also control releases of their feedstocks, intermediates, products, and wastes to the environment. The measures used to achieve this control depend upon the physical form of the material, its chemical properties, level of hazard, and the engineering requirements of the production line.

Mixing and Blending Operations

Operations that disperse solids or gases in liquids, blend liquids, and produce fluid motion are collectively referred to as **mixing and blending processes**. The distinction between mixing and blending is rather vague; both processes physically mix reactants, but blending generally results in a more homogenous product that cannot be separated by physical processes. Successful chemical reactions require good contact between reactants; therefore, mixing and blending are important processes in the chemical production industries.

The vessel in which a chemical reaction occurs is referred to as a **reactor**. There are several types of reactors, described below, used to process liquids. The configuration, the operating characteristics, and the engineering design of reactors make up the heart of a chemical production plant. In order to reduce worker exposure and releases to the environment, nearly all reactions are carried out within closed reactors. Industrial reactor design is influenced by:

— Reaction stoichiometry (molar ratios of the reactants);

— Reaction kinetics (speed of the reaction);

— Material handling requirements (reactant delivery, product removal);

— Supply and withdrawal of heat (heating and cooling);

— Accommodation of phase changes (solid $\leftrightarrow$ liquid $\leftrightarrow$ gas);

—Mixing requirements (assuring adequate contact between reactants);

—Catalyst replenishment or regeneration;

—Energy requirements (electricity and fuel to operate motors, blowers, heaters, coolers, pumps, etc.); and

—Pollution control equipment.

Batch and Semibatch Reactors

A **batch** process calls for all the reactants to be placed in the reactor in the beginning and remain there for a fixed period of time; usually until the reaction is complete or until the product yield reaches an acceptable level. The term **product yield** refers to the amount of product formed from reactants. The yields are often expressed in terms of percentage, which are relative to the maximum amount of product that can be formed from the reactants. A variation of this process – a **semibatch** reactor – is characterized by the continuous addition of a specific reactant to the chemicals that were put in the vessel initially. Batch reactors are more difficult to operate than the other common types of reactor. This is because the flow of raw materials and products from batch reactors is intermittent (i.e., these reactors are periodically loaded and unloaded), and the reactor often needs to be cleaned between batches. Therefore, batch reactors are generally more costly to operate than other types of reactors, and they are not often used in chemical production. Because batch processes are more easily controlled, however, they often achieve better yields of the desired products. Some specialty chemicals, such as pharmaceuticals, are produced in batch processes.

A test tube is an example of a simple batch reactor. Industrial scale batch reactors may hold thousands of gallons.

Continuous-flow Stirred Tank Reactor (CSTR)

CSTRs are the most common type of reactor used for industrial chemical production. As the name implies, materials are continuously flowing through CSTRs. Reactants are continuously added via the influent, and a portion of the reactant mixture is withdrawn continuously (effluent). The influent and effluent flow rates are the same, so the volume in the reactor remains constant. Ideally, reactants that enter the reactor are instantaneously mixed with the contents of the reactor to form a homogenous mixture. The average time that fluid remains in the reactor, or retention time, depends upon the flow rate and the volume of the reactor. The **retention time** is the average time during which chemicals entering the reactor form products.

Because some reactants leave the CSTRs unreacted, product yields are lower in this process than in batch reactors. The effluent contains a mixture of product and reactant molecules. Despite this limitation, CSTRs are frequently used in the chemical industry because they are easier and cheaper to operate than batch systems.

Plug Flow Reactor

In **plug flow reactor** systems, the reaction takes place as the reactants flow through a tubular reactor. Ideally, all chemicals entering the reactor flow at a steady rate through the reactor (i.e., plug flow), and the reaction progress corresponds with the position of the reactants within the reactor. No mechanical mixing occurs in these reactors.

Other Types of Reactors

Multiphase reactors are used for reactions in which the reactants are present in more than one phase. A solid-liquid reactor may suspend the solids in a liquid phase to create a slurry, or the reaction may occur as liquid reactants flow through a solid phase. Liquid-gas reactors usually use **gas diffusers** to bubble small gas bubbles through liquid reactants. (e.g., solid-liquid, liquid-gas, or immiscible liquids). Various other methods are used to increase the contact between liquid phases and facilitate the reactions. For example, **emulsifiers** are often used to aid the dispersion of water-insoluble materials in aqueous solutions. There are numerous other reactor types that combine aspects of these idealized reactors, but they are beyond the scope and intent of this chapter.

Catalysts

A **catalyst** is a substance that changes the rate of a chemical reaction without changing chemical equilibria and without being consumed. Selective catalysts are frequently used during chemical production to speed up reactions, reduce energy requirements, and increase the production of desired products from the reaction mixture by favoring one reaction over another.

In industrial usage, catalysts usually lose activity and selectivity over time due to overheating or poisoning (contamination). Therefore, the production process must be designed to allow catalyst replacement or regeneration. Common contaminants that may poison catalysts include oxygen, nitrogen, and sulfur compounds, polynuclear aromatics (PNAs), and metals present as contaminants in feedstocks.

> Chemical products must usually be isolated from reaction mixtures in order to be useful. The end use of a product will influence the level of purity that is required.

Product Separation

Once a chemical product is formed, it typically needs to be separated from the reaction mixture in which it was produced. These reaction mixtures may contain solvents, co-products, by-products, unreacted chemicals, catalysts, etc. The separation technology employed depends on many factors including the phases of the substances being separated, the number of components in the mixture, and whether recovery of by-products is important. Numerous techniques such as distillation, extraction, filtration, and settling are used either singly or in combination to accomplish the separations. Publications such as *Perry's Chemical Engineers' Handbook* or basic texts on chemical plant design can be consulted if highly specific information is desired. A few of the more common separation methods are discussed below.

Adsorption

Adsorption is a separation method that relies on the

> The success of adsorptive separations depends upon the choice of **adsorbent**. The compound to be removed must have a high affinity for the adsorbent. However, the adsorption process must be reversible so that products can be recovered and adsorbents can be regenerated.

selective collection and concentration of particular types of molecules onto solid surfaces. These types of separations are typically used for liquids or gases. Adsorptive separations are subdivided into either **bulk separations** or **purification** processes. Bulk separations involve the adsorption of 20-50 percent of the process stream; the product of interest will be in either the solid or the liquid phase depending on the process; and, sometimes, useful products will be found in both phases. Purification processes on the other hand, remove impurities that account for less than 3-5 percent of a process stream.

Adsorption technologies are commonly used for the following purposes:

— Odor, taste, or color removal;

— Chemical drying (water removal);

— Product recovery;

— Purification;

— Demineralization; and

— Pollution control (particularly in air scrubbers).

Distillation

Distillation is a common, but energy-intensive method of performing separations. It accounts for 95 percent of all separations made in the organic chemical and petroleum industries. It is also one of the most energy-intensive industrial processes, accounting for approximately eight percent of the total energy use of the U.S. industrial sector. In a distillation process, a liquid mixture is heated inside a vessel or column. The component with the lowest boiling point vaporizes and exits the vessel leaving the rest of the mixture behind. The vaporized component is recovered by cooling it, which causes it to condense. This process can be repeated as many times as needed to separate all of the components of the mixture.

Drying and Evaporation

The terms **drying** and **evaporation** are not interchangeable in the chemical production industry. Drying refers to the separation of volatile liquids from solid or semisolid material by vaporization. Evaporation refers to vaporizing a solvent from a solution or from a relatively dilute slurry. These processes are used for the following purposes:

— Volume reduction;

— Production of a useful form (e.g., salt from brine);

— Resource recovery;

— Waste concentration to minimize disposal costs; and

— Transformation of a waste into a useful product.

Extractions

Extractions are based on the attraction between solvent and solute molecules – following the principle of "like dissolves like." Liquid-liquid extractions transfer components from one liquid phase to another. Liquid-solid extractions – also known as leaching – rely on the solubility of the desired component relative to other components in the mixture.

Filtration

Filtration processes separate solids from fluids. The fluid is typically passed through a porous medium and the solid particles that are too large to pass through the medium are retained. In the chemical industries, filtration is used in a number of ways:

—To purify liquid products by removing solid contaminants;

—To recover solid phase catalysts from reaction mixtures; and

—To recover solid phase product from slurries.

Gravity Separation

Settling tanks, flotation tanks, and immiscible fluid separations are examples of gravity separation techniques. These processes rely on differences in density to achieve the separation.

Energy Requirements

Chemical production is an energy-intensive enterprise. When both inorganic and organic chemical production are considered, the chemical industry accounts for approximately seven percent of the total United States energy consumption and 25 percent of the energy used by all manufacturers. Thus, the chemical production industries are indirectly responsible for additional pollution associated with energy production.

In 1991, the chemical production industry used more than five quadrillion Btu of energy (approximately equivalent to 2.5 million barrels of crude oil per day).

Checking Your Understanding

1. What are three functions served by a catalyst?

2. Name three factors that may influence the choice of technology used to separate reaction mixtures.

11-3 Inorganic Chemical Production Industries

Inorganic chemicals are those that do not contain the element carbon, with the exception of carbon monoxide, carbon dioxide, carbonates, bicarbonates, and cyanides. The inorganic chemical industry produces more than 300 chemicals including alkalis, acids, chlorine, industrial gases (e.g., oxygen, nitrogen, helium, hydrogen, etc.), inorganic pigments, and others. Table 11-1 lists the top 16 inorganic chemicals produced in the United States; note that naturally occurring minerals, which are used but not produced by the inorganic chemical industry, are excluded from this list. A chemical that is produced

at this scale and is subsequently used in other production processes is referred to as a **commodity chemical**. SIC code 281 is used to categorize manufacturing facilities that produce inorganic chemicals.

Raw Materials

Naturally occurring minerals (such as salt, limestone, and gypsum), inorganic by-products of petroleum refining (such as H_2SO_4), air (such as N_2, O_2, and CO_2), and water are the primary raw materials used by the inorganic chemical industry. These raw materials are processed in various ways to extract the chemicals of interest. Mineral ores are typically crushed and processed to extract the compounds of interest. Air is compressed and cooled to separate to atmospheric gases. Process water may be purified to remove contaminants that might poison catalysts or otherwise interfere with chemical reactions. Often these steps occur before the primary feedstocks are delivered to the chemical processing plant. Primary inorganic products are often used to produce other inorganic chemicals.

Compound*	Major Uses
Sulfuric acid, H_2SO_4	Acids
Nitrogen, $N_2(g)$	NH_3
Oxygen, $O_2(g)$	Steelmaking
Ammonia, NH_3	Fertilizers
Calcium oxide, CaO	Bases, cement
Sodium hydroxide, NaOH	Paper
Phosphoric acid, H_3PO_4	Fertilizers
Chlorine, $Cl_2(g)$	Paper, plastics
Sodium Carbonate, Na_2CO_3	Glass
Nitric acid, HNO_3	NH_4NO_3
Ammonium nitrate, NH_4NO_3	Fertilizer
Urea, $CO(NH_2)_2$	Fertilizer
Carbon dioxide, $CO_2(g)$	$CO(NH_2)_2$
Hydrochloric acid, HCl	Varied
Ammonium sulfate, $(NH_4)_2SO_4$	Fertilizer
Carbon, C (black)	Tires

*Minerals that occur naturally and are used but not processed by the chemical industry are excluded from this list.

Table 11-1: Major inorganic industrial chemicals produced in the United States.

Production Processes

Most inorganic production facilities produce a relatively small variety of substances, but each one is produced in large volumes. The production of inorganic chemicals typically involves one or more reaction steps. Redox conditions, pH, and temperature are common variables used to control reaction progress, and water is the primary solvent used. Reactions are typically followed by separation steps to recover the desired product(s).

Large volume inorganic products, such as the chemicals listed in Table 11-1, are typically produced in continuous flow reactors, but specialty chemicals may be produced in smaller batch operations. The production of specialty chemicals usually requires more processing steps. After inorganic products are separated, they are often processed to produce a suitable form for their intended use; for example, inorganic solids may be formed into uniform size pellets or crushed to form fine granules. The final product is then packaged for sale or distributed via rail or pipeline.

Waste Streams

The types of waste streams generated by inorganic chemical production industries obviously depend upon the chemicals produced. Some potentially useful by-products are treated as wastes because they have insufficient market value to justify further handling. As disposal costs and markets change, the economic value of these by-products may also change.

Both process water and water used for housekeeping purposes contribute to aqueous waste streams contaminated with minerals, metals, acids and alkalis. Solid wastes include packaging materials, off-spec products, spent catalysts, and spills. Sources of air pollution include dusts, aerosols, and waste gases generated during plant operation. The hazard associated with these waste streams depends upon the nature of the chemicals being processed.

Checking Your Understanding

1. What element is typically not present in an inorganic reactant or product?

2. Name the most used industrial inorganic compound.

11-4 Organic Chemical Production Industries

The organic chemical production industry uses a large number of diverse processes to make thousands of different carbon-containing chemical products. For the purposes of this chapter, it is not necessary to discuss any particular reaction. Some of the common types of reactions are listed in Table 11-2. A few generalizations about these reactions will suffice. Endothermic reactions require energy to proceed; most of these processes require large amounts of thermal energy to carry out the desired conversions. Exothermic reactions give off energy, and excess heat needs to be drawn off. Many separation techniques also require controlled temperatures; therefore, organic chemical production facilities are often large consumers of energy and water (used for cooling). Catalysts are frequently used to enhance the rate of favorable reactions. Despite temperature controls and the use of catalysts, the yield of many chemical processes is often limited by thermodynamic equilibria and other physical or chemical factors. The yield from these processes is often less than 50 percent of the desired product.

Alkoxylation
Condensation
Halogenation
Oxidation
Polymerization
Hydrolysis
Hydrogenation
Esterification
Pyrolysis
Alkylation
Dehydrogenation
Amination (Ammonolysis)
Nitration
Sulfonation
Carbonylation
Hydrohalogenation
Dehydration
Dehydrohalogenation
Oxyhalogenation
Catalytic Cracking
Hydrodealkylation
Phosgenation
Hydration

Table 11-2: Common reactions used to produce organic chemicals.

Primary Building Block	Secondary Building Block	Tertiary Building Block
Ethylene	Ethylene dichloride Ethylene oxide Ethylbenzene	Vinyl chloride Ethylene glycol Vinyl acetate
Propylene	Propylene oxide Acrylonitrile Isopropyl alcohol	Acetone
Benzene	Ethylbenzene Cumene Cyclohexane	Styrene Phenol Acetone Adipic acid
Methanol	Acetic acid Formaldehyde Methyl t-butyl ether	Vinyl acetate
Toluene		
Xylenes p-isomer	Terephthalic acid	
Butadiene		
Butylene		

Table 11-3: High volume organic chemical building blocks.

Raw Materials

The primary feedstocks for the organic chemical production industry come from the petroleum and inorganic chemical industries. In general, feedstocks are combined in a series of reaction steps to produce both intermediates and end products. Table 11-3 shows that the primary organic chemical building blocks are generated principally from petroleum refining. In general, primary building blocks are used in more reactions than secondary and tertiary building blocks. Most feedstocks can be used in more than one reaction; also, a particular product can be made by more than one reaction sequence.

Waste Type	Potential Sources of Emissions
Air Emissions	Point source emissions – stack, vent (e.g. laboratory hood, distillation unit, reactor, storage tank vent), material loading/unloading operations (including rail cars, tank trucks, and marine vessels) Fugitive emissions – pumps, valves, flanges, sample collection, mechanical seals, relief devices, tanks Secondary emissions – waste and wastewater treatment units, cooling tower, process sewer, sump, spill/leak areas
Liquid wastes (Organic or Aqueous)	Equipment wash solvent/water, lab samples, surplus chemicals, product washes/purifications, seal flushes, scrubber blowdown, cooling water, steam jets, vacuum pumps, leaks, spills, spent/used solvents, housekeeping (pad washdown), waste oils/lubricants from maintenance
Solid Wastes	Spent catalysts, spent filters, sludges, wastewater treatment biological sludge, contaminated soil, old equipment/insulation, packaging material, reaction by-products, spent carbon/resins, drying aids
Ground Water Contamination	Unlined ditches, process trenches, sumps, pumps/valves/fittings, wastewater treatment ponds, product storage areas, tanks and tank farms, aboveground and underground piping, loading/unloading areas/racks, manufacturing maintenance facilities

Table 11-4: Potential environmental releases during industrial chemical manufacturing.

Production Processes

The typical organic synthesis involves combining two or more feedstocks in a series of **unit operations**. Unit operations are the individual processes that make up a manufacturing process. The first process is usually a chemical reaction. Most reactions take place at high temperatures, involve metal catalysts, and include one or two additional reaction components. The reaction may yield one or more useful products. The yield of the reaction will partially determine the kind and quantity of by-products and releases. Commodity chemicals tend to be synthesized in a continuous reactor while specialty chemicals usually are produced in batches. Many specialty chemicals require a series of two or three reaction steps.

Once the reaction is complete, the desired product must be separated from the by-product. The product may also require further processing to produce a saleable commodity. Frequently, by-products are also sold. This reduces the waste-handling burden and may actually contribute to the economic viability of the process.

Most organic chemical manufacturing facilities produce multiple chemicals. Many process units are designed so that production levels of related products can be varied over wide ranges. This flexibility allows the manufacturers to adjust to changes in the market – such as variations in feedstocks and product prices – by changing the processes used and the rate of production.

Waste Streams

The chemical industry generates large amounts of waste, including about 20 percent of the nonhazardous manufacturing wastes, about 90 percent of the hazardous manufacturing wastes, and about 40 percent of the toxic manufacturing wastes produced in the United States. The organic chemical industry managed about 6.3 trillion pounds of production-related TRI waste in 1993. About 90 percent of these wastes were managed onsite through recycling, energy recovery, or treatment.

Organic chemical manufacturers emit chemicals to all media including air, through both fugitive and direct emissions, water (direct discharge and runoff), and land. The wastes that a particular manufacturer generates depend on the feedstocks used, the processes and equipment in use, and the storage and handling practices employed. Emissions can

Of the more than 300 chemicals currently listed on the Toxic Release Inventory (TRI), 204 are reported as released or transferred by at least one organic chemical facility.

product transfer operations), than during the reaction. Some potential sources of pollution are shown in Table 11-4.

Checking Your Understanding

1. List three reasons why the chemical production industry is a primary source of pollution.

2. What is the principal source of the high volume feedstocks used by the organic chemical production industries?

vary from hour to hour depending upon the part of the process that is underway. For example, for batch reactions in a closed vessel, the chemicals are more likely to be emitted at the beginning and end of a reaction step (associated with vessel loading and

11-5 Federal Regulations and the Chemical Production Industry

Nearly all the federal environmental statutes affect the chemical production industry. The industry is also subject to numerous laws and regulations from state and local governments. The major federal regulations affecting the chemical industry include:

— Resource Conservation and Recovery Act

— Clean Water Act

— Storm Water Discharge Rule

— Comprehensive Environmental Response, Compensation, and Liability Act

— Superfund Amendments and Reauthorization Act

— Emergency Response and Community Right-to-Know Act

— Toxic Substances Control Act

— Clean Air Act

These laws were described in Chapter 2. Some of the provisions in these laws that are particularly relevant to chemical manufacturers are described below.

The Toxic Substances Control Act (TSCA) mandates that chemical companies 1) submit premanufacture notices that provide information on health and environmental effects for each new product, and 2) test existing products for these effects.

> The chemical industry pays about $300 million a year in Superfund chemical feedstock taxes.

It further authorizes EPA to require testing of certain substances and to prohibit, limit, or ban the manufacture, process, and use of chemicals, if deemed necessary. These provisions directly affect the activities of the chemical production industries.

The Clean Air Act authorizes EPA to set limits on chemical plant emissions. The chemical production industries must comply with National Emission Standards for Hazardous Air Pollutants (NESHAPs). New facilities may be subject to New Source Performance Standards (NSPSs). Many NSPSs apply to organic chemical manufacturers including those regulating flares that are used to burn off waste gases, storage vessels, equipment leaks, air oxidation processes, distillation operations, reactor processes, and wastewater treatment. Title VI restricts the use and distribution of ozone-depleting chemicals (ODCs). Under this legislation, ODCs have been classified according to the danger they pose to the ozone layer, with Class I compounds generally posing the greatest risk. Production of Class I compounds – including 15 kinds of chlorofluorocarbons (CFCs) – will be phased out entirely by the year 2000, while certain hydrochlorofluorocarbons (HCFCs) will be phased out by 2030.

Checking Your Understanding

1. What are the eight federal regulations that affect the chemical industry?

2. List two provisions of TSCA that apply to the chemical industry.

3. What are ODCs?

11-6 Pollution Prevention Options for Chemical Production Industries

Pollution prevention in the chemical industry is process-specific. Thus, it is difficult to generalize about the relative merits of different pollution prevention strategies. The age and size of the facility as well as the type and number of its processes will determine the most effective strategy. Brief descriptions of some of the more widespread, general pollution prevention techniques used at chemical production facilities are provided below. Note that many of these pollution prevention opportunities, aimed at reducing wastes and materials use, have been used for many years as the primary means of improving process efficiencies and increasing profits in this industrial sector.

Setting up a pollution prevention program does not necessarily require exotic or expensive technologies. Some of the most effective techniques are simple and inexpensive. Others require significant capital expenditures, but may provide a return on that investment by reducing waste disposal costs or converting a waste to a product. Pollution prevention has several advantages over end-of-pipe waste treatment technologies, as discussed in previous chapters.

Research and Development

Pollution prevention techniques can be applied at many stages in the chemical production process. Changes made at the research and development stage generally have the greatest impact. All possible reaction pathways that produce the desired product can be examined during research and development. These reactions can then be evaluated in light of yield, undesirable by-products, and health and environmental impacts. This approach is sometimes referred to as "green synthesis." Chemical production processes usually involve large capital investments in technology and equipment; therefore, pollution prevention at the earliest stages is unlikely unless a company designs a new production line or facility. Luckily, modifying current processes and equipment affords numerous pollution prevention opportunities.

Improved Operating Procedures

Good operating procedures rely not on changes in technology or materials, but on human adaptability. Small changes in personnel practices, housekeeping, inventory control, waste stream segregation, material handling and scheduling improvements, spill and leak prevention, and preventive maintenance can mean big waste reductions. To reduce waste in chemical production it is necessary to:

— Reduce inventory of raw materials. Test materials first to determine whether they can be used in current manufacturing processes.

— Reduce excess and off-spec production. Produce only the amount requested or needed.

— Segregate waste to recover useful materials and cut disposal costs.

— Conserve water. Reuse rinse waters. If possible, clean process equipment with process fluids.

— Prevent contamination of storm water runoff, thereby eliminating treatment of contaminated rainwater. Replace leaking valves, pumps, and seals.

Process Redesign

Chemical engineers have excellent backgrounds to solve environmental problems in the chemical industry. In the past, chemical engineers designed chemical production processes to recover product and unconverted raw materials. They pursued this strategy to the point that the cost of further recovery could not be justified. Now waste disposal costs and environmental regulations have made source reduction an equally good investment. Designs that reduce the amount of waste generated can also reduce energy consumption and maintenance costs.

Box 11-1 ■ Elimination of Mercury – Case Study

A chemical manufacturer used mercury in the manufacture of anthraquinone (AQ) dyes, commonly used for dyeing cotton. AQ was sent through a number of intermediate stages to produce aminoanthraquinone. A sulfonation reaction during this process required the use of mercury as a catalyst; thus, mercury wastes were produced. The company's research group developed a pathway for the production of amino-anthraquinone that avoided the sulfonation step, thus eliminating the need for mercury as a catalyst. By fully implementing this new process, the company eliminated the use of 2,280 pounds per year of mercury at a single plant. The plant also reduced mercury air emissions by 10 pounds per year, eliminated 58 pounds per year of mercury released to wastewater, 325 pounds per year of solid waste, and 39,500 pounds per year of mercury-contaminated material (from process filtering operation).

Recycling

Recycling is the use, reuse, or reclamation of a waste after it is generated. Reuse reduces material and disposal costs. Examples of recycling opportunities in the chemical production industry include:

— Recycle and reuse excess, off-spec materials, and samples taken for quality control testing.

— Use closed-loop recycling to use or reuse a waste as an ingredient or feedstock in the production process.

— Reuse inert ingredients when flushing solids handling equipment.

— Segregate and reuse dust emissions in the production process.

— Distill waste solvents, and regenerate catalysts.

— Develop end-use markets and demonstrate economics for recycled materials

Process Modifications

Modifying a manufacturing process can be a very effective way of preventing pollution. Upgrading the system not only reduces waste, but also improves product quality, saves money by reducing the need for maintenance, and increases control of raw materials used in production. Consider the following process modifications:

— Develop technology to better utilize waste heat.

— Schedule the production of chemicals that use the same production line to reduce cleaning requirements.

— Shift from batch manufacturing to continuous manufacturing to reduce losses that occur during material handling.

— Use dedicated process equipment to reduce the frequency of equipment cleaning and associated wastes.

— Relocate process equipment and change piping configuration to prevent possible cross-contamination.

— Clean equipment with small amounts of cleaning solution. If water is the cleaning agent, use sprays or jets of water to clean tanks or equipment. Where possible, the small amount of concentrated waste collected should be recycled as a raw material. Rinse machinery and tanks less often.

— Use pumps and piping systems to transfer liquid materials; this can reduce spillage.

— Reformulate products to improve handling characteristics; for example, make chemicals in pellet form instead of powder to reduce dust emissions.

— Substitute less toxic or non-toxic materials as raw products.

— Use higher-grade feedstocks to increase catalyst life, improve yields, and reduce off-spec product.

Box 11-2 ■ Process Modification – Case Study

A company that manufactures chemical intermediates used in the production of a powerful antibiotic purified one intermediate via caustic hydrolysis to assure that the intermediate did not degrade while being held for subsequent processing. The purification generated 2,300 gallons of toluene-contaminated wastewater per batch. Three process vessels used for purification could emit as much as 1,600 pounds of toluene per year to the air.

The company determined that by modifying the mole ratios of reagents used to manufacture the intermediate, the one reagent causing the degradation could be significantly reduced. The caustic purification step was eliminated. Product stability and throughput dramatically increased. Manufacturing costs were reduced by $280,000 per year without any capital expenditure for new equipment.

Employee Participation and Management Support

Management commitment as well as employee participation are vital to a successful pollution prevention program. Management can demonstrate its commitment to pollution prevention and encourage employee participation by:

— Training employees in pollution prevention techniques;

— Encouraging employee suggestions;

— Providing incentives for employee participation; and

— Providing necessary resources.

The Responsible Care® Initiative

Even before federal pollution prevention efforts were initiated, several of the large chemical production companies put waste reduction programs into place. Companies that were members of the Chemical Manufacturers Association (CMA) in 1988, for example, began implementing the voluntary Responsible Care® initiative that has the following six components:

— Product stewardship;

— Employee safety and health;

— Community awareness and emergency response;

— Process safety;

— Distribution;

— Pollution Prevention.

As part of the initiative, the Codes of Management Practices laid out 106 specific management practices that all CMA member companies must implement as a condition of membership. The Pollution Prevention Code is designed to improve the chemical industry's performance by seeking an ongoing, long-term reduction of waste released to the environment. Reduction goals give priority to wastes having the highest potential hazard. Participating companies also manage any remaining waste through practicable methods that best protect the environment as well as the health and safety of employees and the public. Obtaining employee and community input regarding pollution prevention programs is another requirement of the Code. Companies must also promote pollution prevention concepts to customers, suppliers, other companies, and the government.

Checking Your Understanding

1. What are the six areas that need to be considered when implementing a pollution prevention program?

2. List three advantages of using process modification as an effective way of preventing pollution.

Summary

The chemical production industries supply both inorganic and organic feedstocks to other industries. Since chemical reactivity is among the criteria used to define a hazardous material, most feedstocks, intermediates, and products of this industry as well as its wastes are hazardous. As a result of the many hazardous substances involved, the chemical industry is one of the most heavily regulated industries.

Because of the diversity of raw materials and production practices, and also because of the fact that spills most often occur during transfers, the chemical industries have developed a variety of techniques for moving solid materials including bulk transfers, conveyor belts, screw augers, and pumps and pipelines for transporting liquids and gases. The industry has also developed many different mixing and blending operations to bring the reactants into contact with each other. Among the common pieces of equipment are the batch, semibatch, and multiphase reactors; the plug flow; and the continuous-flow stirred tanks. Catalysts are frequently used to change the reaction rate in the direction favoring the formation of desirable product(s) and to save both production time and costs. Once the products have been formed they are separated and/or concentrated using a vast array of techniques. Adsorption, distillation, evaporation, similar solvent extraction, filtration, and gravity separation techniques using flotation or settling tanks are some of the more important methods.

More than 300 inorganic chemicals are produced by the chemical industries. They include important inorganic acids and bases as well as chlorine, oxygen, nitrogen, helium, and hydrogen. Many of the compounds are minerals or are extracted from minerals like limestone, sulfur, and gypsum. The major sources of pollution generated by their manufacture include dusts, aerosols, and waste gases. Organic chemical production industries make thousands of different carbon-containing chemical products. Their primary feedstock is petroleum. Most organic synthesis involves combining two or more feedstocks in the presence of a catalyst in a series of unit operations. Most processes are continuous while some – usually specialty chemicals like pharmaceuticals – are produced in batches. Once the reaction is completed, the product(s) must again be separated from the by-products using many of the same separation techniques noted above. Organic chemical manufacturers emit chemicals to all media, including air, water, and land.

TSCA requires the filing of premanufacture notices revealing all known information on the health and environmental effects of a new product; it also requires the testing of existing substances when they are given a new use. The CAA authorizes EPA to set limits on chemical plant air emissions. The CWA sets regulatory standards for all types of wastewater discharged.

Pollution prevention activities in the chemical industry tend to be process-specific. The pollution prevention strategies that are initiated at the research and development stage – an approach sometimes referred to as "green synthesis" – generally have the greatest impact. Small changes in personnel practices, good housekeeping, inventory control, waste stream segregation and material handling can all reduce the amount of waste generated. Whenever a waste cannot be avoided, recovery of the waste and its use or reuse can reduce both material and disposal costs. Finally, process modification and replacement of outdated equipment can also save money by reducing the need for maintenance and by increasing control over raw materials used.

Management commitment and employee participation are integral elements of a successful pollution prevention strategy. Even before federal pollution prevention efforts, a number of large chemical production companies formed the Chemical Manufacturers Association (CMA) and implemented the voluntary Responsible Care® initiative. In one part of the initiative, for example, specific management practices that all CMA member companies must implement are specified. Its Pollution Prevention code is designed to improve the industry's overall performance by seeking ongoing, long-term reductions of the amount of waste being released to the environment.

Critical Thinking Questions

1. Many of the pollution prevention options discussed in this chapter are similar to process modifications for optimizing production. Be-

cause of rising waste disposal costs and stricter environmental regulations, the economic incentives shifted to favor higher investment in pollution prevention. How might chemical production processes be altered in response to substantial increases in energy costs? Discuss how this might affect pollution production and control.

2. Spills are most likely to occur during material transfer operations. Name two ways to reduce the likelihood of these types of spills.

3. From a pollution prevention and marketing perspective, what are the advantages and disadvantages of participation in voluntary programs like the Responsible Care® initiative?

12

The Mining Industries

by Andrew J. Silva

Chapter Objectives

Upon completing this chapter, the student will be able to:

1. **Describe** the major mining divisions and subdivisions.

2. **List** the geological processes responsible for valuable rock and mineral formations.

3. **Explain** the major subsurface and surface mining methods.

4. **Describe** the processes that generate mining wastes.

5. **List** the main environmental laws that govern mining.

Chapter Sections

12-1 Introduction

Mining is an activity pursued by man since prehistoric times. As humans learned to make and use tools, they soon realized certain kinds of **rock** worked better than others. They discovered, for example, they could make a good spear point or knife by chipping and shaping **obsidian**, whereas limestone or sandstone were not good choices. Humans soon began going to specific places where the useful rocks were formed to excavate the rock they wanted. As they evolved, they also found that certain **minerals** – such as copper – found in rocks could be pounded into a shape or melted and poured into molds to make metal tools. This was the beginning of the **Copper Age**. They then discovered that copper and arsenic or tin could be alloyed by mixing and melting them together. This alloy was stronger and more durable, and its widespread use gave the name for the subsequent era in human development: the **Bronze Age**.

In today's world, we use a variety of products that were manufactured with ingredients obtained with mining activities. Most of us know that steel girders are made from iron obtained by mining activities. We also know that mining activities produce gold that is used to make rings for our fingers and circuits in our computers. It may be less obvious that window glass is, in part, made from mined silica sand. Both coal and uranium are also mined and are being used to produce the energy that powers our lights, air-conditioners, and heating systems. The phosphate contained in the fertilizer we apply to the lawn, the gypsum on the living room wall, the gravel used in cement and asphalt paving are all made from mined materials.

Two basic activities are required to produce these products: 1) the desirable rock or mineral must be excavated, and 2) it must be processed into a usable product. The preparation may be very simple, as when quarry equipment crushes rock to make gravel. Conversely, metal production often requires multi-step processing efforts that are complex and involve using chemicals and/or heat to extract, concentrate, and transform the ore into a metal we can use. This chapter will discuss the excavation and initial preparation of ore. Most of these activities are performed at the mine site. Chapter 13 will address the more complex processes required to free metals from their ores.

12-2 Mineral Deposition and Mine Types

Mining is the process of extracting and performing the initial activities required to produce an ore. Ore is a mineral or rock that is excavated, prepared, processed, and sold, returning a reasonable profit to the individual or company engaged in the mining activity. Some mining activities produce ores such as gravel or coal, which are immediately saleable; others produce intermediate products that require additional processing before they are marketed.

Mineral Deposition Methods

Geological conditions, **grade**, size, depth, location, and orientation of a mineral deposit all determine what type of mining method will be used to recover the ore. Ore deposits are formed in several ways. One type is formed when **magma** or **hydrothermal fluids** rise from great depths toward the surface of the earth. When this happens, the elements contained in the magma or in the hydrothermal fluids can no longer remain in solution – due to the temperature and pressure decrease that accompany the process – and precipitation of the solutes occurs. If the solutes include metals such as copper or gold, the metals precipitate or combine with other elements to form minerals rich in these metals.

When precipitation of minerals occurs in highly fractured rock and/or in rock that transmits fluids easily, the result is an amorphous rock zone enriched with the mineral. The zone where this occurs can be several miles in height and width. Fracture density and orientation vary in rock formations as do **permeability** and porosity. These variations can result in certain areas of the ore zone having higher concentrations of the metal than others. If the rock has few fractures or does not transmit fluid well, the zone enriched with minerals will usually consist of isolated cracks or breaks in the earth's crust. The faults and/or fractures in which enrichment has occurred are called ore veins. Large enriched zones or single veins can be mined if the concentrations of mineral/metal are high enough.

Ore deposits are also formed by the action of mechanical and/or physical forces that **weather** and transport rock across the earth's surface. For example, the action of the currents, tides, and waves along an ocean shore can – over time – break down rock rich in **quartz**. The resulting sand is transported by the currents and deposited to form beaches or reefs. Millions of years after the seas have receded, a deposit of relatively pure sand may be found in the center of a continent.

Water also plays an important role in the formation of certain organic deposits. Organic matter, for example, can be picked up and transported by rivers that flow through lush tropical areas. Over time, the organic material is deposited in the river's delta material where it becomes compressed and heated and is transformed into peat, lignite, or bituminous or anthracite coal. For centuries these deposits have been harvested as a valuable fuel source for heating and cooking purposes. Evaporates – such as **gypsum** and salt – were formed when mineral-laden waters concentrated in lakes or seas without any discharge flow. As the water evaporates, a supersaturated condition is reached, and the minerals are precipitated from solution. Such a process is occurring today in Utah's Great Salt Lake and California's Salton Sea. Eventually, mineral deposits of fairly constant thickness will be formed.

Types of Mines

Mining can be divided into two broad categories: subsurface and surface mining. Subsurface mines are subdivided into underground mines and **solution** (injection) **mines**; surface mines are commonly subdivided into **strip** and **open pit mines**. These mining methods can be further subdivided into many other types; however, since this chapter is intended to provide an overview of the mining industry, only the common examples from each of the major categories will be discussed.

Underground mining methods are most likely to be used if the deposit is deep or if it contains several veins of high-grade ore. Underground mining methods usually involve the development of a **shaft** or **adit** in the area of the deposit. **Drifts** and tunnels are then bored from the shaft to access the deposit. Shafts, adits, and drifts have relatively small cross-sections, typically ranging from several feet to several yards in height and width. As a result, small mining equipment – such as excavating machines, drills, and hauling equipment – is used underground. To produce the same amount of ore as a surface mine, underground mining requires a larger number of small pieces of equipment, which, in turn, necessitate more operators and maintenance employees.

In addition to the number of employees required, underground mines usually require the ore to be lifted to the surface in small containers called **skips**. The vertical lift in an underground mine can be thousands of feet long, compared to surface mines where roads, trains, or conveyors are used to haul the ore several hundred feet out of the mine. A single truck, for example, can haul in excess of 200 tons of rock from a surface mine whereas a skip can haul only a fraction of that tonnage.

All these factors result in the cost per ton of ore mined deep underground being much higher than that for ore from surface mines. To offset the higher costs, underground mining must concentrate on high-grade ore, which usually occurs in small deposits. Surface mining, on the other hand, usually produces large quantities of lower grade ore because production costs are lower.

Checking Your Understanding

1. What type of geological process is responsible for today's deposits of copper and gold?

2. What are two types of surface mines?

3. What are two types of subsurface mines?

4. Describe the ore extraction process in surface and in subsurface mines and compare costs.

12-3 Mining Methods

Underground Mining Methods

Underground mining methods are subdivided according to the presence or absence of support for the **back**, or ceiling, in the mining area. Regardless of the method employed, the process of excavating rock underground usually involves the following steps: drilling the rock, loading explosives into the drill holes, blasting the rock, loading the rock onto a conveyor or vehicle, transporting it to the skip, and elevating it to the surface.

Most underground mines require some type of support for the back to remain in place as the work below progresses. The support may consist of drilling holes deep into the back and sides of the working area and inserting long metal rods. The rods, called **rock bolts** (see Figure 12-2), are glued to the rock with resins. As they are tightened, the large blocks of cracked rock immediately adjacent to the mine area are attached to the more **competent rock** high above the back. In mines with bedded deposits such as coal, these rock bolts tie the layers of rock above the back together in a manner similar to laminated beams.

Timbering – the installation of wooden pillars or **cribs** – is also used to support the back in many mines. Sometimes a roof of wooden planks is installed above the work area to support the back and to catch falling chunks of rock before they strike the miners below. Another type of support, called **room-and-pillar** mining, leaves pillars of ore to support the back. Figure 12-3 is an example of this method, which is often used in bedded deposits such as coal mining. Excavating the ore and transporting it from the mine creates the open areas, called rooms.

Different support methods are frequently used in combination, such as using both rock bolts and ore pillars. Other types of supported mining methods include **supported stoping** – which uses rock bolts to support the development of a large underground room, called a **stope** – and **square-set stoping** – which utilizes timbers to support the back of a stope.

Some mines are situated in rock so competent that no support is needed, while others are designed so the ore-bearing back intentionally caves. The broken ore is directed to chutes where it is loaded for transport to the mineshaft. An example of an un-

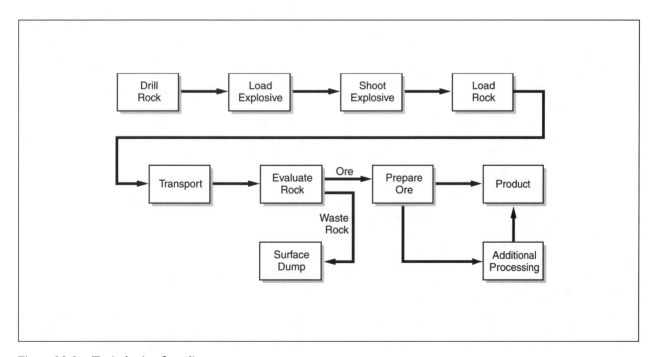

Figure 12-1: Typical mine flow diagram.

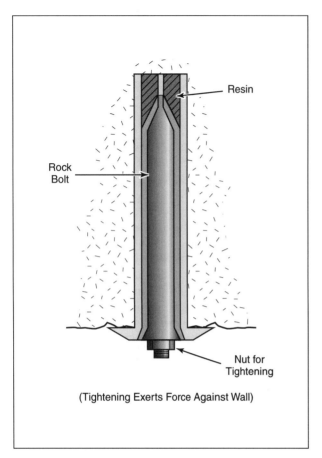

Figure 12-2: Rock bolt cross-section.

supported underground mining method is **shrinkage stoping**.

Shrinkage stoping is used in strong rock that will not break and fall by itself, but requires blasting to cause it to cave into the open area. In a mine that relies on caving methods to **rubblize** (break) the ore, the intent is to use the natural forces of stress and pressure to minimize the amount of blasting required. After blasting, gravity causes the rock to drop into an area where it can be loaded into the transport vehicle or on a conveyor. If shrinkage stoping is employed, the underground structures must be developed to support this method by first providing access to the area to be mined by **haulage drifts**. The haulage drifts are bored below the block of ore that will be stoped. Above the haulage drifts **ore chutes** are excavated in the competent rock and are designed to funnel broken ore from the bottom of the stope to the transport vehicle or conveyor. After these structures are in place, a series

of holes is drilled into the ore above the chutes, loaded with explosives, and shot. The rubblized ore drops to the bottom of the stope and down the ore chutes where it is loaded and transported. As the shot ore is withdrawn from the stope, a space opens between the rubblized rock and the new back. Miners enter this area, drill the next round of holes, and shoot it. This process is repeated until all of the ore is removed.

Other unsupported mining methods include **sub-level stoping**, an adaptation of shrinkage stoping that provides for the intermediate development of access drifts above the haulage level. **Block caving** – like shrinkage stoping – requires the development of haulage ways and ore chutes below the ore body and uses gravity to draw broken rock through chutes and into rail cars or other transport vehicles. The bottom of the ore body is also drilled and shot just like in the shrinkage stopping operation, but differs in that it is used in weak, less competent rock and relies more on gravity, stress, and strain to rubblize the ore. No drilling and blasting occurs after the initial opening at the bottom of the ore body is made.

Solution mining is an *in situ* mining method. It is used to recover minerals or elements that exist underground by dissolving them in water or chemicals. Typically, a circular pattern of holes is drilled into the ore deposit around a center hole. Water or chemicals are pumped into the outer holes and out of the center hole. This creates a **hydraulic gradient** that causes the solution to flow toward the center hole, dissolving the target mineral/metal as it moves through the ore body. This method is obviously cost effective, as it does not involve blasting, loading, and hauling ore and waste to the surface of an underground mine. Unfortunately, underground conditions are not always ideal. Fractures and faults in the ore as well as **country rock** often interfere with the ability to recover the chemicals pumped into the ground. As a result, proposals for this kind of mining are vigorously reviewed by both federal and state agencies and relatively few are given approval.

Surface Mining Methods

As the name implies, surface mining occurs on the earth's surface. Probably the most common type of surface mining is quarrying for rock, gravel, and sand

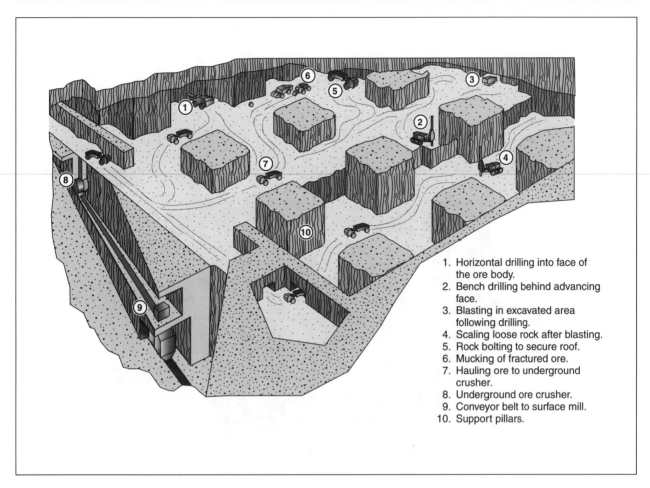

1. Horizontal drilling into face of the ore body.
2. Bench drilling behind advancing face.
3. Blasting in excavated area following drilling.
4. Scaling loose rock after blasting.
5. Rock bolting to secure roof.
6. Mucking of fractured ore.
7. Hauling ore to underground crusher.
8. Underground ore crusher.
9. Conveyor belt to surface mill.
10. Support pillars.

Figure 12-3: Room-and-pillar mining.

used in the construction industry. Every highway, driveway, foundation, and other concrete structures contain quarried materials. Rock **quarries** are usually shallow mining operations that involve blasting the rock into fragments. Once loosened, the rock is loaded onto a conveyor or into a truck, transported to **screens** – where it is sized – and stockpiled. Later it is loaded and transported to the job site or to a mixing facility.

As mentioned earlier, surface mining equipment is typically much larger than equipment used underground. Even with the efficiency of larger surface mining equipment, the mine will eventually reach a depth at which the increased hauling time slows productivity to the point that it is no longer profitable. This point – at which cost equals revenue – is called the break-even point and determines the depth of the mine. Because the materials taken from quarries tend to bring low prices compared with other industrial minerals and metals, the break-even point for a gravel quarry, for example, is reached at a relatively shallow depth.

Another type of surface mine is a **placer mine**, which is the mining of weathered material from rock **outcrops** that was transported by wind or water to a different location. Placer gold deposits, for example, are created when gold from weathered rock is transported downstream by water and deposited. As the water goes around a bend or slows upon entering a larger body of water, the gold – due to its greater density – settles in the bottom sediments. A great deal of placer mining occurred during the Gold Rush – and still occurs today to a certain extent – in an attempt to recover these small amounts of gold from gravel bars in the streams. Placer gold mining requires relatively small amounts of equipment and

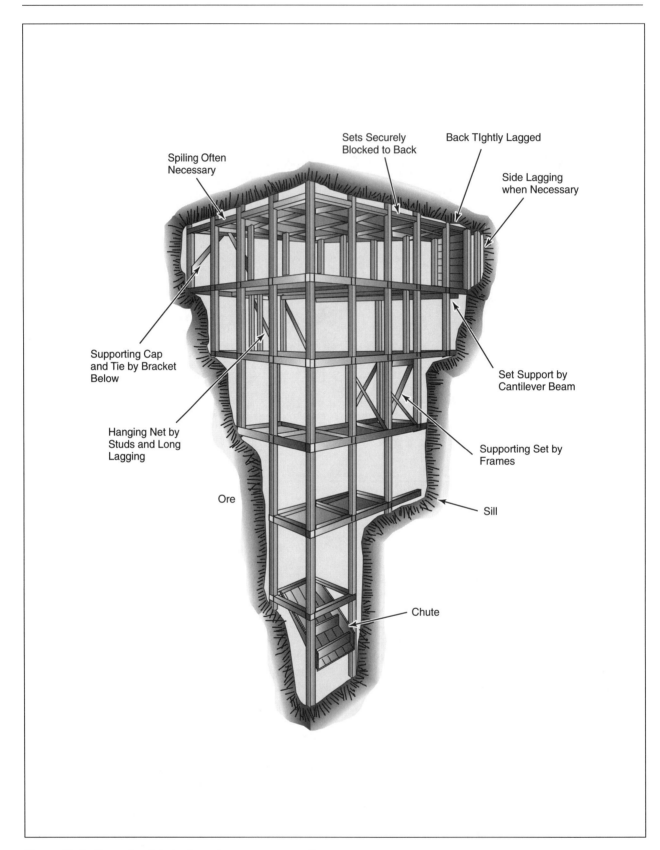

Figure 12-4: Example of timbering using square-set stoping.

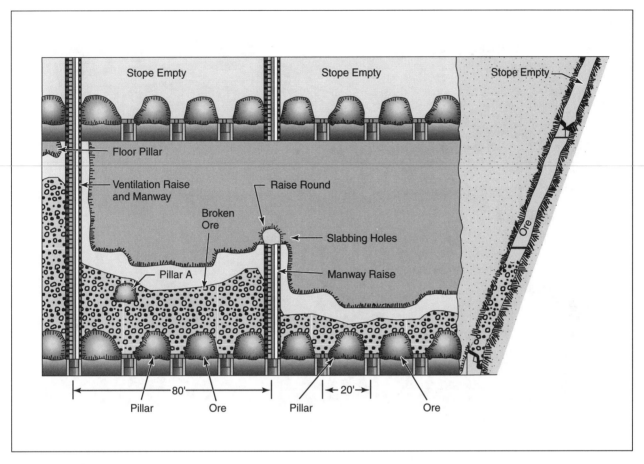

Figure 12-5: Cross-section of a shrinkage stope.

labor since nature has already transported and concentrated the gold. Many recreational miners have staked placer claims or own land containing placers and operate small-scale mining operations on weekends and during vacations. It is believed that most of the rich placer gold deposits have already been discovered and mined; however, some may still exist in more remote areas, such as Alaska.

Strip and open pit mining are the two most common types of large surface mining methods in use today. Strip mining occurs in deposits composed of horizontal linear beds. Coal and **evaporates** such as gypsum that lie near the earth's surface are commonly mined by this method. Generally, the sale price of these products is lower than that for metals. Though these mines have a deeper break-even point than a quarry, they are shallow compared to surface metal mines. The deposit itself is usually covered by a layer of soil, gravel, and/or rock, for which there

is no market. This material, known as **waste, overburden**, or **spoil**, must first be removed or stripped from the deposit. It is then placed in piles that are known as **spoil piles** or **waste dumps**. Since the 1970s, both federal and state regulations require that topsoil be stockpiled separately from the other spoil piles so that it can later be used during reclamation activities.

If the overburden is made of rock, it is removed by drilling and blasting. If it is soft – such as compressed soil, clays, or partially weathered sandstone – bulldozers equipped with spike-like attachments, called rippers, are used to loosen the material. If the overburden is shallow, a **dragline** can be used to expose the ore. Excavating equipment such as electric shovels, front-end loaders, backhoes, and fleets of large trucks are often used in strip mines where the depth is beyond the reach of a dragline. On some occasions, smaller equipment such as a fleet of scrap-

Figure 12-6: A typical strip mining operation.

ers and bulldozers is used to strip the overburden and pile it next to the opened bed. The deposit itself is usually mined with excavating equipment and a fleet of trucks to haul the ore from the pit to the preparation area. Usually, smaller equipment is used to mine the product so that a higher level of quality control can be exercised. Great care is taken to keep the product being mined segregated from the waste rock.

As shown in Figure 12-6, strip mines are shallow and may extend horizontally for great distances. The initial strip or cut can be mined and the overburden from the next strip placed back into the excavation made by the first strip. After the backfill is completed and recontoured, topsoil that was stored separately is applied, and native grasses, shrubs, and trees replanted. In this way, portions of large strip mines are reclaimed at the same time that the mine is developed. Years later, when the last cut is mined, the overburden that was stockpiled from the first cut may be hauled to the final cut and backfilled to complete the reclamation process.

Open pit and strip mining are very similar and, in most cases, use the same equipment. The primary differences between the two lie in the kinds of deposit they mine and in the depth of the excavation. As discussed earlier, a number of minerals are deposited by hydrothermal fluid and magma. These deposits are usually not contained in well-defined beds, but rather in amorphous regions that may extend to great depths; strip mining is used for these kinds of deposits. As shown in Figure 12-7, open

Figure 12-7: Typical open pit mining operation.

pit mines are characterized by vertical excavations in areas of moderately high mineral concentrations, as opposed to the extensive horizontal developments found in strip mines.

Most open pit mines operate for decades and remove millions of tons of ore and overburden. The overburden is typically placed in piles next to the pit, but it cannot be used for backfill, and the area cannot be reclaimed until the mine operation closes. Since it is not economical to the mining companies to reclaim the area, and the overburden would not refill the pit, anyway, the regulatory agencies have not specified any requirements in this matter.

Checking Your Understanding

1. What are the various uses for rock bolts?

2. What supports the back in the room-and-pillar mining method?

3. What steps are repeated when using the shrinkage stoping method?

4. Of the four types of surface mines discussed, which is typically the deepest?

5. What types of material are found in the overburden?

12-4 Mining Wastes

Surface Disturbance

All mining operations result in some degree of surface disturbance and – predictably – the method, size, and location are all contributing factors. Underground mining activities usually result in the least surface disturbance, because – obviously – they are underground and because they are smaller operations. Remote locations require the construction of access roads, power lines, pipelines, etc.; therefore, they have a much greater impact on the environment than mines located in already developed areas.

Disturbance of the earth's surface by clearing trees, building roads, laying foundations for buildings, stripping topsoil, and removing the overburden all accelerate the natural geological weathering processes. Plant removal, for example, increases the chance that wind will blow dust into the air and that surface water **runoff** will contain sediments. The severity of sediment problems on nearby streams will be increased by highly erodible soils, steep slopes, intense rains, and rapid snow melts. To reduce the problem, federal and state agencies require mining companies to collect and divert runoff from disturbed areas into sediment ponds. These ponds are sized and constructed to insure runoff is held long enough for the sediments to settle before discharging. The water quality exiting these ponds is regulated by EPA as a point source discharge under the Clean Water Act and must meet the requirements of the mine's NPDES permit. For runoffs containing clay or other colloidally suspended materials, flocculants are added to aid in their removal and to help meet the discharge requirements.

Acid Mine Drainage and Heavy Metals

Geologic areas that support mining are usually rich in a variety of metals. Metal-bearing ore deposits are generally formed deep underground, under anaerobic conditions. As a result, the sulfur found in the minerals along with the metals is in a reduced oxidation state. These deposits are called sulfide deposits or **sulfide rock**. Under these same anaerobic reducing conditions – but closer to the earth's surface – coal was formed in the **benthic zone** of ancient swamps, lakes, and deltas. Since coal is formed from ancient plant material, it also contains other elements. As a result, coal deposits also contain metals and reduced oxidation state sulfur compounds. The most common metal and sulfur mineral in coal is **pyrite**, commonly known as fool's gold, FeS_2. When mining occurs, these minerals become exposed to air and water. Since thermodynamics favor the oxidation of these sulfur compounds, the mineral crystals are destroyed liberating both the oxidized sulfur compounds and the metal ions. Percolating surface or aquifer waters moving through the fractured rock pick up these metal ions resulting in runoff that can have metal levels exceeding drinking water standards and causing – in severe cases – distress or death to microorganisms, fish, and wildlife. When the oxidized sulfur compounds dissolve into water, sulfuric acid – H_2SO_4 – forms and results in **Acid Mine Drainage (AMD)**.

In addition to the physical and chemical processes at work, there are also bacterial processes that can produce AMD and runoff or **leachate** with elevated metals content. Certain bacteria – such as *Thiobacillus ferrooxidans* and *Thiobacillus thiooxidans* – can oxidize sulfides in their metabolic process. Several oxidation reactions are possible, but a simplified equation for the sulfide oxidation is shown in the following equation:

$$MS + 2O_2 \xrightarrow{\text{Bacteria}} MSO_4$$

In this equation, MS is a metal sulfide, O_2 is oxygen, and MSO_4 is the oxidized metal sulfate. In this reaction, the bacteria use the oxidation reaction to obtain energy in the form of captured electrons. Just as in the example above, the oxidation of the sulfur compound liberates the metal ions and the sulfur-containing ions making them available for AMD and metal ion mobilization. If the metal ion concentration in mine discharge becomes a problem, then chemical treatment technologies, like hydroxide precipitation, can be used to remove it. In recent years, interest has grown in the use of bioremediation technologies to remove metals from mine discharges.

When required, low pH AMD can been neutralized by the addition of **lime** to the sediment ponds. The lime raises the pH of the acid runoff and mine discharges as they flow through the ponds before exiting the property. This method of treatment is effective while the mine is in operation. However, the natural and bacteria-driven oxidation process does not end with mine closure, so the problem can persist for years, decades, and sometimes centuries after the mine has closed. Regular and periodic applications of lime must continue as long as AMD remains a problem.

Particulate Emissions

Disturbed areas are particularly susceptible to weathering by wind. Areas disturbed by mining activities are regulated by the Clean Air Act (CAA) as administered by EPA. In addition to wind-driven particulate emissions, mines using outdoor crushers or screening plants emit dust from these activities. Mobile equipment such as trucks, bulldozers, and graders produce large quantities of dust as they move over mine roads. Dust from the roads is often controlled by the use of water trucks. In some cases, roads that are heavily traveled receive surface treatments. The materials used for the treatment bind fine particles of soil and rock to each other so they are less susceptible to erosion. Dust from crushers, screens, etc. can be controlled by having spray nozzles deliver a fine water spray over the dust-generating points. Nowadays this equipment is often enclosed in a building that contains a bag house or a similar device for capturing the dust before it is emitted into the atmosphere.

Waste Rock

As previously explained, the volume of waste rock produced by underground mines is usually smaller than that produced by surface mines. As a result, surface waste dumps at underground mines are usually small and contain primarily waste from the initial development activities, such as waste rock that resulted from the shaft or adit excavation and from the development of access drifts and tunnels. Although these dumps do not contain economic concentrations of metals and minerals, they often contain higher concentrations than the surrounding weathered rock. Runoff from these piles is regulated by EPA or by individual states under the CWA. It must, therefore, also be monitored and treated in accordance with provisions of the mine's NPDES permit.

Waste rock from surface mines has the same general characteristics as waste from underground mines. The primary difference is the volume, since in surface mines large quantities of waste rock must be removed before the ore is reached. The waste dumps from open pit mines commonly contain million of tons of material in piles that are hundreds of feet deep and extend for thousands of feet. These waste piles have the potential of generating much more contaminated leachate than do the small waste dumps found at most underground mining sites.

Regulatory agencies often require wastes with high metal content or with an acid potential to be segregated in areas of the dump that are less likely to be subject to later disturbance, leaching, or weathering. Therefore, they are typically placed above the water table and relatively deep in the pile. Placement between the earth's surface and the groundwater, in the area called the **vadose zone**, reduces the chances of the waste later coming in contact with water that could leach metals or form acid drainage. If the dump is sited in an arid region, it is unlikely that significant contamination of metals or ADM will result. In areas receiving high levels of precipitation, however, it may be necessary to encapsulate the problem waste with a high-clay content material. The lower permeability of clay-rich soil diverts the water soaking into the ground around the problem waste, thus minimizing the amount of metal or acid leachate formed. By burying problem waste deep within the dump and away from the sides, the chance of this material being susceptible to exposure by erosion is greatly reduced.

It is usually difficult for mining companies to reduce the volume of waste rock generated by the mine. Mining companies usually attempt to eliminate waste rock production before submitting mine designs in their permit applications. The potential for contamination can also be reduced by constructing waste dumps and facilities in areas that are less likely to produce problems with surface or groundwater contamination or with airborne contaminants. Locating waste dumps high on the side of a mountain or on elevated and relatively flat areas away from streams and gullies reduces the amount of water that will come in contact with the pile. This, in turn, reduces the possibility of problems with AMD and heavy metal contamination. Where wind erosion is likely to be a problem, placing the surface dumps in sheltered areas can help reduce particulate emissions. Periodic treatment of dumps with soil binders can also help to reduce dust problems.

Waste from Maintenance Facilities

Hydraulically powered mining equipment is often used to generate the forces necessary to break and load the rock. Mining equipment, even powered by electricity, contains many moving parts that must be constantly lubricated to prevent damage and costly repairs. As a result, most mining operations produce large quantities of waste oils, hydraulic fluids, antifreeze, and other fluids. Maintenance facilities are usually operated on the surface, but they may be excavated from rock at the active mining level underground. Typically, these facilities use large quantities of solvents to clean parts and equipment and although most waste rock at mine facilities is exempt from RCRA regulation, the oil and solvent wastes are not. To reduce these wastes, mining companies have adopted several practices. For example, the replacement of solvents with steam cleaning creates a wastewater stream that can be collected in large sumps where grease and oils can be separated. The oil is later skimmed off and sent to a recycler or disposal site. Mines located in colder climates can install EPA-approved waste oil burning heaters that use the oil to heat the mine facilities.

Large mining trucks cost more than $750,000 each. The purchase price of draglines, electric shovels, and other excavating equipment can exceed millions of dollars. Because of its cost, great care is taken to insure that the equipment is properly lubricated and maintained. In the past, companies established preventive maintenance schedules that required the equipment to be oiled, lubed, parts replaced, etc. on a regular schedule. Today, companies take periodic samples of the oils used in equipment and send it to a laboratory for analysis. Oil changes no longer occur on an arbitrary time schedule, but only when the analysis indicates that the oil is dirty and/or is beginning to break down. The effect of this process change has been to increase the average time between oil changes and, therefore, to reduce the volume of waste oil generated. Regular analysis of the lubricants has a second advantage. By monitoring the amount of foreign material contained in the lubricant, it is often possible to identify the parts that are starting to fail. Component replacement no longer occurs on a prescribed schedule, but only when the evidence suggests that the part needs replacing. Companies can now schedule part replacement before the equipment fails, thereby avoiding the costs associated with damage to other parts and with idle equipment operators.

Sewage and Trash Disposal

Mines are typically located in remote areas; therefore, they must install wastewater treatment facilities or hire private contractors to handle sewage and other wastes. Because the mine workings – whether underground or on the surface – are constantly being changed as the mine is developed, permanent toilet facilities at the active work site is usually not practical. Self-contained portable toilets are typically used so that they can be easily moved and remain convenient to the workers. Permanent mining facilities such as the maintenance shops, offices, and process facilities usually have permanent restrooms and showers. Waste from these facilities must be disposed of in accordance with the CWA, state, and local regulations. Permitted septic systems and, in many cases, small sewage treatment facilities are common at the mine site. In the past, trash generated by the mining operation was placed in the abandoned workings or in waste dumps. Today, RCRA does not exempt this type of waste from regulation; therefore, refuse, garbage, and trash is either picked up by commercial trash haulers or placed in a permitted landfill onsite.

Checking Your Understanding

1. What are two problems associated with the removal of surface trees, shrubs, and grasses in preparation for a mining operation?

2. Describe two problems associated the weathering of sulfide rock ores.

3. Describe a treatment process that can be used to raise the pH of AMD.

4. List a process change that can be used to minimize the amount of waste lubricants generated by a mining operation.

12-5 Other Laws Governing Mining Practices

Earlier in the chapter, it was noted that most waste rock from a mining operation is exempt from RCRA regulation. This is true even though EPA test methods show that its metal content exceeds RCRA hazardous waste standards. Mine waste from **non-locatable mineral** mines – such as coal mines – is regulated under the Surface Mining Control and Reclamation Act of 1977 (SMCRA), but it does not closely regulate **locatable mineral** mines producing metal ores. SMCRA applies to all mines that were abandoned prior to 1977 and that have, or are causing, serious environmental problems or pose a threat to life or property by degrading the quality of life of the local community. This provision also specifically targets areas where collapsing underground mines have caused surface subsidence or abandoned coal mines where fires have produced toxic smoke. Today all operating mines regulated under SMCRA are required to pay a fee or tax on each unit of production to create a fund to pay for reclamation of these abandoned mines.

Mine waste is not exempt from the Comprehensive Environmental Response, Compensation, and Liability Act (CERCLA), which also regulates past and present mining operations that pose a threat to human health and the environment. In addition, many states have enacted laws requiring, at a minimum, stabilization and pollution prevention measures to insure that overburden dumps will not pose a pollution problem in the future.

As a final note, today it is extremely difficult to severely restrict or block the development of a metal mine based solely on environmental arguments. The Mining Law of 1872 gave miners the right to find, acquire ownership, and mine locatable minerals on public lands and it predates all other environmental laws. In environmental settings, such as National Parks, Monuments, and Wilderness Areas, however, mining can be prohibited or at least severely restricted by a requirement to not disturb features such as wetlands or riparian areas. In many cases buffer areas must be established to protect these unique features. Many congressional efforts were made to eliminate or severely modify the 1872 Mining Law, but to no avail. However, representatives of the mining industry have come to recognize the inevitability of change. They have therefore been working with federal agencies to develop environmentally sound legislation that at the same time insures the continuation of mining operations in the United States. Consequently, it seems likely that the old mining law will be changed in the near future.

Checking Your Understanding

1. What are two situations that are regulated by the Surface Mining Control and Reclamation Act of 1977?

2. What rights are granted miners under the Mining Law of 1872?

Summary

Mining is an activity pursued by man since prehistoric times. It includes excavation of the ore and ore processing. The minerals in ores have been deposited by a variety of geological processes. Hydrothermal fluids or magma rise from great depths in the earth and enter porous rocks or highly fractured rock, forming veins of concentrated metals or minerals. Ore deposits are formed by mechanical and/or physical forces – such as weathering – that may transport the rock across the earth and deposit it elsewhere. Some mineral deposits, known as evaporates, are the result of mineral-laden waters evaporating from an ancient lake or sea.

Different types of mining methods are used for different types of mineral deposits. Surface mines are commonly subdivided into strip and open pit mines. Subsurface mines are subdivided into underground and injection or solution mines. The typical underground mining methods involve the development of a shaft or adit in the area of the deposit. Smaller drifts or tunnels are bored from the shaft to access the deposit. Underground mines require many miners and the equipment is small compared to surface mines. Underground mines must, therefore, concentrate on high-grade ore, which usually occurs in small deposits. Surface mines, by comparison, usually produce larger quantities of low-grade ore, because the unit production cost is less.

Underground mining methods always involve drilling, blasting, loading the rock on a transport device, and elevating it to the surface. The methods for developing underground mines can be subdivided according to the presence or absence of support for the ceiling of the mining area. The ceiling may be supported through the use of rock bolts, timbering, or the room-and-pillar method. Some mines have rock so competent that no ceiling support is needed; this ore-bearing ceiling rock is intentionally allowed to cave. These methods are known as shrinkage stoping, sub-level stoping, and block caving. *In situ* mining methods include drilling a series of holes into the mineral deposit and then pumping the water or chemicals that have been injected in the perimeter holes from a center hole. This method has the advantage of eliminating the blasting, loading, and hauling expenses; however, fractures and faults in the country rock sometimes make it impossible to recover the injected solution.

Surface mining operations involve placer, strip, and open pit mines. Most surface mines use a fleet of large trucks and excavating equipment to move millions of tons of overburden and ore.

All mining operations result in some degree of surface disturbance. Underground mining usually result in the least surface disturbance. Metal-bearing ore deposits often contain sulfide rock, which becomes a source of acid mine drainage and metal-laden water once it has been exposed to the atmosphere. Disturbed areas resulting from mining operations are particularly susceptible to weathering by wind.

It is extremely difficult to restrict or block the development of a metal mine. The Mining Law of 1872 – which predates all environmental laws – gave

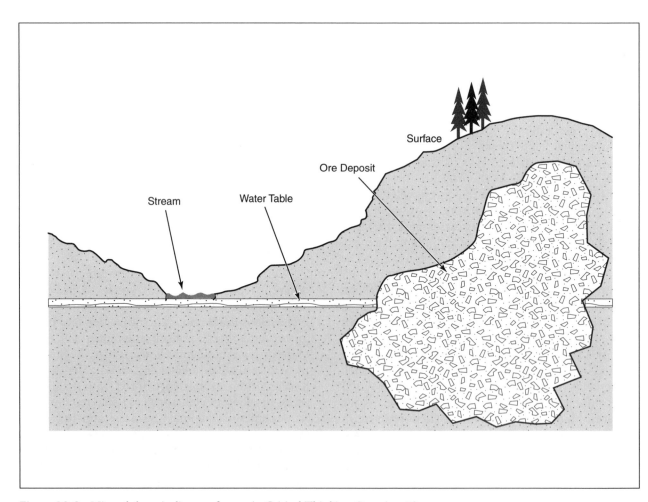

Figure 12-8: Mineral deposit diagram for use in Critical Thinking Question #2.

miners the right to find, acquire ownership, and mine locatable minerals on public lands. A number of Congressional efforts were made to eliminate or severely modify this law, but to no avail. However, due to recent cooperative efforts between legislators and the mining industry, it seems likely that this law will be changed in the near future.

Critical Thinking Questions

1. Mining companies often complain that the burden of environmental regulatory compliance makes mining uneconomical. Discuss some strategies that both subsurface and surface mine operators could use to make mining more profitable, while meeting environmental requirements.

2. Assume that a mining company has plans to develop the ore deposit depicted in Figure 12-8. It is a high grade, relatively shallow deposit that contains sulfide gold ore. It is located in a very heavily fractured rock, and the gold is deposited throughout in millions of tiny rock fractures.
   a. What method (surface or subsurface) would you recommend the company use to mine this deposit? Justify your answer.
   b. Draw a flow diagram depicting the steps in the process you have selected.
   c. What environmental problems might result from the development of this mine?

13

The Metal Production Industries

by Andrew J. Silva

Chapter Objectives

Upon completing this chapter, the student will be able to:

1. **Describe** the two basic activities and the common processes used in ore dressing.

2. **Describe** the thermal, chemical, and biological metal extraction processes that are in common use.

3. **Identify** three categories of waste generated by the metal processing industries and the requirements that are designed to reduce their harmful effects.

13-1 Introduction

The metal production industries are engaged in the extraction of metals from their ores and in forming the metal into **billets**, **ingots**, sheets, cables, etc. to be sold to a **fabricator**. The fabricators use the metal to make tools, machines, structural materials, jewelry, and other useful products. **Metallurgy** is the study of the chemical and physical properties of metal; the technology of extracting metals from ore and forming the metal product is called **extractive metallurgy**.

The typical processes for freeing the metal from the ore use the following three steps:

1. **Ore dressing** prepares the ore for metal extraction by grinding it into very small particles. The metal-bearing particles are then separated from the nonmetal-bearing particles, called **gangue**.

2. An **extraction** process removes the metal from the mineral. Often, the extraction process is accomplished by dissolving the metal and then reclaiming it from the solution by either precipitating or electrowinning.

3. **Forming** is the process in which the metal is shaped to meet the customer's specifications. This process usually requires the metal to be melted; the remaining impurities are often removed during this final process.

Some metals, such as iron, may undergo additional processes to give them special properties. To produce steel or stainless steel, for example, the iron is alloyed with carbon and/or chromium. This process is often performed just before forming the final marketable product.

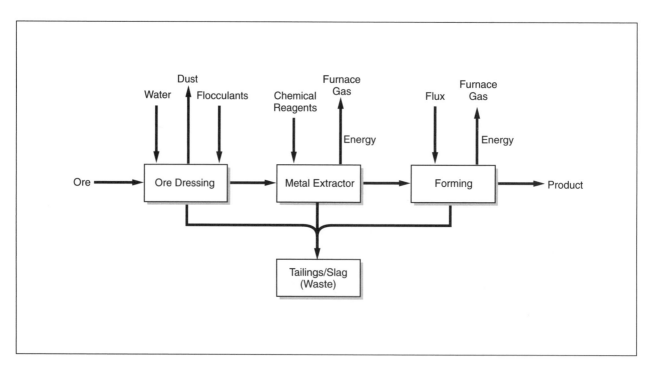

Figure 13-1: Typical metal processing flow diagram.

13-2 Ore Dressing

As defined in Chapter 12, ore is a mineral or rock that is excavated, prepared, processed, and sold, returning a reasonable profit to the individual or company engaged in the mining activity. For example, the mineral galena has the general chemical formula of PbS and is often found joined together with quartz in the form of rock. The **galena** is the metal-bearing component of the ore, and the quartz is the nonmetal-bearing mineral, or the gangue. The term ore dressing refers to the activities undertaken to prepare the ore for extraction of its metal. Basic ore dressing techniques commonly used throughout the industry – regardless of the type of ore being processed – are: 1) **comminution** or liberation of the minerals from the gangue, and 2) sorting and concentrating both the metal-bearing mineral and the gangue.

Comminution

The term comminution refers to the crushing and milling processes used to reduce the size of the metal-bearing ore particles, prior to the extraction process. The crushing process can be done either at the mine or at the metal processing facility, but is always accomplished by forcing the rock through a

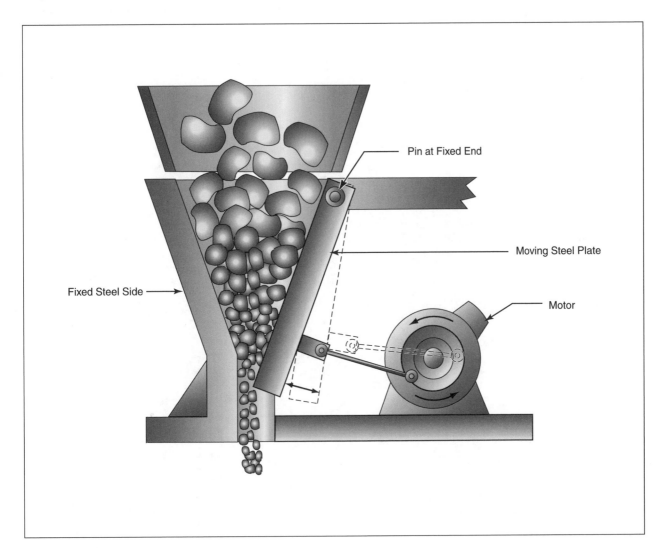

Figure 13-2: Jaw crusher.

crusher that – as shown in Figures 13-2 – has at least one hard steel part (heavy black lines) that is designed to strike and/or break the ore. The size of **run-of-mine rock** can vary considerably, up to boulders that must be broken before they will enter the crusher. The acceptable feed size of a crusher varies from several inches to more than a foot in diameter. A crusher's output varies from about one inch in diameter to particles as small as a grain of sand.

Milling is used to further reduce the particle size. This process involves **grinding** the product from the crusher to a more even consistency, with particle sizes of coarsely ground grain or smaller. In the process, the minerals containing the target metal and occasionally the metals themselves are liberated from the gangue. The metal-containing mineral can then, hopefully, be separated from the gangue using only physical methods. Grinding the ore to an extremely fine powder, like flour, is often necessary before this separation can occur.

The mills used for this grinding process are usually cylindrical in shape and mounted so that they can be rotated. Several different kinds of mills are used, but only the three most common ones will be discussed: **ball**, **rod**, and **autogenous mills**.

The Ball Mill

As shown in Figure 13-3, ball mills are cylinders partially filled with hardened steel balls. Crushed rock is fed into the end of the rotating cylinder. The **centrifugal force** resulting from the drum's rotation causes the hardened steel balls and rocks to move up the side of the cylinder. Eventually, gravity overcomes the centrifugal force and the balls and rock fall back the bottom striking other rocks and causing them to break. The efficiency of rock to rock breakage is dependent on the density and hardness of the rocks as well as on the rotational speed

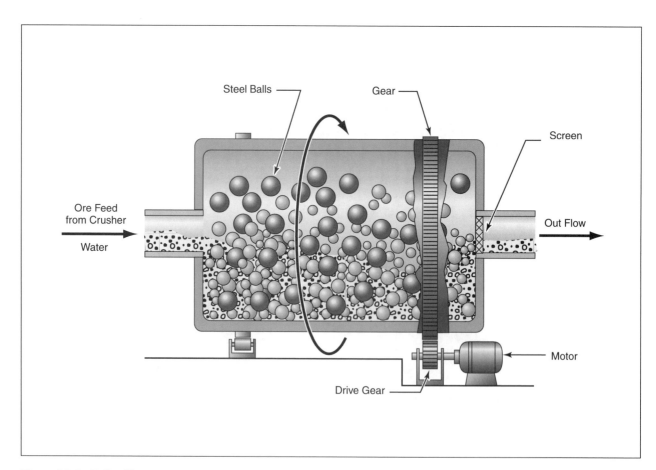

Figure 13-3: Ball mill.

and diameter of the mill. Dense, hard rocks break better than less dense, softer rocks. The diameter and the rotation speed determine the height reached before the balls and rocks fall. Obviously, the higher their position along the wall when they fall, the greater their impact when they strike the rocks below. The repeated impacts from the falling balls and rocks reduce the size of the ground material to sand- or powder-sized particles.

The exit end of the cylinder is covered with a grate, which retains the steel balls and also determines the size of the material leaving the mill. Outflow from the mill is usually screened to insure that the optimum sized particle has been achieved and the oversized material is routed back for additional grinding. To reduce dust losses and make the movement of the finely ground material to other processing circuits more efficient, water is added to the ground ore converting it into slurry. The slurry is then pumped through the remaining milling circuits.

The Rod Mill

As is shown in Figure 13-4, rod mills are very similar to ball mills, except that they contain free moving steel rods that extend from one end of the cylinder to the other. As with the ball mill, the discharged product from the crusher is fed into the mill. Again, as the cylinder is rotated, centrifugal force causes the rocks and rods to move up the side of the cylinder and then fall under the force of gravity. The impact of the steel rods and the rocks above falling on the rocks below cause more rocks to break. The ground rock exiting through the grate at the end of the mill is screened and oversized material returned to the mill.

The Autogenous Mill

The autogenous grinding method is really a combination of crushing and milling in the same machine. If the ore-bearing rock is dense and hard, it can be used to crush and grind itself in an autogenous mill. As can be seen from Figure 13-5, the autogenous mill is short but large in diameter, while ball and rod mills tend to be longer and smaller in diameter. As run-of-mine ore or material from a primary crusher is fed into the this type of mill it again moves up the side of the rotating cylinder until it eventually falls and strikes the rocks below.

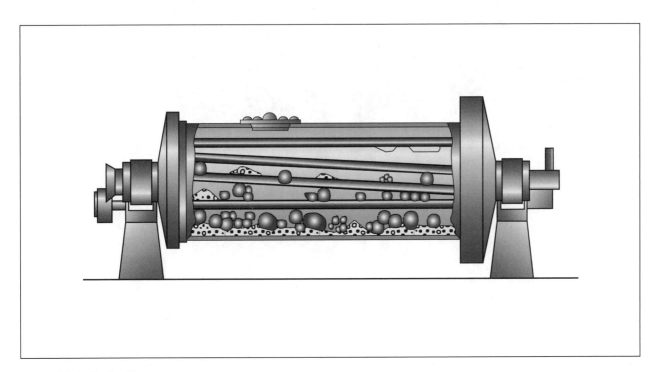

Figure 13-4: Rod mill.

Because of its large diameter, the falling rocks gain more speed and strike the ones below with a greater impact. If the impact is great enough, both the falling rock and the rock below will break. As the particle size is reduced, air being pushed through the cylinder will sweep the smaller particles out; these can then be fed into the remaining processing circuits. Often steel balls are used to supplement the rock used as the primary grinding agent in an autogenous mill. When this is done, the mill is called a semi-autogenous mill.

Sorting Mineral Fractions

Once the ore has been ground to a size that liberates the minerals and/or metal, it then enters the sorting and concentrating circuit. The nonmetal-bearing mineral wastes called **tailings** are pumped as slurry to large evaporation ponds called **tailings ponds**. The remaining ore fraction is pumped to the **extraction circuits**. Within the extraction circuits, the differences between the physical proper-

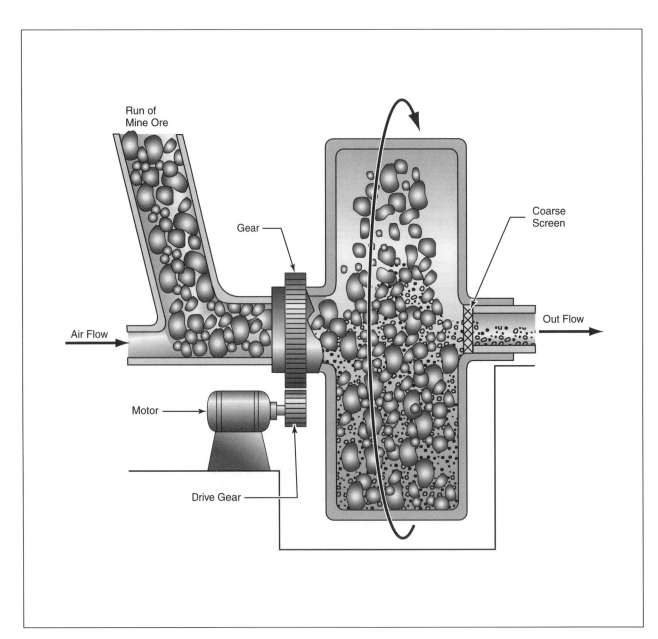

Figure 13-5: Autogenous mill.

ties of the minerals and the gangue are used to effect the sorting/concentrating process in several ways. Three of the most common methods will be discussed here: **classifiers**, **flotation cells**, and **magnetic separation**.

Classifiers

Classification makes use of the differences in a mineral's density, size, and shape to separate the target mineral or metal from the rock as it moves through water. One example of a classifier is the **sluice box** that was widely used in the nineteenth and early twentieth centuries to recover gold from the sand and gravel bars of a stream. The miners would set up their boxes containing a series of perpendicular ridges – called riffles – and direct the flow of the stream over the ridges and through the box. They would shovel the stream deposit into the upper end of the box and allow the flowing water to carry the lighter and less dense gravel through the box and out the opposite end. The denser gold and gold-containing materials, however, would start to sink and would strike one of the riffles, which would prevent them from exiting the box. Periodically, the miners would stop shoveling and deflect the water flow from the box so they could examine the material retained behind the riffles. If they were lucky and gold was present, they would find it trapped behind one of the riffles.

The bowl classifier – commonly called **thickener** – is another type of classifier commonly used by POTWs for sewage separation. As Figure 13-6 shows, the watery mixture to be separated is pumped into the center of the bowl classifier, which is a large circular tank. The denser and larger particles quickly sink to the bottom where large rotating rakes move the settled material down a sloping bottom into the center drain. The water, the less dense materials, and the **fines**, however, are carried across the tank and overflow into the surrounding trough. Depending on their composition, the materials entering the drain are either pumped into the tailings pond, if they are a waste, or into the remaining milling circuits, if they contain the valuable mineral.

A third type of classifier is called a **cyclone**. Cyclones are most commonly used to dewater miner-

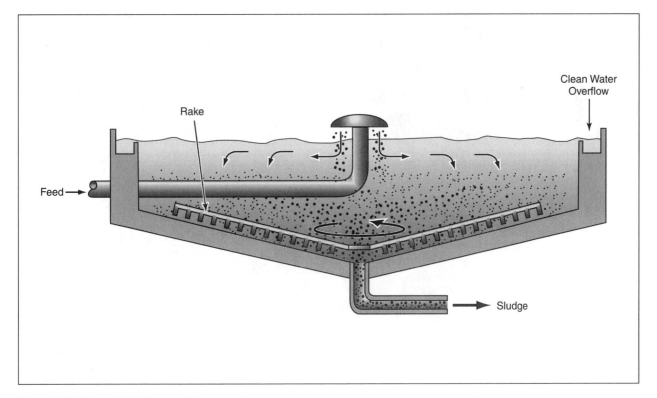

Figure 13-6: Bowl classifier or thickener.

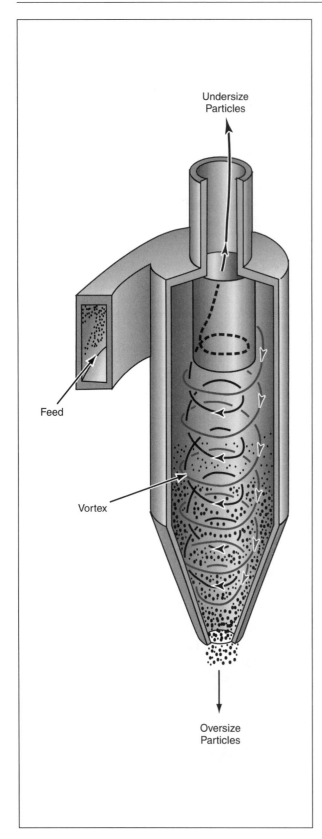

Figure 13-7: Cyclone separator.

als before entering the extraction circuits. They can be adjusted to separate particles according to both density and size. Cyclones, as shown in Figure 13-7, are cone-shaped tanks, in which the inflow mixture is injected at high velocity at the top and tangentially to the outside wall. As the fast-moving mixture of milled ore and water starts to spin around the cyclone, it creates a vortex in the center of the classifier. Gravity pulls the denser and larger particles down the outside edge of the tapering classifier. This reduction in circumference causes the water-rock particle mixture to accelerate. The water and the less dense materials are drawn to the vortex and are swept out of the top of the classifier. The denser particles continue to move down the walls of the cyclone and are then removed from the bottom.

Cyclones are frequently used in the copper industry. Many other classifying methods exist, with names like jigs, concentrating (Wilfley) tables, and Humphries Spirals. The one thing all classifiers have in common is that they rely on the physical properties of the mineral particles to separate them from each other and from the water.

Flotation Separation

Another type of sorting/concentrating method used in the metal production industries is **flotation**. Although this method is now in common use for the treatment of various waste streams, it was initially developed in the metal production industry to separate finely ground mineral particles. In the process, a mixture of small mineral particles and water is placed in a tank called a **float cell** or flotation cell. Small air bubbles are created by diffusers and float up from the bottom of the tank. The charge on the surface of the minerals attracts the bubbles attaching them to the particles. The rising air bubbles then carry the attached mineral particles to the surface. Upon reaching the surface, the floating particles overflow the cell and are collected as a concentrate. Particles with little or no attraction for the air bubbles fall to the bottom and are removed from the cell.

On occasion, flocculants – developed to selectively coat target mineral particles – are added to the float cell. By controlling inflow rates, flocculant feed rates, and other operating conditions, only particles containing the target mineral can be made to float and overflow the cell as a concentrate. The use of flocculants allows for a great deal of selectivity in

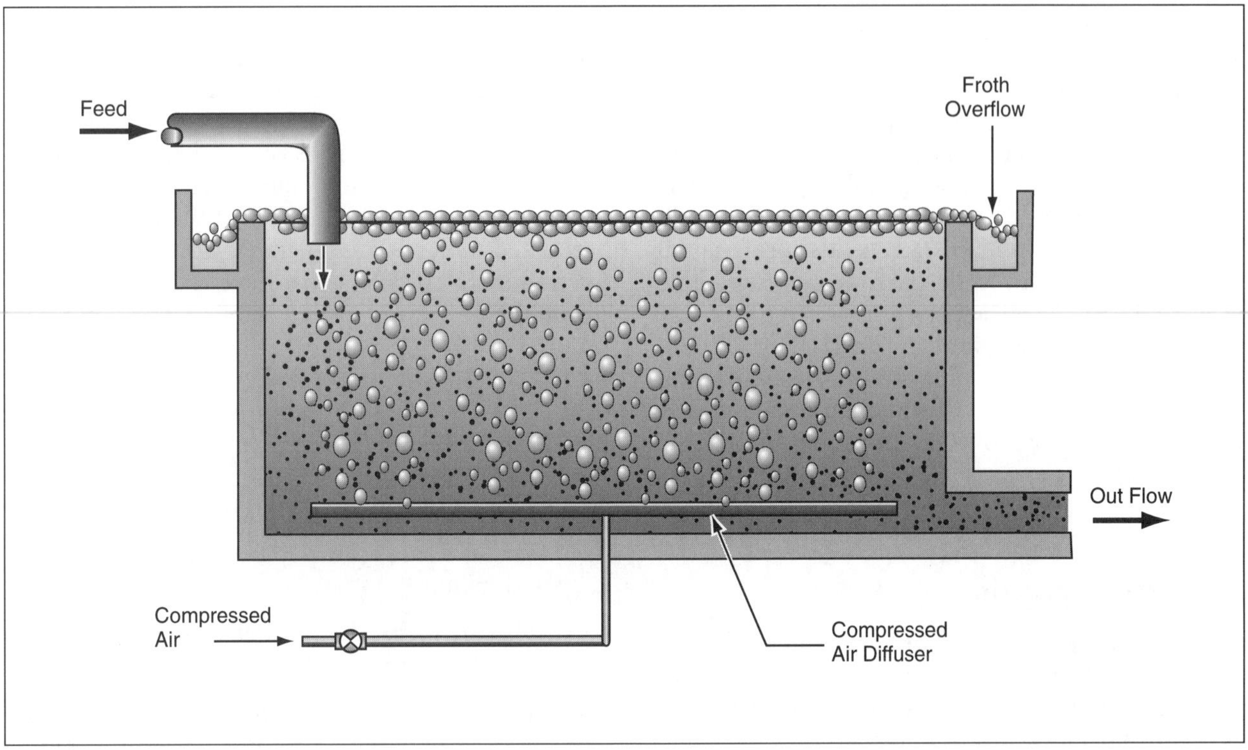

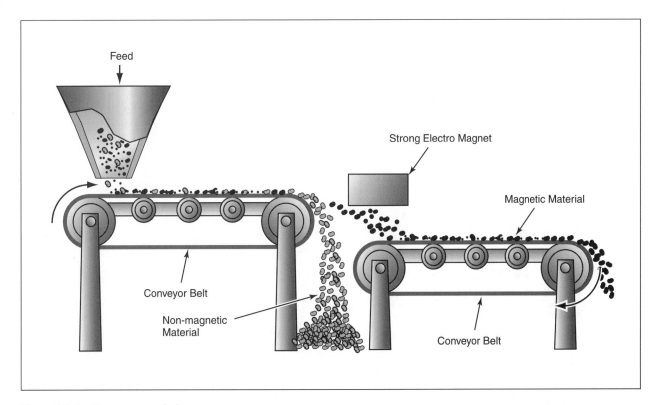

Figure 13-8: Flotation cell.

Figure 13-9: Dry magnetic belt separator.

determining which particles float and which drop as waste.

Phosphate for fertilizer production is a good example of a substance that is produced by flotation. Apatite is the mineral name for naturally occurring calcium phosphate and its mineral impurities. The rock is first pulverized and then put into a float cell. Through the addition of fatty acids, preferential wetting of the different particles occurs. The unwetted particles attract and are carried to the surface by the rising air bubbles where they are collected as a phosphate concentrate. The remaining minerals that were wetted by the fatty acids are hydrophobic and are moved from the bottom of the cell as a waste.

Magnetic Separation

Magnetic separation is a technique that can be used only on metals that are attracted to a magnet. Iron-bearing ores, for example, such as magnetite and hematite, are inexpensively and effectively separated in this manner. Separation of magnetic minerals is a simple process accomplished by passing the crushed ore either under or over strong magnets. The magnets attract the magnetic mineral and the remaining materials pass unaffected. Processes using this separation technique often place the magnets near the end of a conveyor belt, as illustrated in Figure 13-9. As the materials reach the end of the conveyor, the attracted minerals are separated from the nonmagnetic minerals by falling in an arced path farther away from the conveyor.

Checking Your Understanding

1. Name three ore dressing processes.

2. What is the relationship between comminution and milling?

3. What are the factors that determine the efficiency of rock-to-rock breakage?

4. What are the basic design and operational differences between a ball mill, a rod mill, and an autogenous mill?

13-3 Metal Extraction and Forming Processes

The degree of difficulty in extracting metals that are chemically bound in minerals is dependent on the chemical nature of both the metal and the mineral. Extraction processes may vary from relatively simple processes, such as acid leaching, to very complex ones involving many steps. No attempt will be made here to survey all of the possible metal extraction processes; rather, representative examples from thermal, chemical, and biological extraction techniques will be discussed.

Thermal Extraction Techniques

Three common thermal extraction techniques are **calcination**, **roasting**, and **smelting**. They are typically selected for use to achieve one of the following goals:

1. To remove volatile compounds from the minerals,

2. To convert nonvolatile elements in the mineral into an oxidized state that is volatile,

3. To facilitate the dissolution of a metal by changing its oxidation state, or

4. To melt the mineral so that the elements can be separated from the cool melt.

Calcination

The elimination of volatile products from a mineral is generally accomplished by thermal decomposition. Calcination, for example, is used on carbonate minerals such as **magnesite**, $MgCO_3$, and **rhodochrosite**, $MnCO_3$, to remove carbon dioxide gas. At temperatures above 400°C, the carbonate radical is decomposed according to the following equation:

$$MgCO_3 \rightarrow MgO + CO_2$$

Thermal decomposition can also be used to remove the water from hydrated ores such as **baux**-ite, $Al_2O_3 \cdot 2\ H_2O$ according to the following formula:

$$Al_2O_3 \cdot 2\ H_2O \rightarrow Al_2O_3 + 2\ H_2O$$

Roasting

Roasting is another thermal treatment process commonly used in metal extraction. Roasting is most often used to oxidize the metals in the mineral. Substances such as **arsenic** and **antimony** – which can pose problems in some of the extraction processes discussed later – can be removed by being converted into volatile oxides in the roasting process. Scrubbers can then be used to reclaim the oxides before discharging the furnace gases into the atmosphere.

Roasting can also be used to convert the target metal into a more soluble compound. Since metal sulfides are typically much less soluble than their oxides, the roasting of a sulfide ore can make the later chemical extraction of the metal oxide possible.

Smelting

Smelting is a process that uses first heat to melt metal-containing minerals and then rapid cooling to separate the constituents into layers. Usually, metals are more **fusible** than nonmetals and tend to form one or more layers. The less fusible components, called **slag**, form another layer, which is generally considered waste.

On occasion, **fluxes** are added to the melt to facilitate metal/slag or metal/metal separations. Fluxes are substances that help separate the more fusible materials from the less fusible ones or that react with a specific element in the melt. Different fluxes can be used to separate one component of the melt from the rest. For example, fluxes can be added to separate the small amounts of cobalt sometimes found in copper ore. The non-target portion of the melt – in this case the cobalt – is called the **species**, while the target portion is the copper. In lead smelting, copper is often present as a secondary metal along with sulfides. Again, fluxes are added to the melt to separate lead from the copper, which in this case is the non-target portion of the melt. This copper-sulfide impurity is called **matte**. In both processes the species and the matte are sent to secondary process circuits to recover the cobalt or the copper.

Chemical Extraction Techniques

Hydrometallurgy

The process of extracting metals from their ores by dissolving them in a solution is known as **hydrometallurgy**. The primary hydrometallurgical technique used today is **leaching**, which involves using various aqueous solutions to dissolve the metals. Some of the more common leaching solution/metal combinations are sulfuric acid/copper, sodium cyanide/gold, and sodium hydroxide/aluminum oxide. Other leaching reagents that are used include ammonia, ammonium carbonate, sodium carbonate, and in some applications just plain water.

Typically, leaching involves placing crushed and/or milled ore into vats filled with one of the aqueous solutions described above. Often the vat's contents are agitated to increase its oxygen content, which aids in the dissolution of some metals, and to insure adequate solution/ore contact. Contact with the solution causes the metal to separate from the

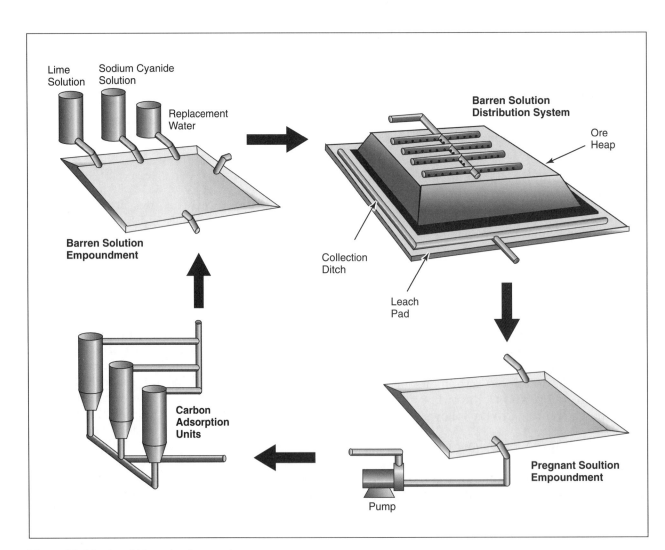

Figure 13-10: A gold heap leach operation.

waste rock and to dissolve, creating what is known as a **pregnant solution**.

In recent years, it has become increasingly common to use **heap leaching** to extract metals from undressed run-of-mine ores. By using undressed ores, the high costs associated with building, operating, and maintaining the crushing and grinding circuits can be avoided. The basic procedure involves creating an earthen pad that is covered with an impermeable liner. The ore is placed on the liner surrounded by a pregnant solution collection system, consisting of perforated pipes covered by a protective layer of gravel. A leaching solution that is selective for the target metal is applied to the top of the heap and allowed to percolate downward. At the time of application, the leaching solution is referred to as a **barren solution** because it does not contain any dissolved metal.

As the barren solution percolates through the ore pile, it comes into contact with and dissolves the target metal, thereby becoming a pregnant solution. The impermeable liner and collection system at the base of the heap intercepts this solution and drains it into one of the nearby collection ponds. Metal contained in the pregnant solution is concentrated and removed from the solution in subsequent process circuits.

A diagram for a typical gold heap leach operation is shown in Figure 13-10. Because gold and silver are often found commingled and react similarly with the cyanide extraction solution, this operation produces a gold/silver pregnant solution, which is extracted by the use of carbon adsorption columns. When the columns become loaded with the metals, they are washed with a strong caustic solution, which redissolves the gold and silver producing a much more concentrated solution than the original pregnant one. The caustic-gold/silver solution is pumped to an electrowinning cell where the metals are plated out on the cathode. The cathode from this process is later sent to a refinery where it is smelted and where the impurities are removed, before the metal is melted and poured into a mold. The product of this process is called **doré**, which is still a mixture of gold, silver, and a trace of remaining impurities. The doré, in turn, is sent to another refinery, where the remaining impurities are removed, and the gold and silver are separated. Both silver and gold are poured into bars after this process, with the gold bars typically assaying 99.99 percent pure.

Solution or *In Situ* Mining

Successful attempts have been made to leach metal from ore without mining it. This process is called **solution** or *in situ* **mining**. Functionally, the process is similar to heap leaching; however, in this method a series of holes must first be drilled into the ore deposit and the barren solution pumped down selected holes. As this solution migrates through the ore, it dissolves the target metal. The resulting pregnant solution is removed by being pumped out of the remaining holes; the metal is then recovered in the extraction circuits.

Bioextraction Techniques

Many metallurgists consider **bioextraction** to be yet another type of hydrometallurgical process, because it also results in a pregnant solution. In bioextraction procedures, bacteria use a part of the mineral as an energy source. As the mineral is being used by the bacteria, the metal is set free. From the ore pile or vat a pregnant solution can be recovered and sent on to other circuits for metal recovery.

Distillation

Metals with low boiling points, such as mercury, zinc, and magnesium, can be extracted from their ores by distillation. In this process, the ore is heated until the metal is vaporized in a vessel called a **retort**. The hot metal vapors pass from the retort into a condenser where they are cooled and recovered as pure metal.

Metal Recovery

Chemical Precipitation and Electrowinning

The two most common metal extraction methods are chemical precipitation and **electrowinning**. Chemical precipitation involves adding a reagent to the solution – which causes the metal ions to form an insoluble precipitate – and then separating the precipitate from the liquid.

Electrowinning uses electricity to convert the metal ions in the pregnant solution into their elemental form. In this process, DC electricity is introduced into the solution through the use of an anode and a cathode. At the cathode, metal ions are reduced by gaining electrons and are plated from the solution as a nearly pure metal. Copper, for example, is often recovered from pregnant leach solutions using this method. The half-cell reaction occurring at the cathode is:

$$Cu^{2+} + 2e^- \rightarrow Cu^o$$

where Cu^{2+} is the copper(II) ion present in the leach solution and Cu^o is the copper metal that plates out on the cathode.

Electrowinning and precipitation techniques usually result in an impure product. In the case of chemical precipitation, the metal is removed from the solution as an insoluble substance. With an electrowinning product, other metals present in the pregnant solution also plate onto the cathode. The final step in the metal extraction therefore involves a refining process before the formation phase, in which the pure metal is sized and shaped according to the fabricator's specifications.

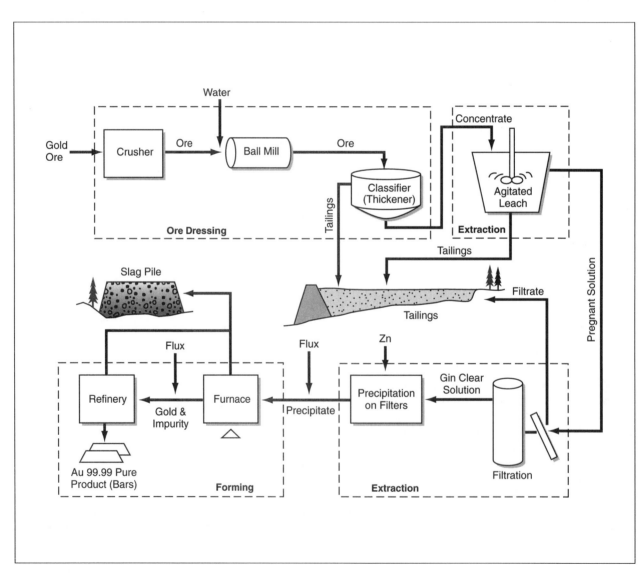

Figure 13-11: Flow diagram of a gold processing facility.

Refining

Refining is a smelting process that removes impurities from the metal or that separates one metal from another. It can be as simple as allowing entrapped hydrogen to escape the melted metals through diffusion; or, it may require a more complex procedure involving the introduction of elements or compounds that form precipitates along with the impurities to be removed. Reactions involving oxygen are most commonly used for refining, but other substances such as sulfur and lime are also chosen.

Nonetheless, some level of contamination always remains after refining. When the amounts are negligible, no attempt is made to further purify the melt, but in very specialized applications additional purification procedures must be used, which inevitably increase production costs.

Forming

The final step in metal processing is forming the metal into a marketable shape and size. Most often the metal is simply poured into a mold, allowed to cool, and sold as an ingot or a bar, as in the case of gold, for example. Gold bars are sold to a variety of customers including jewelry makers, computer makers, and gold speculators. Increasingly, metals are molded into shapes – such as billets, plates, sheets, cables, and tubes – that can be directly used by the foundry or by the fabricator.

In conclusion, the various processes described above are often used in several different mill circuits. Most metal producers find it necessary to combine and repeat processes to achieve good product quality. The flow diagram in Figure 13-11 illustrates this multiple processing with the example of a traditional gold processing facility, in which the starting material is a sulfide mineral containing gold.

While only a few of the more common processes are described here, other processes will be encountered at individual metal processing facilities; however, nearly all of them use the same basic circuits: ore dressing, metal extraction, and product forming.

Checking Your Understanding

1. What are three thermal metal extraction techniques?

2. What are two functions served by a flux?

3. How do pregnant and barren solutions differ?

4. What types of metals can be successfully extracted from their ores by distillation?

5. Why is bioextraction sometimes considered to be another hydrometallurgical process?

6. What physical property allows a metal to be extracted using distillation?

7. What is the purpose of all metal refining processes?

8. What is metal forming?

13-4 Waste from the Metal Processing Industries

Airborne Waste

There are several different types of airborne wastes generated by the metal production industries. All mining, conveying, crushing, grinding, and dry screening circuits designed to move the ore, reduce its particle size, or sort the ore are potential dust sources. Besides dust, other airborne wastes may include gases, such as SO_2 from roasting, acid mists from extraction, or metal fumes from distillation processes. The Clean Air Act (CAA) regulates all discharges to the atmosphere and, as a result, most metal processing facilities are required to have a permit regulating their stack gases and particulate emissions. Bag houses, scrubbers, and electrostatic precipitators are commonly used to capture the pollutants before they are discharged. The wastes captured by these devices are regulated under the Resource Conservation and Recovery Act (RCRA) as administered by EPA or by the state in which the facility is located.

Liquid Waste

Tailings

Ore dressing circuits produce tailings, which are typically aqueous/solid suspensions or slurries composed of water and waste rock. Tailings from both the ore dressing processes and the mining often contain high levels of sulfide rock, which is primarily responsible for the formation of acid mine drainage (AMD). AMD is characterized by both a high metal concentration and a low pH. The following series of chemical reactions shows, for example, the origin of this combination from a pyrite ore, FeS_2, commonly called fool's gold. In the first step, the pyrite undergoes oxidation in the presence of oxygen and water producing hydrogen ions, $H^+_{(aq)}$, according to the following half-reaction:

$$2\,FeS_2 + 2\,H_2O + 7\,O_2 \rightarrow 4\,H^+_{(aq)} + 4\,SO_4^{2-}_{(aq)} + 2\,Fe^{2+}_{(aq)}$$

As the pH falls below 4.5, bacteria (*Metallogenium*) begin to act as a catalyst to accelerate the

process. When the pH falls below 3.5, the bacteria *Thiobacillus ferrooxidans* and *Thiobacillus thiooxidans* start to oxidize ferrous to ferric ions in their metabolic processes. The half-reaction for this reaction is the following:

$$4\,Fe^{2+}_{(aq)} + O_2 + 4\,H^+_{(aq)} \rightarrow 4\,Fe^{3+}_{(aq)} + 2\,H_2O$$

The Fe^{3+} produced by this process, in turn, causes the generation of additional hydrogen ions in the following way:

$$8\,Fe^{3+}_{(aq)} + S^{2-}_{(aq)} + 4\,H_2O \rightarrow 8\,Fe^{2+}_{(aq)} + SO_4^{2-}_{(aq)} + 8\,H^+_{(aq)}$$

In addition to metal and hydrogen ions, tailings impoundments also commonly contain chemical wastes from other processes, such as cyanide from the metal extraction circuits. In the past, tailings impoundments were unlined and abandoned when the processing facility closed. Over the years, plumes of heavy metals – occasionally also containing toxic reagents – have entered aquifers leading to many Superfund sites in the Western United States. It was common practice to place these impoundments in low-lying areas near streams or on flood plains. This presented the potential for a breach during flooding or seepage into surface or groundwater.

To correct these problems, permitting agencies now require tailings impoundments to be located outside flood plains, to be designed for 100-year storm events, and – if the tailings contain toxic reagents such as cyanide – a zero discharge requirement is mandated. Ideally, the impoundment should be designed so that no discharge occurs and be lined with clay and/or synthetic liners to assure groundwater protection. Under these conditions the impoundments function like large evaporation ponds and are built to accommodate both precipitation and the daily inflow from the process facility. When the volume of the impoundment cannot meet these requirements, some of the inflow must be discharged. These discharges are regulated by the Clean Water Act (CWA), and the companies must obtain an NPDES permit.

Depending on the location, the NPDES permit is issued and administered by the state or by EPA; in either case the facility must meet the requirements specified in the permit. To do so, usually metal processing facilities must install wastewater treatment plants to remove both metals and toxic reagents and adjust the pH to meet the requirement of their discharge permit. One exception exists to the requirements for receiving a CWA permit before discharging liquids from a tailings impoundment: facilities that do not discharge into waters of the United States are not required to obtain an NPDES permit. Such discharges are not covered under the CWA. The courts have broadly defined "waters of the United States" to include almost any drainage, ephemeral stream, seep, lake, creek, or flowing stream. Nevertheless, certain discharges occurring in the Western United States, particularly in Nevada, are not covered by the CWA and do not require NPDES permits. These discharges occur in areas where the existing topography produces a natural evaporation pond. The discharge goes into a basin from which no additional discharge can occur.

Runoff

In addition to AMD and tailings impoundments, metal processors also produce waste from runoff of disturbed areas around the process facility. Due to the need to keep and maintain stockpiles of ore, these areas may be very large, sometimes up to hundreds of acres. This runoff is also regulated under the CWA and requires an NPDES permit for discharges from the property. Because this runoff often contains elevated metal levels and a low pH, it is generally treated before discharge by being passed through sediment ponds; there its pH is adjusted by adding lime or any other inexpensive base. Often the pH adjustment will result in enough metal being precipitated that additional treatment is not necessary.

Other Process Solutions

Depending on the metals being extracted and the processes being used, waste process solutions from quenching activities, electrowinning cells, etc. may require an NPDES permit for discharge, which may require cooling, metal removal, pH adjustment, or other treatments. If the facility is in a town or city, it may be allowed to discharge liquid waste into the local POTW under the community's own NPDES

permit. In this case, however, CWA mandates that the processing facility pretreat its effluent before introducing it into the POTW sewer system.

Solid Waste

The primary solid waste produced at metal processing facilities is the solid portion of the tailings. This waste often poses an environmental hazard because it contains high concentrations of heavy metals and is subject to wind erosion, leaching, and runoff. Many old tailings piles have been cleaned up under the provisions of Superfund or the Comprehensive Environmental Response, Compensation, and Liability Act (CERCLA).

To prevent this in the future, most states now require mining companies to stabilize the tailings when a processing facility is closed. Stabilization may include capping with a liner and commonly involves the reapplication of topsoil and re-vegetation of the impoundment area. An exemption from RCRA exists for tailings produced at a metal processing facility; however, if the ore was transported from the mine site for treatment, the tailings would then be regulated under RCRA.

To insure compliance with these regulations, reclamation bonding is usually required of a metal processing company to insure funds will be available. Any processing activity taking place on public lands – such as National Forests or land administered by the Bureau of Land Management – would also require an operating permit. The permit would include provisions for bonding and reclamation of the tailings and facility sites when the metal processing activities are complete.

Slag from thermal processes is another solid waste produced by the metal extraction and processing circuits. Like mine tailings, slag is a solid waste that usually contains elevated quantities of heavy metals. Because it is the waste product of a melt, slag is generally a massive waste and not a fine dust like the one from tailings. As a result, the metals contained in the slag are usually less mobile than those contained in tailings. Nevertheless, leaching of heavy metals can and does occur from slag piles. Companies are now required to stabilize slag piles – as well as tailings piles – to prevent metal contamination from entering groundwater and surface waters.

Checking Your Understanding

1. What are three sources of air emissions associated with the metal processing industries?

2. What is the source and composition of AMD?

3. List two cases in which a metal processing facility is not required to obtain an NPDES permit for its discharges.

4. When closing a metal processing facility, what are three things that must be done to stabilize the tailings impoundment?

Summary

The metal production industries are engaged in the extraction of metals from ores and in the forming of the metals into billets, ingots, sheets, cables, etc. to be used by a metal fabricator. There are three steps involved in freeing the metals from the ores: ore dressing, extraction, and forming. Ore dressing prepares the ore for metal extraction by grinding it into very small particles. Extraction processes remove the metal from the mineral wastes. Forming shapes the metal to meet the customer's specifications.

The steps in ore dressing include comminution (or liberation of the minerals from the gangue by some type of crushing or grinding), sorting, and then concentrating both the metal-bearing mineral and the gangue by using classifiers, flotation cells, and magnetic separation techniques.

Metal extraction processes include thermal techniques (calcination, roasting, and smelting), chemical techniques (hydrometallurgy and leaching), bioextraction techniques, and distillation techniques. The two most common methods for recovering metals from pregnant solutions are chemical precipitation and electrowinning. The recovered metals always contain some degree of impurity. Refining is the process used to further purify the metals. The final step in metal processing is forming, which shapes the metals into a marketable form.

Critical Thinking Questions

1. Identify the three types of processes used in the metal production industries and describe how each of them transforms the minerals into metals.

2. Identify and describe metal processing equipment, procedures, or circuits that are also employed in the treatment of hazardous waste.

3. Select a specific metal extraction and identify the tailings produced. Design a process that could be used to reduce its potential environmental damage.

4. Analyze the processes of recycling aluminum and extracting aluminum ore. Compare the two in terms of costs and environmental impact.

14

The Metalworking Industries

by Melinda Trizinsky

Chapter Objectives

Upon completing this chapter, the student will be able to:

1. **Describe** the following unit operations used in the metalworking industry: shaping, surface preparation, bonding, heat treating, and surface treatment.

2. **Identify** common waste streams generated by the metalworking industry.

3. **Identify** regulations that impact the operation of metalworking facilities.

4. **Describe** various opportunities for waste minimization and source reduction available to the fabricated metal industry.

5. **Develop** a process flow diagram for a metal fabricator and identify opportunity points for waste minimization.

14-1 Introduction

Several steps are necessary to convert metal ores into useful metal products. This chapter will review the processes used and the waste streams generated by the metalworking industries; specifically the manufacturing industries identified by **Standard Industrial Classification Codes (SIC Codes)** 3400-3499.

The metalworking industries commonly use five processes or **unit operations** to change the shape or properties of metal stock materials:

1. **Shaping** – Processes that change the shape of metal stock materials to produce the basic form of the metal parts (e.g., milling and bending).

2. **Surface Preparation and Cleaning** – Usually cleaning operations that prepare the metal surface for subsequent operations such as welding, painting, electroplating, or packaging (e.g., degreasing).

> The term unit operations refers to individual processes that are part of a manufacturing process.

3. **Bonding** – Processes that join two metal objects or metal to some other material (e.g., welding and adhesive bonding).

4. **Heat Treating** – Processes that use changes in temperature to affect the properties of the metal (e.g., hardening steel).

5. **Surface Treatment** – Processes that alter the surface characteristics of the metal (e.g., polishing and electroplating).

14 - 2 Metalworking Industry Profile

The fabricated metal industry produces metal products, parts, or packaging; for example, metal containers and cans, cutlery, hand tools, hardware, heating devices, structural metal products, nails, metal forgings, fabricated pipe and fittings. Metal fabrication supports other industries including the appliance, automotive, defense, electronics, furniture, and other assembly industries. Industries that do not use metal parts in their products may have in-house **machine shops** to repair or manufacture metal parts for their production facilities. Figure 14-1 presents a flow diagram of the unit operations that

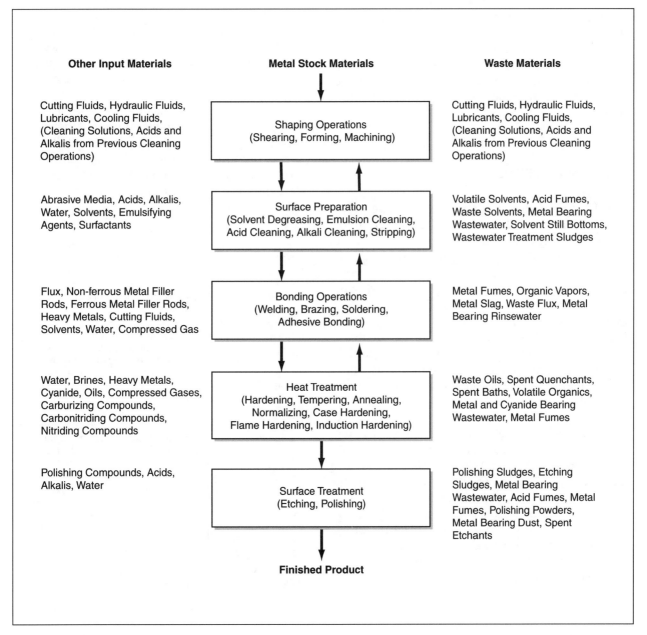

Figure 14-1: Process flow diagram for the metalworking industry showing material inputs and waste products for the unit operations discussed in this chapter.

turn raw metal stock materials into finished products; it also includes other material inputs and outputs (wastes). The flow diagram does not represent operations of an actual manufacturer; the order of unit operations may vary considerably from one manufacturer to another.

Process flow diagrams are useful tools for businesses conducting waste management audits and attempting to identify opportunity points for pollution prevention. Every material input stream identifies an opportunity for material substitution. Connections between unit operations are potential spillage sites; thus waste caused by spills can be reduced at these points. Wastes can be segregated at points where waste streams converge. Unit operations may provide opportunities for process modification or elimination. These opportunity points can then be evaluated to determine the best management options.

Raw Materials

Metal usage varies according to the specific needs of a given industry. Most of the metals used in industry are **alloys** (mixtures of two or more metallic elements). **Metallurgical properties** are the inherent properties of a given metal or alloy; these properties affect the suitability of a metal for a specific purpose. Important properties of metals include: density, melting point, toxicity, corrosion resistance, hardness, tensile strength, hardenability, ductility, malleability, brittleness, elasticity, weldability, fusibility, machinability, and hardening temperature. Factors such as cost and availability also affect the choice of metals for a given purpose.

For the purposes of this discussion, it is assumed that the stock materials used in the metalworking industries have the proper metallurgical properties and are in a form suitable for shaping operations. The metal production processes used to supply the stock materials are covered in Chapter 13. In addition to the stock materials, the metalworking industry uses various fluids, oils, solvents, gases, abrasives, paints, coatings, acids, alkalis, and adhesives. The unit processes and the wastes they generate are discussed below.

Unit Processes and Wastes

Shaping

Metals may be worked by **heat treating** or **cold pressure** methods. Heat treating modifies the physical properties of the workpiece through controlled heating and cooling cycles. Cold pressure methods shape the workpiece by applying direct physical pressure to cold metal. Shaping operations include **shearing, forming**, and **machining processes**.

Shearing processes (e.g., piercing, punching, blanking, and trimming) produce holes, openings, blanks, or parts by cutting the workpiece into a desired shape or size. These operations produce metal scraps. Shearing processes are frequently used to prepare materials for subsequent shaping or bonding processes.

Forming processes alter the shape of a metal object without reducing the amount of metal in the piece. Examples of forming processes include **bending, extruding, drawing, rolling, spinning, coining**, and **forging**. Table 14-1 presents brief descriptions of some forming operations.

Bending	Bends part to form a specific angle or curve.
Extruding	Blank is forced through a die to form a specific shape.
Drawing	Metal sheet stock is forced into a die by a punch. The shape is formed in the space between the punch and die.
Rolling	Metal is passed through a set or series of rollers that bend and form the part into the desired shape.
Spinning	Pressure is applied to a shape while it spins on a rotating form and forces the metal to acquire the form's shape.
Coining	Part is stamped to produce three-dimensional relief on one or both sides.
Forging	Produce a specific shape by applying external pressure that either strikes or squeezes a heated blank into a die.

Table 14-1: Forming operations.

Shaping operations require the use of metalworking fluids to lubricate the contact between the tools and the workpiece.

Machining operations use machine tools to refine the shape of a workpiece by removing small amounts of material from it. Examples of machining processes include drilling, milling, turning, shaping/planing, sawing, and grinding.

Most shaping operations involve high-pressure, metal-on-metal moving contacts between tools and workpieces. To reduce the heat and friction caused by these operations, **metalworking fluids** are circulated over the working surfaces. These fluids are used to:

— Control and reduce the temperature of tools and working surfaces;

— Reduce friction and vibration by acting as a lubricant;

— Wash away chips and metal debris;

— Improve the quality of the machined surface; and

— Inhibit corrosion or surface oxidation on newly exposed metal surfaces.

The five most common types of metalworking fluids include:

— **Mineral oils** – Oils derived from crude oil (e.g., liquid petrolatum).

— **Natural oils, fats, and derivatives** – Water-insoluble substances of animal or vegetable origin (e.g., lanolin).

— **Synthetic lubricants** – Chemical compounds manufactured to provide specific working properties. Most synthetic lubricants are water based. The water helps cool the tools and workpiece, while water-soluble chemicals provide the lubrication (e.g., silicone).

— **Compounded metalworking fluids** – Fluid mixtures that generally combine one of the previous three types of fluids with additives that enhance properties missing in the parent fluid.

— Aqueous metalworking fluids – Water-soluble oils mixed with water to produce cutting fluids with 90 to 98 percent water content (e.g., ethylene glycol mixtures).

In addition to the metalworking fluids, many of the machines used in this industry require hydraulic fluids, lubricating oils, and coolants. Ideally, these fluids should remain within the machines, but hydraulic fluids and oils frequently leak into the work area where they may contaminate the metalworking coolants or soil the surface of the workpiece.

Shaping Wastes – Metalworking fluids eventually become spoiled or contaminated and must be replaced. The heat generated during the shaping operations can degrade oils or cause a loss of water and depletion or degradation of additives. Bacterial growth may cause contamination and lead to fluid degradation. Waste fluids may contain high levels of metals. Other contaminants may include solvents, hydraulic fluids, acids, alkalis, and waste oils.

The hydraulic fluids and lubricating oils used in the machines may also degrade with use and time. These waste oils generally do not contain hazardous wastes. Petroleum products, including oils, are specifically excluded from RCRA and CERCLA regulation unless they are contaminated with hazardous substances (e.g. heavy metals, PCBs). Therefore disposal of waste oils is currently less problematic than some other types of wastes.

Other waste streams associated with shaping operations include metals and oily wastewaters. Shearing and machining operations may generate scrap metal, metal chips, or metal dust. These waste metals are usually contaminated with traces of metalworking fluids and solvents. The washdown of machining areas produces oily wastewater.

Fluids that no longer meet the needs of their intended applications become waste products.

Surface Preparation and Cleaning

The surface of the metal may require preparation prior to bonding operations, finishing operations, painting operations, packaging, and shipping. Common surface preparation techniques include **stripping**, **solvent cleaning**, **emulsion cleaning**, **alkaline cleaning**, and **acid cleaning**. These processes are used before and after metalworking operations such as welding, heat treatment, machining, applying protective coatings, assembling parts, inspection, and packaging.

Stripping – Stripping is often the first step in reworking out-of-specification products. Stripping processes use either mechanical or chemical action to remove surface coatings or surface layers. Conventional stripping processes include abrasive blasting, manual grinding, and chemical stripping. Chemical strippers – which frequently contain volatile organic chemicals (VOCs) and ozone-depleting chemicals (ODCs) – react with surface coatings to break them down for easy removal.

Solvent Cleaning – Cleaning processes remove contaminants such as oil, grease, flux, grit, oxides, and metal chips from metal parts. Solvent cleaning (degreasing) uses organic solvents to remove unwanted grease, oil, or other organic films from metal surfaces. These processes do not require rinsing or forced air drying. The fabricated metal industry uses two types of solvent cleaning:

1. **Cold Cleaning** – Solvent is applied at or below ambient temperature. The solvent may be applied by brush or the part may be dipped into the solvent. The most common solvent used in cold cleaning is a highly flammable mineral spirit.

2. **Vapor Degreasing** – Nonflammable, chlorinated hydrocarbons are heated to produce a vapor phase solvent. The part is suspended in the vapor phase solvent and is cleaned as the solvent condenses on the cooler metal surface. This method is faster than cold cleaning because of the higher temperatures. Common solvents used in vapor degreasing include methyl chloroform, perchloroethylene (PCE), trichloroethylene (TCE), Freon, and methylene chloride.

Semi-Aqueous Cleaning – Semi-aqueous cleaning uses an organic solvent dispersed in an aqueous solution. This type of cleaning uses smaller quantities of chemicals than solvent degreasing because the concentration of the solvent is lower. Emulsion cleaning is a type of semi-aqueous cleaning frequently used in metalworking operations. Emulsion cleaning uses common organic solvents (e.g., kerosene, mineral oil, and glycols) dispersed in water with the aid of emulsifying agents.

Aqueous Cleaning – Aqueous chemical treatments use acids, alkalis, and surfactants to prepare metal surfaces for subsequent processes. Alkaline cleaning removes organic residues. The composition of these cleaning solutions includes: builders, such as alkali hydroxides and carbonates; organic or inorganic additives, which promote better cleaning or affect the metal surface in some way; and surfactants. Alkaline cleaning is often assisted by mechanical action, ultrasonics, or by electrical potential (e.g., electrolytic cleaning).

Acid cleaning (sometimes referred to as **pickling**) prepares the surface of metal products by chemically removing oxides and scale. Several types of strong acids are used, including sulfuric, nitric, chromic, hydrochloric, and hydrofluoric acids. The metal generally passes from the acid bath through a series of rinses.

Wastes from Surface Preparation Activities – The primary air emissions from cleaning are due to the evaporation of chemicals from chemical stripping, solvent degreasing, and emulsion cleaning processes. These VOC and ODC emissions are the result of volatilization of solvents during storage, fugitive losses during use, and direct ventilation of fumes. Other air emissions such as acid fumes from acid cleaning processes and metal-bearing dusts or aerosols from abrasive cleaning processes are also possible.

Aqueous wastes generated during surface preparation activities include solvent-contaminated wastewater from vapor degreaser-water separators, rinse

> Surface preparation activities can result in 1) air emissions of VOCs, ODCs, and acid fumes, 2) contaminated rinsewater and spent aqueous and semi-aqueous cleaning solutions, and 3) solid wastes consisting of metal bearing sludges, waste solvents, stripping debris, contaminated abrasives, and still bottoms.

waters, and spent semi-aqueous and aqueous cleaning solutions. Solvent-bearing wastes are typically pretreated to comply with any applicable National Pollutant Discharge System (NPDES) permits and then discharged off site. Aqueous wastes from alkaline and acid cleaning that do not contain solvents are often treated onsite.

Other wastes may include liquid waste solvent containing organic contaminants, wastewater treatment sludges, still bottoms, cleaning tank residues, stripping debris, and contaminated abrasives. Many of these wastes contain heavy metals. Depending upon the nature of the coating and the underlying metal, the waste generated may be hazardous.

Bonding

Bonding processes are used to fuse metal parts, assemble finished products, and attach metal parts to nonmetallic parts. Common bonding processes include **welding**, **soldering**, **brazing**, and **adhesive bonding**. In welding, the pieces to be joined are melted and fused together. The filler rod used for welding is basically the same metal as the pieces being joined; this is not true for soldering or brazing. Soldering is a process of fastening metals together with a nonferrous metal that has a low melting point (< 800∞F). Surfaces to be bonded must be physically and chemically clean. **Flux** chemically cleans the metal, inhibits corrosion, and helps solder flow. The choice of flux depends upon the type of metal being joined. After soldering, the workpiece is washed to remove traces of the flux because flux can discolor the metal. Brazing is a similar process that uses a nonferrous metal with a higher melting point (> 800∞F). In both brazing and soldering the temperatures used are below the melting point of the metals being joined.

Adhesive bonding also requires clean surfaces, but differs from the other three processes in three major ways: 1) it uses a nonmetallic compound to join parts, 2) it does not require high temperatures, and 3) it can join metal to nonmetallic parts. The fabricated metal industry uses various adhesives, many of which emit VOCs as they cure.

Bonding Wastes – Air emissions are the primary wastes associated with bonding operations. The high temperatures used in welding, brazing, and soldering can volatilize organic residues and fluxes. The formation of toxic metal vapors is also a concern, so it is important to use hoods or vents to draw fumes away from the work area.

Heat Treatment

Heat treatment involves the heating and cooling of metals in the solid state to change their mechanical, microstructural, or corrosion-resisting properties. All heat treatment processes involve heating and cooling metal according to a time-temperature cycle that includes the following three steps: 1) heating the metal to a specific temperature, 2) holding the metal at that temperature for a specific time, and 3) cooling the metal at a certain rate. The principal types of heat treatment processes are outlined below.

— **Hardening** – This process makes steel harder but also increases its brittleness. Pre-cleaned medium-carbon and high-carbon steels are hardened by slowly heating to the hardening temperature and then rapidly cooling in brine, water, oil, or air. The choice of quenching solution affects the rate at which the metal cools, and the rate of cooling influences the hardness of the finished product. Brines generally cool the fastest and produce the hardest results.

— **Tempering** – This heat treatment process is used after hardening to relieve the internal strain in hardened steel and to increase its toughness.

— **Annealing** and **Normalizing** – These are similar processes that soften and improve the machinability of hard or hardened steel. They relieve internal stresses that may have been caused by machining, previous heat treatment, or cold-working operations.

— **Case Hardening** – This is a surface-hardening process used with alloys that are not amenable to normal hardening procedures. Case hardening involves two heat treatment steps. In the first step,

The principal properties changed by heat treatment include: hardness, brittleness, toughness, tensile strength, ductility, malleability, machinability, and elasticity.

the part is heat treated with chemical compounds that react with the surface metal to improve the hardening properties of the alloy. This surface treatment is performed by any of the following three methods: carburizing, which adds carbon to the metal, **carbonitriding**, which adds carbon and nitrogen, and **nitriding**, which adds nitrogen. Once the surface treatment is completed, the second step takes place, which is the hardening of the alloy.

—**Flame Hardening** – A surface hardening procedure that uses an oxyacetylene flame to rapidly heat the metal surface to the hardening temperature. The surface is quenched with a spray of water or other coolant.

—**Induction Hardening** – A surface hardening procedure that is similar to flame hardening except that the heat source is a high-frequency electric current.

Heat Treatment Wastes – Like other liquids used in the manufacturing process, quenching solutions and baths eventually become contaminated or lose their effectiveness. Spoiled quenching solutions and spent baths are the primary wastes associated with heat treatment. Some of these wastes require special handling. Quenching solutions may include hazardous brines and waste oils, and some compounds used in the first step of case hardening are hazardous (e.g., cyanide).

Surface Treatment and Metal Finishing

The metalworking industries frequently treat metal surfaces to inhibit **corrosion**, enhance appearance, add value, or increase the usefulness of finished pieces. Polishing and etching will be discussed here; additional metal and surface finishing processes are discussed in separate chapters about the electroplating and metal finishing industries (Chapter 15) and the paint and surface coating industries (Chapter 16). **Polishing** uses abrasives to remove or smooth out surface defects (scratches, pits, or tool marks) that affect the appearance or function of a part. Area cleaning and washdown after polishing operations can produce metal-bearing wastewaters and sludges. **Etching** uses chemical reagents or **etchants** to produce a specific design or surface appearance through the action of controlled dissolution. Strong acids or bases serve as etchants. The spent solutions usually contain high concentrations of spent metals.

Surface Treatment Wastes – Surface treatment processes may generate solid, liquid, or gaseous wastes. Polishing and etching sludges are the primary solid wastes. Metal and acid fumes are potential air pollutants from etching operations. Spent etchants (acidic or alkaline) and metal-bearing rinse waters are the most common aqueous wastes from these operations. Wastes from other surface treatments are covered in Chapters 15 and 16.

Checking Your Understanding

1. What is the primary role of the metalworking industry?

2. Name five types of unit operations used in the metalworking industry. Give two examples of wastes associated with each of these processes.

14-3 Federal Regulations and the Metalworking Industry

This section briefly outlines the federal statutes and regulations that may apply to the metalworking industry. A more complete discussion of each of these regulations is given in Chapter 2.

Resource Conservation and Recovery Act (RCRA)

RCRA addresses solid (Subtitle D) and hazardous (Subtitle C) waste management activities. The wastes listed in Table 14-2 are all subject to regulation under RCRA. Most RCRA requirements are not industry specific but apply to any company that transports, treats, stores, or disposes of hazardous waste. Some of the RCRA provisions that impact the management of metalworking wastes include:

> The primary environmental compliance problems faced by the metalworking industries include:
>
> — Increasingly restrictive discharge limitations for all media.
>
> — Product phaseout of ODCs mandated by the Clean Air Act Amendment of 1990.
>
> — The use of hazardous raw materials regulated by OSHA.
>
> — Wastes regulated by RCRA.
>
> — EPCRA requirement for manufacturing industries to report onsite release and off site transfer data for more than 600 toxic chemicals.
>
> — The Pollution Prevention Act of 1990 requirement for manufacturers to report information about the management of chemicals subject to TRI reporting.

— (40 CFR 268) Land Disposal Restrictions apply to solvents, electroplating wastes, heavy metals, and acids.

— (40 CFR Part 279) Used Oil Management Standards impose management requirements affecting the storage, transportation, burning, processing, and re-processing of used oil.

— Tanks and Containers used to store hazardous waste with a high volatile organic concentration must meet emission standards under RCRA (40 CFR Part 264-265, Subpart CC).

— If underground storage tanks (USTs) are used, then Subtitle I regulations (40 CFR Part 280) apply.

— Boilers and industrial furnaces that use or burn fuel containing hazardous waste must comply with strict design and operating standards (40 CFR Part 266, Subpart H).

Clean Water Act (CWA)

Metalworking facilities that intend to discharge wastewater into U.S. waters must obtain a permit prior to discharge. To obtain a permit, the metal fabricator must provide quantitative analytical data identifying the types of pollutants present in the facility's effluent. The permit sets forth the conditions and effluent limitations under which a facility may make a discharge.

The Effluent Guidelines and Standards for Metal Finishing (40 CFR Part 433) are applicable to wastewater generated by any of these operations: electro-

---

> Wastewater discharges are regulated by provisions of the Clean Water Act.

EPA #	Hazardous Waste
D006-D011	Wastes that are hazardous due to the toxicity of the constituents: cadmium, chromium, lead, mercury, selenium, and silver.
F001	(1) Halogenated solvents used in degreasing: tetrachloroethylene, methylene chloride, 1,1,1-trichloroethane, carbon tetrachloride, and chlorinated fluorocarbons. (2) All spent solvent mixtures/blends containing, before use, 10% (by volume) of one or more of these solvents or those listed in F002, F004, and F005. (3) Still bottoms from recovery of these spent solvents and spent solvent mixtures.
F002	(1) Spent halogenated solvents: tetrachloroethylene, methylene chloride, trichlorethylene, 1,1,1-trichloroethane, chlorobenzene, 1,1,2-trichloro-1,2,2-trifluoroethane, ortho-dichlorobenzene, trichlorofluoromethane, and 1,1,2-trichloroethane. (2) All spent solvent mixtures/blends containing, before use, one or more of the above halogenated solvents or those listed in F001, F004, F005. (3) Still bottoms from the recovery of these spent solvents and spent solvent mixtures.
F003	(1) Spent non-halogenated solvents: xylene, acetone, ethyl acetate, ethyl benzene, ethyl ether, methyl isobutyl ketone, n-butyl alcohol, cyclohexanone, and methanol. (2) All spent solvent mixtures/blends containing, before use, only the above spent non-halogenated solvents. (3) All spent solvent mixtures/blends containing, before use, one or more of the above non-halogenated solvents, and, a total of 10 percent or more (by volume) of one of those solvents listed in F001, F002, F004, F005. (4) Still bottoms from the recovery of these spent solvents and spent solvent mixtures.
F004	(1) Spent non-halogenated solvents: cresols and cresylic acid, and nitrobenzene. (2) All spent solvent mixtures/blends containing, before use, a total of 10 percent or more (by volume) of one or more of the above non-halogenated solvents or those solvents listed in F001, F002, and F005. (3) Still bottoms from the recovery of these spent solvents and spent solvent mixtures.
F005	(1) Spent non-halogenated solvents: toluene, methyl ethyl ketone, carbon disulfide, isobutanol, pyridine, benzene, 2-ethoxyethanol, and 2-nitropropane. (2) All spent solvent mixtures/blends containing, before use, a total of 10 percent or more (by volume) of one or more of the above non-halogenated solvents or those solvents listed in F001, F002, or F004. (3) Still bottoms from the recovery of these spent solvents and spent solvents mixtures.
F006	Wastewater treatment sludges from electroplating operations except from the following processes: (1) sulfuric acid anodizing of aluminum; (2) tin plating on carbon steel; (3) zinc plating (segregated basis) on carbon steel; (4) aluminum or zinc-aluminum plating on carbon steel; (5) cleaning/stripping associated with tin, zinc, and aluminum plating on carbon steel; and (6) chemical etching and milling of aluminum.
F007	Spent cyanide plating bath solutions from electroplating operations.
F008	Plating bath residues from the bottom of plating baths from electroplating operations where cyanides are used in the process.
F009	Spent stripping and cleaning bath solutions from electroplating operations where cyanides are used in the process.
F010	Quenching bath residues from oil baths from metal heat treating operations where cyanides are used in the process.
F011	Spent cyanide solutions from metal heat treating operations.
F012	Quenching wastewater treatment sludges from metal heat treating operations where cyanides are used in the process.
F019	Wastewater treatment sludges from the chemical conversion coating of aluminum.
K090	Emission control dust or sludge from ferrochromiumsilicon production.
K091	Emission control dust or sludge from ferrochromium production.

Table 14-2: Hazardous wastes relevant to the metal finishing industry.

plating, electroless plating, anodizing, coating, chemical etching and milling, or printed circuit board manufacturing. If any of these processes are used, then the metal finishing standards will apply to discharges from a number of additional processes including cleaning, polishing, shearing, hot dip coating, solvent degreasing, and painting. The standards include daily maximums and maximum monthly average concentration limitations. The standards are based on milligrams per square meter of operation and determine the quantity of wastewater pollutants that may be discharged.

The national pretreatment program – CWA 307(b) – controls the indirect discharge of pollutants to publicly owned treatment works (POTWs) by "industrial users." Metalworking industries that discharge to a POTW must meet certain pretreatment standards.

Storm Water Discharges

The NPDES regulations apply to storm water discharges directly related to runoff from manufacturing, processing, or raw materials storage areas at industrial plants – 40 CFR 122.26(b)(14). Metalworking industries are subject to these regulations.

Comprehensive Environmental Response, Compensation, and Liability Act (CERCLA)

CERCLA regulations require that all environmental releases of hazardous substances that exceed certain limits be reported to the National Response Center (40 CFR 302.4). A release report may trigger a response by EPA or by one or more Federal or State emergency response authorities. EPA implements hazardous substance responses according to procedures outlined in the National Oil and Hazardous Substances Pollution Contingency Plan (40 CFR Part 300).

Emergency Planning and Community Right-to-Know Act (EPCRA)

Many fabricated metal operations must comply with EPCRA 313. Manufacturing facilities with more than ten employees that do not qualify for an exemption and produce at least 25,000 lbs. or use at least 10,000 lbs. of a listed compound are subject to Toxic Release Inventory (TRI) reporting. EPCRA requires these industries to submit an annual toxic chemical release report (Form R). Form R covers releases and transfers of toxic chemicals to various facilities and environmental media, and allows EPA to compile the national database. Other EPCRA reporting obligations may also apply. All information submitted pursuant to EPCRA regulations is publicly accessible, unless protected by a trade secret claim.

Safe Drinking Water Act (SDWA)

Facilities that use underground injection of wastes are subject to SDWA regulation. The SDWA Underground Injection Control (UIC) program (40 CFR Parts 144-148) is a permit program that protects underground sources of drinking water by regulating five classes of injection wells. Wells used to inject hazardous wastes must also comply with RCRA corrective action and land disposal standards in order to receive a RCRA permit.

Toxic Substances Control Act (TSCA)

Facilities that use asbestos, chlorofluorocarbons (CFCs), or polychlorinated biphenyls (PCBs) need to adhere to TSCA standards. TSCA standards may apply at any point during a chemical's life cycle.

Clean Air Act (CAA)

The fabricated metal industry is subject to CAA regulations including National Emission Standards for Hazardous Air Pollutants (NESHAPs). New facilities may be subject to New Source Performance Standards (NSPSs).

Title VI restricts the use and distribution of ozone-depleting chemicals. Production of Class I substances, including 15 kinds of chlorofluorocarbons (CFCs), will be phased out entirely by the year 2000, while certain hydrochlorofluorocarbons (HCFCs) will be phased out by 2030. In addition to the CAA requirements discussed above, EPA is currently working on several regulations that will directly affect the metal finishing industry. Various potential standards are described below.

EPA has proposed a NESHAP (58 FR 62566, November 19, 1993) for the source category of halogenated solvent degreasing/cleaning that will directly affect the metal finishing industry. EPA is specifically targeting vapor degreasers that use the following haz-ardous air pollutants (HAPs): methylene chloride, perchloroethylene, trichloroethylene, 1,1,1-trichloroethane, carbon tetrachloride, and chloroform.

Hydrochloric acid (HCl) and chlorine are among the pollutants listed as hazardous air pollutants in Section 112 of the Clean Air Act Amendments of 1990. Steel pickling processes that use HCl solution and HCl regeneration processes have been identified by the EPA as a source category for which air emission standards may be warranted.

Checking Your Understanding

1. Under the Clean Water Act, what kind of data must be provided about the facility's effluent to obtain a wastewater discharge permit?

2. Which regulations impact the use and disposal of hazardous solvents (such as methylene chloride) that are used as degreasers by the metalworking industry?

14 - 4 Pollution Prevention Options for Metalworking Industries

The first step toward preventing pollution is to understand the processes that create the wastes. Constructing process flow diagrams (see Figure 14-1) and material balances (see Chapter 1) can help to identify opportunities for waste minimization and pollution prevention. Every industrial operation has material inputs and outputs. Material that is not incorporated into the finished product eventually becomes a waste product. Implementing pollution prevention can increase industrial efficiency and potentially decrease costs. Minimizing hazardous waste production has additional benefits such as improved safety, decreases in liability, less regulatory burden, lower disposal costs, and improved public relations. The following pollution prevention strategies can be used to minimize waste production in the metal fabricating industry.

Production Planning and Sequencing

Production planning and sequencing ensures that only necessary operations are performed and that no operation is needlessly reversed by a subsequent one. There are no capital costs associated with this action; on the contrary, material costs will be reduced. Additional staff hours may or may not be needed to implement these options.

— Improve scheduling of processes that require use of varying fluid types to reduce the number of cleanouts.

— Eliminate unnecessary cleaning steps.

— Inspect parts for quality before each manufacturing step.

Process Modification or Elimination

Process modification or elimination reduces the amount of waste generated. Waste disposal costs will be reduced, and material costs may decrease. Automation may reduce labor costs, but process modification may require capital outlay for equipment changes.

— Standardize fluids to reduce variety (this minimizes cross contamination, simplifies stocking, and aids recycling).

— Air emissions from vapor degreasing equipment can be reduced by: increasing the freeboard height above the vapor level; covering the degreasing unit; installing refrigerator coils above the vapor zone; rotating parts before removal to allow all condensed solvent to return to degreasing unit; controlling the removal rate (10 feet or less per minute is desirable) so as not to disturb the vapor line; and installing thermostatic heating controls on solvent tanks.

— Add automatic oilers to avoid excess oil applications.

— Use demineralized water to increase rinse efficiency, and keep the rinsing counter-current.

— Reduce rinse contamination via drag-out by: slowing and smoothing removal of parts (rotating them if necessary); maximizing drip time; using drainage boards to direct dripping solutions back to process tanks, installing drag-out recovery tanks to capture dripping solutions; using techniques such as air knives or squeegees to wipe bath solutions from the part; and changing bath temperature or concentrations to reduce the solution surface tension.

— Use abrasive cleaning instead of solvents for oxide removal and stripping operations.

— Replace solvent cleaning with aqueous or emulsion cleaning.

Raw Material Substitution or Elimination

Raw material substitution or elimination replaces existing raw materials with other materials that are less hazardous, produce less waste, or produce a non-toxic waste.

— Use easy-clean or no-clean fluxes, rust inhibitors, machine oils, etc.

— Use less hazardous acid and alkaline compounds in aqueous cleaning operations.

— Substitute alkali washes for solvent degreasers.

— Replace oil with lime or borax soap as the drawing agent in cold forming.

— Use metalworking fluids that can remain on the piece through heat treatment processes, where they can burn off. This eliminates the need for hazardous degreasing solvents and alkali cleaners.

— Use less hazardous degreasing agents such as petroleum solvents or alkali washes.

— Replace barium and cyanide salt heat treating with a carbonate/chloride carbon mixture, or with furnace heat treating.

Loss Prevention and Housekeeping

Preventive maintenance as well as equipment and materials management minimize opportunities for leaks, spills, evaporative losses, cross-contamination, and releases of potentially toxic chemicals. This reduces waste disposal costs as well as wear on machines and helps maintain metalworking fluid quality. These procedures also aid recycling.

— Maintain equipment by scheduling regular maintenance of gaskets, wipers, and seals on machines, scheduling regular sump cleaning and machine cleaning, and inspecting equipment for cracks, rust, and leaks.

— Use drip pans to recover leaking fluid.

Waste Segregation and Separation

Waste segregation and separation avoid the mixture of different types of wastes and the mixture of haz-

ardous and non-hazardous wastes. This facilitates recycling, reuse, resale, and disposal.

— Use specific lines for each set of metals or processes with similar inputs.

— Segregate scrap metal by metal type.

— Segregate different kinds of used oils and avoid solvent contamination.

— Keep solvent waste streams segregated and free from water contamination.

Recycling and Reuse

Closed-loop recycling uses or reuses a waste as an ingredient or feedstock in the production process. Reuse reduces material costs and disposal costs. Recycling may require capital outlay to purchase equipment.

— Centrifuge oil and scrap mixtures to recover oil for reuse.

— Use metal scrap as feedstock or recover for resale.

— Extend the life of process fluids through chemical, physical, and biological methods (precipitation, settling tanks, filtration, skimmers, dissolved air flotation, coalescing, centrifugation, ultrafiltration, biocides, etc.).

— Recycle spent solvents onsite using batch stills.

— Arrange a cooperative agreement with other small companies to centralize solvent recycling.

— Recover metals from solutions for resale.

— Recycle metal sludges through metal recovery vendors.

— Install a recovery system for solvents contained in air emissions.

Training and Supervision

Training and supervision provide employees with the information and the incentive to minimize waste generation in their daily duties.

— Ensure that employees and management are

aware of, understand, and support the company's pollution prevention goals.

— Ensure that employees know and practice proper and efficient use of tools and supplies.

Checking Your Understanding

1. What are two strategies that can be used to analyze waste streams to identify pollution prevention opportunities?

2. Name five strategies for pollution prevention.

Summary

The metalworking industries serve the functions of shaping metals into useful products, packaging materials, and parts manufacturing. These parts, in turn, supply other assembly industries including the appliance, automotive, defense, electronics, and furniture manufacturing. These parts and products are produced using five basic unit operations: shaping, surface preparation and cleaning, bonding, heat treating, and surface treatments. The metals used in these operations vary according to the specific need, but most are alloys that give the mixture certain metallurgical properties like strength, hardness, ductility, malleability, machinability, and corrosion resistance, to name a few.

Shaping operations take the form of heat treating, cold pressure, shearing, forming, or a machining process. Forming processes include bending, extruding, drawing, rolling, spinning, coining, or forging. During most of the shaping processes, metalworking fluids are circulated over the work surfaces to wash away the metal debris, to cool the cutting tools, and to otherwise improve the quality of the machined surface. These fluids eventually become contaminated and must be replaced, thus becoming industry waste sources.

During storage, transportation, and shaping metals are subject to surface changes such as rusting. Frequently the metal surface must undergo some type of stripping, solvent, alkaline, emulsion, or acid cleaning processes prior to such operations as welding or heat-treating. Many of these operations produce air emissions as well as liquid and sludges that require hazardous waste disposal due to their metal contents. Bonding operations may take the form of welding, soldering, brazing, or adhesive bonding. The joining of the surfaces of many metals requires the use of fluxes to chemically clean the metals and inhibit corrosion. The later removal of these fluxes creates another waste stream in addition to the air emissions resulting from the volatilization of organic residues in the fluxes during the bonding process.

To improve the metallurgical properties of the metals, several heat treatment processes are used including hardening, tempering, annealing, flame and induction hardening, and case hardening, the latter including both carbonitriding and carburizing. The wastes from these processes are mostly in the form of quenching solutions that may be oils, hazardous brines, or some other toxic compounds. In the finishing phase, a metal product is often subjected to a final surface process that provides it with protection against corrosion, improves its appearance, adds value, and/or increases its usefulness. Polishing abrasives, for example, are used to smooth surface defects, and etchants are used to produce specific designs. All surface finishing processes generate some wastes that contain metal residues.

Most environmental laws and regulations have some application in the metalworking industries. For example, RCRA addresses solid and hazardous waste management activities that are particularly applicable because of the generation of electroplating wastes and heavy metal wastes as well as the use of acid etchants, cutting oils, and various surface-cleaning materials. The CWA regulates the discharges of wastewater from electroplating, anodizing, coating, chemical etching, and milling operations as well as from printed circuit board manufacturing processes. The SDWA regulates the disposal of various hazardous materials by underground injection. The CAA regulates all air emissions and has restricted the continued use and distribution of ozone-depleting chemicals including CFCs and certain HCFCs.

Over the years the metalworking industries have constructed process flow diagrams and material balances to identify opportunities for waste minimization and to implement pollution prevention activities. Some of these are: improving production sequences to eliminate unnecessary cleanups; making process modifications to reduce the amount of waste generated by drag-out, by cross-contamination of fluids, by using abrasives rather than stripping operations, or by replacing solvent cleaning operations with aqueous cleaning methods. The elimination or substitution of raw materials has helped reduce the amount of toxic substances, and the use of good housekeeping practices has eliminated needless waste. For those wastes that are unavoidable, waste segregation, reuse, and recycling techniques have reduced the amounts of waste and recovered valuable raw materials.

Critical Thinking Questions

1. Develop a list of compliance issues that may be affected when a metal fabricator eliminates organic solvents from the production process.

2. Develop a simple process flow diagram and identify the opportunity points for waste minimization in the following manufacturing process. Sheet steel and aluminum castings are cut to the appropriate size by shearing operations. The metalworking fluids used in the shearing operations are emulsion type. To produce the final product form, the cut metal undergoes additional bending and milling operations. These operations use synthetic cutting oils. Residual cutting fluid is removed from the parts after each shaping process using petroleum naphtha. Components are welded, subjected to abrasive cleaning, and are then annealed for stress relief. Components are polished and then sent off site for assembly.

15

PRESERVING
THE LEGACY

The Electroplating and Metal Finishing Industries

by Azita Yazdani

Chapter Objectives

Upon completing this chapter, the student will be able to:

1. **Describe** the overall electroplating process.

2. **Explain** the function of each of the essential parts in an electroplating system.

3. **Discuss** the pollution prevention techniques being used within the industry.

4. **Identify** and apply the regulations that govern wastes produced by the electroplating process.

5. **Identify** and discuss pollution prevention activities that are used by the electroplating industries.

Chapter Sections

15-1 Introduction

Metals are shaped, formed, forged, drilled, turned, cast, and acid etched into appropriate shapes according to their intended use: these are metalworking processes and the topic of Chapter 14. After they have been shaped and before they can be considered ready for their intended application, most metals must also be finished to be protected from **corrosion**.

In the 1830s the British scientist Michael Faraday elucidated the principles of **electrochemistry**. It had been previously discovered that a metal in solution could be caused to accumulate on the surface of another – the workpiece – through the application of **direct current (DC)** electricity. By making the workpiece the **cathode** (negative electrode), the **cations** (positively charged metal ions) in the solution are attracted to it. Once in contact, the electrons are gained (**reduction**) by the cations and are converted into neutral atoms. The insoluble metal atoms accumulate on the surface of the workpiece, eventually forming a thin layer of the metal. This process is known as **electroplating**.

Some electroplated finishes are applied to improve appearance, either temporarily or for extended periods, while others are for purely decorative reasons. Steel, for example, is the cheapest structural material available for countless uses. However, it must be protected from corrosion by the surrounding atmosphere. Copper is an excellent conductor of electricity and is used for such items as printed circuits and communications equipment. It does, however, quickly form tarnish films that interfere with joining operations such as soldering, and cause unacceptably high contact resistance in relays and switches. In general, electroplating allows the use of relatively inexpensive metals for the production of a part, while affording a more expensive metal for the exterior coating according to the desired physical and chemical properties.

15-2 The Electroplating Industry

Types, Locations, and Functions

Electroplating facilities fall into two major categories: **job shops** and **captives**. Job shops provide a service, taking parts manufactured by others and electroplating them with any combination of more than one hundred different metallic coatings. These shops tend to be very small businesses, averaging less than 50 employees and with annual sales of less than $5 million, although larger shops exceeding 100 employees and with sales of more than $10 million per year do exist. Job shop electroplaters in the United States number about 3,000 companies, most of them located in or near the major metropolitan areas; notably Chicago, Detroit, Cleveland, New York, and the New England states.

Captive electroplating facilities perform electroplating operations for parts manufactured in-house. Captive facilities can be found throughout the United States and include numerous large (Fortune 500) manufacturing and service corporations, such as major airlines, aerospace firms, computer and electronics manufacturers, hardware manufacturers, and all automotive firms.

Electroplating operations are utilized in the manufacture of both strategic and consumer products. Some of the items that are electroplated include printed circuit boards as well as hundreds of automotive parts such as piston rings, dashboards, electronic sensors and controls, air bags, metallic and plastic hardware, fuses, lights, and most engine components. Each aircraft contains hundreds of electroplated components, such as engine parts, landing gear, and cockpit instrumentation. In general, most metallic components in a manufactured item are electroplated to enhance corrosion resistance, appearance, **solderability**, and/or **weldability**.

Electroplating is not just limited to application over metallic substrates. While other nonmetallic materials have been electroplated since the mid-nineteenth century, the electroplating of plastics began in 1963 with the introduction of ABS (acrylonitrile-butadiene-styrene) plastic. In one method, the plastic is first acid etched, then sensitized and activated by being dipped into a stannous chloride solution followed by a palladium chloride solution. It is then allowed to form an **electroless copper** or **electroless nickel** conductive coating, before entering the remaining electroplating steps to add the final finish. Today, nearly any plastic, ceramic, or other substrate can be electroplated using specialized techniques, solutions, and equipment. The variety of chemicals used in an electroplating process obviously depends on the substrate used and the type of metallic coating that is applied.

The Electroplating Process

Regardless of the composition of the workpiece, the electroplating method to be used, or the type of protective or decorative coating that is to be applied, the basic steps involved in electroplating remain the same. As shown in Figure 15-1, the first step is to clean the surface of all foreign materials including grease, loose particles like **smut** or dust, metal turnings or grinding debris, and adherent soils including scale, rust, paint, or burned-on oil. The removal of each type of material presents its own challenge, but four methods are typically used: 1) solvent action (**vapor degreasing**), 2) detergent action (**surfactants**), 3) chemical reaction (pickling), and 4) mechanical action (sandblasting). Each of these cleaning methods is then followed by a water rinse to remove chemicals or particulate matter.

The cleaned and rinsed workpieces are subsequently placed piece by piece on a manual or automated **rack plating** system. The purposes of the rack are to bring each workpiece into contact with the **process bath** and to serve as the electrical contact, making each workpiece a part of the cathode. While in contact with the process bath, the cations in the solution are attracted to the negatively charged surface, where they are converted into metal atoms. Depending on the amount of electrical current and the length of time the workpiece is left in contact with the process bath, a metal coating of varying depths can be deposited on its surface.

If the workpieces are small and numerous, it is likely that the final cleaning, rinsing, and electroplating steps will be carried out in a **barrel plating** system, as shown in Figure 15-2. In this method, the barrel becomes the electrical contact that trans-

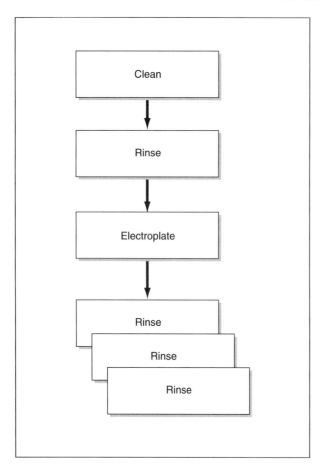

Figure 15-1: Process steps in electroplating.

mits the electrical charge to each of the smaller pieces as they tumble within the rotating barrel. The barrel plating method is used for such smaller items as nuts, bolts, screws, and electrical junction boxes, where some surface imperfections on the final part are acceptable.

Once the finish layer is of the desired depth, the workpiece is removed from the process bath and allowed to drip for a few moments before entering the first of a series of rinse baths. The number of rinses varies but, as shown in Figure 15-3, most plating operations now use a minimum of three **countercurrent rinses**. As the workpiece leaves each rinse, it is allowed to drip a few moments before it is moved to the next bath. By having fresh (deionized) water entering the last of the rinse baths and flowing in a direction opposite the movement of the workpiece, the workpiece becomes progressively free of process chemicals as it is removed from the final rinse. After leaving the final rinse, the workpiece is complete unless it is sent on to a polishing process.

The Plating Bath

There are four essential parts to an electroplating system: 1) the workpiece, 2) the process solution, 3) the anodes, and 4) the external electrical circuits, including the source of the electrical current, the wires, and the associated instrumentation. The process bath contains the cations to be deposited and is where the workpiece is placed as the cathode. The total composition of the process bath is often a closely guarded industrial secret. In general, it contains a soluble salt of the metal to be deposited and several other substances, such as a buffer to control the pH, intended to insure a desirable finish.

Ingredients of a Plating Bath – Every plating bath contains ingredients that serve one or more of the following functions. They:

1. Provide a source of the metal or metals being deposited;

2. Form complexes with ions of the metal being deposited;

3. Provide the conductivity of the plating bath;

4. Stabilize the plating bath solution against hydrolysis;

5. Act as a buffer and stabilize the pH;

6. Modify or regulate the physical form of the finish deposited;

7. Aid in the dissolving of the anodes;

8. Modify other properties, either of the bath or of the finish, that are application-specific.

This does not mean that all plating baths contain eight ingredients, since some compounds perform more than one function and in some instances not all of the listed functions are necessary. Consider the following points:

1. The electroplating bath must always contain ions of the metal that is to be electroplated. The addition of a substance to cause a complex of this ion to be formed, however, is not always required. It is true that in many cases the finish obtained from the use of complex ions is superior to those from the use of simple ions.

2. Any ionic solution conducts electricity, but many metal salts produce ions with low mobility and

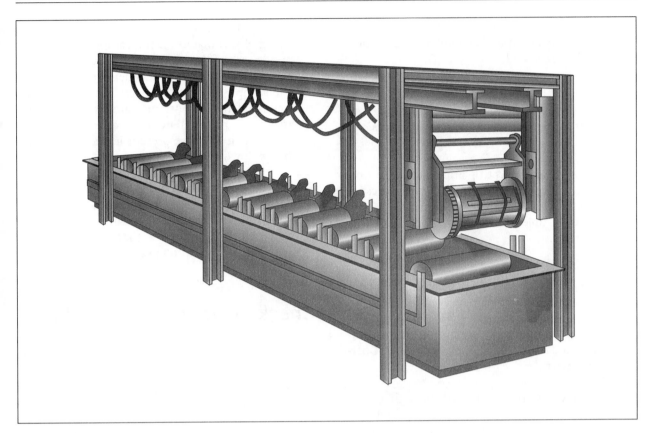

Figure 15-2: Barrel method for electroplating many small workpieces.

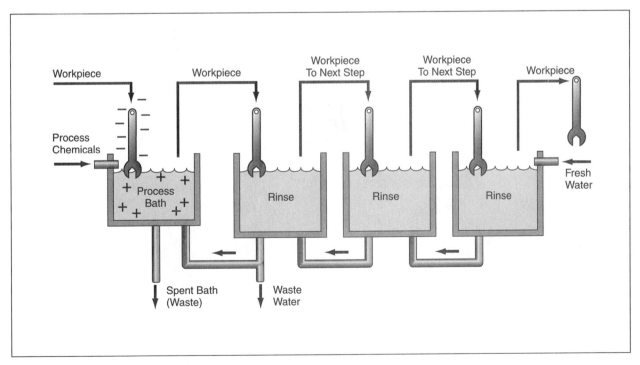

Figure 15-3: Rack plating method with countercurrent rinse system.

are therefore rather poor conductors. To avoid the necessity of using higher voltages, **conducting salts** are added.

3. Most metal hydroxides, M(OH), have low water solubilities, therefore making the metal ions subject to precipitation due to hydrolysis as shown: $MX + H_2O \rightarrow M(OH) + HX$.

4. In some alkaline electroplating baths, absorption of carbon dioxide from the air will cause precipitate of the metal ions as a metal carbonate, MCO_3, unless an acceptor for carbon dioxide is present.

5. Many plating solutions are either highly acidic or highly alkaline. For these solutions, pH control is a minor concern. However, for those solutions known as neutral, (pH = 5-8), control of the pH within these prescribed limits is important and is usually accomplished using a **buffering agent**.

6. When DC electricity is passed through a solution containing a metal ion, the metal will deposit on the cathode. On occasion, the metal finish will be useless if additives are present to control its deposition.

7. Unless the **anodes** used in an electroplating process are deliberately inert and only being used to complete the circuit, it may be desirable for them to be made of the same metal as the one being electroplated on the cathode. When the anodes are of the same metal, they can be used to replenish the metal ions being deposited, so the composition of the electroplating bath remains relatively constant. Anodes used in this manner, however, tend to become **passive**, unless specific ions are present in the solution to break down the passivity.

8. Finally, some baths require specific additives for specific purposes. An example is the addition of sodium polysulfide to most zinc cyanide baths to precipitate impurities that would interfere with the formation of a satisfactory finish.

Checking Your Understanding

1. What is the difference between a job shop and a captive electroplating facility?

2. What are two purposes for using counterflow rinse tanks?

3. What are the four essential parts of an electroplating system?

4. What is the purpose of a buffer?

5. Why are substances sometimes added to electroplating baths to form a complex with the ions being electroplated?

15-3 Regulatory Overview and Pollution Prevention

Regulatory Overview

The activities of electroplaters are subject to a variety of regulatory requirements. Due to the large volumes of wastewater generated, the most significant of these is the Clean Water Act (CWA). Many electroplating facilities also generate concentrated wastes that contain high levels of toxic metals such as lead, cadmium, or hexavalent chromium as well as other toxics such as chlorinated solvents and cyanide wastes. These wastestreams typically originate from stripping or other pretreatment operations, unrecoverable contamination of electroplating solutions, or use of processing solutions with a finite utility life. The resulting sludges typically fall under RCRA solid (Subtitle D) and hazardous (Subtitle C) waste management regulations. Most plating shops also operate an air scrubbing system to comply with the Clean Air Act Amendments of 1990 (CAAA) and local air discharge regulations.

The discharge of metal-containing wastewaters is regulated by EPA through NPDES guidelines (40 CFR 403) and standards established by the local agencies. As a consequence, the discharge limits may vary throughout the country. Since most metal finishers must operate their own wastewater pretreatment system to meet CWA standards, they also generate metal sludges as by-products. The metal sludges generated by wastewater pretreatment processes are a relatively low hazard and are regulated as an F-006 (hazardous wastes from non-specific sources) RCRA waste. Regulations under Subtitle C of RCRA (40 CFR Parts 260-299), however, establish a "cradle-to-grave" system for governing listed hazardous wastes; off site management of these metal sludges must therefore be accomplished according to the regulations.

The air quality regulations also place requirements on metal finishing facilities. The emissions from boilers and other combustion sources at these plants as well as the emissions of solvents from cleaning operations are regulated. At this time, air emissions of heavy metals from plating operations are not closely regulated, except for certain metals such as chromium (40 CFR Parts 9 and 63, Subpart N, 60 FR 498, January 1995). It is expected that the heavy metal air emissions will become more regulated.

Wastes and Pollution Prevention Measures

Waste minimization is a policy specifically mandated by Congress in the 1984 Hazardous and Solid Waste Amendments (HSWA) to RCRA. As the federal agency responsible for writing regulations under RCRA, EPA has an interest in ensuring that new methods and approaches are developed for minimizing hazardous waste. In the following sections, waste minimization practices that can be initiated in the electroplating and metal finishing industries to meet this mandate are presented.

Operational Improvements

The most commonly practiced pollution prevention option in electroplating is the utilization of **drag-out rinses**, as displayed in Fig 15-3. These are the first rinse tanks that are used after the plating bath. The water and chemicals collected in the drag-out rinses are returned to the process tank by **evaporative recovery** to make up for the fluid losses in the tank. The system works only on chemical processes that operate with sufficient evaporative losses to make room for the returned drag-out volume. This set-up also allows for installation of reuse and recovery technologies, which will be discussed later.

Chemical Substitution

The second most popular pollution prevention alternative is product substitution. For example, there are alternative zinc plating methods available, including some that use cyanide. By switching to a noncyanide zinc plating process, cyanide can be eliminated from the entire plating process.

Electroplaters considering such a process substitution will need to know as much as possible to make an intelligent decision. Popular substitutions for a few other common plating processes are discussed in the following sections.

Trivalent Chromium – The driving forces behind substituting trivalent for hexavalent chrome are 1) the CAA regulations, 2) local constrains on hexavalent chromium emissions to the air and water, and 3) the health hazard associated with the use of hexavalent chromium. Because trivalent chrome lacks the hardness of hexavalent chrome, it is currently used only in decorative applications; however, even in those applications it is not entirely satisfactory since it fails to produce the desired bright chrome finish.

Trivalent chromium processes can be expected to generate between five and 10 percent solid waste from rinse water treatment, due to the lower chromium content in the drag-out, when compared to hexavalent baths (assuming no recovery of drag-out). The use of trivalent chrome eliminates the need for the reduction of hexavalent to trivalent chrome in the wastewater treatment process, thereby leaving only the pH adjustment step. The exhaust from trivalent baths does not need to be controlled, which means there is no need for scrubbing equipment, nor treatment or disposal of scrubbing solutions. This translates to lower capital costs for waste treatment and no hexavalent discharge violations due to malfunctions of the wastewater treatment system.

For companies that must comply with Chemical Oxygen Demand (COD) regulations, trivalent baths can increase COD loading because they contain a significant amount of organic additives. At this time there are no commercially viable hard chromium plating processes using trivalent chemistry on the market.

Substitute Processes for Hard-Chromium Plating – While a viable substitute for conventional hard chromium finishes is presently not on the market, a considerable amount of research is being conducted to develop coatings that could function in place of heavy chromium deposits. In one such process, an alloy containing nickel and tungsten metals with fine particles of silicon carbide dispersed throughout is produced by **electrodeposition**. This alloy process can deposit a finish with the same thickness in half the time due to an electrical current efficiency of 24-35 percent vs. 10-12 percent for the conventional hard chromium plating process.

The successful substitution of the alloy plating process would eliminate hexavalent chromium from the wastewater and air discharges of the metal finishing facility. None of the ingredients in the alloy bath are oxidants, so storage of the chemicals is simplified, and the hazards of unwanted reactions between oxidizing and reducing agents are eliminated. However, the alloy process utilizes nickel and tungsten compounds that have also been linked to cancer. The net effect, therefore, is to replace one potentially carcinogenic material with two that are suspected carcinogens. The plating process also requires exhaust and scrubbing to reduce air emissions. High concentrations of ammonium ions in the wastewater stream would cause serious wastewater treatment problems, due to the ammonium ion tendency to **chelate** metals such as copper and nickel.

Presently there are no good techniques for stripping the alloy or performing spot repairs. Chromium, on the other hand, is easily stripped and some parts can even be spot repaired. The alloy process is also sensitive to metallic contamination, while the chromium process is very tolerant. The alloy process utilizes citrates at a relatively neutral pH, creating an environment that is conducive to biological activity, such as mold growth. Other substitute nickel alloy baths such as nickel-tungsten-boron have been investigated and reported in literature, but have not been commercialized.

Alkaline Noncyanide Copper for Cyanide Copper – Alkaline noncyanide copper plating processes have been available as a substitute for the cyanide process since 1990. The obvious benefit is the elimination of cyanide from the wastewater stream. The cyanide and alkaline noncyanide processes typically function at elevated temperatures, making recovery through drag-out control and other recovery systems viable. The noncyanide process contains only 25-50 percent as much copper as a full-strength cyanide copper bath, which translates into lower sludge generation. Waste treatment is accomplished by pH adjustment with lime, CaO, or magnesium hydroxide, $Mg(OH)_2$, thereby eliminating the two-stage chlorination system in the wastewater treatment process and the use of more dangerous chemicals such as chlorine, Cl_2, or sodium hypochlorite, NaClO. The noncyanide process does not create a cyanide-bearing waste. The cost of disposing of a cyanide copper solution is astronomical compared to the alkaline noncyanide process. If copper plating on steel is the only source of cyanide in the waste, substitution of the noncyanide process may create a waste that qualifies for delisting as a hazardous waste by EPA.

Cyanide-based copper plating solutions are very tolerant of impurities and typically last many years before requiring treatment and disposal. Noncyanide baths, on the other hand, are very sensitive to impurities and frequently become contaminated requiring disposal. Other disadvantages of the alkaline noncyanide copper plating process are 1) higher operating costs, 2) inability to use the process on zinc surfaces, like die castings and **zincated** aluminum or magnesium, 3) greater sensitivity to impurities, and 4) a chemistry that is more difficult to control.

High pH Nickel for Copper Strike – High pH nickel plating solutions have been available for a long time as a substitute for cyanide copper **strike** on zincated aluminum surfaces. Cyanide copper has fewer cleaning and analytical requirements, but the nickel process allows for the elimination of one more cyanide bearing process.

To obtain optimum results, electroplaters must balance the ratio between the nickel sulfate, $NiSO_4$, and the sodium sulfate, Na_2SO_4. Parts with complex shapes require higher sodium sulfate concentrations. Zinc contamination should be continuously removed through the use of a purification cell.

The higher the sodium content in the nickel-plating bath, the more brittle the deposit becomes. The bath should, therefore, be used only as a strike, before conventional nickel plating. Parts that undergo **fatigue cycles** or extreme temperature changes may experience fatigue failures and less corrosion resistance.

The presence of ammonium ion in the high pH nickel formulation may cause waste treatment problems unless its concentration can be minimized through drag-out recovery techniques. This bath contains a higher metal content than the cyanide copper process and twice the metal content of the

alkaline noncyanide process. Sludge volume from wastewater treatment would be affected accordingly.

Recovery and Recycling

Closed Loop and Open Loop Evaporative Recovery Systems

The electroplating industry utilizes numerous recovery and recycling techniques to return a portion or all of the process chemicals to the original plating bath. A typical electroplater will first evaluate the effectiveness of rinsing; determine if a viable, less polluting or nonpolluting substitute exists; and make those changes before investing in recovery and recycling equipment.

Recovery and recycling equipment is generally expensive; it requires reduction of water usage to be economically feasible, and increases the maintenance workload and operational complexity of the electroplating facility. However, such systems can reduce the amount of solid waste generated by the electroplater and can often yield a net saving in chemical costs that can pay for the equipment in a matter of months.

Recovery involves direct reuse without further processing. Recycled chemicals are used or reused in other industrial processes, or used as substitutes for other chemicals. Some of the electroplating operations in which reuse and/or recycling is commonly employed are:

— Reuse of concentrated rinse water in the electroplating process baths;

— Recovery of metal or metal concentrates;

— Sale of by-product sludges;

— Regeneration and reuse of process solutions;

— Recycling of treated wastewater.

The use of closed or open loop recovery units, as displayed in Figure 15-4, can recover process chemicals from the first rinse of a counterflow rinse system. The water can then be reused, depending on process requirements. Any chemicals that are not recovered are typically treated by the wastewater pretreatment system.

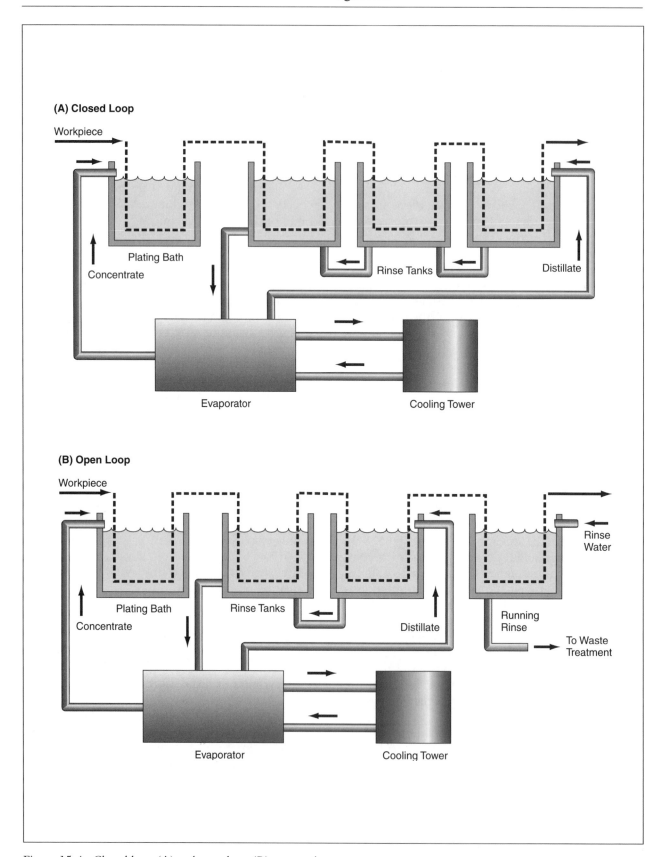

Figure 15-4: Closed loop (A) and open loop (B) evaporative recovery systems.

Atmospheric Evaporators

Atmospheric evaporators operate by spraying the wastewater over packing media, grids, or plates, and then blowing ambient air over the packing to promote evaporation. Advantages include low capital cost and simple maintenance. A major drawback is the inability to evaporate on days when air humidity levels approach 80-90 percent. In areas of generally high humidity, the solution will require some heating to work effectively. Atmospheric evaporators generally have low capacities, ranging from 10-40 gallons/hour, although several can be installed in parallel to attain the desired evaporation rate. With cyanide-based plating solutions, use of any evaporative system can increase the rate of carbonate buildup in the plating bath, due to carbon dioxide absorption from entrained air, and cause thermal breakdown of cyanide.

Vacuum Evaporators

There are several different **vacuum evaporator** designs on the market, normally differing in how the vacuum is achieved, the amount of energy required, or the amount of vacuum used. The higher the vacuum, the lower the temperature at which the water will boil. By using enough vacuum so that the water will boil at temperatures as low as 110-130°F, vacuum evaporators protect some of the delicate ingredients of the processing solutions that might otherwise decompose at higher temperatures. Advantages of these systems include recovery of both the concentrate and the condensed water as well as the capability of operating in all weather conditions. Its disadvantages include high energy and maintenance costs as well as foaming of some process solutions.

Single-effect Evaporators – A single-effect evaporator usually uses steam or high-temperature water passing through a steam coil or jacket to heat the liquid to its boiling temperature. The vapors produced by the boiling liquid are drawn off and condensed. The concentrated liquid is then pumped from the bottom of the vessel and returned to the process bath. This process requires about 1,200 BTU per pound of water evaporated.

Multiple-effect Evaporators – A multiple-effect unit consists of a series of single-effect evaporators. Vapor from the first evaporator is used as the heat source to boil liquid in the second evaporator. The process can continue for several evaporators (effects). Depending on the number of effects, multiple effect units can require as little as 200 BTU per pound of water evaporated.

Vapor Recompression Evaporators – The **vapor recompression evaporators** initially use steam to boil the liquid. The vapors produced are then compressed to a higher pressure and temperature. The compressed vapors are then directed to the jacketed side of the next evaporator and, instead of using more steam, they are used as the heat source to vaporize more liquid. The unit requires as little as 40 BTU per pound of water evaporated.

Electrolytic Recovery

Electrolytic recovery, shown in Figure 15-5, is a technology that uses special electroplating equipment to lower the concentration of dissolved metals in drag-out rinse tanks. Benefits of electrolytic recovery include reduction of sludge generation, some electrolytic destruction of cyanide, and reuse or sale of scrap metal plated out. Disadvantages include incomplete recovery (some waste is generated), tendency for spontaneous combustion of plated metal, and high energy costs.

Ion Exchange

Ion exchange systems, as shown in Figure 15-6, have been in use in the metal finishing industry for decades. A conventional metal finishing ion exchange system has a fixed bed of resin with the ability to exchange or remove cations or anions from rinse waters. In general, divalent and trivalent ions are easier to remove than monovalent ions using ion exchange.

The chief advantage of ion exchange is that it can be selective in what it removes. In recycling applications, this means that undesirable impurities need not be recovered along with the desired materials. Ion exchange can also be used to remove specific ions from a rinse tank. A typical application may be to remove nickel, copper (from an acid bath), or chromium from their respective rinses and then to return the regenerated metal ions to the plating bath as concentrates.

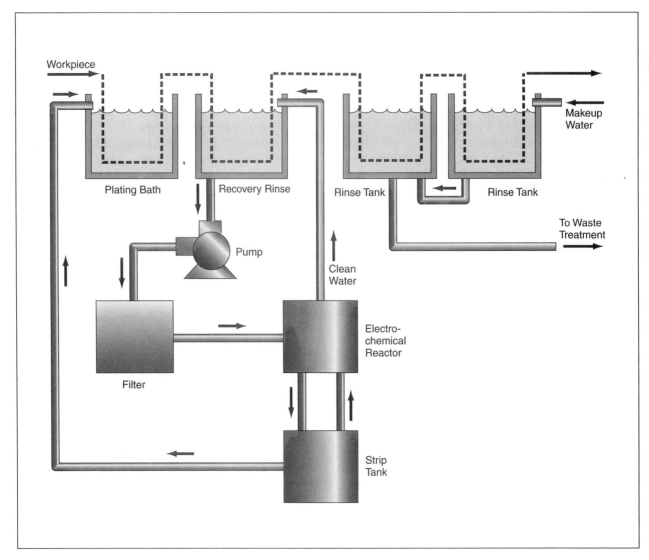

Figure 15-5: Electrolytic recovery.

Ion exchange is also frequently utilized to remove **tramp** heavy metals from hard-chromium-plating solutions. The presence of heavy metals such as iron, copper, and nickel can significantly reduce the efficiency of the chromium-plating solution, creating excessive misting and solution loss through the exhaust stack. Such systems produce a regeneration waste that needs to be treated to remove the heavy metals prior to discharge.

When the useful capacity of the ion exchange column is exhausted, it is regenerated using dilute acid (sulfuric or hydrochloric) for cationic resins and sodium hydroxide solution for anionic resins. The hardware of prepackaged units consists of pressure vessels ranging from two to six feet or more in di-

ameter, handling flow rates up to 300 gallons per minute (gpm) or more. Custom-made units, with 12 feet-diameter columns, have been built to handle flows as high as 1,150 gpm. The loading capacity is typically 10 gpm per square foot of resin cross-section.

Close control of the pH of the wastewater is important; lower pH reduces the capacity of the resin while high pH tends to clog the resin with solids (metal hydroxides). The systems have the drawback of not having suitable instrumentation to indicate when the resin is saturated. Saturated resin will discharge an effluent containing high concentrations of the pollutant it is supposed to remove. Operators typically compensate for this drawback by taking ion

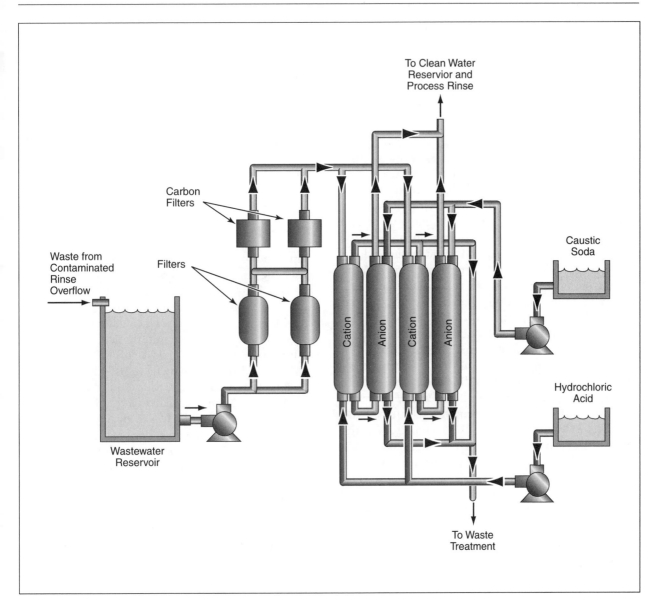

Figure 15-6: Conventional ion exchange system.

exchange cylinders off line and replacing them according to a schedule based on a resin loading rate study.

Additional disadvantages of ion exchange recovery systems may include the requirement of additional treatment equipment for modifying the regenerated stream to a chemistry suitable for reuse. An example is a nickel recovery system, which requires acid neutralization of the regenerated solution before it can be reused in the plating bath. The regenerated stream is often not very highly concentrated (2-4 ounces per gallon is typical), and evaporative systems are needed to concentrate the recovered stream to a usable level. Ion exchange is considered an expensive recovery method, which requires a lot of care and knowledge of operation.

Reverse Osmosis

Reverse Osmosis (RO) can be used to separate water from inorganic salts. Through use of a selectively permeable membrane, water – the permeate – is allowed to pass, but the salts are retained. The system utilizes pumps to generate pressures of 400-800 psi that force the permeate through the membrane, leaving a concentrated residual liquid behind. To prevent fouling of the membrane, feed solutions must be pretreated to remove materials such as magnesium, calcium, lead, iron, carbonates, particulates, oils, and other fouling materials.

The membranes may be made of cellulose acetate (similar to the clear plastic covering on cigarette packages) and aromatic polyamides or cross-linked polyamides (nylons). The membrane cost can often represent 50 percent or more of the total equipment cost. The membrane comes in various configurations, including spiral wound (see Figure 23 in Chapter 3), tubular, and hollow fiber. The tubular-type membrane is inserted into a porous tube. This type of RO is used for low-volume (low-pressure) applications. The spiral-wound membrane is a flat sheet separated by a mesh spacer that is spirally wound around a perforated plastic tube that acts to channel the permeate flow. The hollow-fiber membrane consists of millions of membrane fibers.

The most common (99 percent) application of RO is for the rinse water tanks. RO systems were developed to be used on alkaline solutions, such as cyanide copper and brass. To be effective, however, the rinse water must be extremely well filtered and have a low carbonate level; the water used in the plating and rinse tanks must be deionized. Aside from the fouling, some RO systems are very temperature sensitive. For example, a cellulose acetate membrane cannot withstand temperatures higher than 96°F. The membrane can also rapidly deteriorate if the waste stream has a pH less than three or higher than 10. Nickel plating is the most popular process in the electroplating industry that utilizes RO recovery equipment.

Electrodialysis

Electrodialysis (ED) is a process that makes use of a stack of closely spaced ion-selective membranes through which ionic materials are selectively transferred or rejected (see Figures 24 and 25 in Chapter 3). The driving force causing the ions to migrate within the stack is an electrical potential applied between two electrodes. In plating operations, ED is typically applied to the rinse water and an ED system used to separate the dissolved metal and associated anions. Because each pass through the system does not create a very concentrated product stream, the system is usually connected to a re-cir-culating drag-out rinse tank. Typical applications are on the first drag-out tank after the process bath of gold, silver, nickel, and acid tin plating systems. On a nickel plating system, ED does not recover organic contaminants from the rinse water, thereby eliminating organic contaminant buildup that would have existed had an evaporative system been used. ED equipment is very expensive and involves the same level of membrane fouling and replacement cost problems as RO.

Other Pollution Prevention Measures

There are several other pollution prevention options available to metal finishers. One unique proprietary process is a resin-based system for purification of acid solutions contaminated with heavy metals. A typical application would be on aluminum-anodizing solutions based on sulfuric acid and stripping solutions containing nitric acid. The resin retains the acid and allows the metal to pass through. Upon backwashing with water, the acid is released by the resin, purified, and returned to the process.

Crystallization, or freeze-drying, systems are available and used in a number of large manufacturing firms that have electroplating facilities. Crystallizers are commonly used in printed circuit board manufacturing facilities to recover copper sulfate crystals from the etching and plating wastes. Freeze-drying systems have been pilot tested on nickel plating solutions, yielding mixtures of crystals of nickel sulfate and nickel chloride. These systems are typically too expensive for most job shops to justify.

Checking Your Understanding

1. What is the major federal legislation that affects electroplaters?

2. What are three pollution prevention activities practiced by electroplaters?

3. What is the function performed by each of the different types of evaporators?

4. What are two advantages of the electrolytic recovery method?

5. What types of materials are used for the membranes in RO?

Summary

Metal parts are usually not ready for application until they have had a finish added to protect it from corrosion. Electroplating is a method for adding a protective coating to a workpiece by making it the cathode in an electrolytic bath. Through the application of direct current, the cations in the bath are attracted to the surface of the workpiece where they are converted to atoms. By regulating the time and current, protective coatings of varying thickness can be deposited on the workpiece.

There are two types of electroplaters: job shops that provide electroplating as a service to other companies, and captives that only plate items that are also produced by the company. Electroplating is no longer limited to just conducting workpieces. Using a variety of specialized techniques, solutions, and equipment, today even plastics and ceramics can be electroplated. Regardless of the substrate, all electroplating processes involve the same four basic steps: cleaning the surface of all foreign materials, rinsing, electroplating, and a second series of rinses. The four methods most often used for surface cleaning include solvent action, detergent action, chemical reaction, and mechanical action. In each case, the cleaning process must be followed by rinsing to remove chemical or particulate matter. The electroplating process may be of the manual or the automated rack type if the objects are large and the finishes must be flawless, and of the barrel plating type when there are many small items on which surface imperfections are acceptable. In both cases the plating process is followed by a series of washes, typically using a three tank countercurrent system. The purposes of the rinse tanks are twofold: 1) they remove process chemicals from the workpiece and 2) they recover the process chemicals.

The major regulatory requirements affecting the electroplating industry are the CWA, RCRA, and CAA. Wastewaters typically contain high concentrations of toxic metals or other toxics such as chlorinated solvents and cyanide wastes. Discharges of wastewaters, after pretreatment, are regulated by EPA through NPDES guidelines and local agencies. Solid wastes resulting from wastewater pretreatment processes fall under RCRA solid and hazardous waste management regulations. Due to acid mists created by the electroplating process, most plating shops must operate an air scrubbing system to comply with CAAA and local air regulations.

Pollution prevention activities center on the use of three practices: operational improvements, substitution, and recovery and reuse. The most common operational improvement is the use of countercurrent drag-out rinse tanks. Substitution activities center on the replacement of more hazardous chemicals and processes with less hazardous ones. In particular, the replacement of hexavalent with trivalent chrome and the elimination of cyanide-containing process solutions. Recovery and recycling efforts center mainly on the recovery of valuable process chemicals, thereby not only saving on additional purchases, but also saving on waste disposal costs. Some of the methods used to recover the process chemicals are the use of various types of evaporators, electrolytic recovery techniques, ion exchange, reverse osmosis, and electrodialysis. Several other unique recovery measures have been demonstrated; however, they are often too expensive for most small electroplaters to consider.

Critical Thinking Questions

1. In Chapter 16, other surface coating technologies that can be used to protect metal surfaces from corrosion are discussed. In Chapter 8, it is suggested that as a part of life cycle design, consideration be given to the surface finish applied because of recycle and/or reuse considerations. Describe the type of surface finish you would suggest for a brass bathroom water faucet and defend your answer in terms of both environmental and recycle/reuse concerns.

2. Research and compare/contrast the reverse osmosis and ion exchange methods specifically for metal ion recovery. Which of the systems provides the greatest flexibility and is the least expensive to install and operate?

3. Compare the amount of chrome used on a car manufactured 20 years ago to a similar model today. What are the principal factors that account for the differences found?

16

The Paint and Surface Coating Industries

by Richard Marty and Larry McGaughey

Chapter Objectives

Upon completing this chapter, the student will be able to:

1. **Discuss** the purpose of and methods used in surface preparation, coating application, and in curing paint and surface coatings.

2. **Describe** the common waste streams generated during surface preparation, coating application, and curing operations.

3. **Discuss** the economics of pollution prevention practices in the paint and surface coating industries.

4. **List** and discuss the various specific regulations that apply to the paint and surface coating industries.

Chapter Sections

16-1 Introduction

Paints and surface coatings are widely used to protect homes, automobiles, appliances, boats, furniture, toys, bridges, and roads. According to the National Painting and Coating Association, the Unites States alone produced almost 500 million gallons of paint and surface coatings in 1996; almost half of this production was **architectural paint** used to protect residences and commercial properties. Paints and surface coatings fall into four major types: 1) **latex** or water-based coatings; 2) **solvent-based** coatings; 3) 100 percent solids coatings, which includes **epoxy paint, polyurethane coatings**, and similar materials; and 4) **powder coatings**.

Water-based (or latex) paints contain synthetic resins and **pigments** dispersed in water by surfactants; these paints dry and cure as the water evaporates. Solvent-based paints include pigments thinned with organic solvents; these paints dry and **cure** as the solvent evaporates. The 100 percent solid coatings cure through chemical reactions, while the solid particles in powder coatings are fused to form a coating by the application of heat.

This chapter describes the processes involved in the paint and surface coating industries; it identifies the common waste streams generated and describes opportunities for waste minimization and reduction; it evaluates the economics of source reduction, recycling, and treatment; it identifies the elements of a waste audit protocol and cites specific regulations and permits pertaining to the industry.

16-2 Preparation, Coating, and Curing Processes

The basic purpose of paint and surface coatings is to preserve unstable materials such as steel and wood by preventing direct contact with the air and water, which would cause them to rust or decay. Painting and surface coating involves three basic stages: preparation of the surface, application of the coating, and curing.

Surface Preparation

Surface preparation is intended to provide a clean stable base for the paint or coating. Preparation includes cleaning unstable materials from the surface, improving the adhesion of primer and paint, and masking areas where paint is not desired.

Cleaning removes dampness, grease, dust, loose rust, salts, and mineral scales before the coating is applied and commonly generates large amounts of waste compared to other aspects of surface coating. Cleaning can be accomplished by using chemical solutions such as alkaline cleaning solutions and **acidic pickling solutions**, degreasing agents such as trichloroethene, mechanical abrasion, or thermal treatment. When chemical cleaning and surface preparation are used, rinses are placed between the chemical stages to minimize commingling of the chemicals. These rinses can generate large quantities of wastewater.

The chemical solutions employed to remove surface contaminants and prepare surfaces for painting include pickling baths, which use chemical reactions or electrolysis to remove scale, rust, and foreign matter. Pickling baths range in size from continuous pickling lines in steel mills to small batch processes for small-scale manufacturing. Pickling baths are preferred over mechanical cleaning in many applications because they produce a cleaner base, which allows a more durable coating. The most widely used pickling solution is hydrochloric acid, which will **etch** or cut into the metal's surface as it removes oil, grease, and oxide coatings. The pickling bath usually follows the chemical cleaning stage of the surface preparation process.

Sulfuric acid baths were widely used at one time, but have been almost entirely displaced by hydrochloric acid because they required high operating temperatures and generated large amounts of waste, resulting in an overall higher cost. If hydrochloric acid is not sufficiently aggressive to clean and etch acid-resistant materials such as stainless steel, hydrofluoric or nitric acids may then be added to the baths. Pickling efficiency falls dramatically when metallic contamination in the solution reaches 5-15 percent; the solution must then be either regenerated or replaced.

Metal phosphate and phosphoric acid solutions are also frequently used to prepare the surface of steel, aluminum, and other metals prior to surface coating. The phosphate coatings are typically applied after the surface has been cleaned and pickled. These solutions react with the metal producing a metal phosphate coating, which promotes adhesion of paint and surface coatings and limits corrosion. Commonly the solution contains metal ions such as zinc, which impart desirable properties to the phosphate coating. The coating acts as a physical barrier to air and water and can serve as a **sacrificial layer** protecting the metal from oxidation.

Chlorinated and nonchlorinated organic solvents are also used to degrease, **depaint**, and clean items to be painted or coated. Tetrachloroethene (perchloroethylene or PCE), trichoroethene (TCE), toluene, and acetone all have been widely used for these purposes. Large amounts of these chemicals were used, for example, to depaint and strip coated surfaces of airplanes and other military equipment. Historically, chlorinated solvents were inexpensive and therefore used in large amounts. These spent solvents were commonly discharged to the environment where they polluted air, soil, and water. Cleanup of the sites that have been extensively contaminated by such operations is difficult or impossible, and remediation efforts are frequently quite expensive.

Solvent cleaning continues to be used in some applications; plastic parts, for example, are commonly washed in alcohol-based solutions to remove grease and prepare the surface before they are coated. Similar results, however, can be achieved in many applications using caustic cleaning solutions. Caustic solvents are used to clean, degrease, and prepare surfaces and have replaced chlorinated solvents in many degreasing applications. Caustic solutions do not, however, etch most metallic components; the solutions may be augmented by acid pickling baths in applications where etching is important. High-volume water rinses are commonly used between stages of different chemical treatments. Rinse water from these processes is regulated under the Clean Water Act (CWA) while sludges may be regulated under RCRA.

Sandblasting, abrasives, wire wheels, and other mechanical techniques are widely used to remove surface coatings and oxides from surfaces before they are recoated. Paint chips removed by these processes commonly contain lead and chromium and may be a RCRA characteristic hazardous waste. Conventional sandblasting uses cheap silicate abrasives and produces a large volume of waste. Recyclable or biodegradable abrasives including **plastic media**, steel shot, wheat starch, sodium bicarbonate, and water blasting produce smaller volumes of more concentrated waste. Some substrates are not compatible with the alternate abrasives: bicarbonate blasting, for example, corrodes aluminum.

Thermal techniques can also be used to prepare surfaces for painting and coating. **Dry ice** (carbon dioxide) pellets produce a thermal shock that weakens the bonding between the coating and the substrate. **Sublimation** of the CO_2 pellets also releases gas at high velocity, which helps remove the paint chips. The remaining carbon dioxide pellets are lost from the waste by sublimation leaving a waste stream containing only paint chips.

High temperatures can also strip existing coatings and prepare surfaces for recoating. Fluidized-bed paint strippers are similar to fluidized-bed incinerators and can be used to remove coatings from forged steel and other heat resistant materials. The fluidized-bed furnace pyrolizes existing coatings and treats the off-gases with afterburners and scrubbers. The technique can obviously not be used with plastic, wood, aluminum, or other materials that burn or degrade at high temperatures; the fluidized-bed media may need to be regulated as a RCRA characteristic waste. Lower-temperature topical heat-stripping can be used to remove surface coatings from wood and other surfaces.

Application of Coatings

Paints and coatings can be applied through **dip coating**, directed mechanical application, air spray paint guns, **electrostatic paint guns**, or airless paint guns.

Masking is used in many applications to control the application of paint and prevent any overspray from affecting areas where it is not desired.

Most application techniques can be adapted for water-based paints, solvent-based paints, powder coatings, or **dual-component paints**. Dual-component systems that include epoxy paints, 100 percent solids polyurethane paints, and similar coatings present some special application problems. Dual-component systems mix two liquid components to produce a desirable reaction. In epoxy-based paints these components are a base and a catalyst while in polyurethane systems the components are isocyanate and **polyol-rich liquids**. Conventionally, the two components can be mixed, allowed to sit for a specific time, and then applied. Once mixed, however, the reaction begins to set up the paint and, if the **pot life** of the mixed paint is exceeded, the paint must be cleaned from the equipment. The excess paint and cleaning solvents may be hazardous wastes.

Alternatively, the components can be mixed as they are applied. Such mixing/application systems can increase throughput and reduce cleanup labor, but it can also pose problems if components are incorrectly measured when they are mixed. Off-spec coatings can fail, making it necessary to prepare the surface again and reapply a fresh coat of paint.

All types of paint can be applied through a number of mechanisms. Dip coating is widely used to apply paint and coatings to new products and involves immersing the item to be coated directly into the coating. Coatings can be applied either hot or cold, but this process is limited by the thickness of the coating that can be applied and the need to control VOC emissions from the tank to meet regulatory limits. Brushes, paint rollers, and similar tools also are widely used to apply paints directly.

Paints and surface coatings are sprayed in many applications. Conventional paint spray systems atomize paint with high velocity air jets that force paint through small holes. Such systems can apply solvent-based paints, latex-based paints, powder coatings, and multi-component paints that are mixed prior to application. Air pressures generally range between 40 and 80 psi and impart high velocities to the paint particles. Some paint particles arrive at the intended target, but others miss it or bounce off, forming **overspray**. According to the Naval Facilities Engineering Service Center, transfer efficiency – the fraction of applied paint that adheres to the surface – for conventional paint spray systems is typically 15-30 percent.

Several refinements have been developed to increase the efficiency with which paints and powder coatings are transferred to items. **High Volume/ Low Pressure (HVLP)** paint spray systems apply paint at low velocities by atomizing the paint with high volumes of air at less than 10 psi pressure. The atomized particles have low velocities; this reduces overspray, bounce-back, and blow-back, resulting in transfer efficiency reaching up to 65 percent. Airless spray painting systems atomize paint by hydraulically or mechanically forcing paint through small openings at pressures of up to 3,000 psi. An airless nozzle atomizes the paint and produces a fan of particles at high pressures (approximately 800 psi). These systems deliver large quantities of coarsely divided paint particles that can cover large areas quickly with transfer efficiencies up to 40 percent. Pressure-atomized, air-assisted systems combine air atomized and airless systems. A stream of low-pressure air is injected after the nozzle and improves atomization and the spray pattern. These systems can have transfer efficiencies of up to 40 percent while allowing for the control of the paint spray pattern that is not available with the airless systems.

Application efficiency can also be increased through the use of electrostatic paint systems. These systems place a negative charge on atomized paint droplets or powder particles, which are drawn to the positively charged target. The transfer efficiency for electrostatic paint spray systems can reach 75 percent, reducing overspray to levels much less than with conventional systems. Air-atomized, airless, and rotating disc and bell systems all can be combined with electrostatic spraying. Electrostatic spray systems, however, cannot be used with insulating materials such as plastic, rubber, ceramic, or glass. The technique is also difficult to use on textured surfaces because charge builds up on the outer surface of concave areas, pulling paint to these areas and preventing it from reaching into the cavities.

Even the most efficient spray painting and powder coating systems produce significant quantities of overspray. This overspray is generally controlled by applying the paint in booths which control particulate emissions and protect both workers and the environment. A water curtain paint **spray booth** uses water (often with chemicals added) to remove particulates and soluble organics from the air. This generates a wastewater and paint sludge. The wastewater may require off site treatment and the paint sludge may have to be disposed of as a hazardous waste.

Dry filter booths use fiberglass cartridges, multi-layer honeycombed paper rolls/pads, accordion-pleated paper sheets, or cloth rolls/pads to remove paint particles from the air. These systems typically remove particles more efficiently than water curtain booths and operate on the principle that paint particles carried by air change direction more slowly than the air itself. As particle-laden air flows around the media, the inertia of the particles causes them to continue toward the media and be deposited on the filter. The dry filters are replaced when they become plugged with paint; such booths typically generate only 10-50 percent of the waste generated by water curtain booths. Dry filters are not used with powder coatings, however, because they do not allow the coating to be recycled.

Buildup of paint in booths must be regulated to limit the potential for fire. This can be accomplished by scraping or sandblasting the booth after the paint has hardened or by applying easily strippable coatings before the booth is used. The strippable coating and accumulated paint can then be removed relatively easily.

Curing

Curing produces a durable layer from fresh coatings. As previously noted, latex-based paints cure by losing water while solvent-based paints cure by losing solvents. The water evaporating from latex paints is innocuous, but the volatile organic compounds given off by solvent-based paints may be flammable or toxic, and must be controlled in large-scale applications. The 100 percent solids polyurethane and epoxy paints cure through chemical reactions in the paint and do not emit large quantities of compounds. Some latex or solvent coatings may cure by air-drying, while others may require oven curing.

All common powder coatings (epoxy, hybrid, aromatic urethane, aliphatic urethane, and TGIC polyester) are cured by heat. The powder particles fuse into an impervious coating upon heating. Powder coatings generally are formulated to minimize emissions during heating, but the emissions from some systems require control. The ovens used to bake powder coatings are typically fired with natural gas or oil and emit carbon dioxide, carbon monoxide, nitrogen oxides, and unburned hydrocarbons. All of these compounds are subject to regulation.

Checking Your Understanding

1. What are the major categories of paints and surface coatings?

2. What is the purpose of surface preparation?

3. What techniques can be used to remove old coatings from surfaces?

4. How do electrostatic systems affect overspray?

5. Why aren't dry paint filter systems commonly used with powder coatings?

16-3 Common Waste Streams

As shown in Figure 16-1, waste streams are generated throughout the surface coating process. Toxic and regulated waste streams can be produced in all phases of painting, but high toxicity surface coating and paint formulations that were used in the past have been reformulated to reduce toxicity and the ensuing production of regulated waste. The following subsections discuss the types of wastes generated during the preparation of surfaces, the application of coatings and paints, and the curing processes.

Waste Generated in Surface Preparation

Surface preparation activities generate a variety of waste streams. If the acid/alkaline waste stream generated is a RCRA listed hazardous waste, it must be treated to meet Land Disposal Restrictions (LDRs) before it can be disposed of; the treated residue must

Preparation of Surfaces				
Spent Acid Pickling Media/ Alkaline Cleaning Solutions	Rinse Water	Spent Organic Solvents	Paint Chips and Spent Abrasives	
Application of Paint and Surface Coatings				
Overspray and Overspray Collection Media	Leftover Paint	Organic Solvents Used in Cleanup	Off-spec Coatings	Waste-water
Curing of Paint and Surface Coatings				
Volatile Organic Compounds From Paint	Furnace Gases			

Figure 16-1: Wastes generated during the stages of painting and surface coating process.

be disposed of to a RCRA Subtitle C landfill. If the waste is simply a RCRA characteristic waste due to pH or toxicity, it must be treated to remove the characteristic – usually accomplished through neutralization – and to reduce any underlying hazardous constituents to promulgated treatment standards (unless metal content drives the characteristic). Once this is accomplished, treatment residues from characteristic wastes can be disposed of at a RCRA Subtitle D landfill as a nonhazardous solid waste. Figure 16-2 presents the treatment options for RCRA hazardous waste streams.

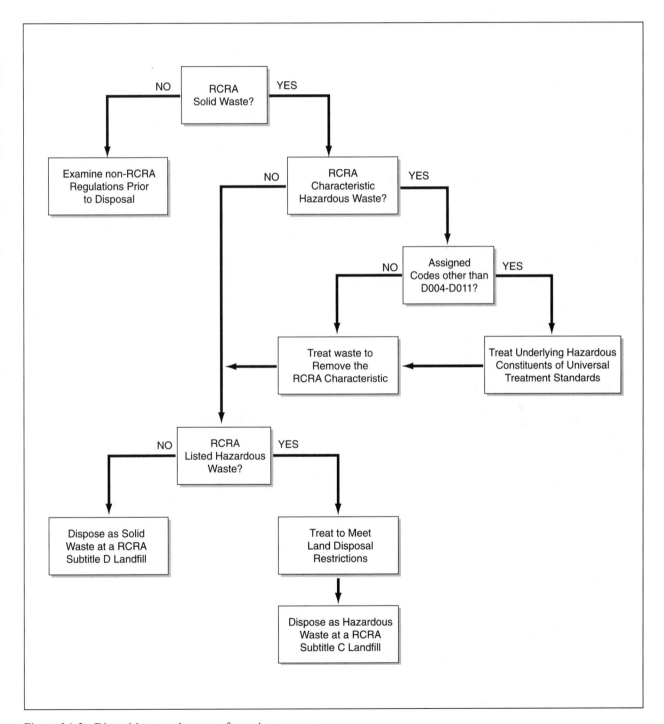

Figure 16-2: Disposition requirements for various wastes.

In the past, degreasing and depainting solvents have been major waste streams produced by painting and coating operations. Chlorinated and nonchlorinated solvents were commonly disposed of indiscriminately, and in some locations such disposal has contaminated large volumes of soil and groundwater. Solvent stripping and solvent cleaning have been replaced by mechanical stripping or by degradable solvents, but solvents continue to be used for some purposes. The waste solvents are RCRA listed hazardous waste if they have been used in certain applications. If this is the case, then the solvents must be treated to meet LDRs before treatment residues are disposed of at a RCRA Subtitle C landfill. Nonhalogenated solvents such as toluene may exhibit the RCRA characteristic of ignitability (D001). If this is the case, the solvent must be treated to remove this characteristic, and the waste must be treated to reduce any underlying hazardous constituents to required levels.

Mechanical preparation of surfaces produces paint chips, spent abrasives, protective clothing, and plastic sheeting. The properties of these wastes are determined by the properties of the original coating and of the surface contamination that is removed. Heavy metals are the most common contaminants driving regulation because lead was used widely as a paint pigment and as a corrosion inhibitor in metal primers, while chromium was used extensively in zinc chromate-based paints. Other metallic contaminants are also present in some paints. Mercury, for example, has been incorporated into paints in the past as a biocide, and highly toxic copper and tin compounds are used in marine anti-fouling paints.

Wastes that are RCRA characteristic hazardous waste because of their metal content require special handling prior to disposal. Generally the RCRA characteristic can be removed through stabilization procedures, such as mixing the waste with a specially formulated grout.

Wastes Generated in Application

Overspray, overspray collection media (including strippable paint booth coatings), personal protective equipment, respiratory cartridges, air filtration media, waste generated during color changes, paint-contaminated water and solvents, and unopened and partially used leftover paint are among the waste streams generated by paint application. Some materials, such as the leftover paint, can be reused, but most of the waste must be stored, managed, and disposed of as a solid waste, a RCRA hazardous waste, or a **household hazardous waste (HHW)**.

Several practices have been adopted to reduce the quantities of waste generated and to ensure that the waste is nonhazardous. Recycling of materials is one example. Large-scale powder-coating operations, for example, commonly use dedicated equipment for each color that is applied. Overspray can then be collected and reused, eliminating one potential waste stream. Solvents used in post-application cleanup are also being extensively recycled, and less toxic compounds are being substituted. Latex-based paints that are not RCRA hazardous, for example, have replaced solvent-based paints in many applications, and water-based washing solutions have been substituted for solvent washing in some cleanup operations.

Most overspray, overspray collection media, personal protective equipment, respiratory cartridges, air filtration media are considered nonhazardous solid waste. Leftover paint, paint-contaminated solvents, and paint contaminated wastes, however, may be RCRA hazardous waste requiring special handling.

Wastes Generated in Curing

In general, the amounts of solid waste generated during curing are relatively small in comparison to waste generated during surface preparation and the application of paint/surface coatings. Off-gases, however, can be produced both by the coating itself and by furnaces used to cure the coating and can be a major air emission concern.

Checking Your Understanding

1. Why are lead and chromium commonly found in paint chips?

2. What is the most common type of household hazardous waste generated during surface coating applications?

3. Name three waste streams that can be generated during the application of paint.

4. Why was mercury widely used in paints and surface coatings?

16-4 Pollution Prevention Opportunities and Economics

Reuse, reduction, recycling, substitution, and volume reduction have been widely applied to the painting and surface coating industry and have profoundly reduced the amount of waste generated.

Minimization of Waste During Surface Preparation

Examples of reuse and waste reduction in preparation of surfaces include development of hydrochloric acid-based pickling operations that are closed or nearly closed loops. In this type of closed-loop system, the metal – i.e., iron chloride – can be recovered as a saleable product, while the spent hydrochloric acid can always be regenerated and reused. This allows hydrochloric acid-based systems to be run with little or no waste. Solid wastes that cannot be reused or recovered are commonly RCRA characteristic wastes (D002) because of their low pH; they may also contain high levels of some underlying hazardous constituents. The hydrochloric acid

system has largely displaced sulfuric acid systems, which do not have the capability to regenerate the acid and also produce large volumes of precipitated sulfates requiring disposal.

Substitution of materials and techniques is widely practiced to lower toxicity levels. The chlorinated solvents formerly used to strip paint and grease from surfaces, for example, have been replaced in large part by mechanical preparation or by relatively nontoxic chemical strippers. Substitution of less toxic compounds and in-place management of old paint has also reduced lead and chromium waste generated in surface preparation. Lead and chromium were widely used in primers and paints to increase durability and prevent metal corrosion, but these are being supplanted by formulations based on relatively nontoxic organic and inorganic zinc compounds.

Reduction of waste volume is behind many of the innovative particle blasting schemes that have been developed. Recyclable steel blasting grit, carbon dioxide pellets, plastic abrasives, and similar techniques yield a waste stream of concentrated paint chips with a much lower volume than the combined paint chip-blasting media waste stream generated by conventional blasting. The waste stream must be managed as a hazardous waste in many cases, but the lower volume of material to be disposed of substantially reduces the cost of managing the waste.

Innovative blasting schemes may also reduce the toxicity of paint-chips and eliminate the production of hazardous waste. Specially formulated steel shot, for example, stabilizes paint chips against the Toxic Characteristic Leaching Process (TCLP) by converting soluble lead compounds into insoluble forms. The conversion/stabilization may be partially reversible, but silicate-based additives can increase the long-term stability of the waste.

Minimization of Paint Application Waste

Low toxicity paints and surface coatings have widely replaced higher toxicity materials in recent years. Latex paints, for example, have been substituted for solvent-based paints in a variety of applications and have significantly reduced hazardous waste generation. The solutions generated in cleanup of latex paints are not hazardous wastes and can be treated as wastewater, while the solid paint waste generated can be used in waste-to-energy plants or disposed of directly as nonhazardous solid waste. However, low toxicity paints and coatings cannot replace higher toxicity formulations in all cases, because low-toxicity paints often have less chemical and abrasion resistance, decreased temperature ranges, higher sensitivity to humidity, higher dependence on surface cleanliness, and poorer adhesion. Other reformulations to reduce toxicity include the removal of lead from house paints, the decreased use of lead primers on steel structures, and the removal of mercury-based biocides from architectural paint.

Recycling is also practiced more widely than in the past. Leftover paint can either be reused via donation/exchange programs or blended for reuse. Reblending, however, requires stringent quality control, and the reblended paint is of limited usefulness for the following three reasons: solvent-based paints generally cannot be reblended, latex-based paints are commonly contaminated with bacteria, and reblended paint can generally be used only on exteriors. Powder coats also can be recycled. Overspray from coatings applied with dedicated equipment for each color can be reclaimed for reuse. Mixed color and limited volume powder coating waste can be used for surface coating jobs where color is unimportant; residues can be used as fillers in roofing materials, cement, and sealer/caulking compounds. Recycling can also be used during post-application cleanup. Automatic paint gun washers recycle and reuse solvents in a process similar to conventional dishwashers. Impurities are filtered from the solvent and concentrated into a sludge, which is then collected and disposed of, reducing both worker exposure and hazardous waste generation.

Minimization of Waste Generated During Curing

Relatively little solid waste or wastewater is generated during curing, but reuse, recycling, substitution, and volume reduction can also be applied to these wastes. The amounts of air emission allowed during curing are strictly regulated, and strategies to control these emission have included reformulation of coatings and paints to reduce volatile off-gases and addition of pollution control devices to off-gas streams and furnace exhausts. The requirements for emissions controls vary by location of the operation and are generally most stringent in the most polluted locations.

Conducting Waste Audits

Waste audits are intended to assure compliance with hazardous waste laws and best management practices during the disposal of wastes. The basic elements of an audit protocol are the same as for other industries:

1. Document types, volumes, and weights of waste generated from the operation;

2. Determine the regulatory requirements and best management practices for each solid, aqueous, gaseous, and RCRA hazardous waste stream;

3. Audit the operation to identify shortcomings in storage, transportation, and disposal;

4. Develop suggestions to correct shortcomings; and

5. Correct the shortcomings.

OSHA-required Material Safety Data Sheets (MSDS) that accompany paints and other materials are good starting points for identifying potential waste types and problems. The manufacturers of the materials should also be contacted for information on the material itself. It may also be necessary to analyze waste streams to obtain defensible data concerning their regulatory status.

Economics of Pollution Prevention

When materials and disposal costs for polluting substances are minimal, economic pressure can be applied through regulations that demand high costs for their continued use. These regulations are the most important driving force for source reduction, recycling, and treatment technologies. For example, chlorinated solvents are inexpensive and would continue in wide use if it were not for legal requirements that increase the expense of their use and disposal. Or, the requirement that spent halogenated solvents used for degreasing and surface preparation be handled as RCRA listed hazardous waste, rather than being disposed of to the land, has greatly increased disposal costs and driven the industry to-

ward product substitution, source reduction, recycling, and treatment.

Regulatory pressures are less important for non-hazardous solid wastes. The cost of the raw material becomes the driving force behind source reduction and recycling, because the cost of the raw material is generally considerably higher than the cost of its waste disposal. Powder coatings, for example, cost $10-20 per pound while residues from its application can be disposed of for pennies per pound.

Checking Your Understanding

1. How can leftover latex paints be reused or recycled?

2. What steps have been taken to reduce the toxicity of paint chips?

3. Why have sulfuric acid solutions largely disappeared from pickling operations?

4. What are the basic elements of a waste audit protocol?

5. What are the major driving forces behind the decreased use of chlorinated solvents in the painting and surface coating industry?

16-5 Industry-specific Regulations

A variety of local, state, and federal regulations affect the paint and surface coating industries. The most important federal regulations are discussed below.

The Federal Clean Air Act

The Clean Air Act, as amended in 1990, sets stringent standards for emissions from industrial facilities. Overspray particulates must be virtually eliminated from the air prior to discharge to the atmosphere, and volatile organic compounds are strongly regulated.

CAA standards and requirements regulate the discharge of VOCs from industries applying surface coatings. 40 CFR Part 60 sets performance standards for surface coating of metal furniture (Subpart EE), automobile and light-duty truck operations (Subpart MM), large appliances (Subpart SS), metal coil surface coating (Subpart TT), beverage cans (Subpart WW), plastic parts for business machines (Subpart TTT). EPA is currently working on even more restrictive regulations for air emissions from metal finishing operations. Standards based on Maximum Achievable Control Technology (MACT) will decrease allowable emissions. The proposed NESHAP for methylene chloride, tetrachloroethene, trichloroethylene, 1,1,1-trichloroethane, carbon tetrachloride, and chloroform used as degreasers/cleaners, for example, would require facilities to use a designated pollution prevention technology along with proper operating procedures to control vapors (58 FR 62566, November 19, 1993).

The Occupational Safety and Health Act

OSHA requirements of 29 CFR 1910.107 contain numerous regulations regarding the application of surface coatings. The regulations require that particulate-laden air be vented from the booth away from the operator. Air must move past the operator at 60-100 linear feet/minute, depending on the application. The regulation also requires spray booths to be free of combustible residue buildup and the floor of paint booths to be noncombustible.

Hazardous Waste Regulations

The hazardous waste codes most likely to be relevant to the painting and surface coating industry are specified in 40 CFR 261 and include the D001 code for flammability; the D002 code for pH; D codes for metals that may be present (especially D007 for chromium and D008 for lead); D codes for halogenated and nonhalogenated organic compounds; F codes for organic solvents used in degreasing and cleaning (F001 etc.); and K codes for specific chemicals such as spent pickling liquor (K062).

RCRA has the following requirements: that wastes that are characteristic for metallic constituents be treated to remove the characteristic prior to disposal; that wastes that are characteristic for properties or constituents other than metals be treated to remove the characteristic and reduce underlying hazardous constituents to regulatory levels; and that RCRA listed hazardous wastes be treated to meet LDRs prior to disposal. In general, the treated residues of listed hazardous waste must be disposed of at a RCRA Subtitle C landfill while treated residues of characteristic hazardous wastes can be disposed of at RCRA Subtitle C or D landfills.

Paint chips that fail the TCLP test for lead (D008), for example, must be stabilized in some manner before they can be disposed of, but once that is done they can be disposed of as non-RCRA hazardous solid waste. Ignitable paint thinner (D001) must be treated in order to be made nonignitable and also for underlying hazardous constituents (such as toluene and xylenes) before it can be disposed of as non-RCRA hazardous solid waste. Moreover, degreasing solvents regulated as F001 must be treated to meet LDRs prior to disposal. Except under special circumstances, listed hazardous wastes continue to be RCRA hazardous waste after treatment and must be disposed of as such.

Effluent Guidelines and Standards

Effluent guidelines and standards for metal finishing (40 CFR Part 433) apply to wastewater generated by coating, cleaning, polishing, shearing, hot dip coating, and solvent degreasing operations. EPA has recently developed new effluent guidelines under the CWA for electroplaters (40 CFR 413), metal finishers (40 CFR 433), and the metal products and machinery industry (40 CFR 438), all of which may be applicable to the painting and surface coating industry. New regulatory options for the metal finishing industry are integrated into this new guideline. Under Phase I of the regulation, EPA proposed effluent limits for facilities that generate wastewater from processing metal parts, metal products, and machinery during manufacture, assembly, rebuilding, repair, and maintenance. Industries making aircraft, aerospace hardware including machine tools, screw machines, metal forgings and stampings, metal springs, heating equipment, and fabricated structural metal, ordinance, stationary industrial equipment including electrical, mobile industrial equipment, and electronic equipment including communication equipment are all affected.

Phase II had an implementation deadline of December 31, 1997 and will limit effluents from facilities that process metal parts, metal products, and machinery. This will include motor vehicles, buses and trucks, household equipment, business equipment, instruments, precious and nonprecious metals, shipbuilding, and railroads.

Community Right-to-Know

Community right-to-know laws apply to hazardous materials used in the paint and surface coating industries.

Checking Your Understanding

1. What waste code would apply to paint chips that failed the TCLP test for lead?

2. What is the proper means for disposing of spent degreasing solvents (F001)?

3. What Act sets effluent guidelines for electroplaters and metal finishers?

4. What Act sets standards and requirements on the discharge of volatile organic compounds (VOCs) from industries applying surface coatings?

Summary

Paints and surface coatings are widely applied to unstable materials such as metal and wood to prevent them from rusting or rotting. The four major types of paints and surface coatings applied are 1) latex or water-based coatings, 2) solvent-based coatings, 3) 100 percent solids coatings, which include epoxy paint, polyurethane coatings, and similar materials, and 4) powder coatings. Water-based and solvent-based paints dry and cure by evaporation of water and solvents. In contrast, 100 percent solids coatings undergo a chemical reaction to cure, while powder coatings are heated to fuse the powder particles into a tough smooth coating.

Regardless of the use, all paint and surface coating processes involve surface preparation, application of the coating, and curing. Surface preparation may be accomplished by alkaline solutions, acidic pickling solutions, degreasing agents, mechanical abrasion, or thermal treatment. Each of these processes generates large quantities of waste, especially due to the water used between the various phases. The paint or surface coating may be applied by either dipping, direct mechanical application (brush or roller), air spray and airless paint guns, and electrostatic paint guns. In many types of application it is necessary to mask various areas of the object to prevent overspray from affecting other areas. The switch to HVLP paint spray systems has greatly reduced overspray, improving transfer efficiencies to nearly 65 percent. The highest transfer efficiencies – 75 percent – have been reached through the use of electrostatic paint and powder coating systems.

The common waste streams generated by the paint and surface coating industries include spent acid pickling media and alkaline cleaning solutions, degreasing and depainting solvents, spent abrasives, protective clothing, plastic sheeting, respiratory cartridges, air filtration media, paint-contaminated water and solvents, unopened and partially used leftover paint or powder, and air emissions resulting from the curing process or curing furnace emissions. The industry has implemented a number of pollution prevention activities including: 1) the use of closed or nearly closed-loop pickling solution recovery systems; 2) the elimination of lead-, chrome-, mercury-, and tin-based paints; 3) the switch to steel blasting grit, carbon dioxide pellets, or plastic abrasives to reduce the volume and toxicity of blasting media wastes; 4) the recycling of paints, particularly water-based paints; and 5) the reuse of powder paints.

To be more effective in reducing the generation of wastes, the industry conducts audits in which the types, volumes, and weights of waste generated are gathered; the best management practice for each waste are determined; and shortcomings in storage, transportation, and disposal are identified and corrected. Regulatory pressure continues to be the most important driving force behind pollution prevention activities involving hazardous wastes, while raw material costs are what drives nonhazardous solid waste reduction efforts. Due to the nature of the processes and materials involved, the paint and surface coating industries are regulated by most major environmental regulations.

Critical Thinking Questions

1. Lead is a highly toxic element that was a common component of surface coatings applied in the past. Under what circumstances should these coatings be left in place and when should they be removed prior to the application of a new coating?

2. Government regulation has driven waste minimization and the development of less toxic formulations in the painting and surface coating industries. Should additional regulations be applied to the industry and how should these new regulations be structured?

3. Contact a local paint manufacturer and obtain information about the composition of an indoor household paint and the recycling opportunities the company provides. Make a list of recycling and reuse applications for latex paints.

Integrated Circuits and Electronics Assembly Industries

by Azita Yazdani

Chapter Objectives

Upon completing this chapter, the student will be able to:

1. **Identify** and describe the steps involved in the fabrication and assembly of an IC.

2. **Identify** waste streams generated by the fabrication and assembly processes.

3. **Describe** pollution prevention options that are available to eliminate the waste generated by the IC fabrication and assembly processes.

17-1 Introduction

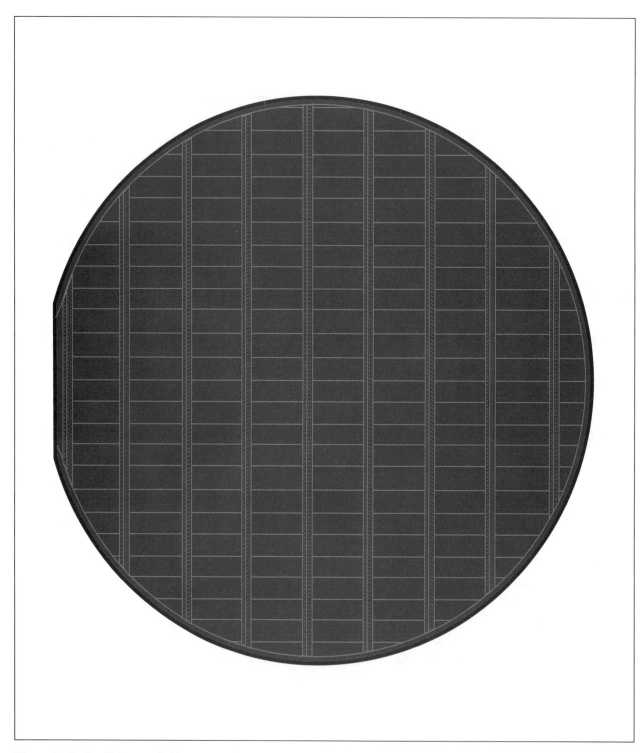

Figure 17-1: Parallel rows of chips on a wafer.

The electronics industry manufactures components and electronic packages. The industry is comprised of three major sectors: **semiconductor** manufacturing, **Printed Wiring Board (PWB)** fabrication, and PWB assembly. Semiconductors are used either as insulators or as conductors of electric current. Printed wiring boards are composed of a sheet of non-conducting substrate onto which the interconnecting conductors and some of the circuit components have been printed, electroplated, or etched. Also known as printed circuit boards (PCBs), they serve the additional function of providing the surface onto which other electronic components are mounted. PWB assembly involves the process of attaching, often by soldering, the tested electronic components to the PWB, thereby creating an electronic package.

Semiconductors are made from solid crystalline materials – typically silicon – that are formed into circuits, which are either as simple as one diode or as complex as an **Integrated Circuit (IC)**. A diode performs one single on/off function that controls the flow of the electrical current. Integrated circuits are composed of hundreds, or even thousands, of diodes formed on a single **wafer**. Each of the grid cells on the wafer is called a **chip** or a **die** and is the area onto which the integrated circuit will be built (see Figure 17-1).

ICs are manufactured by the semiconductor industry for use in microelectronic applications such as computer systems, telecommunications, automobiles, industrial machinery, military hardware, and consumer electronics products. Typical functions of semiconductors in these applications are: information processing, LCD displays, electrical power handling, data storage, signal conditioning, and photocells.

The semiconductor industry is comprised of both small and large facilities. The industry is kept in a constant state of flux due to constant updating and process modifications demanded by the competitive nature of the business. The production of semiconductors requires high standards of cleanliness, extensive capital investment in facilities, and sophisticated mechanical equipment. The growth of the electronics industry has been significant in recent years and continues to use many toxic chemicals. This chapter describes the steps used in the production of semiconductors and their assembly on PWBs as well as pollution prevention activities that are used to reduce environmental harm.

17-2 IC Manufacturing Processes

Semiconductor Manufacturing Process

ICs are the major products of the semiconductor industry. The process of manufacturing semiconductors or ICs typically consists of more than one hundred steps, during which hundreds of copies of an integrated circuit are formed on a single wafer. The primary reason for failure in semiconductors is contamination, particularly the presence of microscopic residue on the surface of the base material or circuit path. For this reason, cleaning operations are performed before and after many of the manufacturing steps.

Unlike discrete semiconductor devices, such as diodes, resistors, capacitors, or transistors, ICs are combinations of such devices on a single semiconductor chip. Semiconductor manufacturing processes employ 200 different materials and 100 proprietary methods. The two overall steps in the production process are wafer fabrication and wafer assembly. Figure 17-2 shows the steps involved in the semiconductor manufacturing process.

Wafer Production

The first step in the wafer production process is to grow a single crystal or **ingot** of **silicon** from a seed

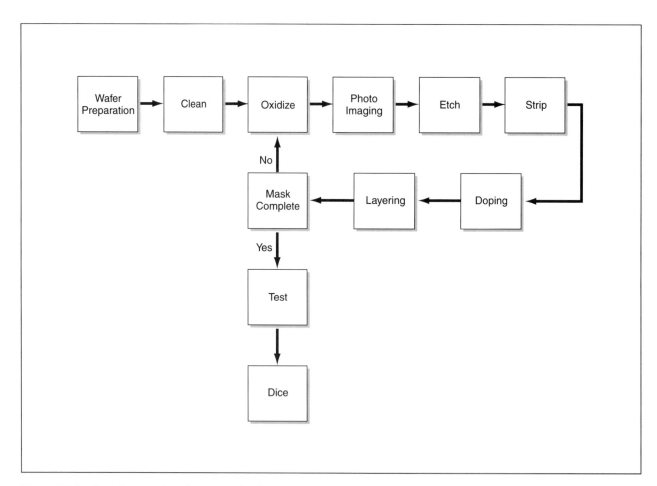

Figure 17-2: Steps in a semiconductor production process.

crystal. Starting with sand in an electric furnace, this process results in purified silicon as a molten liquid. A small piece of pure silicon is used as a **seed crystal** and brought into contact with the molten liquid. As the seed crystal starts to grow, it is slowly pulled from the melt forming a single ingot as the liquid cools and crystallizes. These ingots often measure six to eight inches in diameter and may be up to two feet in length.

Once the ingot is grown, it is smoothed and polished in a process called **lapping**. The ingot is then sliced into thin (0.76 mm or 0.03 inch) individual wafers (see Figure 17-1) with a diamond saw, much the same as a butcher would slice a stick of bologna. Various cooling waters and cutting oils are used in the slicing process. The wafers are now ready to be sent to the wafer fabrication area where they are used as the starting material for the manufacture of semiconductors and integrated circuits.

The sliced wafers are mechanically lapped to further flatten the surfaces of the wafer and to reduce saw marks. After lapping, wafers are etched in nitric acid/acetic acid solution or a sodium hydroxide solution to remove microscopic cracks and surface damage created by the lapping process. Reverse osmosis is used to produce high purity **DI water**, which is used to remove the solutions from the etched wafers.

The wafers are then polished in a series of chemical and mechanical polishing steps. The wafers are buffed with a slurry of silica powder, DI water, and sodium hydroxide. The polishing steps utilize progressively finer slurries and intermediate cleanings with DI water. The final cleaning step uses a cleaning method developed by RCA in 1970. The three-step process starts with an **SC1 solution** (ammonia, hydrogen peroxide, and DI water) to remove organic impurities and particulate matter from the wafer surface. Next natural oxides and metal impurities are removed with hydrochloric acid, and finally the **SC2 solution** (hydrochloric acid and hydrogen peroxide). This solution causes super-clean new natural oxides to form on the surface of the wafer.

Wafer Fabrication

The following steps in semiconductor manufacturing are the **wafer fabrication processes** that result in the integrated circuitry being formed in and on the wafer. This portion of the wafer fabrication process, which takes place in a clean room, involves a

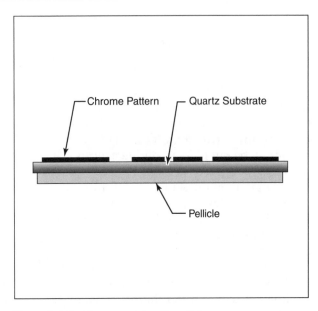

Figure 17-3: Cross-section of a reticle

series of steps, which are described below. Typically, it takes from 10 to 30 days to complete the fabrication process.

The first step in wafer fabrication is oxidation. A silicon dioxide layer is grown on the wafer to provide a base for the **photolithography** process to follow. The silicon dioxide layer is formed on the surface of the wafer by exposing it to high temperatures and oxygen in a furnace. This layer will later prevent diffusion of the **dopants** into the silicon substrate below. After oxidation, the wafer is cleaned with strong acids and bases.

In the second step, a photoresist is spun onto the wafer as a uniform film. Photoresist is a photosensitive liquid that is applied in small quantities to the wafer. The wafer is then spun at 3,000 rpm to spread the photoresist puddle out and produce a uniform layer between two and 200 μm thick. There are two basic types of photoresist: a **negative photoresist** containing a polymer and a **positive photoresist** containing resin. When a negative resist is used, unexposed portions of the photoresist are removed leaving a negative image. This method is only capable of producing features down to about 2.0 μm. Most semiconductor processes now use a positive resist where exposed portions are removed, leaving a positive image of the mask pattern on the surface of the wafer.

Semiconductor devices are made up of as many as 50 individual layers of silicon, polysilicon, silicon dioxide, metal, and **silicides**. The pattern for each layer is contained on a mask called a **reticle**, which is an optically clear quartz substrate with a chrome pattern. To keep the surface of the reticle clean, a thin plastic sheet called a **pellicle** is mounted a short distance away from the surface of the reticle. This allows the wafer to be cleaned without directly contacting the chrome mask surface, while also insuring that any microscopic dust that settles on the reticle will be out of focus during exposure and will therefore not create defects. Reticles are 1 to 10 times larger than the actual size of the pattern they produce. The design pattern is developed for each layer using CAD software. The CAD pattern is transferred to the reticle using either a laser pattern generator or an **e-beam** (see Figure 17-3).

After exposure of the photoresist, as determined by the pattern on the reticle, wafers are developed in either an acid or base solution to remove the exposed areas of photoresist. Once the exposed photoresist is removed, the wafer is then **soft-baked** at a low temperature to harden the remaining photoresist.

As shown in Figure 17-4, the third step is etching. Removing selected areas of silicon dioxide from a wafer involves the use of many different types of acid, base, and caustic solutions. Much of the work with chemicals takes place at large wet benches where special solutions are prepared for specific tasks. For instance, **buffered oxide etch (BOE)**, prepared from hydrochloric acid buffered with ammonium fluoride is used to remove silicon dioxide without etching away an underlying silicon or polysilicon layer.

As previously noted, the wafers must be constantly cleaned. Since this is such a frequent and important operation, a special tool called a **Spin, Rinse, and Dryer (SRD)** is used. The SRD cleans the wafer with DI water and dries it with nitrogen.

The fourth step in the sequence is **doping**. The purpose of doping is to change the electrical conductivity of the silicon. This process is different from other semiconductor fabrication processes in that it does not create a new layer on the wafer. Instead, a process called **ion implant** changes the electrical characteristics of a precise area within an existing layer on the wafer. The dopant ions are implanted into the top layer of the wafer, just below the surface, changing the electrical conductivity of a precise region. To create a **p-type** insulating (nonconducting) region, an **acceptor ion** such as boron, gallium, or indium is implanted. To create an **n-type** conducting region, a **donor ion** such as antimony, arsenic, phosphorus, or bismuth is implanted.

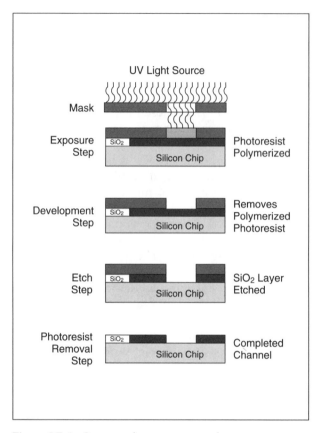

Figure 17-4: Steps used to create an etch pattern.

The fifth step is **layering**. In this step, a thin layer of material that may be a conductor, semiconductor, or insulator is deposited over the whole surface of the wafer. Prior to **Chemical Vapor Deposition (CVD)**, the wafer is usually cleaned with a **dry plasma etch** process using either sulfur hexafluoride or a combination of tetrafluoromethane and oxygen. (These recipes are some of the most closely guarded secrets among the semiconductor manufacturers.) Using nitrogen and hydrogen as carrier gases, CVD can produce a variety of layer types. For example, ammonia and dichlorosilane will produce a silicon nitride layer. Silane and oxygen are used to create silicon dioxide layers. Metals like aluminum, gold, and tungsten are used to create conducting layers on a device.

Once a silicon, metal, or silicide layer has been created, the remaining photoresist is removed. This is done with a procedure called **ashing** – the sixth and final step – where a high temperature plasma is used to selectively remove photoresist without damaging the layers of the device. The processes of oxidation, photolithography, doping, and layering steps are repeated until the desired number of circuit layers is achieved. The circuits then go to wafer assembly.

Wafer Assembly

In this step, the back of the wafer is mechanically ground to remove unnecessary material. A film of gold may be applied to the back of the wafer to aid in the connection of its leads to the bonding pad during a later process step. Tests are performed to insure that the circuitry performs as designed. Any chips failing this test are marked and will be discarded during later assembly operations.

The wafer is cleaned after testing, masked off, and sawed into single chips. As shown in Figure 17-5, the chips are assembled by being mounted on an **attachment pad** and being connected to the lead frame using **bonding wires**. The connection of several chips will ultimately form an IC that will perform the desired electronic function.

In the final step of assembly, the stack of chips is enclosed in a plastic or ceramic case to protect it against mechanical shock and any damage from the external environment. Because of its external appearance, the completed IC is often referred to as a bug. The lead frame or legs of the bug are often used to connect the completed IC to the PWB.

PCB Assembly

In the assembly process, electrical components are placed on the PWB, flux is added, and the components are welded to the boards using **solder**. The flux is used to reduce the surface tension so the solder will flow evenly. In wave, dip, and drag soldering techniques, molten solder acts as both a heat source and a supply of solder that is deposited on the board. In the wave soldering technique, for example, the board is fluxed and a wave of solder is allowed to pass over the board. The solder adheres only to the metal leads and any other tin, tin/lead, or gold-plated parts of the board.

The various types of flux and their characteristics are summarized in Table 17-1. The flux residues as well as other contaminants need to be removed from the board after the soldering pro-

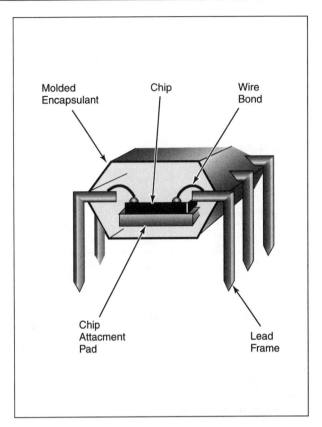

Figure 17-5: Completed semiconductor showing various parts.

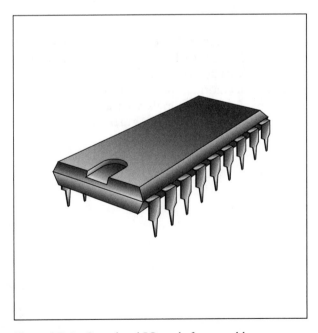

Figure 17-6: Completed IC ready for assembly.

Flux	Common Reference	Composition	Approved by Military	Solubility
Rosin	R	Abietic acid and other isomeric acids	Yes	S*, WS
Mildly Activated Rosin	RMA	Abietic acid and other isomeric acids; amine hydrochloride activators	Yes	S*, WS
Activated Rosin	RA	Abietic acid and other isomeric acids; amine hydrochloric activators	No	S*, WS
Super Activated Rosin	RSA	Abietic acid and other isomeric acids; amine hydrochloric activators	No	S*, WS
Organic Acid	OA	Abietic acid and other isomeric acids; amine hydrochloric activators	No	W
Synthetically Activated	SA	Alkyl acid phosphates; halide activators	No	S*
Resin	Resin		No	W
Low Solids	Low solids	2 to 5 percent solids `	No	**

S* = chlorinated solvent soluble
W = water soluble
WS = water soluble with saponifier
** = Cleaning may not be required

Table 17-1: Solder types.

cess. In general, the other contaminants include: 1) nonpolar materials such as oils, greases, **rosin**, and waxes; 2) polar materials such as rosin flux activators, sodium chloride, and plating and etching salts; and 3) particulate materials such as dust, machining, drilling and punching fragments.

Checking Your Understanding

1. What is the function served by a silicon seed crystal?

2. What are the six steps used in the wafer fabrication process?

3. What is the function of a dopant?

4. What is the purpose of layering?

5. What substance is typically used to connect ICs to the PWB?

17-3 Waste Streams and P2 Options

Wastes Options for Semiconductor Manufacturing

There are many processes utilized in the semiconductor production and assembly industries. Table 17-2 summarizes several of the wastes that can be generated by each of the processes. The two major environmental concerns for the electronics industry are: 1) the use of chlorinated solvents for cleaning processes, and 2) the presence of lead in the solder during assembly.

Process	Materials Used	Possible Types of Environmental Releases		
	Semiconductor	**Hazardous Waste**	**Wastewater**	**Air**
Wafer Production and Fabrication	1. Acid Cleaners 2. Chlorinated Solvents or CFCs 3. Alkaline Cleaning 4. Deionized Water	1. Acidic Waste 2. Chlorinated Solvents or CFC Wastes 3. Tank Cleaning Waste from Alkaline Cleaning 4. Deionized Water Baths cleaning	1. Acidic Rinses 2. None 3. Alkaline Cleaning Rinses 4. Deionized Bath Discharge	1. None 2. Solvent Emissions 3. None 4. None
Photoresist Stripping	1. Chlorinated Solvents for Negative Photoresist 2. Organic Solvents for Positive Photoresist	1. Chlorinated Solvent Waste 2. Organic Solvents Waste	1. None 2. None	1. Solvent Emissions 2. Solvent Emissions
Etching	1. Liquid or Dry Strippers 2. Chlorinated Solvents 3. Acids	1. Liquid or Dry Stripper Waste 2. Chlorinated Solvent Waste 3. Acidic Wastes 4. Contaminated Application Equipment	1. Rinse Wastewater 2. None 3. Rinse Wastewater	1. None 2. Solvent Emissions 3. Acidic Emissions
Wafer Assembly	1. Chlorinated Solvents or CFCs 2. Alcohol 3. Methanol and CFC-113 for Nomenclature Testing	1. Solvent Waste 2. Solvent Waste 3. Contaminated Application Equipment and Rags	1. None 2. None 3. None	1. Solvent Emissions 2. Solvent Emissions 3. Solvent Emissions
PCB Assembly				
Component Fluxing and Soldering (Wave, Dip, and Drag)	Soldering 1. Molten Solder 2. Flux Cleaning 3. Chlorinated Solvents or CFCs 4. Aqueous and Semi-aqueous Cleaners 5. Mechanical Cleaning	1. Solder Waste Containing Lead 2. None 3. Solvent Waste 4. Bath Clean-up Waste and Contaminated Bath Disposal 5. Contaminated Application Equipment	1. None 2. None 3. None 4. Rinse Water Discharge 5. None, unless the Application Equipment is Rinsed with Water	1. Lead 2. None, unless it contains VOCs 3. Solvent Emissions 4. None 5. None

Table 17-2: Typical wastes generated in the electronics industry.

N-Methyl Pyrrolidone (NMP) Recycling

NMP is an organic solvent commonly used in the fabrication process, i.e., in the stripping of **polyimide**, and as a **passivator** between layers on the wafer. The waste stream containing this material can be recycled onsite, by removal of the water in a single-stage packed column followed by flash distillation of the NMP to separate it from the polyimide. The condensed water, containing some residual NMP, can be discharged or further processed. The NMP is condensed and transferred to a separate tank for analysis and then recycled back into the process. The polyimide is accumulated and sold to an off site reclamation facility.

Automated Photoresist Dispensing System

Photoresist is one of the most expensive virgin chemicals demanding a price of $300-400 per gallon. The use of an automated dispensing system can reduce photoresist waste by as much as 50 percent. Several different automated dispensing system designs are available. All systems provide for maximum usage of photoresist from the supply bottles; they allow for uninterrupted operation with minimal operator assistance; and also provide point-of-use filtration resulting in higher yields. In addition, the dispensing systems do not require any modifications or adjustments be made to existing dispensing mechanisms. Photoresist accounts for less than two percent of the total annual hazardous waste generated by the semiconductor industry. As a waste, it also requires the use of expensive disposal options such as incineration.

PCB Assembly Cleaning

Several solvents and cleaners are used in the cleaning of PCB after assembly. The phaseout of CFC-113 and TCA has led many users to search for suitable alternatives. Among the various types of solvents being used are alcohols, petroleum solvents, **terpenes** and heavy hydrocarbons, as well as detergents and **saponifiers**.

HCFCs – The chemical substitutes for TCA and CFC-113 are the newly proposed **hydrochlorofluorocarbons (HCFCs)**. Three major HCFCs were proposed initially as replacements to CFCs for the electronics industry. The HCFCs have been shown to be suitable substitutes for a limited number of uses. HCFC 123 has shown chronic toxicity on male rats, which eventually led to its removal as a solvent from the electronics and cleaning markets. HCFC 141b has shown ozone depletion potential equal to that of TCA, thus it is not approved at this time for general cleaning uses. HCFC 141b will be available for limited critical cleaning applications only. HCFC 225 is in the process of being toxicity tested.

HCFCs can be used in conventional vapor degreasing systems, in which high solvent emissions could occur. A number of equipment manufacturers have introduced totally enclosed or vacuum-cleaning systems for the use of HCFCs to control and minimize solvent emissions. Operations from initial cleaning to drying occur in single chambers, in which the parts are placed. The solvent vapor is drawn to the storage vessel once cleaning of the subassembly is completed. Through the use of additional vacuum, the residual vapors are also recovered.

Heavy Hydrocarbons – The heavy hydrocarbon solvents, such as terpenes and N-methyl-2-pyrrolidone (NMP), have been introduced to the electronics market as substitutes for the CFCs. These solvents are used in either batch or continuous semiaqueous cleaning systems; in both systems the solvent is regenerated, as shown in Figure 17-7. In this type of system, for example, flux residues would be dissolved in the solvent in the first tank. The dragout solvent being removed by the parts can be replaced with fresh solvent. The rinse water in the second tank removes the solvent and the flux residue by forming an emulsion, which is transferred into a decanter. Since the solvent is not soluble in water, the emulsion separates, and the residue droplets float on the surface of the water. The residue level of the emulsion is monitored, and the separated solvent is returned to the cleaning chamber.

Perfluorocarbon Compounds – A new process uses perfluorocarbon compounds in conjunction with the heavy hydrocarbons in traditional vapor degreasing equipment. The process called the aqueous vapor degreasing (AVD) process, uses terpene-based compounds as solvating agents. The rinsing agent is generally a perfluorocarbon. The equipment used for this purpose is commonly just a modified vapor degreaser with 150 percent freeboard, and a −20°F chiller coil above usual condenser coils in the freeboard area. Perfluorocarbon compounds, how-

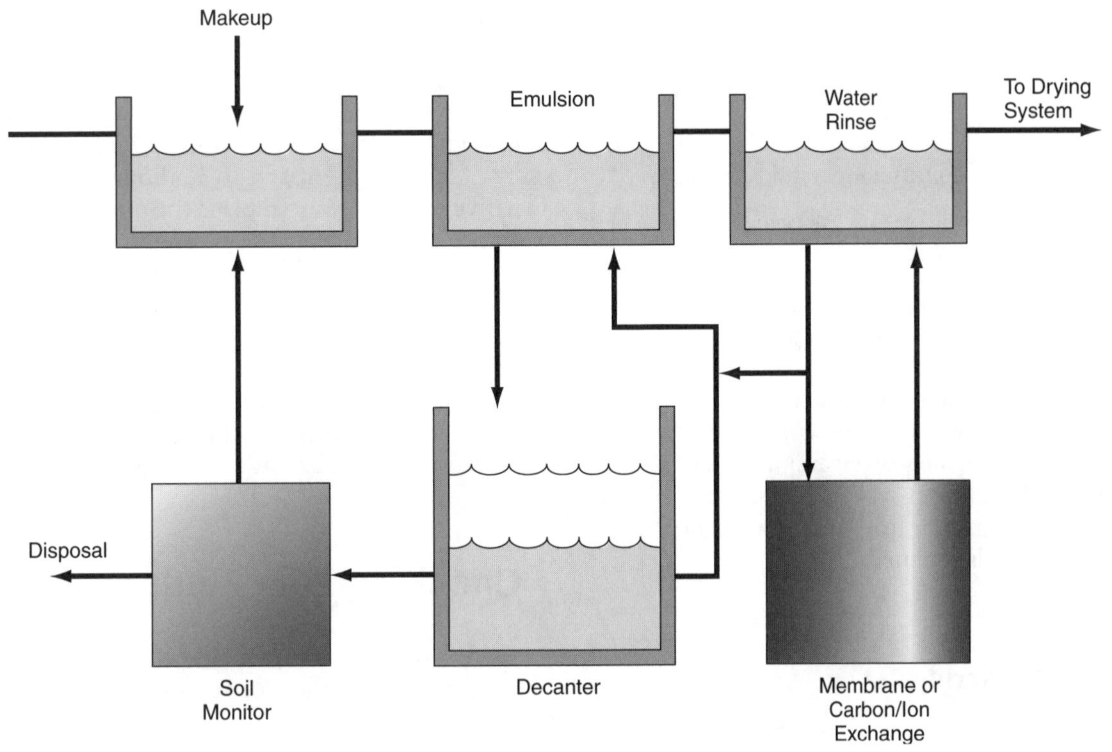

Figure 17-7: Semiaqueous cleaning process.

ever, are quite expensive and believed to be greenhouse gases. Thus, they are not approved for use by EPA for standard cleaning applications.

Process Substitution

Surface Mount Technology

On a traditional PCB, devices are connected to the board by drilling holes in the board, inserting and crimping the leads, and then soldering the components to the boards. Today, there is a movement toward **Surface Mount Technology (SMT)**. Surface-mounted devices are small and have no connector leads. Because no holes are needed with these devices, the components may be more densely

packed on the board. The movement toward SMT means that cleaners used must be able to penetrate the smaller crevices and spacing between the devices and the board. The recent phaseout of ozone-depleting substances such as TCA and CFCs has led to higher use of aqueous and semi-aqueous cleaning systems.

Aqueous Cleaning Process

Cleaning with water has its pros and cons. Aqueous cleaning systems have been used in electronics production for many years. Water-based cleaning agents, with the addition of saponifiers, can be used to remove both non-ionic contaminants (such as rosin flux, grease, and oil) and ionic contaminants (i.e., flux activators). However, the use of water in cleaning applications will result in wastes that must be managed. These wastes consist of:

— Spent wash solutions that contain the materials removed from PWBs/assemblies;

— Rinse water containing impurities dragged over from the wash;

— Energy lost by conductance and radiation;

— Energy lost by disposing of heated rinse water into a drain;

— Filter media (mechanical, carbon, or ion exchange) used for the purification of tap/raw water.

The volume of wastewater generated from aqueous cleaning processes can be considerable. A significant reduction in wastewater is achievable by employing closed loop systems, where process water is purified continuously, and a major portion of the heat energy is retained.

No Flux Soldering

Nitrogen-controlled reflow soldering machines, when used in conjunction with **formic acid** or **abietic acid** as activators, can eliminate the need for soldering flux. Because the oxygen content in the soldering machine is controlled, no oxides form on the boards. The residues remaining from the activator on the board do not have to be cleaned. Several technical and operational issues need to be considered in adoption of this technology; however, it should be noted that this process eliminates, in a single step, the need for liquid cleaning systems.

Soldering

The widespread use of solder containing lead has become a concern within the electronics industry. In recent years, several European and U.S. suppliers have researched the possibility of making a lead-free solder. Finding a suitable substitute for lead-tin solder, however, is an extremely complex proposition. Considered as a possible replacement for lead is bismuth, a by-product of lead and copper refining, but various technical and product issues remain to be resolved before bismuth can be considered as a viable substitute.

Other alloys such as tin-antimony or tin-silver have also been considered. However, they pose other technical problems, such as: 1) having melting points higher than conventional solder, 2) under certain conditions, tin-antimony alloys form intermetallic needles that can create structural problems, and 3) copper must be added to the alloys to achieve wetting rates comparable to conventional solders.

A second approach to finding a suitable alternative for solder is to replace the metallurgical bond with a connection made by a conductive adhesive. Use of conductive adhesives – typically silver-filled epoxies – as a replacement for solder is not new. A number of process variables must be considered before full adoption of conductive adhesives is possible. It should be noted that the use of these adhesives will eliminate the need for cleaning, representing a substantial cost savings.

Checking Your Understanding

1. What are the two major environmental problem areas in the electronics industry?

2. What are two reasons for using an automated photoresist dispensing system?

3. What are two of the problems preventing the use of HCFCs that have been discovered?

4. What are two substances used as activators for no flux soldering?

Summary

The electronics industry manufactures components and electronic packages that are used in telecommunication devices, automobiles, industrial machinery, military hardware, and consumer electronic products. The industry is composed of three major sectors: semiconductor manufacturing, printed wiring board fabrication, and printed wiring board assembly.

Semiconductors are made primarily from silicon. The silicon is grown from a seed crystal into an ingot measuring six to eight inches in diameter and about two feet in length. The ingot is lapped to smooth its outer surfaces, and thin wafers sliced from its length. After polishing and preparing the surface of the wafer, it is coated with a photoresist, and the design for a circuit is transferred by photolithography. The circuit may be simple, containing a single diode, or it may become an integrated circuit, which is a combination of many diodes performing thousands of functions. The exposed photoresist is removed and the circuit etched into the surface. Dopants may later be added by ion implantation to change the electrical characteristics of an area, making it either a nonconducting or a conducting region. The completed circuit may next be covered with a thin layer of conductor, semiconductor, or insulator material; after that, the following layer is added. This sequence of steps is repeated until all of the circuits have been built.

When the circuits are completed, each one of them is tested; then the wafer is masked off and sawed into single chips. After a final cleaning, they are mounted on attachment pads and placed in a plastic or ceramic case that protects them against mechanical shock and any external damage. The completed ICs are then attached to a PWB, typically by soldering, thus forming the completed electronic assembly.

There are two major environmental concerns in the electronics industry: 1) the use of chlorinated solvents for cleaning processes, and 2) presence of lead in the solder used during assembly. Several replacements for chlorinated solvents have been found, including N-methyl pyrrolidone (NMP) and terpenes. Several HCFCs have been developed to replace CFCs, but all pose other environmental and health problems. The use of aqueous and semiaqueous cleaning systems have met with some success; however, if not operated in a closed loop system they generate high volumes of wastewater and great heat losses.

In an attempt to improve the soldering process, both no flux and lead-free solders have been used. Nitrogen-controlled reflow soldering machines make use of acid activators in an oxygen-controlled atmosphere to eliminate the need for flux and, therefore, the need for a liquid cleaning system. Several solder alloys that do not contain lead have been tested, but they pose other problems that have not yet been resolved. The use of conductive adhesives, such as silver-filled epoxies, is not new, but it has not yet gained wide acceptance.

Critical Thinking Questions

1. Considering that the circuitry currently being etched on wafers is about 2.0 μm, what are the factors that will most likely limit further reductions? Be sure to consider such items as contamination, chemical purity, and all of the circuit transfer processes in your answer.

2. Conduct a literature search and prepare a class report on how ion implantation can be used to create p-type and n-type areas.

3. Research the composition of photoresist used in IC manufacture and determine why it results in a waste that must be incinerated.

4. Electronic devices become obsolete very quickly and are replaced at a rapid rate. The discarded devices contain both valuable metals and toxic substances. Develop a list of the different materials found in a personal computer and propose a system for the collection, disassembly, and recovery of these materials.

18

The Printed Wiring Board Industries

by Azita Yazdani

Chapter Objectives

Upon completing this chapter, the student will be able to:

1. **Identify** and describe the processes used to prepare a PWB.

2. **Describe** methods used to reduce and recover drag-out.

3. **Discuss** various methods used to recover or remove chemicals from the process and rinse baths.

4. **List** and describe the four process used for chemical recovery in the PWB industry.

Chapter Sections

18–1 Introduction

18–2 Overall PWB Processes

18–3 Regulatory and Pollution Prevention Measures

18-1 Introduction

It would be hard to imagine living in a world without telephones, radios, TVs, VCRs, video games, computers, and the many other electronic wonders that are a part of our daily lives. At the heart of these devices is a printed wiring board (PWB). The PWB is the foundation upon which the electronic circuitry has been built and, at the same time, onto which integrated circuits (ICs) are subsequently attached.

At the most basic level, a PWB is composed of a nonconductive substrate upon which an electrically conductive pattern has been formed. The conductive layers are then connected to each other by electroplated through-holes that are also used to mount and connect other components, such as ICs. After the components have been mounted on the PWB, the combination of the PWB with its components is known as an **electronic assembly** or a **printed wiring assembly (PWA)**. This assembly – and perhaps several others – is the basic building block for all larger electronic devices, from wristwatches to telecommunication networks.

The PWB and PWA industries are, in turn, the cornerstones for the overall electronics industry. The domestic interconnect industry consists of approximately 700 PWB manufacturers (90 percent of which have annual sales of less than $10 million) and 1,200 electronic manufacturing service providers. The industry employs more than 200,000 workers in the United States and, in 1997, had annual sales that exceeded $17 billion.

The PWB industry consists of large facilities totally dedicated to printed circuit boards; large and small captive facilities; small job shops doing contract work; and specialty shops doing low-volume and high-precision work. Approximately half of the printed circuit boards produced are by independent producers, while the rest are by captive producers. More than 65 percent of all printed circuit board manufacturing sites are located in the northeastern states and in California.

The Far East is the leader in manufacturing capabilities, including the development of small holes and fine conductors. Development and production of PWB is dominated by the United States, Asia, and Western Europe.

18-2 Overall PWB Processes

PWB Manufacturing

PWBs, also called **printed circuit boards (PCBs)**, consist of carefully designed patterns of conductive material that have been formed on a nonconductive base material. Nonconductive base materials include pressed epoxy paper, phenolic, epoxy-glass resins, teflon/glass, and many others. The conductive patterns are most typically made with copper, although aluminum, chrome, nickel, and other metals have also been used. The metal is fixed to the nonconductive base using adhesives, pressure/heat bonding, and sometimes even screws.

There are three basic types of PWBs: single-sided, double-sided, and multilayer. Single-sided boards have a conductive pattern on only one side, while double-sided boards have conductive patterns on both sides. Multilayer boards consist of a sandwich of alternating layers of conducting and insulating materials that are bonded together. As shown in Figure 18-1, as the complexity of the PWB increases, so does the number of process steps required. The raw materials most commonly used for PWB manufacture are shown in Table 18-1.

The main processes, common to all PWBs, are drilling, image transfer, and electroplating. Through-

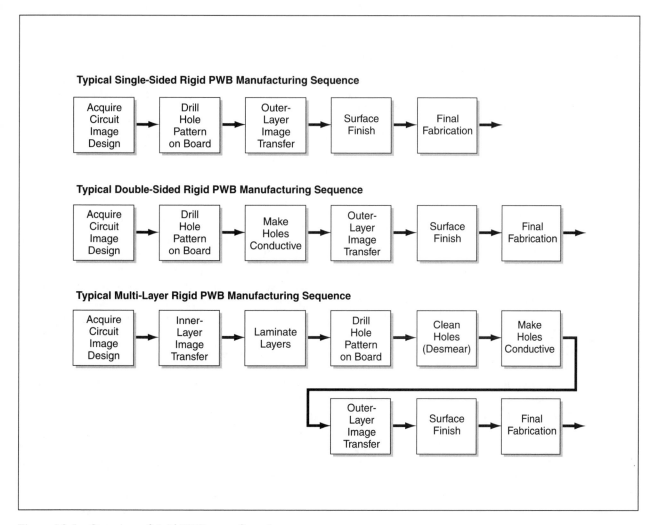

Figure 18-1: Overview of rigid PWB manufacturing process sequences.

Component	Common Material Used
Board Materials	Glass-epoxy, ceramics, plastic, phenolic paper, copper foil
Cleaners	Sulfuric acid, fluoroacetic acid, hydrofluoric acid, sodium hydroxide, potassium hydroxide, trichloroethylene, 1,1,1- trichloroethane, perchloroethylene, methylene chloride
Etchants	Sulfuric and chromic acid, ammonium persulfate, hydrogen peroxide, cupric chloride, ferric chloride, alkaline ammonia
Catalysts	Stannous chloride, palladium chloride
Electroless Copper Bath	Copper sulfate, sodium carbonate, sodium gluconate, Rochelle salts, sodium hydroxide, formaldehyde
Screen	Silk, polyester, stainless steel
Screen Ink	Composed of oil, cellulose, asphalt, vinyl or other resins
Resists	Polyvinyl cinnamate, allyl ester, resins, isoprenoid resins, methacrylate derivatives, polyolefin sulfones
Sensitizers	Thiazoline compounds, azido compounds, nitro compounds, nitro aniline derivatives, anthones, quinones, diphenyls, azides, xanthone, benzil
Resist Solvents	Ortho-xylene, meta-xylene, para-xylene, toluene, benzene, chlorobenzene, cellosolve and cellosolve acetate, butyl acetate, 1,1,1-trichoroethane, acetone, methylethylketone, methylisobutyl ketone
Electroplating Baths	Copper pyrophosphate solution, acid-copper sulfate solution, acid-copper fluoroborate solution, tin-leas, gold, and nickel plating solutions
Resist Stripping	Sulfuric-dichromate, ammoniacal hydrogen peroxide, solutions metachloroperbenzoic acid, methylene chloride, methyl alcohol, furfural, phenol, ketones, chlorinated hydrocarbons, non-chlorinated organic solvents, sodium hydroxide.

Table 18-1: Raw materials most commonly used in PWB manufacture.

holes are drilled into PWBs (or punched, in the case of paper-based substrates) to provide layer-to-layer interconnection on double-sided and multi-layer circuits. The interiors of these holes are subsequently made conductive by electroplating copper into the **hole barrels** (the interior surfaces of the cylinder). Most rigid PWB substrates consist of epoxy-resin and glass, which makes direct electroplating of hole barrels impossible, since this material will not conduct electricity. Therefore, a **seed layer** or coating of conductive material (usually electroless copper) must be deposited into the hole barrels before the electrolytic copper plating can occur. Holes are also drilled into PWBs – including single-sided boards – to accommodate component leads that are inserted through the hole and soldered to the board. Holes drilled for this purpose do not require electroplating.

Image transfer is the process by which the image of a circuit layer is transferred from film, glass, or directly from image data files, to the copper foil of the PWB material. For inner layers, this includes the application of a photoresist (a photosensitive film that also serves as the etch resist), imaging, developing, and etching, (see Figure 18-2). For out-layers, image transfer may include the electroplating of copper, tin, tin-lead, or nickel/gold coatings.

The predominant way of accomplishing image-transfer and layer-to-layer interconnection is called the **subtractive plating process** or **print-and-etch** method. With this process, a copper foil layer of uniform thickness is selectively etched to create a circuit pattern. Copper foil is an integral part of PWB base laminate and is applied by PWB manufacturers during the **laminate stage,** or it is bonded to substrate by laminate makers.

In an alternative process, called **additive processing**, the manufacturer forms the copper image by selectively plating electroless copper onto a sensitized substrate (fully additive), or by plating a thin layer of electroless copper non-selectively onto a substrate, then applying a photoresist and selectively electroplating additional copper onto the circuit areas (semi-additive). The fully additive process does not require etching; semi-additive processing requires etching of only the thin electroless copper layer. Additive processing, although attractive in terms of efficiency and waste reduction, is not a common choice among manufacturers producing the basic or low-cost PWBs. This is due to the complexity of the process compared with subtractive processing, the electrical properties of electroless copper deposits, and certain other physical properties of electroless copper (e.g., the low copper-to-substrate peel strength).

PWB Production Steps

Each box in Figure 18-1 is a logically distinct process or group of processes. For some of the boxes, a set of competing processes exists. In this chapter, only the most common processes utilized in rigid multilayer PWB manufacturing are discussed.

Computer Aided Design (CAD)

Currently, the PWB shops utilize personal computers, modems, and inexpensive circuit layout software to receive and manipulate circuit images. The circuit image files are edited to produce a customized **photo-tool**. The CAD department responsible for these data may also perform design checks and other engineering functions.

Inner Layer Image Transfer

Image transfer for inner layers is accomplished through a series of steps that transfer the image of the desired circuit from film, glass, or data to the copper foil layer of the PWB base material. Inner layer image transfer processes transfer an image of the circuit from a film photo-tool onto the copper foil of the PWB base laminate, using either a subtractive or the previously mentioned additive technique. In the commonly used subtractive image transfer method, the transfer is accomplished through a se-

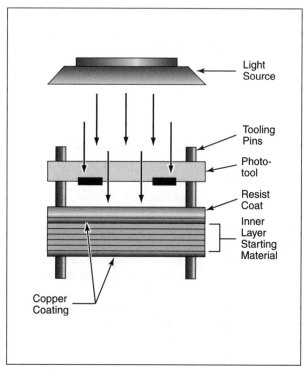

Figure 18-2: Use of photo-tool for inner layer exposure.

ries of processes that together are referred to as "print-and-etch."

To accomplish the "print" step, a photoresist, which is a light sensitive organic coating, is applied to the copper foil of the PWB. The image is next transferred to the photoresist using a photo-tool and a light source. The exposed photoresist is then developed and becomes the **etch resist**. When the panel enters the etch process, therefore, the etch resist layer selectively protects the circuit areas from the etchant, whereas the remaining copper foil is etched away. The etch resist is then stripped away, using hot potassium hydroxide, KOH, or other solvent, revealing the copper circuitry.

In the past, ferric chloride, $FeCl_3$, and chromic acid, H_2CrO_4, were the commonly used etchants. Both have fallen into disfavor, leaving cupric chloride, $CuCl_2$, and ammoniacal etchants as the preferred. Ammoniacal etchants, composed mainly of ammonium hydroxide and ammonium chloride, have the additional advantage of being compatible with both organic and metallic etch resists, therefore making it the only choice for shops where two etching systems are not economically feasible. Although onsite regeneration methods are available for both resists, cupric chloride etchant is more frequently processed onsite, which is a major consideration when the volume of etchant used is taken into account.

Oxide

The oxide step is used to enhance the copper-to-epoxy bond strength of the multi-layer panel. The oxidation may be performed in a variety of ways, but the conventional method relies on the use of sodium hydroxide, NaOH, and sodium chlorite, $NaClO_2$, at elevated (140-150°F) temperatures. The need for the oxide step can be eliminated by purchasing "double-treated" oxide-coated laminate from the manufacturer. All shops producing rigid multi-layer PWBs, however, use the conventional oxide step for at least some of their multi-layered products, and they purchase some double-treated material for other products.

The Desmear Process

Vias are holes drilled through a PWB for the purpose of layer-to-layer interconnections. During the drilling process, the drill bits may become heated, resulting in some melting and smearing of the epoxy-resin base material across the copper interface of the hole barrel. If the smeared resin is not removed, it may lead to a nonconductive circuit. **Smear**, which is the name given to smeared epoxy-resin, is not a concern on single-sided circuits because there is no interconnection in the holes. It is not a major concern on double-sided circuits, either, because the copper-to-copper interface extends out of the hole barrel and onto the horizontal surface of the PWB. In the case of multi-layer panels, however, the only interface is provided by the copper plated to the inner layers of the hole barrels. If the hole barrels are covered with smear no connection occurs, and the PWB will be defective.

The **desmear** process is often grouped and sometimes confused with **etchback**. Desmear is the removal of smear from the copper surfaces within the hole barrel to facilitate interconnection between the inner-layer copper and the electroless copper that is plated next. Etchback is the removal of a significant amount of epoxy-resin and glass fibers from the hole barrel. Etchback is performed to expose a greater copper surface area, which will enhance the

bond between the inner-layer copper and subsequently added electroless copper. The problem of smear is now considered to be only a result of poor drilling practices.

Two distinct methods of desmear currently exist: wet chemical processing is most common, but **plasma etching** is seeing increased use. With the wet process, sodium or potassium permanganate, $NaMnO_4$ or $KMnO_4$, are the oxidizers of choice; they replace the use of chromic acid, which is a known carcinogen. Permanganate solutions alone are poor glass etchants; therefore, if etchback is required, a concentrated sulfuric acid step is used first.

The cost of a plasma etchback system for some PWB applications is around $100,000. A combination of oxygen, O_2, and carbon tetraflouride, CF_4, must be present in the plasma gases to oxidize the epoxy and remove fiberglass. This may result in some fluorocarbon molecules as by-products. Plasma etching is employed at approximately 30 percent of multi-layer shops, but only very few use plasma etching exclusively.

Through Hole Metalizing

In this step, relatively thick (0.001 inches) electroless copper is plated in the hole barrels to form an interconnect between the various inner layers and the surface copper. This process is also known as making holes conductive. The conventional electroless copper plating process used for this operation includes at least seven different solutions, accompanying rinse tanks, plus the electroless copper line, which may include as many as 20 to 25 tanks. The electroless copper process presents problems with respect to waste treatment, due to significant concentrations of ethylenediaminetetraacetic acid (EDTA) or other **chelating agents** found in all electroless copper baths and associated wastewaters. Therefore, most PWB facilities use special treatment steps to separate or destroy these compounds. These treatment steps increase chemical reagent use, operating costs, and sludge production. Furthermore, most electroless copper baths contain significant concentrations of formaldehyde, which in turn produce air emissions.

Until the late 1980s, no commercially viable alternative for electroless copper process existed. However, recently several alternatives have been developed and are being used by a limited number of shops. These alternatives include carbon-based, graphite-based, and palladium-based processes. In addition, there are alternative processes that use conductive polymers and inks and one that is a non-formaldehyde electroless copper process. All of the alternative methods reduce copper discharges and eliminate formaldehyde, and most of them require fewer process baths than the conventional electroless copper method.

Outer Layer Etch Resist

Outer layer imaging is a process that is quite similar to inner layer imaging. Typically, the same type of photoresist is used for both. When used for the outer layers, however, its thickness becomes critical. To prevent the copper or tin-lead applied later from **mushrooming**, the photoresist is generally applied to a thickness equaling that of the metal to follow. When the pattern plating of metal is to follow, the photo-tools are positive images of the circuit. Therefore, when developed, the circuit resist is developed away, exposing the underlying copper. The photoresist remaining on the panel becomes the plating resist for the pattern plating process.

Immediately after copper pattern plating, an etch-resist metal is plated over the copper. The elimination of lead metal plating has been a goal of many PWB manufacturers, partially due to strict lead discharge limitations. Tin-lead is most often plated as an etch resist on panels subsequently processed with **solder-mask-over-bare-copper (SMOBC)**. Therefore, when the SMOBC process is specified, tin-lead is easily replaced by tin, as panels must be processed with a tin-lead etch resist, which is subsequently fused into solder. For economic reasons, many shops cannot maintain both tin and tin-lead plating lines and are, therefore, unable to employ tin-only plating on the portion of their product that is SMOBC.

In short, the transition from tin-lead plating to tin-only has been slow.

Many of the alternatives available to this process are based on end-user specifications. For example, if electrolytic gold is specified by the end user, gold must be plated as an etch resist. If gold is not specified, it is almost never considered as an alternative to tin-lead due to cost.

Outer Layer Etchant

The choice between cupric chloride and ammoniacal etchants for outer layer etching is largely determined by compatibility with the various etch resists employed. Metallic etch resists are generally incompatible with cupric chloride, which greatly limits their use on outer layers. Ammoniacal etchants are generally compatible with all etch resists and are therefore very common as an outer-layer etch resist. Sulfuric-peroxide is also compatible with metallic resists.

Checking Your Understanding

1. What are the three basic types of PWBs and how do they differ from each other?

2. What is the purpose of hole barrels?

3. What is a photoresist?

4. What is the purpose of the oxide step?

5. What is the source of smear and what does smear interfere with?

6. What is the purpose for the through-hole metallizing step?

18-3 Regulatory and Pollution Prevention Measures

Good Operating Procedures

Pollution prevention (P2) methods listed under good operating procedures include various administrative and equipment-related topics that can affect waste generation. Many facilities have established a formal P2 program, in which employee education is a major component of the program. Because of the similarities in the processes used in the electroplating and PWB industries, many of the P2 procedures are the same or address similar problems.

In general, the PWB facilities make more frequent use than plating shops of pollution prevention methods that relate to chemical inventory, process bath control, and use of chemicals. For example, nearly all of the PWB manufacturing facilities in a 1995 EPA survey indicated that they perform in-house process bath analysis and maintain records of this work. Other areas in which the PWB shops have a higher percentage of application methods include preventative maintenance of racks and tanks and use of overflow alarms and leak detection. Surprisingly, only about one-fourth of the shops indicated that they employ **statistical process control (SPC)** for bath maintenance. SPC is a potential cost reduction and a P2 method that relates to analytical work and record-keeping practices already employed by most of these facilities.

Drag-out Reduction and Recovery Methods

Drag-out is the process solution clinging to a workpiece when it is removed from a tank. This solution is usually small in volume, but high in chemical concentration. The primary purpose of rinsing is to remove the drag-out from the workpiece and prevent it from contaminating the subsequent baths. The quantity of rinse water needed to sufficiently clean the workpiece is directly related to the quantity and characteristics of the drag-out. When the quantity of drag-out is reduced, its characteristics changed, or the drag-out solution recovered and reused, the overall quantity of pollution from a facility is reduced.

To reduce drag-out, PWB manufacturers most often use workpiece motion control (e.g., slow withdrawal from tank and/or long drip tanks), rather than adjusting the bath chemistry, its temperature, or drag-out recovery methods. Plating shops, on the other hand, tend to rely more on drag-out recovery than other methods to reduce drag-out losses, and they also make significant efforts to control process chemistry and bath temperature.

The flat surfaces of PWBs improve the performance of motion control methods for drag-out reduction as compared to the range of workpiece shapes processed by plating shops. Flat surfaces, like PWBs, will more completely drain when held for an extended time period as compared to cup-shaped workpieces with internal surfaces, such as auto bumpers. Although many plating processes require close chemical control, there are some (e.g., chrome plating) that can be operated over a significant range of concentrations. By reducing the chemical concentration of the bath, electroplaters are able to reduce **viscosity** and, in turn, reduce drag-out. PWB manufacturers may have less latitude with regard to varying the bath concentration.

To a moderate extent, plating shops also increase bath temperatures as another way of reducing viscosity. In general, there is a smaller percentage of heated process tanks in PWB shops than in electroplating shops and, therefore, less opportunity to take advantage of this method of reducing drag-out. Heated process tanks in PWB facilities where this method could be used include cleaners, permanganate desmear, and micro-etches.

The most common method of drag-out recovery employed in the electroplating industry is the use of drag-out tanks. Drag-out tanks are initially filled with clean water and are located immediately after the process bath. Workpieces exiting the process tanks are immediately rinsed in the drag-out tank. The contents of the drag-out tank are then used to replace evaporate loss in the preceding tank. **Drip tanks** – a less sophisticated method of drag-out recovery – are not initially filled with water. A drip tank is simply a tank over which racks are hung to drip, and the content of which is periodically returned to the process tank. Both drag-out tanks and drip tanks require the use of heated process tanks, because enough liquid must evaporate from the process tanks to make room for the returned solution. As such, these recovery methods are only applicable to heated process solutions.

The PWB industry uses drag-out tanks on a limited basis, and only a few companies use drip tanks. The low use of these methods in PWB shops is possibly due to the permissible latitude in process control and differences in bath temperatures, as discussed above. However, some PWB heated process tanks (cleaners, permanganate, micro-etches) do lend themselves to the use of drag-out tanks for drag-out recovery.

Rinse Water and Chemical Use Reduction

The vast majority of wastewater generated at PWB and electroplating facilities is the result of rinsing. The use of several important rinse water reduction methods can lead to a significant reduction in water usage. The primary way to reduce the volume of rinse water is through countercurrent rinsing methods. A countercurrent rinse system is a series of two or more rinse tanks piped together in such a way that clean water enters the final tank of the series and flows into the next tank in a direction opposite to that of the workpiece flow. Far less water is required to maintain sufficiently clean water than in a system with a single rinse tank.

A large percentage of PWB shops employ countercurrent rinse methods as a means of reducing total water usage. Flow controllers, used in conjunction with countercurrent rinsing, are also very common in PWB shops. Automated water use controllers that monitor such variables as conductivity, pH, or time are also more commonly used in PWB shops than in plating shops. Workpiece sensors are used primarily on conveyorized equipment to turn the water flow on when a PWB reaches the rinse area and turn it off until the next part arrives. The fact that rinse water control devices are more commonly used in PWB production than in electroplating operations is due to the higher requirements for rinse water purity in the manufacture of PWBs.

Porous Pot

Electrolytic regeneration of permanganate desmear baths using a **porous pot** (or similar ion transfer designs) is a pollution prevention technology employed by a number of PWB facilities. This relatively inexpensive and simple technology is used for bath maintenance (i.e., extending the useful life span of permanganate desmear baths). In the conventional permanganate process, the permanganate ion, MnO_4^-, is reduced by heat and contact with PWBs and is replaced by chemical addition. Also, during operation of this bath, by-products including the manganate ion, MnO_4^{2-}, accumulate in concentration causing a sludge to form, which makes frequent disposal necessary. The porous pot can be used to maintain a sufficiently low concentration of contaminants to reduce the frequency of disposal.

The common porous pot design consists of a rectifier, a ceramic pot that houses a cathode (protecting the cathode from direct contact with the process solution), and an anode, which surrounds the pot and is in direct contact with the bath. At startup, the pot is immersed into the bath (with the top remaining above the solution, preventing it from flowing into the cathode compartment) and filled with an electrolyte, usually sodium hydroxide. With the bath shielded from the cathode, the primary reaction that occurs is the anodic re-oxidation of the manganate ion back to permanganate. The use of the porous pot can realize a bath-life extension of ten-fold or more.

Ion Exchange

Ion exchange is a versatile technology applied by PWB manufacturers for various and sometimes overlapping purposes, including chemical recovery, water recycling, solution maintenance, and waste treatment. In general, most of the waste streams discharged from the PWB processes are compatible with ion exchange technology. Many shops mix several similar rinse streams and treat them with a single ion exchange unit. The ion exchange effluent may be discharged, and the **regenerant** processed using electrowinning, thereby making ion exchange both an end-of-pipe waste treatment and a component of a metal recovery system.

Ion exchange resins operate by exchanging an H^+ ion for a cation or, in the case of anion resins, an OH^- ion for an anion in the waste stream. When most sites in the resin have exchanged their ions, the resin must be regenerated. During the regeneration phase, an acid is passed through the cation resin, and a base is passed through the anion resin. The metal ions recovered in the regenerant are in concentrations of a few grams per liter and are removed by either using electrowinning or by being sent to a conventional precipitation treatment system.

Besides the application of ion exchange as a raw water treatment method (i.e., water softening or deionization of city water), ion exchange is most often used by PWB shops in one of two configurations. If wastewater is to be recycled for re-use as rinse water, both cation and anion removal is required. A second strategy is to remove only certain cations from the wastewater and discharge the treated water. For many streams, the only regulated ions are typically Cu, Ni, and Pb, and these may be selectively removed by special cation resins. This cation-only configuration can be employed as a stand-alone treatment system or as a polishing step for effluents from a conventional precipitation system. To provide for continuous operation, multiple columns of resin are used, with one regenerating while the other is operational.

Electrowinning

Electrowinning is a technology employed by PWB manufacturers to remove metallic ions from spent process fluids, ion exchange regenerants, and concentrated rinse waters. Although the technology is not yet widely used, some facilities are using electrowinning as a part of their end-of-pipe treatment, i.e. in conjunction with ion exchange.

An electrowinning unit consists of a rectifier and a reaction chamber that houses anodes and cathodes. In the simplest design, cathodes and anodes are set in the reaction chamber in an alternating series, and the electrolyte is allowed to flow past them. Positively charged metal ions are reduced onto the negatively charged cathode. The rate at which metals can be recovered depends on several factors, including 1) the concentration of metal in the electrolyte (the higher, the faster), 2) the size of the unit in terms of current and cathode area, and 3) the predominant species of metal being recovered.

Electrowinning is commonly applied to spent gold solutions, gold drag-out and drip tanks, silver-bearing photographic developer and fixer solutions, and other copper-bearing or lead-bearing spent solutions, such as strippers or acid dips. Recovery of precious metals rapidly recuperates the cost of electrowinning equipment, but most PWB shops produce only small quantities of gold-bearing or silver-bearing spent solutions. Solutions containing hydrochloric acid, or the chlorine ion in general,

are usually not processed using electrowinning since electrolysis of these fluids can result in the evolution of deadly chlorine gas. The concentration of metal ions in a solution can be readily reduced below one gram per liter using electrowinning; lower concentrations (as low as one mg/l) are possible with some electrolytes. However, the efficiency of the electrowinning process (i.e., mass of metal removed per unit consumption of energy) steadily decreases as the metal ion concentration is depleted.

Environmental Challenges

The increasing cost of compliance is the most pressing environmental and occupational health challenge facing most businesses. There are also several specific environmental and occupational health challenges that the PWB industry has identified. The most common are: 1) changing environmental regulations; 2) consistently meeting effluent discharge limits; 3) reducing worker exposure to chemicals; and 4) eliminating solvent use. Other less frequently cited challenges are: 1) inconsistent enforcement of regulations; 2) hazardous waste transportation liabilities; 3) meeting air emission standards; and 4) the lack of hazardous waste disposal sites.

Most environmental requirements impact the operations of PWB facilities. These include, but are not limited to, the CAA, CWA, RCRA and Superfund, and EPCRA. The details of the impact these regulations have on PWB, electroplating, and other related industries are discussed elsewhere in this book.

Checking Your Understanding

1. Define drag-out and list two methods that can be used to recover it.

2. What is the most common method used in PWB manufacture to reduce drag-out?

3. What is the most common method used to reduce rinse water usage?

4. What are three common methods used to recover or remove spent chemicals?

5. List the environmental and health challenges that have been identified by the PWB industry.

Summary

Printed wiring boards (PWBs) – also known as printed circuit boards (PCBs) – along with mounted components, are collectively known as printed wiring assemblies (PWA) and make up the heart of most telephones, TVs, VCRs, and other electronic devices we use daily. The most common products of the PWB industry are the single-sided, double-sided, and multi-layer rigid boards. All PWBs consist of a nonconductive base material and a conductive pattern. Copper is the most commonly used metal for the conductive patterns, although other metals are also used.

The main processes used in PCB production are drilling, image transfer, etching, and electroplating. Multilayer boards consist of a sandwich of alternating layers of conducting and insulating materials bonded together. The image for the inner layers is transferred by using a photo-tool, a photoresist, and a light source. The photo-tool determines the areas where light strikes and interacts with the photoresist. Once the photoresist is developed, the part struck by light becomes the etch resist. By placing this panel in an etchant, the unprotected areas of copper foil are etched away, leaving the circuitry under the etch resist. This process is known as the print-and-etch method. The various inner and nonconductive layers are then sandwiched together and prepared for the hole-making process.

The holes, known as hole barrels, are drilled through the sandwich to form the layer-to-layer interconnections. Due to heat created by the drilling process, nonconductive base material may line the hole barrels with smear that has to be removed by a desmear process. The conventional desmear process involves the use of potassium permanganate, which prepares the copper surfaces within the hole barrels for copper plating. The hole barrels may also have to undergo an etchback procedure to remove some of the nonconductive base material that would otherwise interfere with the plating process. An electroless copper plating process is typically used to form the interconnect between the various inner layers and the surface copper.

The final outer imaging processes are carried out in much the same print-and-etch way as the inner layers. After the copper pattern plating and etch resist removal, an etch-resist metal is plated over the copper pattern. Tin-lead is most often used on the panels that subsequently will be processed with solder-mask-over-bare-copper (SMOBC).

Pollution prevention methods employed within the industry involve various administrative and equipment-related activities. Employee education is a major component of most administrative P2 programs. Reduction and/or recovery and reuse of process chemical drag-out remain some of the most important chemical and wastewater reduction tactics. Because PWBs have flat surfaces, motion control of the workpiece has been one of the most cost-effective methods to control drag-out. The use of countercurrent rinse baths that recover process chemicals by returning the rinse waters to the process bath has been effective and also provides a source of fresh rinse water for the rinse tanks. PWB shops also make use of automated water controllers, workpiece sensors, and flow controllers in conjunction with the countercurrent rinse systems.

The usefulness of process baths is also prolonged through the use of a porous pot electrolytic regeneration of the permanganate desmear baths as well as through recovery of metals from waste streams using ion exchange and/or electrowinning.

The PWB industry still faces several challenges, including meeting the effluent discharge limits and changing environmental regulations, reducing worker exposure to chemicals, and eliminating solvent usage. Due to the nature of the processes and chemicals used, PWB facilities are very similar to electroplaters and are affected by nearly every major piece of federal environmental regulation.

Critical Thinking Questions

1. Of the four environmental and occupational health challenges identified in this chapter, which is the most pressing and how and why would you attempt to meet the challenge?

2. Design a closed-loop electroless copper line, using a combination of the technologies described in this chapter.

3. Analyze the process steps used to make a multilayer PWB and describe what effect the development of a hypothetical electrically conducting polymer would have on the industry, if it did replace the use of copper.

19

The Pulp and Paper Industries

by Christopher J. Biermann

Chapter Objectives

Upon completing this chapter, the student will be able to:

1. **Describe** the operations used to produce thermomechanical and kraft chemical pulps.
2. **Explain** the advantages and disadvantages of various bleaching agents.
3. **Describe** the papermaking process.
4. **List** the tools used to recover fiber from recycled paper.
5. **List** the major sources of solid, liquid, and gaseous pollutants for each of the pulp and papermaking areas.

Chapter Sections

19-1 Introduction

Paper products are present in many aspects of our everyday life: newspapers, grocery bags, packing boxes, junk mail, writing paper, and many other products. In fact, there are more than 400 types of paper, made by a variety of processes, used for printing and writing, construction, packaging, sanitary products, and even capacitors. It is not surprising that the pulp and paper industry is large and diverse. In 1995, for example, sales for eight of the larger U.S. paper companies reached $5 billion each, while the largest of them – International Paper – had sales of $20 billion.

Paper is a mat or web of **pulp**. Pulp consists of cellulose-containing fibers that have been separated from each other. Along with the pulp, printing papers often contain fillers such as clay, calcium carbonate, and titanium dioxide, which enhance the printing properties of paper and usually cost less than the pulp they displace.

In 1900, the United States produced only 2.3 million tons of paper – about 70 percent of the total world production. By 1990, the U.S. production was 80 million tons, representing about 30 percent of the total world production. Since most paper is used for packaging and printing, the use of paper in any country is closely related to its economic status.

The Nature of Wood

The fiber source for about 93 percent of paper manufactured in the world is wood, with the remainder consisting of fillers and non-wood vegetable fibers. Wood consists of plant cell walls or fibers. The visible grain in wood is the result of the fiber orientation. More than 90 percent of the fibers are aligned

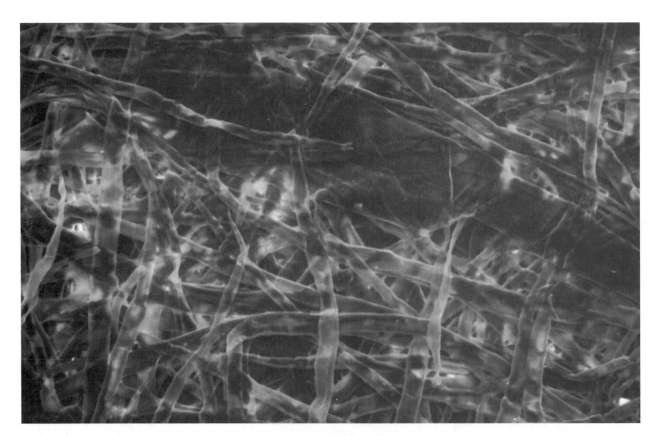

Figure 19-1: Scanning electron micrograph of a paper sample.

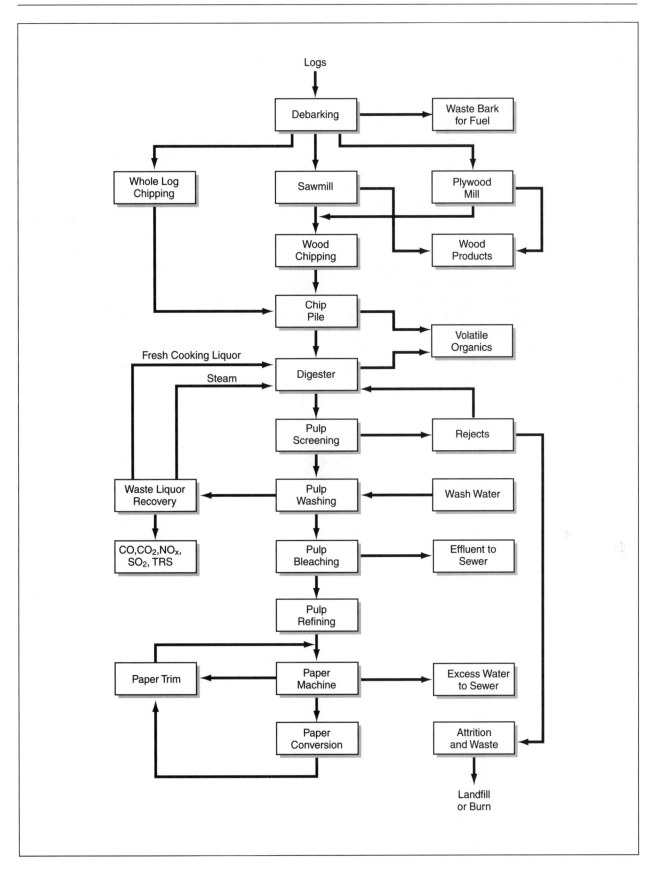

Figure 19-2: Flow diagram for the papermaking process.

in the main direction of the tree branch or stem.

As shown in Figure 19-2, wood must be debarked and chipped prior to making paper; chips are usually screened by size prior to pulping so that cooking time for the chips is uniform. The cellulose fibers derived from the pulping process are then washed and, depending on the desired product, may or may not be bleached before entering the papermaking process.

Papermaking fibers from **softwoods** – conifers such as pine, spruce, and fir – are 3-5 mm long and 0.03-0.05 mm in diameter, while those from **hardwoods** – deciduous trees such as aspen, birch, and maple – are 0.7-1.2 mm long. The length-to-diameter ratio is about 100 to 1 in both cases. Softwood pulp is used in papers that require high strength such as printing papers and structural papers used for items such as bags and boxes. Hardwood pulp is also used in printing papers with the specific purpose to impart smoothness, opacity – which is necessary for the print not to show through on the backside of the paper sheet – and even formation, which gives the paper uniform thickness.

Chemically, wood consists of three major components: **cellulose** (about 45 percent of the wood), **hemicellulose** (25-35 percent), and **lignin** (20-35 percent). Cellulose and hemicellulose, which are white polysaccharides, are desired in paper. Lignin, a complex polymer of phenylpropane units, is concentrated between the fibers and bonds the fibers together. To form pulp, the lignin must be altered (mechanical pulp) or removed (chemical pulp) by a pulping process.

Wood also contains about 1-8 percent **extractives**, which are made up of thousands of chemical compounds. Some extractives are volatile and are lost in the wood chip pile over time. The volatile extractives of fresh softwood chips are recovered from the digester vent gases of chemical pulping operations as turpentine, fatty acids, and rosin. Turpentine is used as a solvent or in pine oils for cleaning; it consists of **pinenes** and other chemicals, each with the empirical formula of $C_{20}H_{32}$.

19-2 Pulping and Bleaching Processes

Pulping Processes

There are four major pulping process categories used to separate the cellulose fibers. Listed in order of increasing reliance on mechanical energy and decreasing chemical action they are: chemical, semichemical, chemi-mechanical, and mechanical. These processes are summarized in Table 19-1 according to type and to the extent – in percentage – each one of them is used to produce various types of paper.

Chemical and Semichemical Pulping Processes

Chemical pulps are formed by using materials that de-polymerize and dissolve the lignin from the wood fibers. About 80 percent of pulp manufactured in the U.S. is kraft chemical pulp, while four percent is acid sulfite chemical pulp. The kraft chemical process uses sodium hydroxide, NaOH, and sodium sulfide, Na_2S, to remove the lignin and will be described in more detail in the next section.

Acid sulfite pulping was the dominant process in the 1930s, but its use is currently limited because the pulp has only 60 percent of the strength of a kraft pulp. All acid sulfite processes rely on mixtures of sulfurous acid, H_2SO_3, and/or its alkali salts (Na^+, Mg^{2+}, Ca^{2+}, and NH_4^+) to solubilize the lignin. The traditional salt used was calcium carbonate, $CaCO_3$, which reacts with the sulfurous acid producing a calcium bisulfite, $Ca(HSO_3)_2$, liquor that is used at a temperature of 140°C and a pH of 1-2. However, there are two disadvantages of using a calcium bisulfite liquor: 1) it has a reduced lignin solubility and precipitates as calcium sulfite, $CaSO_3$, at higher pHs, and 2) there is no recovery process for the spent liquor chemicals.

When magnesium oxide is allowed to react with sulfurous acid, magnesium bisulfite, $Mg(HSO_3)_2$, is the cooking liquor, which offers several advantages. The pulping can be successfully carried out at a lower temperature (130°C in four hours) and at a pH of 3-4. A chemical recovery cycle can also be used to recover the spent liquor chemicals. In this process, the spent liquor is burned at 1,350°C and both the sulfur dioxide gas, SO_2, and the magnesium oxide, MgO, are recovered.

Semichemical pulping uses sodium sulfite, Na_2SO_3, and sodium carbonate, Na_2CO_3, at pH 7-10 as the cooking liquor. The resulting pulp is used to make corrugating medium, which is the inner, fluted layer of paper found in corrugated boxes; it accounts for only six percent of U.S. paper production.

Mechanical and Chemi-mechanical Pulping Processes

Mechanical pulp is produced by grinding the wood against a rough surface or metal plate with metal bars. The pulp yield is 90-95 percent (dry pulp mass based on dry wood mass). Lignin is retained in the pulp, which lowers brightness and strength. Furthermore, the paper darkens with age and turns yellow in strong sunlight. It is used in temporary papers such as newsprint and catalogs. If a mild chemical pretreatment (with Na_2SO_3 and Na_2CO_3) is used to enhance mechanical pulping, the process is termed chemi-mechanical pulping.

Type of Process	Example	Importance	Types of Products
Chemical	kraft	80%	brown and white papers
Semi-chemical	acid sulfite	4%	white printing paper
	neutral sulfite	6%	corrugating medium
Chemi-mechanical	CTMP	< 1%	newsprint, catalogs
Mechanical pulp	TMP	10%	newsprint, catalogs
	ground-wood	< 1%	newsprint, catalogs

Table 19-1: Pulping processes used in the United States.

Until the 1970s, most mechanical pulp was made by grinding the sides of wood logs against large grinding wheels with aluminum oxide or silicon carbide surfaces in the groundwood process. This treatment resulted in cutting many of the fibers into small fragments known as **fines**. Mechanical pulp is still often called groundwood.

Kraft Pulping Process

Kraft chemical pulps are the strongest of all pulps. The **kraft pulping process** – sometimes called the sulfate process – cooks the wood chips in an aqueous sodium hydroxide, NaOH, and sodium sulfide, Na_2S, liquor at 170-180°C for 1-2 hours in a pressurized (120 psi) reactor. Once the delignification process has reached the desired point, the mixture is abruptly discharged into a blow tank. The sudden pressure drop results in the partially delignified chips bursting into a pulpy mixture with an oatmeal consistency. The pulp yield is only about 50 percent, but the lignin content of the cooked fiber is about three percent compared to 30 percent in the wood source. The pulp, in its brown form, is used to make linerboard, which is used for the top and bottom layers of corrugated boxes, or for brown bags.

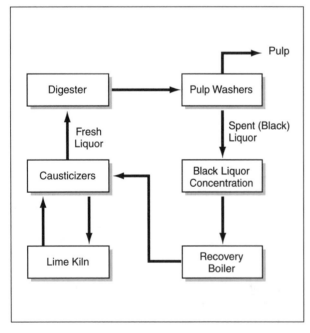

Figure 19-3 The kraft chemical pulping process.

To produce white printing paper, the brown pulp is bleached. The bleaching process uses powerful oxidants to remove the remaining lignin. Paper made from bleached kraft pulp remains white, since the lignin, which tends to darken with age, is no longer present. The spent liquor from the pulp and bleach washing processes are collected and concentrated into a strong **black liquor** solution.

Kraft Liquor Recovery Cycle – Kraft mills always have a chemical recovery process for the spent black liquor. Figure 19-4 is a diagram of the process, in which the weak black liquor is first concentrated in multiple-effect evaporators from about 15 percent to 65 percent solids. This concentrate is then fed into the recovery boiler where it is burned; the inorganic materials are recovered as a solid called **smelt**. The smelt is then dissolved in water to make **green liquor**. This green liquor, which contains sodium carbonate, Na_2CO_3, and sodium sulfide, Na_2S, is treated with calcium hydroxide, $Ca(OH)_2$, in a process called **causticizing** to regenerate **white liquor** to be used in the next pulping cycle. The calcium carbonate, $CaCO_3$, that is generated in the causticizing step is washed and heated in a lime kiln at 1,200°C to generate calcium oxide, CaO, which is **slaked** to produce $Ca(OH)_2$ when dissolved in water. The organic materials in the concentrated strong black liquor – generated by the multiple effect evaporators – are burned to produce carbon dioxide and water. The heat produced is trapped by water-filled tubes in the walls of the recovery boiler and used to produce process steam. At many mills, the steam is also used in a cogeneration plant to produce electricity.

The recovery boilers produce the usual combustion byproducts: H_2O, CO_2, CO, and nitrogen oxides, NO_x. The amount of oxygen entering the combustion process is critical since there is a tradeoff between the formation of NO_x when relatively high levels of oxygen are used and CO when relatively low amounts of oxygen are used. Sulfur from the various organic compounds is emitted as sulfur dioxide, SO_2, and reduced sulfur compounds such as methyl sulfide, CH_3-S-CH_3, or methyl disulfide, $CH_3-S-S-CH_3$. Unfortunately, the human nose can detect the disagreeable rotten-egg odors of these reduced sulfur compounds at the parts per billion level. The kraft liquor recovery cycle also produces large amounts of suspended solids when the green liquor is regenerated by dissolving the smelt into water.

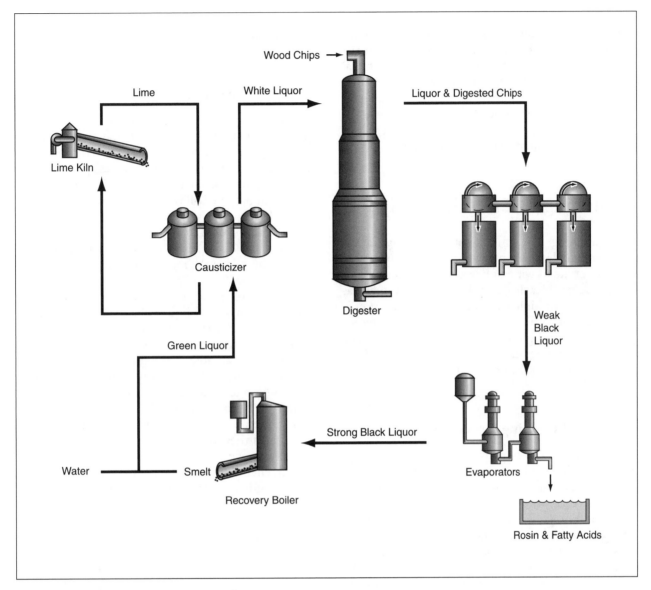

Figure 19-4: The kraft liquor recovery cycle.

Thermomechanical Pulping

Thermomechanical pulp (TMP) is the dominant mechanical pulping process and accounts for an additional 10 percent of U.S. pulp production. In this process, wood chips are first introduced into a pressurized **disk refiner** at elevated temperatures to free the fibers. The refiner consists of two large round metal plates – up to two meters in diameter – with a series of bars on their surface; the plates are separated by less than a millimeter and turn in opposite directions. Steam at 130°C softens the lignin and allows fairly good fiber separation while generating much fewer fines than in the groundwood process. A second **refining** step, using a disk refiner, is carried out at room temperature to further alter the surface of the fibers. This process uses large amounts of steam and electricity – about 2,500 kWh per ton. Hydrogen peroxide, H_2O_2, and/or sodium dithionite, $Na_2S_2O_4$, are used to brighten the pulp by masking some of the color of the lignin; removal of lignin from mechanical pulps would require vast amounts of bleaching agents. The United States imports large amount of newsprint made with TMP from Canada.

The chemi-thermomechanical process (CTMP) uses 1-5 percent sodium sulfite, Na_2SO_3, as a chemical pretreatment for the wood chips. The pretreatment produces an effluent high in biochemical oxygen demand (BOD). This process is not used appreciably in the United States, but it does have some use in Canada. A new CTMP process uses alkaline hydrogen peroxide in the pretreatment and produces no effluent that leaves the mill.

Bleaching of Chemical Pulps

The bleaching of chemical pulps has a significant environmental impact, primarily because it produces aqueous effluents with high levels of color and BOD. Brown pulp is bleached in at least four to as many as nine stages. Multiple stages are actually more efficient because different chemicals attack different portions of the lignin, and overall chemical usage actually decreases. Many changes in this area of the paper industry can be expected in future years.

Chlorine Bleaching Processes

Until 1990, the first step in bleaching chemical pulps at most mills involved the use of elemental chlorine, Cl_2. This was followed by an alkaline extraction where about 75 percent of the lignin in the pulp was removed. The resulting effluent was highly colored, and secondary biological treatment processes removed less than half of its color. The use of chlorine also caused the production of large amounts of chlorinated organic molecules, including a small amount of dioxin. The effluent from the bleaching and alkaline extraction stages cannot be sent to the recovery boiler since the chlorine would build up in the system and cause widespread corrosion problems.

Most mills have replaced at least half of the elemental chlorine with chlorine dioxide, ClO_2, while others have eliminated the use of elemental chlorine completely. In 1994, the use of chlorine dioxide surpassed that of elemental chlorine in the paper industry. Chlorine dioxide oxidizes the lignin, but does not add chlorine to organic material, thereby virtually eliminating dioxin formation. Chlorine dioxide is also very specific for lignin removal and is often used in later stages of bleaching when only small amounts of lignin remain. There is increasing

pressure today for all pulp mills to become elemental chlorine free (ECF) in their bleaching practices.

Oxygen Bleaching Processes

Many mills are now using oxygen, O_2, and NaOH at pressures of 80-100 psi for delignification in the first bleaching stage. While oxygen is not overly specific for lignin removal and therefore causes appreciable damage to the remaining fiber, it can be used to remove about one-half of the lignin from the pulp. Two advantages of oxygen delignification are: 1) the effluent can be used to wash the pulp after exiting the digester, and 2) the lignin removed from the pulp can then be burned in the recovery boiler, which is an environmentally sound practice.

Sodium hypochlorite, NaClO – the ingredient found in household liquid bleach – is also used to bleach pulp; however, this chemical produces small amounts of chloroform, $CHCl_3$, which may be considered unacceptable in the future. Although technically the hypochlorite ion is an oxygen-based bleaching agent, there is pressure for mills to become totally chlorine free (TCF), which means they would use no chlorine or chlorine-containing compounds in their bleaching practices.

Hydrogen peroxide is also used at high pH to bleach chemical pulp. Ozone, O_3, has recently been used commercially at several mills around the world. The use of these materials may increase as pressure increases for mills to abandon the use of chlorinated compounds. However, these materials are not as efficient as chlorine dioxide for lignin removal. Therefore, finding a balance between efficiency and environmental impact is the challenge in the years to come.

Fiber from Recycled Paper

Between 1980 and 1995 the amount of pulp from recycled (recovered) sources grew from about 25 to 40 percent. Recycling paper not only decreases the need for harvesting wood, but it also keeps a valuable resource from entering rapidly filling landfills. In 1990, about 30 to 40 percent of all material in municipal landfills was paper. Most of the increased production of paper in the 1990s was the result of the use of recycled paper as a resource. This diminished the use of trees as a resource, while at the same time extending landfill capacities. Prices for fiber from recycled paper vary according to availability in different areas. In 1995, for example, repulpers in some locations paid as low as $37 per ton while others paid as high as $202 per ton for recycled newsprint.

When paper recycling is mentioned many of us think only of newsprint. Old corrugated containers (OCC), however, are recycled at the highest rate – about 50 percent – since they are composed of the heaviest type of paper produced and can be collected from relatively few points, such as grocery and retail stores. Post consumer waste paper is not necessarily the best source for recycling since 1) it tends to be made up of mixed types of paper that must first be sorted, 2) it is often contaminated with non-paper materials, and 3) its collection is typically labor-intensive.

Recovering useful fiber from recycled paper involves many tools, each of which is designed to separate the fiber from various contaminants that may interfere with the papermaking process. At each stage there are tradeoffs between the reject rate and the efficiency of the contaminant removal process.

The first step uses a pulper to disperse the paper into water creating a fiber slurry. In this process the paper is changed back into individual fibers that can be further processed.

Two types of screens are used to recover fiber from the slurry and separate it from contaminants. Coarse screens are used early in the process to remove the larger debris that might otherwise damage the processing equipment; they are followed by fine screens (mesh size of 0.20 mm or smaller) to remove the smaller contaminants. Recent advances in fine screens have improved the overall quality of fiber derived from recovered paper.

Two types of centrifugal or cyclone cleaners are used on the slurry. First, forward centrifugal cleaners remove materials that are heavier than fiber – such as metal, dirt, and sand – by letting them fall out the bottom of the cyclone. Next, throughflow centrifugal cleaners are used to remove lightweight materials – such as plastics and polymers – by collecting and rejecting the suspension at the center of the vortex.

Most paper is not **deinked**, but this trend is changing as more and more newsprint and office papers are recycled. Three types of deinking processes are commercially used. Air flotation makes use of air bubbles to bring relatively large ink particles (5-100 μm) to the surface where they are removed by skimmers. Pulp washing is a process quite similar to washing clothes where ink is dispersed into small particles (~ 1 μm) that are removed with the wash water. Ink agglomeration is a process where ink is made to form large particles (5 mm or larger) that are removed by screening systems.

When paper is recycled, shrinkage occurs. OCC has a shrinkage of only about five percent, while white paper might shrink 30-50 percent. The material lost in shrinkage – ink, fillers, and other contaminants – usually ends up as a solid waste with 80 percent moisture that must be sent – at high cost – to landfills. Research is currently being focused on trying to find uses for this material.

Pulp from recycled paper may be bleached. Recovered mechanical pulp is bleached in a manner similar to that used for virgin mechanical pulp. White chemical pulps are sometimes bleached using one percent sodium hypochlorite; this bleaching helps mask residual dyes that may be in the pulp.

Checking Your Understanding

1. What is the function of fillers that are added to printing papers?

2. What component of wood is removed or altered in the pulping process?

3. What are the advantages and disadvantages of the thermomechanical and kraft pulping processes?

4. How does a mill achieve decreased amounts of aqueous effluents by using oxygen delignification instead of elemental chlorine as the first bleaching stage?

5. What does each stage in fiber recovery from recycled fiber accomplish?

19-3 The Papermaking Process

As previously noted, there are more than 400 types of paper made for a wide array of uses. The papermaking industry typically divides these into broad categories based on weight and on the types of fibers used. Some of the lighter grades are used for newsprint, stationery, tissue, bags, towels, and napkins, whereas the heavier grades – the so-called paperboard grades – include such items as linerboard, corrugating media, tubes, drums, milk cartons, and recycled board used in shoe and cereal boxes. Anything that weighs up to 200 grams per square meter is designated as paper, while any product exceeding this weight is classified as a paperboard. Printing paper, for example, is about 20 pounds per ream (3,000 square feet), whereas a lightweight linerboard is 78 pounds per 3,000 square feet, or nearly four times heavier.

Paper is made from a dilute slurry of pulp fibers. Before papermaking, chemical pulps are refined between two rotating metal plates with raised bars on their surfaces in a process that is similar, but milder, than the second step of thermomechanical pulping. The refining process exposes the cellulose strands contained within the fiber – called microfibrils – to optimize the papermaking properties. The exposed microfibrils increase the number of hydrogen bonds between them, which in turn increases the strength of the final sheet of paper.

Paper additives are used to control the papermaking process and to improve the quality of the final sheet. Additives include retention aids; strength agents; biocides to prevent slime growth on the paper machine; and sizing agents, so the paper does not rapidly absorb water and ink like paper towels do. Printing papers often contain **paper fillers** such as clay, calcium carbonate, and titanium dioxide. All fillers increase the opacity (hiding power) of the paper so ink on the backside does not show through; they often improve the brightness as well. As previously noted, clay and calcium carbonate have the additional benefit of being less expensive than the pulp they replace, whereas the more expensive titanium dioxide is used

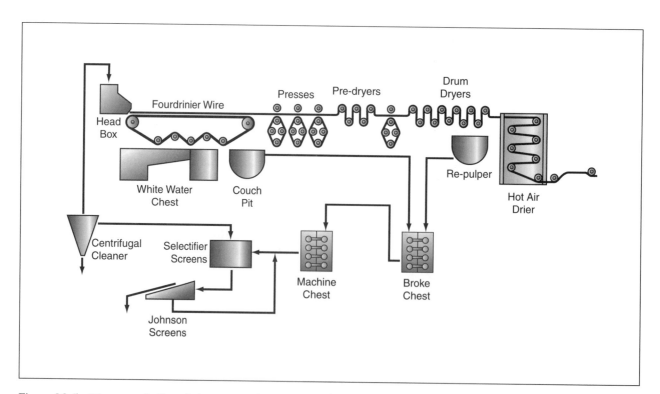

Figure 19-5: Diagram of a Fourdinier papermaking machine.

to make very thin papers with high opacity, such as those used in many editions of the Bible.

Paper Formation

To produce a mat or web of pulp with uniform fiber distribution – called good formation – requires a dilute slurry (about 0.5 percent dry fibers and other materials). The basic functions of the paper machine are to form the mat of fibers, vacuum the water from the mat, press additional water from the sheet, and finally use steam-heated rollers to remove the remaining water.

To accomplish these steps, the dilute slurry of fibers is applied from the headbox (A) of the paper-making machine onto a continuously moving screen (B) – also known as the forming fabric – to form the mat of fibers. Light grades of paper are usually made by applying the slurry between two screens, whereas heavier papers are applied to a single flat wire screen in the Fourdinier machine. (The heaviest grades of paper, such as chipboard used in cereal boxes and shoeboxes, are made in a series of layers with cylinder machines.) Gravity and vacuum boxes against the screen remove and recycle much of the water, known as **white water**, by using increasing levels of suction.

The mat of fibers, now at about 20 percent solids, goes to the press section (C), which usually contains three presses, where additional water is squeezed from the mat. The mat leaves the press section at about 40 to 50 percent solids and goes to the drier section (D), where a series of steam-filled metal rolls (usually six feet in diameter) supplies heat to evaporate much of the remaining water. Light grades of paper, such as tissue, use one large dryer since they cannot withstand numerous transfers between dryer rolls. Even after the dryer section the paper reaches only 90-93 percent solids, retaining 7-10 percent bound water.

The paper produced by the paper machine is wound onto large rolls, which must later be rewound to smaller rolls. These smaller rolls are then either shipped to customers, converted into products at the mill, or subjected to additional processes. To smooth their surfaces, for example, some printing papers are coated with pigments like clay or calcium carbonate to fill the voids between fibers.

Checking Your Understanding

1. What is the purpose of refining prior to the papermaking process?

2. What improvements are provided by the use of paper fillers?

3. What is white water?

4. How much bound water is retained by finished paper?

19-4 Paper Mill Regulations

Water and Air Regulations

The pulp and paper making processes require the use of a large amount of water. In the papermaking processes most of the white water is collected from the pit of the paper machine and reused. Even though any excess white water produced is filtered prior to discharge from the mill, it is still a large source of suspended solids. These and the other water discharges from the paper mill fall under the Clean Water Act (CWA) and depend upon the type of material being produced (there are 26 classifications). For example, the pH of liquids discharged from the mill must be between 6 and 9 pH, which the mills find easy to control. Mills making brown paper such as linerboard or brown bags are allowed to discharge up to 1.8 kg of BOD and 6.0 kg of total suspended solids (TSS) per ton of air dry pulp, whereas semichemical mills are only allowed to discharge 4.4 kg BOD and 5.5 kg TSS per ton of production. Many states also regulate the color in liquid effluents.

New Source Performance Standards for particulate matter and total reduced sulfur emissions for the kraft process are regulated under §111 of the 1978 Clean Air Act (CAA). Total reduced sulfur (TRS) must be below five ppm (corrected to 10 percent oxygen) for most sources, but if the overall emissions are below five g/ton air-dried pulp, then point sources may be higher than five ppm. Individual states may regulate industries by allowing rates of discharges lower than those allowed by EPA. For example, air emission standards, such as those for NO_x and SO_x, are usually covered by state regulations. State regulations are enforceable at the federal level if they are established as State Implementation Plans (SIPs).

EPA is in the process of combining air and liquid effluent requirements for the paper industry. This combination of regulations has been nicknamed the "cluster rules." The combined regulations, however, have not been finalized at this time.

Opportunities for Waste Reduction

The chemical pulping and papermaking industry is capital intensive. The cost of a new 1,500 ton per day kraft mill – which is the smallest size considered profitable – is at least $2 billion. The high capital costs make it difficult for the industry to upgrade existing equipment. Therefore, pollution prevention in the industry continues to rely on source reduction and process modification. The continued use of chlorine dioxide, for example, will decrease the amount of chlorinated organic materials and dioxin generated. The increased use of oxygen delignification will allow more of the lignin-containing wastes to be burned in the recovery boiler, also allowing recovery of its heat content.

Some new technologies are also being explored. Ozone delignification offers many of the same advantages as oxygen delignification; however, at this time it has only been implemented in a few U.S. mills. Enzymes are also being introduced to assist in the bleaching process. Even farther on the horizon are new methods of processing black liquor, for example gasification. The use of organic solvents to pulp wood is now being practiced at one small mill. If they prove to be economically viable, such new technologies have the potential to accomplish a dramatic decrease in pollution due to the pulp and paper industry.

Checking Your Understanding

1. What range of pH is allowable in the discharge from a paper mill?

2. What section of the 1978 CAA regulates the particulate matter and total reduced sulfur emissions from the kraft process?

3. What is meant by the "cluster rules"?

Summary

Paper is made from fibers isolated from wood or other cellulose-containing material by the pulping processes. Wood is pulped by chemical means – usually the kraft process – or by mechanical means. Chemicals used to produce chemical pulp are recovered within the mill. Pulp from recycled paper is made by dispersing the paper in water using agitation followed by several steps designed to separate the useful fiber from ink, fillers, and contaminants. Paper is made from a dilute slurry of fiber applied to a continuously moving screen. The wet fiber mat is pressed and then dried in steam-filled dryers. The paper is then converted into useful products.

Sources of pollutants include CO, NO_x, SO_2, and reduced sulfur compounds from the kraft chemical recovery process, BOD from most processes, and chlorinated organic compounds from bleaching chemical pulp. Solid wastes are produced from chemical pulping, from the kraft liquor recovery cycle, and from recovery of pulp from recycled paper. These are usually sent to a landfill, although pulp wastes may be applied to farmland since they contain appreciable amounts of calcium. Total suspended solids are produced from the paper machine and liquor recovery cycle.

Due to the large amounts of wastewater and air emissions generated, the primary regulations affecting the pulp and papermaking industry are the CWA and the CAA. As a result, EPA is in the process of implementing combined air and liquid effluent requirements for the industry. This combination is known as the "cluster rules." High capital costs make it difficult for the industry to upgrade existing equipment; as a result, the industry is relying on source reduction and process modification to meet pollution prevention goals.

Critical Thinking Questions

1. What are three benefits of the kraft liquor recovery cycle and how does each help the environment?

2. Some people want to see the use of chlorine dioxide eliminated only because it is a chlorine-containing compound. Research has shown that the levels of chlorinated organic compounds produced from bleaching with chlorine dioxide are much lower than when elemental chlorine is used. In other words, ClO_2 is much less environmentally harmful than Cl_2 and the two should not be lumped in the same category. Chlorine dioxide is very specific for lignin removal. It has been suggested that the use of less specific chemicals may, on the other hand, have other disadvantages such as: higher costs; additional energy usage, which in turn contributes to increases in carbon dioxide in the atmosphere; and the use of more trees for the same amount of paper produced. After researching these two arguments, select a position and defend it based on your research.

3. A closed carbonated beverage with contents at 35°C is much more apt to bubble out the top when opened than one at 5°C. Why is this true and what does it have to do with BOD emissions from a paper mill? Hint: Allowable BOD emissions are sometimes lower in summer months than in winter months.

4. If a mill discharges five parts per quadrillion of dioxin in 20 million gallons of effluent every day, what is the mass of dioxin discharge in one year?

20

The Graphics and Printing Industries

by Lisa Cummins

Chapter Objectives

Upon completing this chapter, the student will be able to:

1. **Describe** the five methods that currently dominate the printing industry.

2. **Identify** wastes generated during each of the three steps of the printing process.

3. **Explain** how the Clean Air Act, Resource Conservation and Recovery Act, and the Clean Water Act impact printers.

4. **Identify** ways printers can minimize or eliminate air emissions, hazardous waste generation, and wastewater discharges.

5. **Explain** how printers are affected by Department of Transportation regulations.

6. **Explain** how printers are affected by Occupational Safety and Health Administration regulations.

20-1 Introduction

One of the oldest known printed illustrations was found on a Buddhist scroll and is believed to have been printed around 750 BCE. This print was made using a flat wooden block on which the non-image areas had been carved out, leaving a raised image area to be printed. Of the five modern printing methods, two still rely on the principle of removing materials from the substrate in the nonprinting areas: letterpress and flexography. Three additional printing methods are also in use today: gravure, lithography, and screen printing. Each of these printing methods is subdivided in three steps: prepress, press, and post press operations, each of which produces different kinds of waste. This chapter will discuss the methods used to produce the printing surface, the image transfer methods, the wastes they generate, and the federal regulations that impact the printing and graphics industries.

20-2 Printing Industry Overview

Printing Methods

The goal of all printing is a simple one: to make a copy of an image by pressing paper or another substrate against an inked printing surface. All printing occurs on one of two standard types of printing press: the web presses and the sheet-fed presses.

—Web Press – The image prints on a **web** – which is a continuous roll of paper or other substrate – at speeds ranging from 1,000 to 1,600 feet per minute.

—Sheet-Fed Press – The image prints on separate sheets of paper or other substrate and can produce up to three impressions per second.

> Printing is a process that transfers an inked image to paper or another substrate.

On both types of presses, a printing plate transfers ink to the substrate. The printing plate – also called the image carrier – may be either a flat plate or a cylinder. Rollers apply ink from an ink fountain to the printing plate, forming the pattern of the image. Currently, two types of ink are used throughout the printing industry: the heat set ink and the non heat set ink.

—**Heat Set Ink** – This type of ink requires heat to be cured (dried) and can contain up to 45 percent **VOCs**, which are organic solvents that have relatively high vapor pressures and therefore evaporate easily.

—**Non-heat Set Ink** – Ink of this type cures by absorption of the ink's carrier solvent – either water or volatile organic solvent – into the substrate. Volatile organic solvent-borne inks contain less than 35 percent VOCs.

As shown in Table 20-1, there are currently five printing methods that dominate the industry.

Method	Application	
1. Letterpress	magazines newspapers books	stationery advertising
2. Flexography	packaging newspapers	magazines directories
3. Gravure	packaging newspapers	greeting cards art books
4. Lithography	magazines newspapers books	stationery advertising containers
5. Screen Printing	signs electronics greeting cards wallpaper	ceramics decals banners textiles

Table 20-1: The five printing methods dominating the industry.

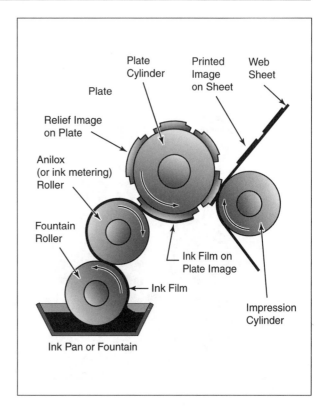

Figure 20-2: Flexographic printing.

Letterpress

Letterpress is the oldest printing technique. It is a relief printing process, which means the **image area** is raised above the **non-image area** on the printing plate. The image area is inked by rollers and pressed against the substrate, directly transferring ink from the plate to the substrate.

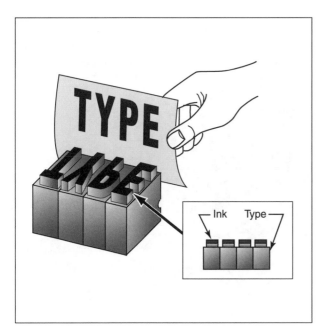

Figure 20-1: Letterpress printing.

Flexography

Flexography is also a relief printing process, in which the image area is raised above the non-image area on the printing plate. The image area is inked by an **anilox roller**, which is a screened roller in which cells of uniform size and depth are engraved. The size and depth of these cells control ink film thickness during printing. Once again, the inked image is transferred directly to the substrate from the plate.

Gravure

Gravure printing uses a plate on which the image area is set below the plate's surface. The entire plate is flooded with ink and the excess is wiped away by a metal blade. The ink-filled areas transfer the image directly to the substrate.

Lithography

Lithography is a planographic means of printing, meaning that both the image and non-image areas

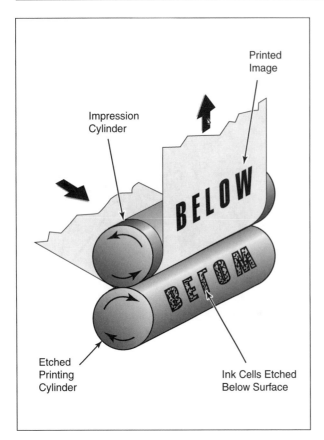

Figure 20-3: Gravure prinnting.

are on the same plane on the printing plate. Lithography is based on the immiscibility of water and oil. The plate is chemically treated, making the image area ink-accepting and water repellent, while the non-image area is made water-accepting and ink-repellent. During the printing process, the plate is first dampened with **fountain solution**, which keeps non-image areas clean and ink-free. Most fountain solutions contain a mixture of isopropyl alcohol with a high water concentration.

After dampening, the water forms beads on the water-repellent image areas of the plate while forming a continuous film on non-image areas. The ink roller easily pushes away the beaded water, replacing it with an oil-based ink. Lithography is referred to as an offset process because the inked image is first transferred from the plate to a rubber blanket and then from the rubber blanket to the substrate.

> Lithographic printing is based on the immiscibility of water (fountain solution) and oil (ink).

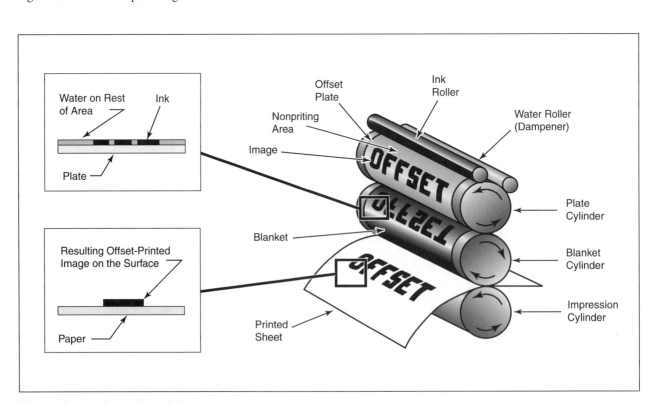

Figure 20-4: Lithographic printing.

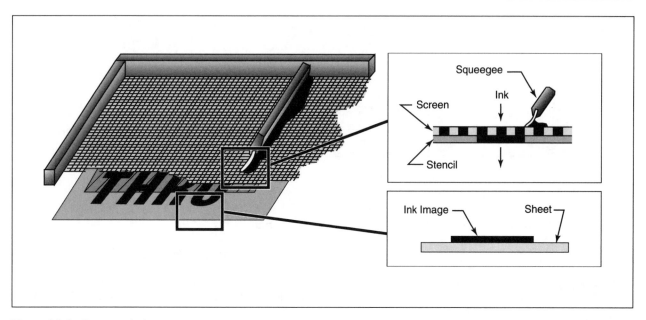

Figure 20-5: Screen printing.

Screen Printing

During **screen printing**, ink is forced through the image area of a screen. The screen is composed of a mesh made from either stainless steel or finely woven fabric stretched over a supporting frame. The image area, or stencil, allows ink to pass through the mesh, while it blocks passage in the non-image areas. By drawing a rubber squeegee across the stencil, ink is mechanically forced through the unblocked mesh onto the substrate below.

Checking Your Understanding

1. Describe the five methods that currently dominate the printing industry.

2. In what way are the letterpress and flexography printing methods similar?

3. Why is lithography called an offset printing process?

20-3 Waste Generation Steps

Wastes are generated during three steps in all five printing methods.

In each of the five printing methods described earlier, wastes are generated during the three steps of prepress, press, and post press operations.

Prepress

Image Process

The image is photographed to produce a transparency. This transparency is used to prepare the printing plate. The wastes generated by this process are the photochemical solutions (fixer, developer, and wash) and the film.

Proof

The proof copy enables the printer and client to review the image for correctness prior to platemaking. Off press proofing by photography is the most common technique used, especially by lithographic printers. The wastes from this step include photographic film and paper.

Plate Process

Plates are usually prepared and processed by photochemical means, involving coatings whose physical properties change after exposure to light. Typically, the exposed areas become insoluble in developer solution, while non-exposed areas become soluble and wash away in the developer solution. The result is an image or a stencil for an image. The wastes generated include photochemical and platemaking solutions.

Makeready

All necessary adjustments are made to the press just prior to printing. The wastes from this process include paper, air emissions from the ink, and alcohol-containing fountain solution.

Press

Actual Printing Operations

The wastes generated by the printing operation include empty ink containers, unused ink from the

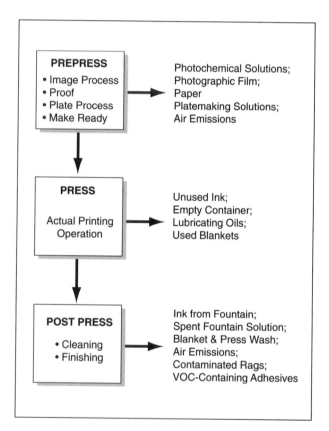

Figure 20-6: The three waste-generating steps in all five printing methods.

fountain, lubricating oils, used blankets, spoiled prints, air emissions from the inks, and solvents.

Post Press

Cleaning

The press requires cleaning between jobs. The wastes generated by cleaning are the ink from the fountain, the spent fountain solution, the solvent-containing blanket and press washes, the air emissions from the solvent-containing blanket and press washes, and the contaminated rags.

Finishing

Finishing involves cutting, folding, collating, binding, and otherwise preparing printed materials for distribution and use. Waste generated by these processes may include paper trimmings and VOC-containing adhesives from the binding operations.

Checking Your Understanding

1. What are the five methods that currently dominate the printing industry?

2. Identify at least two types of wastes generated during the prepress, press, and post press printing steps.

20-4 Federal Regulations

Due to increasingly strict environmental regulations, the printing industry has, in recent years, sought ways to use fewer raw materials and to recycle or reuse materials as much as possible.

Printers are regulated under three provisions of the Clean Air Act.

All five printing methods use materials that may adversely impact land, air, and water quality. Due to increasingly strict environmental regulations, the printing industry has, in recent years, sought ways to use fewer raw materials and to recycle or reuse materials as much as possible. Printers are modifying operations to produce less waste by reducing ink and solvent use, by using low-alcohol or alcohol-free fountain solutions, and by switching to new, lower emission inks and solvents. Environmental regulations that impact printers – discussed below – are aimed at protecting air, land, and surface waters from pollution.

Air Regulations

The biggest environmental issue facing most printers is controlling air emissions. Therefore, most waste minimization efforts in the printing industry are aimed, either directly or indirectly, at reducing use of materials that generate emissions. Several provisions of the Clean Air Act have direct applications to the printing industry.

Title I National Ambient Air Quality Standards (NAAQS)

Of the six established NAAQS, printers are primarily affected by ozone standards. Although printers are not major sources of ozone, they are sources of

The biggest environmental issue facing most printers is controlling air emissions.

VOC emissions, which interact with other pollutants and sunlight to produce ground level ozone (smog).

Part 112 Emissions of Hazardous Substances

Many chemicals used by printers are listed as Hazardous Air Pollutants (HAPs). Chemicals used in the printing industry that are listed as HAPs in the Clean Air Act Amendments are given in Table 20-2.

Benzene	Isophorone
Cadmium Compounds	Lead Compounds
Carbon Tetrachloride	Methanol
Chromium Compounds	Methyl Ethyl Ketone
Cobalt Compounds	Methylene Chloride
Cumene	Perchloroethylene
Dibutylphthalate	Polycyclic Organic Matter
Diethanolamine	Propylene Oxide
Ethyl Benzene	Toluene
Ethylene Glycol	2,4-Toluene Diisocyanate
Formaldehyde	1,1,2-Trichloroethane
Glycol Ethers	Trichloroethylene
Hexane	Vinyl Chloride
Hydrochloric Acid	Xylenes

Table 20-2: HAPs chemicals used in the printing industry.

> Printers can minimize and eliminate air emissions by process changes and by product substitution.

> If an ink is RCRA hazardous, it is due either to ignitability or to toxic metals.

> Printers can minimize and eliminate hazardous waste by better inventory control and by recycling used ink into reusable ink.

Title VI Ozone-depleting Substances

The **stratospheric ozone layer** is important to us because it protects the earth from much of the harmful ultraviolet light emitted by the sun. Carbon tetrachloride and 1,1,1-trichloroethane – cleaning solvents historically used by printers – are regulated under Title VI of the Clean Air Act because they are ozone-depleting substances (ODS).

To reduce harmful air emissions, printers must first assess what they are releasing into the air. The primary releases are VOCs from alcohol-containing fountain solutions, inks formulated with volatile solvents, and press-cleaning solvents. In addition, some printing chemicals contain HAPs and/or ozone-depleting substances.

To minimize and/or eliminate these emissions, printers can use low-alcohol fountain solutions; replace alcohol in fountain solutions with low-VOC substitutes (such as soaps and detergents); chill alcohol-containing fountain solutions, making them less volatile; substitute less hazardous materials for products containing HAPs or ODS; and use **automatic blanket washes** that can decrease overall solvent use and emissions

VOC emissions from inks can be lessened by switching to lower-VOC formulations, such as water-based inks or inks made with soy or vegetable oils rather than petroleum. Also, in certain situations, **UV-curable inks** – which are inks that cure upon exposure to ultraviolet light – may be used. UV-curable inks have the additional benefit of being able to remain in the fountains and on the press for much longer periods of time.

Hazardous Waste Regulations

Most solid waste generated by the printing industry is nonhazardous. By volume, paper is the largest waste

> Most solid waste generated by printing is nonhazardous

stream generated by printers. Much of it is recycled, and the rest is either incinerated or disposed of as trash. Sources of paper waste include rejected print runs, paper wrappers, start and end run scraps, web roll ends, and cardboard web cores.

Resource Conservation and Recovery Act (RCRA)

By RCRA definition, most waste printing inks are nonhazardous. If an ink is RCRA hazardous, it is either due to its ignitability (flashpoint <140°F) or its levels of toxic metals. Whether hazardous or nonhazardous, many printers drum waste inks and ship them to companies to be used for fuel blending. As an alternative, printers are starting to recycle or reformulate their ink wastes. With the use of new computer software, leftover inks can be reblended into **Pantone** (industry standard) colors or recycled into a general purpose black ink. The quantity of waste generated can be minimized by better estimating the amount needed for a job and controlling the amount of supplies ordered.

Printers are required to inventory their wastes and determine if their wastes are a RCRA hazard. They must also determine their hazardous waste generator status, according to the criteria listed in Table 20-3.

In addition to determining their generator status, all hazardous waste generators must:

1. Use structurally sound containers for their hazardous waste;

2. Label hazardous waste containers according to EPA requirements;

3. Comply with the storage and accumulation restrictions according to their generator status;

Hazardous Waste Generator Status	Hazardous Waste Generation Limits	Hazardous Waste Accumulation Limits	Hazardous Waste Storage Time Limits
Conditionally Exempt Small Quantity Generator (CESQG)	<220 lbs/month	<2,200 lbs	No Limit
Small Quantity Generator (SQG)	220-2,200 lbs/month	13,200 lbs	<180 days*
Large Quantity Generator (LQG)	>2,200 lbs/month	No Limit	<90 days
* or <270 days if TSD facility is greater than 200 miles away			

Table 20-3: Printer hazardous waste generator status.

4. Apply for the appropriate identification numbers as needed;

5. Comply with applicable manifest and land disposal restrictions;

6. Comply with any and all reporting requirements.

Water Regulations

Wastewater discharges from printers typically consist of spent photochemicals and platemaking solutions. Photochemicals have traditionally been discharged into the sanitary sewer under a permit from the local **publicly owned treatment works (POTW)**. However, due to the high biological oxygen demand (BOD) of these chemicals, many local POTW authorities require treatment prior to discharge.

Photochemical wastes typically contain significant amounts of silver, which is strictly regulated

Printing wastewater discharges typically consist of spent photochemicals and platemaking solutions.

Printers can minimize and eliminate wastewater discharges by product substitution and by recycling spent solutions.

under the Clean Water Act, due to its toxic effects on aquatic life. Therefore, printers may opt for silver recovery prior to discharge in order to meet POTW requirements. The benefits of silver recovery are both environmental and economical, since the reclaimed silver can be sold as a precious metal.

Platemaking solutions typically consist of acids or caustics, along with heavy metals. These solutions, therefore, must either undergo wastewater treatment or be placed in drums for later disposal. Any solutions containing mercury or other hazardous chemicals may require special handling.

Many printers now opt for presensitized plates to avoid generation of these wastes. Presensitized plates are treated by the manufacturer with light-sensitive polymer coatings, which allow for the creation of image and non-image areas without the use of platemaking solutions.

Agency	Regulation	Provisions
OSHA	Hazard Communication	Employee Right-to-Know training; Material Safety Data Sheets; container labels
DOT	Hazardous Materials Transportation Act	shipping papers; shipping descriptions; container labels; packaging requirements; placarding
EPA	Clean Air Act	National Ambient Air Quality Standards; Hazardous Air Pollutants; Ozone Depleting Substances
EPA	Clean Water Act	waste water discharges
EPA	Resource Conservation and Recovery Act	hazardous waste generation and storage; container labeling; generator status; manifesting requirements

Table 20-4: Summary of federal regulations applicable to the printing industry.

Process/Operation	Materials Used	General Type of Waste Generated
Plate Preparation		
counter etching to remove oxides	phosphoric acid	acid/alkaline wastes
deep etch coating of plates	deep etch bath	acid/alkaline wastes, heavy metal wastes
etch baths	multimetal plate and platecoating	acid/alkaline wastes, heavy metal wastes
applying light-sensitive coating	resins, binders, emulsifiers, photosensitizers, gelatin, photoinitiators	photographic processing wastes
developing plates	developer	photographic processing wastes
applying lacquer	resins, solvents, vinyl lacquer, lacquer developers	solvent wastes
using ink (lithography, letterpress, screen printing, flexography)	pigments, dyes, varnish, drier, extender, modifier, fountain solutions	toxic waste ink with solvents/chromium/lead/barium. ink sludges with lead/barium
making gravure cylinders	acid plating bath	plating wastes
Stencil Preparation for Screen Printing		
lacquer stencil film	solvents, polyester film, vinyl film, dyes	solvent wastes
photographic stencil film	organic acids, gelatin (pigmented), polyester film base	acid/alkaline wastes
photoemulsion	resins, binders, photosensitizers, dyes	photographic processing wastes
blockout (screen filler)	pigmented polymers, solvents, acetates	solvent wastes
Photoprocessing		
developing negatives and prints	developer, cleaning agents, wetting agents, fixers, bleaches	photographic processing wastes
Printing		
using ink (lithography, letterpress, screen printing, flexography)	pigments, dyes, varnish, drier, extender, modifier, fountain solutions, inks, solvents, plates, shellacs	heavy metal wastes (dust and sludge) ink-sludges with chromium or lead ink-toxic wastes with metals or organic constituents solvent wastes
Clean Up		
wash/clean plates, type, die, press blankets and rollers	alcohols, solvents, rags, alkaline cleaner	acid/alkaline wastes ink-toxic wastes with metals or organic constituents solvent wastes

Table 20-5: Typical printing and allied industries operations: materials used and hazardous wastes that may be generated.

PRINTER COMPLIANCE CHECKLIST	YES	NO	N/A
EPA			
Generator has determined whether wastes are RCRA hazardous			
Hazardous waste containers are in sound condition			
All hazardous waste containers are marked per RCRA requirements:			
1. "Hazardous Waste- Federal Law Prohibits Improper Disposal. If Found, Contact The Nearest Police Or Public Safety Authority Or The U.S. Environmental Protection Agency"			
2. Generators name and address			
3. Manifest document number			
Accumulation start date indicated on each hazardous waste storage container			
Hazardous waste generator status has been determined:			
• CESQG <220 lbs (100 kg)/month			
• SQG 220 - 2,200 lbs (100-1000 kg)/month			
• LQG >2,200 lbs (1000 kg)/month			
Accumulation and storage limits based on generator status are followed:			
• CESQG: no more than 1000 kg; no storage time limit			
• SQG: no more than 6000 kg; 180 days (270 days if TSD >200 miles away)			
• LQG: no accumulation limit; 90 days storage limit			
All necessary identification numbers have been obtained			
Hazardous waste shipments are manifested			
Hazardous wastes transporters have all necessary identification numbers			
Hazardous waste is shipped to a permitted TSD for treatment/disposal			
Generator is in compliance with any and all reporting requirements			
Any/all sewer discharges are in accordance with local POTW requirements			
Measures are in place to eliminate/minimize air emissions			
Measures are in place to eliminate/minimize hazardous waste generation			
Measures are in place to eliminate/minimize sewer discharges			

Table 20-6: Printers compliance checklist.

PRINTER COMPLIANCE CHECKLIST	YES	NO	N/A
DOT			
Materials are shipped in DOT-approved containers			
Container type (X,Y,Z) corresponds to Packing Group (I, II, III)			
PG Container Type I X II X or Y III X or Y or Z			
Materials are shipped on proper DOT shipping papers			
Materials are shipped on proper DOT shipping descriptions:			
1. Proper shipping name			
2. Technical name (if needed)			
3. Hazard class number			
4. UN/NA identification number			
5. Packing Group (I, II, or III)			
Containers are labeled per DOT hazard class			
Hazardous materials containers (<110 gallons) offered for shipment are marked per DOT requirements:			
1. Proper shipping name and UN/NA identification number			
2. Technical name (if needed)			
3. Name and address of either shipper (consignor) or receiver (consignee)			
OSHA			
All product-containing containers are labeled			
Current MSDSs are readily available for all products in the workplace			
Employees have received Hazard Communication training			

Table 20-6: Printers compliance checklist (continued).

> Printers are also affected by regulations issued by the Occupational Safety and Health Administration and the Department of Transportation.

To minimize discharges of certain hazardous chemicals, many printers have switched from solvent-based to water-based plate development solutions. Also, equipment has been developed that allows printers to reuse these chemicals, enabling them to purchase less chemicals and further minimize discharges. The trend toward computerized, direct-to-press printing will also decrease the amount of these wastes generated.

Other Regulations Affecting Printers

In addition to environmental regulations, printers are also impacted by transportation and worker health and safety regulations. Table 20-4 provides a summary, by agency, of other federal regulations affecting the printing industry.

Printing involves many materials and processes that are affected by federal regulations. Printers must inventory their materials and processes to determine how they can most effectively make the changes and/or substitutions mandated by increasingly strict environmental regulations. In addition, printers must comply with regulations covering worker health and safety as well as proper transportation of hazardous materials.

As a reminder, Table 20-5 provides a list of the different printing operations and the types of waste generated. Table 20-6 is a compliance checklist that can be used to determine whether a printer is complying with all of the applicable regulations.

Checking Your Understanding

1. List two ways printers can minimize/eliminate air emissions.

2. List two ways printers can minimize/eliminate hazardous waste generation.

3. List two ways printers can minimize/eliminate wastewater discharges.

Summary

The goal of all printing is to make a copy of an image by pressing paper or other substrate against an inked printing surface. All printing occurs by using one of two standard types of printing presses: the web press, which uses a continuous roll of paper, and the sheet-fed press, which uses individual sheets of paper. The inks used in printing are of the heat set or non heat set type. Both of these inks rely on the use of VOCs.

The letterpress is the oldest printing technique. It uses a relief printing process, where the image is raised above the non-image area on the printing plate. Flexography is another relief printing process, but one in which the image area is inked by an anilox roller. Gravure printing uses a plate on which the image areas are below the plate's surface. The entire plate is flooded with ink and the excess is wiped away from the image area with a metal blade. The remaining ink-filled areas transfer the image to the paper. Lithography is a planographic printing process, where the image and non-image areas are on the same plane; however, the plate is chemically treated making the image area ink-accepting and water repellent, while the non-image areas are made water-accepting and ink-repellent. Lithography is referred to as an offset process, because the inked image is first transferred from the plate to a rubber blanket and then from the rubber blanket onto the paper. Finally, screen printing makes use of a screen that has blocked and non-blocked image areas. By drawing a rubber squeegee across the stencil, ink is forced through the unblocked mesh onto the substrate below.

In each of the five printing methods, wastes are generated during the prepress, press, and post press operations. Prepress wastes include photochemical solutions, film, proofs, and platemaking solutions. Press wastes include paper, air emissions from ink and alcohol-containing fountain solutions, empty ink containers, unused ink, lubricating oils, used blankets, spoiled prints, and air emissions from the inks and the solvents. Post press wastes are ink and fountain solutions, solvent-containing blanket and press washes, contaminated rags, and air emissions from organic solvents. Finishing processes involving cutting, folding, collating, and binding to prepare the materials for distribution and use also generate waste paper trimmings and adhesives from the binding operations.

The printing industry is attempting to reduce the kind and amount of waste it generates. Fountain solutions are being replaced with low-alcohol solutions that are chilled to reduce VOC emissions. UV-curable inks and water-based inks made with soy or vegetable oils are being used to replace high VOC inks. Much of the waste paper generated is now being recycled. The colors of printing inks have been standardized, so leftover inks can be re-blended into other colors or recycled into general purpose black ink. Wastewater discharges from printers consist mostly of spent photochemicals and platemaking solutions. These must either be treated or placed in drums for later disposal. The use of presensitized plates avoids generation of most of these wastes.

Critical Thinking Questions

1. Develop a revised block flow diagram of the printing process that incorporates computerized, direct-to-press technology as one of the steps. From a regulatory compliance perspective, what are the benefits printers may achieve by switching to computerized direct-to-press printing?

2. Develop a list of all the compliance issues that may be affected when a printer switches from a petroleum-based to soy-based ink.

21

The Textile Industries

by Thomas D. Shahady

Chapter Objectives

Upon completing this chapter, the student will be able to:

1. **Identify** and describe the fibers and processes used to produce fabrics.

2. **Describe** the major waste streams generated by the textile industries.

3. **Identify** and explain the environmental regulations impacting each of the waste streams produced within the textile industries.

21-1 Introduction

The evidence of human textile making disappears into ancient history. Remains of woolen textiles dating as far back as 6,000 BCE have been found, and it is known that the Egyptians were wrapping mummies in fine linen cloth around 2,500 BCE. The Chinese began cultivating silkworms and developed special looms for making fine silk garments as early as 2,700 BCE.

The modern textile industry manufactures upholstery, carpeting, draperies, linens, towels, and many other products in addition to clothing. As shown in Table 21-1, all textile products are made from some type of natural or synthetic **fiber** that is either used directly or spun into a yarn. In the manufacturing process, the yarn or fiber is either woven, knitted, or interlocked; it is then dyed to add color, and finished to provide a desired appearance or function.

Natural			Synthetic		
Name	**Source**	**Use**	**Name**	**Source**	**Use**
Wool	Sheep Hair	Clothing, Blankets, Rugs	Nylon	Polyamide Fibers	Carpeting, Clothing, Hosiery, Sporting Goods
Silk	Silk Worm	Home Furnishings, Clothing, Dental Floss	Rayon	Cellulose	Apparel, Home Furnishings
Linen	Flax Plant	Clothing, Linens, Household items	Acetate	Cellulose Acetate	Carpeting, Clothing, Home Furnishings
Cotton	Cotton Plant	Clothing, Household Items, Upholstery	Polyester	Polyester Fibers	Apparel, Filling Material
Jute	Jute Plant	Burlap, Sportswear	Acrylic	Acrylonitrile Fibers	Wash & Wear Fabric
Hemp	Hemp Plant	Rope	Polyvinyl	Polyvinylidene Chloride	Cloth, Upholstery, Rope
Camel Hair	Bactrian Camel	Clothing			

Table 21-1: Types of fibers used in textile production.

21-2 Textile Production

Fibers

Many factors must be considered when producing a textile product. What is the desired quality, appearance, durability, and cost? How strong, elastic, or resilient must it be? What type of **dyeing** or finishing procedure will be used? Specific fibers are selected for specific uses based on their characteristics, as shown in Table 21-2. To produce a soft, luxurious garment, for example, silk may be the choice. When multiple characteristics are needed, fibers may need to be combined. Construction of a long-lasting work garment may demand the use of a cotton-synthetic blend.

Fibers used in textile production originate from either natural or synthetic sources. Natural fibers may come from plants or animals, such as the cotton fibers produced by the cotton plant. Once the cotton pod has matured, it is picked and separated from the seeds and surrounding plant material. This discarded material represents a large solid waste problem for the cotton manufacturing industry. Cotton fibers are very valuable in textile production because they are durable, comfortable, and easily dyed and manipulated. Other examples of plant fibers include jute, which is used in the production of burlap products; hemp used for rope production; and flax used in the production of linen fabrics.

Animal fibers originate from several sources. Silk fiber is the secretion of the silkworm moth larvae as they spin their cocoon. The fiber is unwound from the cocoon providing a luxurious, comfortable, and easily dyed product. Wool fiber from sheep or lamb fleece produces textiles that are warm, comfortable, and naturally water-resistant. Its processing will be discussed in more detail later in this chapter. Other animal fibers used in specialty products include camel and goat hair as well as the fur from a wide variety of other animals.

Synthetic fibers are produced from chemical substances such as natural or synthetic polymers as well as fiberglass. To produce a synthetic fiber, a chemically generated liquid is forced through small holes and then drawn out to form a thread. Synthetic fibers are typically resilient, strong, and easy to care for. At the same time, garments made from these fibers may be uncomfortable to wear or have a tendency to cling to the body. To reduce some of these problems, synthetic and natural fibers are often blended to produce a more desirable product. Examples include wool and rayon or cotton and polyester blends. The chemicals used and the waste generated in the manufacture of synthetic fibers present obvious environmental concerns. While there are many types of synthetic fibers, acrylics, nylons, and polyesters are the most popular ones currently used throughout the industry.

Characteristic	Effect	Example
Density	Weight of textile	Wool fibers produce heavy garments
Strength	Resistance to tearing	Flax produces very strong fibers
Durability	Resistance to wear	Acrylic is a very durable fiber
Chemical Properties	Coloring, flammability, reactivity	Polyester resists stains
Elasticity	Flexibility of textile	Cotton fibers relatively inflexible
Moisture Adsorption	Coloring, end use	Wool highly absorptive giving it warming properties
Resilience	Resistance to stretching, wrinkles	Acetate resistant to stretching
Luster	Appearance of fiber	Silk has high luster
Biological Properties	Susceptibility to mold, mildew or insect damage	Acetate resistant to biological damage

Table 21-2: Fiber characteristics and examples of uses.

Yarn

Yarn is a thread made of natural or synthetic fibers that is formed from a twisted aggregate of fibers.

To create a yarn, fibers are interlocked by arranging or twisting the fibers into strands. Sometimes short fibers must be subjected to a **spinning** or twisting process to be useful, while longer fibers may be used without modification. However, even long fibers are sometimes spun into yarns to provide specific textures, color patterns, or elasticity. Greater twisting, for example, increases the strength and smoothness of the yarn, but decreases its luster and makes it a shorter yarn. Multiple yarns can be combined to produce particular colors or textures. Different fibers create yarns with unique qualities. Yarns may also be blended to capitalize on the good qualities of one fiber, while minimizing those of a weaker or less expensive one.

To **finish** the yarn, a wax or oil may be added to allow for easy manipulation during fabric production. The disadvantage of this use lies in the stripping of these wax or oil products later during the dyeing process, which adds to the wastewater pollutants generated.

Fabric

Fabric is produced by the interlacing of yarns or fibers. **Knitting** and **weaving** are the two common methods used for producing a fabric. As shown in Figure 21-1, knitting forms a series of interlocking loops, generating rows or chains that create a continuous piece of cloth.

Because knitted garments are easy to care for, naturally elastic, and move easily with the body, knitting is used to produce socks, underwear, and many form-fitting garments.

As shown in Figure 21-2, weaving interlaces two or more sets of yarns at right angles. **Warp** yarns run the length of the fabric and **weft yarns** run the width. Fabrics are typically constructed on some type of **loom**, as shown in Figure 21-3. Many interesting effects are created by varying weaving techniques. Weaving constitutes the most widely used method of fabric construction and results in a wide variety of fabric types.

Other methods of fabric creation include **felting**, in which fibers are manipulated so as to become entangled and permanently interlocked; **netting**, in which the yarn is coarsely interlaced producing fishnets, for example; and **sewing**, in which a threaded needle passes through the structure of the yarn stitching the yarn into a solid fabric.

Finishing

Once the fabric is constructed, a finish is applied.

Weft Knitting

Warp Knitting

Figure 21-1: Knitting is the construction of fabric by interlocking loops of one or more yarns.

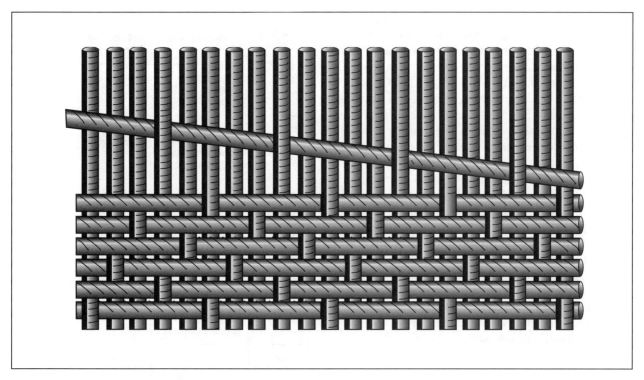

Figure 21-2 A typical pattern used to interlace warp (length) and weft (width) yarns to form a woven fabric.

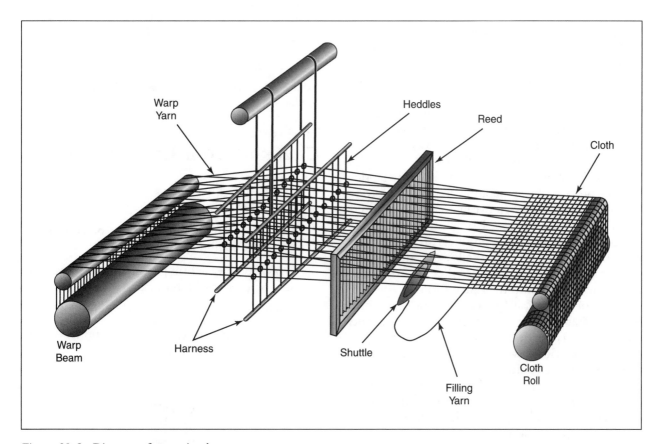

Figure 21-3: Diagram of a weaving loom.

Scouring removes impurities and **bleaching** whitens the product. Finishes make the fabric more acceptable, serviceable, attractive, and predictable; they also provide such characteristics as shrinkage control as well as stain and wrinkle resistance. Some finishes affect the feel, texture, weight, and flexibility of the fabric. Starch, for example, stiffens the fabric, while softeners give it greater flexibility. Other finishes provide the fabric with specialized anti-static, flame retardant, and mildew-resistance qualities. Finish may also aid in the dyeing of the fabric.

Finishes are applied both mechanically and chemically. Mechanical finishes manipulate the fabric in its dry state. These finishes cause physical changes in the fabric and result in minimal environment impact. Chemical finishes use acids, alkalis, bleaches, detergents, resins, and other chemical substances on the fabric. During the process, excess finish material is discharged in the wastewater. Along with dyeing, finishing creates a majority of the pollutants in the textile industry.

Dyeing

Dyeing provides a fabric with its most influential attribute: color. The dyeing process involves penetration of dyestuffs into the textile fiber. The characteristics of the fiber determine the class of dye to be used. Dyes may be reactive, direct, basic, acidic, metallic, or a combination specific to the fiber. Table 21-3 shows a comparison between some dye classes, along with the typical fiber application as well as the rate of fixation to each of the fibers.

Products may be dyed at any time in the fabric-making process. In preparation for dyeing, products may be scoured to remove impurities that would interfere with the process. Using either a continuous or batch method, the products are submersed into the dye long enough to allow the dyestuff to penetrate the fiber. Products are then rinsed, dried, and manufactured into their final form.

Each of these finishing processes produces waste. Scouring transfers the impurities such as waxes and oils from the fabric to the wastewater. Rinsing dyed product adds acids, bases, bleaches, metals, solvents, color, or other chemicals to the wastewater. Even drying the product produces lint waste. Each of these pollutants creates a challenge for the textile industry to control, treat, or prevent through application of pollution prevention strategies.

Dye Class	Description	Fibers Typically Applied To	Fixation
Acid	Water soluble anionic compounds	Wool Nylon	80-93%
Basic	Water-soluble, applied in weakly acidic dye baths, very bright dyes	Acrylic, some polyesters	97-98%
Direct	Water soluble, anionic compounds	Cotton Rayon Cellulostics	70-95%
Disperse	Not water soluble	Polyester Acetate Synthetics	80-92%
Reactive	Water soluble, anionic compounds; largest dye class	Cotton Cellulostics Wool	60-90%
Sulfur	Organic compounds containing sulfur or sodium sulfide	Cotton Cellulostics	60-70%
Vat	Oldest dyes; more chemically complex; water insoluble	Cotton Cellulostics	80-95%

Table 21-3: Characteristics of dyes used on textile products.

Checking Your Understanding

1. What are the two major types of fibers used in the textile industry?

2. List the seven factors that must be considered when producing a textile product.

3. Which type of fabric production process consists of a series of interlocking loops?

4. What is the purpose of a fabric finish?

5. Which two processes create the majority of pollutants in the textile industry?

21-3 Waste Streams and Environmental Regulations

Waste Streams

As noted earlier, each process in textile production generates some type of waste. Since these wastes come from specific processes, one way to categorize them is by the industry operation that generates them.

Wool Scouring

Raw wool must be cleaned by wet processes before it can be transformed into fibers, yarn, or fabric. This process, which is not necessary for cotton or synthetic fibers, is called wool scouring. The steps in the process are shown in Figure 21-4.

The wastes generated by this process include wastewater that contains the following concerns: biochemical oxygen demand (BOD), chemical oxygen demand (COD), total suspended solids (TSS), and oil and grease (O & G); solid wastes including vegetable matter and wastewater sludge; and air emissions in the form of vapors and particulates.

Wool Finishing

Wool finishing processes are used to complete the wool product. The wool finishing process is unique, due to the chemicals used and wastewater produced. Figure 21-5 shows the steps involved in the process.

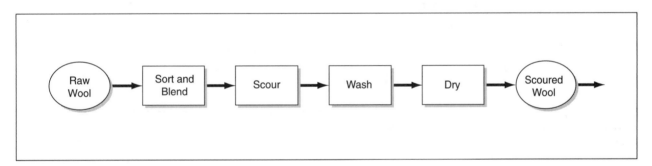

Figure 21-4: Steps involved in wool scouring.

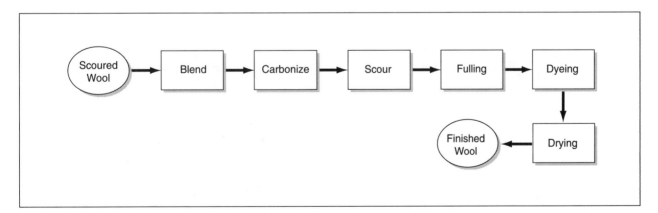

Figure 21-5: The steps involved in wool finishing.

The wastes generated in wool finishing include high volumes of wastewater containing BOD, COD, and TSS concerns; solid waste in the form of wastewater sludge; and vapor and particulate air emissions.

Low Water Use Processes

Yarn production, fabric coating, laminating, and carpet manufacturing use very little water in their production steps. Some waste is produced due to spillage and cleanup. Figure 21-6 shows the steps used in these processes.

The wastes generated include wastewater concerns (BOD, COD, TSS, and color), various solid waste materials, and some vapors and particulate air emissions.

Woven Fabric Finishing

The greatest contributor to wastewater effluent in the textile industry is woven fabric finishing. These processes provide fabrics with color and finishes such as soil repellency and flame proofing. Figure 21-7 shows the processes involved in finishing woven fabrics.

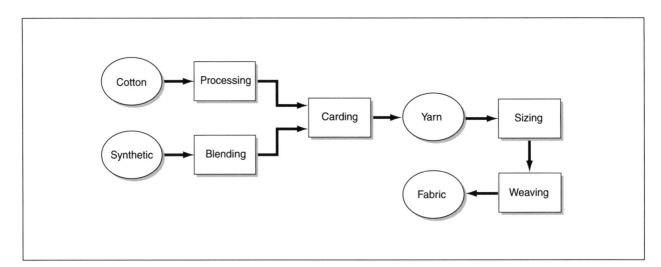

Figure 21-6: Various low water use steps in fabric production.

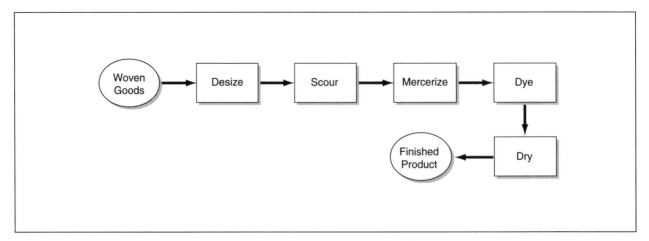

Figure 21-7: Steps involved in finishing woven fabrics.

These processes generate wastewater containing the typical concerns (BOD, COD, TSS, and O & G); phenol, chromium, sulfide, and color are additional problems. The air emissions remain vapors and particulates.

Knit Fabric Finishing

Knit fabric finishing includes knit fabric piece goods, hosiery items, and other specialty products such as underwear. Figure 21-8 shows the steps involved in the finish of knit fabrics.

The wastes generated in these processes are the same as those produced by the finishing of woven fabrics.

Carpet Finishing

Figure 21-9 shows the steps involved in finishing synthetic fibers for the production of carpet material.

Once again the wastewater and air emissions are the same as those produced by the knit and woven fabric finishing processes; however, in producing carpeting, the backing material becomes a new solid waste source.

Stock and Yarn

The processes used to dye and finish stock and yarn products are shown in Figure 21-10.

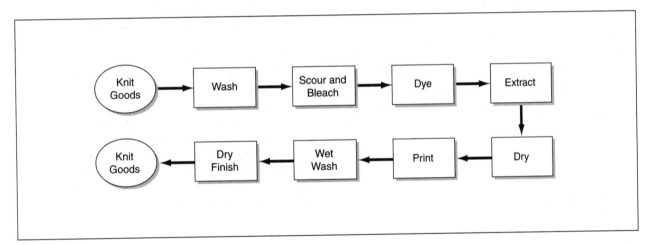

Figure 21-8: Steps involved in finishing knit fabrics.

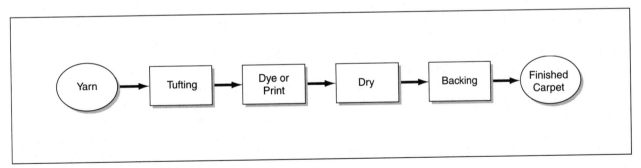

Figure 21-9: Steps involved in finishing carpet material.

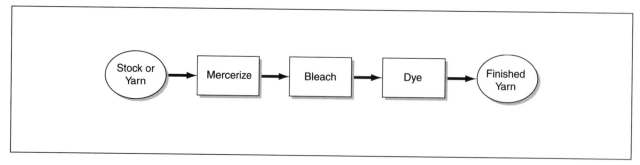

Figure 21-10: Steps used to finish stock or yarn.

Concerns for BOD, COD, TSS, O & G, phenol, chromium, sulfide, and color are again the major wastewater concerns and vapors air emissions.

In addition to these processes, non-woven fabric and felted-fabric processes also contribute to the waste streams generated by the textile industries. The felted fabric finishing process releases effluents high in phenols and COD. Non-woven fabric finishing is a relatively low impact segment; however, its wastewater concerns remain BOD, COD, and TSS. In summary, it is apparent that air and solid waste generation is of far lesser proportion than wastewater, the latter being the industry's greatest problem.

Regulations

Air

Air emissions have traditionally received minor focus in the textile industry due to the emphasis on wastewater. Yet, air emissions and indoor air pollution are significant sources. Air emissions are generated by dryers and curing ovens that emit volatile organic compounds and residues from the finishes applied to the fabric. Solvent vapors are also emitted in the finish process. Many textile mills still operate steam boilers that produce toxic air emissions.

Air pollutants typically emitted by the textile industry are:

—Acetic acid from drying and curing oven emissions;

—Formaldehyde from finishes;

—Oil and acid mists;

—Volatile chemicals from dyes and finishes;

—Nitrogen and sulfur oxides from boilers;

—Dust and lint from the spinning of fibers.

Indoor air quality is also a concern in textile manufacturing. Many of the finishes that are applied to the products have an effect on the quality of indoor air, not only in the factory itself but also in the areas in which those products are used. Up to 90 percent of all formaldehyde used in textile products, for example, takes up to three years to volatize. Also, fabrics provide an attractive home for microorganisms, such as mites and molds, which emit biological contaminants into the air.

Solid Wastes

The textile industry makes extensive use of hazardous chemicals. Chemical use may, in fact, amount to 10-100 percent of the weight of the goods produced. Depending on the operation, varying amounts of hazardous waste are produced and must be disposed of following RCRA guidelines. The sources of these solid hazardous wastes include 1) waste solvents from machine cleaning; 2) unused finish materials; and 3) dyeing and printing materials. Some of the hazardous materials these wastes may include are metals, chlorinated solvents, nondegradable surfactants, and volatile organic substances.

Water

As previously noted, the textile industry makes extensive use of water. As a result, water pollution is the single largest environmental and regulatory concern of the industry. The use can vary from 10 to 40 pounds of water per pound of fabric produced. Textile manufacturers must comply with Clean Water Act (CWA) regulations and NPDES permit requirements. If a manufacturer discharges wastewater or storm water directly into surface waters, an NPDES permit is required. Typically, manufacturers discharge wastewater indirectly into surface waters through a POTW. This requires a local permit with specific pretreatment requirements. Wastewater must meet permit requirements set by these local regulators. The chemical and physical characteristics of the wastewater are largely determined by the type of process involved in the manufacturing process.

Textile wastewater effluents may also contain shop chemicals, biocides, pesticides, and boiler treatments. These chemicals may enter the waste stream from spills, leakage, cleanup, or poor housekeeping. Pollutants found in textile wastewaters – subdivided on the basis of process – are the following:

— *Organics from dyeing and finishing* – Include phenol, naphthalene, chloroethylenes, chloroethanes, and chloroform;

— *Metals from sizes, dyeing, printing* – Include zinc, copper, chromium, lead, and nickel;

— *Conventional* – BOD and COD from sizes, dyeing, printing, finishes; TSS from dyeing, printing, and finishes; aquatic toxicity from finishes; lint from dyeing, preparation, and washing operations; and oil and grease from scouring processes.

Health and Safety

Worker health and safety is taken very seriously in textile operations. OSHA regulates operations and working conditions. Of particular concern in textile operations is cotton dust, which causes **byssinosis** or brown lung disease. Many of the flame-retardant chemicals and finishes applied to textiles are also hazardous to health. Noise levels expose textile workers to the highest levels of any industry. Consequently, the typical health and safety programs for the textile industries include:

— MSDS training;

— Hazard communication training;

— Lockout/tagout of machinery training; and

— Hearing conservation programs;

With special concerns for:

— Cotton dust;

— Noise;

— Proper air ventilation – for heat and lint.

Pollution Prevention

Textile pollution prevention practices include material substitution, process modification, inventory control, better management techniques, recovery, and reuse. Source reduction may also be effective for many hard-to-treat wastes such as color, metals, phosphates, phenol, organics, and surfactants.

Selected pollution prevention efforts in the textile industry include:

— The development of dyes that are treatable, less persistent, and less offensive. Dyes currently used in the textile industry are designed to be stable and resist fading, which makes them difficult to treat in waste management.

— The removal or substitution of sizes and knitting oils added to yarns before and during fabric production. This is one of the greatest intentionally created pollution streams in the industry.

— Machine cleaning between dye baths produces some of the most toxic and offensive chemicals used in the industry. Scheduling batches and sequencing dye baths according to similar colors helps minimize waste during this step.

— Lobbying for review of the current regulatory fee structures – associated with surcharges based on the concentration of pollutants in the wastewater– often applied to textile operations. As a result, commendable efforts such as water conservation or equipment improvements can backlash by triggering higher surcharges due to the increased concentration of these constituents.

Box 21-1 ■ Hosiery Industry in North Carolina

The hosiery industry in North Carolina is a significant producer of socks and women's pantyhose throughout the world and accounts for almost 60 percent of all hosiery for men, women, and children in the United States. For the past 40 years, North Carolina has shipped more hosiery products than all other states combined.

Speed and adaptability are essential to hosiery manufacturers. Major retailers such as Wal-Mart and K-Mart demand customization in the socks and hosiery they put on the shelves. Customization requires exact color matches on request as well as repeat orders. In many instances, socks and hosiery must be reworked to meet exact demands. This involves stripping improperly matched colors and redyeing.

Emerging technologies use less water in the dyeing process, thereby creating a more concentrated wastewater effluent – placing a strain on the facility's ability to meet its NPDES and POTW permit requirements. In addition, manufacturers may purchase raw materials based on price rather than quality. Poor quality yarn, for example, adds pollutants to wastewaters because it requires a more thorough cleaning. While saving money on the purchase of yarn, wastewater pollution costs increase.

To meet these seemingly conflicting demands while minimizing pollution, hosiery manufacturers are turning to new technology and more training. By training dye operators to use computer technology, the need for rework of improperly dyed batches is diminished. Computers not only match colors exactly, but also list the chemicals needed for the dye baths and indicate the environmental impacts of the materials in use. Therefore, less toxic materials are used in the dye bath formulations, which previously overused or unnecessarily used these materials. This technology also improves the operator's ability to track wastewater pollutants in the system and develop techniques to reduce their impact.

Checking Your Understanding

1. Why is it convenient to classify waste streams by industry operation?

2. Which type of waste stream do most finishing processes produce?

3. What are six air pollutants that are typically emitted from the textile industries?

4. Which finishing processes generate solid hazardous wastes?

5. What is the ratio of pounds of fabric produced to weight of water used?

6. Name six organic water pollutants resulting from dyeing and finishing operations.

7. What is another name for byssinosis?

8. What are three special health and safety concerns in the textile industries?

9. What is one way that pollution can be minimized in the textile industry?

Summary

Evidence of human textile making dates back as far as 6,000 BCE. The modern textile industry manufactures upholstery, carpeting, draperies, linens, and towels, in addition to clothing. The fibers used for textile construction are natural or synthetic in origin, woven or knitted, and are typically dyed to improve appearance and function. The natural fibers come from plants (cotton, linen, and hemp) or from animals (silk, wool, and hair). Synthetic fibers include rayon, polyester, and nylon. Many fabrics produced for clothing today are blends, containing specified amounts of both natural and synthetic fibers.

Yarns are twisted threads made from natural or synthetic fibers. The greater the twisting the greater the strength and smoothness of the yarn. By mixing threads, yarns can be produced with a variety of colors and textures. Finishes, such as waxes or oils, are often added to the yarn to facilitate fabric production. These substances must be stripped prior to dyeing or finishing the product, thus creating a considerable amount of pollution.

Knitting and weaving are the two most common methods used for producing fabric. Knitting forms a series of interlocking loops generating rows or chains that create a continuous piece of cloth. Knitted garments are easy to care for and naturally elastic, which makes them well suited for socks, underwear, and other form-fitting garments. Woven fabrics are constructed on a loom. Many different weaving patterns result in a variety of fabrics. Fabric can also be created by felting, netting, and sewing.

After the fabric is constructed, a finish is applied. Scouring removes impurities, while bleaching whitens the products. Other finishes alter the feel, texture, weight, and flexibility of the fabric. Some finishes aid in the dyeing of the fabric or add specialized anti-static or flame-retardant qualities. Dyeing adds the most influential attribute to the fabric, which is its color. Dyes may be reactive, direct, basic, acidic, metallic, or a combination specific to the fiber. Each of the finishing processes produces waste, which is a major challenge for the textile industry. Wool scouring, for example, generates large amounts of wastewater with BOD, COD, and TSS, in addition to sludge and air emissions. Stock and yarn dyeing and finishing generate major wastewater concerns including BOD, COD, TSS, O&G, phenol, chromium, sulfide and colors.

Air emissions are a minor, but important focus in the textile industry. Indoor air pollution is significant due to the VOCs and residues released by dryers and curing ovens. The major sources of solid hazardous wastes are from machine cleaning, unused finish materials, and dyeing and printing materials. Water use can vary from 10 to 40 pounds of water per pound of fabric produced. Typically, producers discharge their wastewater to a POTW, which has established acceptable chemical and physical discharge characteristics by NPDES permit. The pollutants generally found in textile wastewaters can be subdivided based on the process: organics from dyeing and finishing; metals from sizes, dyeing, and printing; and BOD, COD, and TSS from dyeing, printing, and finishes.

Worker health and safety is also a major concern, since many of the chemicals and finishes are hazardous to health. For example, cotton dust is responsible for brown lung disease. The industry is addressing health and safety as well as environmental problems by providing proper worker training, with special concerns for cotton dust, noise, and proper air ventilation. It is instituting pollution prevention practices by developing environmentally safer dyes and finishes and by scheduling dye jobs to keep batch changes to a minimum. Of much concern to the present industry is the regulatory agency's practice of surcharging wastewater based on concentration – a practice that has the counterproductive effect of discouraging water conservation.

Critical Thinking Questions

1. Name as many products as you can that are produced by the textile industry. Of those named, how many are you wearing today? How many are in your home?

2. Describe the process of making fibers into fabric. Include the reason for each step.

3. What are the differences between natural and synthetic fibers? From an environmental point of view, compare/contrast the advantages and disadvantages of the use of synthetic over natural fibers.

4. Describe the finishing processes that are used to give fibers and yarns desirable characteristics. What function does each of these processes serve?

5. What are some of the significant water pollutants from textile production and how might they be minimized?

6. How are pollution problems in the hosiery industry similar to or different from those of the rest of the textile industry? What pollution prevention strategy is being used to address those problems?

22

The Agricultural Industries

by Melinda Trizinsky

Chapter Objectives

Upon completing this chapter, the student will be able to:

1. **Describe** production processes and agricultural activities that influence productivity.

2. **Identify** common waste products resulting from agricultural activities.

3. **Differentiate** between agricultural activities that generate point and nonpoint source pollution.

4. **Describe** opportunities for waste minimization and better resource management.

5. **Describe** control measures that will minimize off-farm impacts of agricultural wastes.

6. **Identify** specific regulations that impact agricultural activities.

22-1 Introduction

> Agriculture is the primary industry of most of the world's population.

The agricultural industry produces basic commodities such as meats, grains, fruits, vegetables, mushrooms, dairy products, and eggs. Agricultural products are also processed for use as animal feeds, fertilizers, fiber, raw materials for other industries (e.g., consumer products, pharmaceuticals, pulp and paper, textiles, chemical production, etc.). They are also used to produce biofuels – a renewable energy source. This chapter discusses the basic production of plant and animal products. Common agricultural activities, the wastes they generate, and pollution prevention strategies will be covered along with a brief overview of regulations and government policies that affect agriculture.

In the past, farms typically produced both crops and livestock, but agricultural trends since the 1940s have favored large-scale production of crops and intensive factory farming of livestock. The family farm no longer dominates U.S. agriculture. Large farms generate most of the agricultural products produced in the United States. The average large farm in the United States is 1,542 acres in size, has more than $1 million in assets, and has annual sales in excess of $400,000. Today's farms are not only larger and more productive, but also more dependent upon technology than the farms that existed prior to World War II. Productivity has soared due to new technologies, mechanization, increased chemical use, higher specialization, and government policies that have favored maximizing productivity.

Increases in agricultural efficiency have allowed fewer farmers to produce the majority of food and fiber in the United States. These changes are dramatically illustrated with just a few statistics from the most recent (1992) farm survey. In 1969 there were 2.7 million farms in the United States; by 1992 that number had dwindled to 1.9 million. In 1969, there were 51,995 large farms (i.e., farms with sales in excess of $100,000); by 1992, the number of large farms had increased sixfold to 333,865. Large farms accounted for 17 percent of the farms in 1992, yet they produced more than 83 percent of the agricultural products. Thus, chances are most of the food we eat was produced on a large farm. Agriculture truly is an industry today.

The basic goal of all agricultural activities is to foster the growth of plants, animals, or fungi. The agricultural industry differs from most of the industries discussed so far because the production process is largely controlled by biological and environmental systems. This limits the degree to which production processes can be altered. For example, to plant earlier in the season a farmer might be able to choose a hybrid corn that can withstand colder soil temperatures, but he cannot do anything to change the weather on his farm. Agricultural activities are largely directed toward influencing and controlling natural cycles. A basic understanding of biology and ecology is necessary to understand the methods and goals of the agricultural industry.

22-2 Overview of Agricultural Activities

Crop Production

Plants obtain carbon from carbon dioxide in the atmosphere and energy from the sun, but they obtain other nutrients from the soil. More precisely, nutrients are dissolved in the water contained in the soil pores and absorbed through the root systems of the plants. As plants grow, they gradually deplete the nutrients that are most essential for their growth. If the plants die or are eaten in the field, a portion of the nutrients may return to the field in the form of crop residues or animal excrement; otherwise, the nutrients are lost from the field's **ecosystem** when the plants are harvested. This process eventually leads to nutrient depletion, a factor in **soil depletion**. In addition to plant uptake of nutrients, soil becomes depleted through processes such as microbial biodegradation of soil organic matter. Soil requires regular influxes of organic matter to remain healthy and productive. To counteract nutrient losses, farmers apply fertilizers to the soil.

Both soil structure and composition influence the soil's ability to retain and release moisture and nutrients. Soil composition is defined by characteristics such as particle-size content, organic matter content, and pH. Soil structure refers to qualities such as aggregate size and degree of compaction. Cultivation can be used in conjunction with soil amendments to improve soil characteristics.

Plants need water to survive. They obtain water from the soil through their root systems. Water application rates for each crop are dependent upon plant uptake rates, local soil conditions, and evapotranspiration rates.

The availability of air and sunlight is not a problem in outdoor agriculture, but plant growth is a seasonal phenomenon. It is dependent upon average daily temperatures, hours of sunlight, and precipitation. Each season calls for special work. Exceptions occur for **greenhouse** crops that are grown under artificial conditions. The basic farm activities that are specific to crop production are discussed below.

Land Cultivation (Tillage)

These procedures physically work the soil. Tillage can be used to aerate soil, improve soil structure, prepare fields for planting, control weeds, control runoff, mix soil additives into the soil, or plow under crop residues to enrich the soil. Land cultivation is almost entirely mechanized in the United States.

Plant Propagation

Crops may be grown from seed, cuttings, or grafts. Sometimes seedlings (i.e., young plants grown from seed) are planted rather than seeds. Perennial crops such as grapes, citrus fruits, nuts, and apples grow only on mature plants. New plants for these crops are frequently propagated by grafting new shoots onto mature rootstock.

Farmers typically plant a single crop in each field (**monoculture**). These uniform fields allow the farmer to increase production and reduce labor costs by using a high degree of mechanization. Although labor costs may be reduced in these systems, other costs have risen because of the ecological instability of monoculture ecosystems.

Fertilizer Application

Fertilizers are usually applied mechanically with the rate of application controlled by variables such as type of crop, soil condition, previous cropping, and hydrologic conditions. Fertilizers enhance plant growth and replenish soil nutrients, but if applied in excess, these beneficial agricultural chemicals can have negative environmental impacts.

Weed and Pest Control

Weed control has always been an issue in agriculture. Weeds compete with crops for resources

Crop pests come in many forms:

— Weeds – Non-crop plants that compete with crops for nutrients, water, and sunlight.

— Insects – Many insects are herbivores; their predation on crops reduces yields and affects produce quality.

— Molds – These fungal parasites weaken plants and ruin produce. Some forms are toxic.

— Larger Herbivores – Rodents, birds, and other wildlife also damage crops.

(e.g., water, nutrients, and even sunlight). They can also disrupt mechanical harvesting or hinder processing. In labor-intensive agriculture, weeds are often controlled physically by pulling or cultivating.

Most U.S. farms rely heavily on chemical means to control pests and weeds. The chemicals used to control these pests are collectively referred to as **pesticides**. Pesticides are applied at critical times in the growth cycle. The application may be directed to the soil or to the plants themselves. Both aerial and truck-mounted sprayers are used to apply these agricultural chemicals. Pesticides may be selective (affecting only a very specific population) or broad spectrum (affecting many organisms). Some classes of pesticides and their targets are given below:

—**Herbicides** – Used to kill or control plants, especially weeds. These products are the most commonly applied pesticides in the U. S.

—**Insecticides** – Used to kill insects or interfere with their reproductive cycle.

—**Fungicides** – Used to kill or control fungi, including molds, yeast, and rot.

—**Rodenticides** – Used to kill rodents (e.g., rats, mice, gophers).

—**Biocides** – Nonspecific pesticides that kill many types of organisms, including bacteria.

Integrated Pest Management (IPM) offers an alternative approach to pest control. One of the goals of IPM is to reduce pesticide use. IPM uses techniques such as choosing pest-resistant crops, rotating crops to avoid recurring outbreaks of host-specific pests, and using tillage practices to control pests and weeds. It also stresses the use of biological controls such as introducing host-specific **parasites** to control pests, using traps baited with insect attractants, and creating environmental conditions that favor proliferation of beneficial species. Because IPM uses fewer chemicals, many farmers are able to save money when they use these techniques.

Irrigation

Plants differ in their water requirements, and rainfall is rarely a predictable event. Therefore, farmers often use **irrigation** to supply water to their crops. Irrigation provides a reliable and controllable source of water that can be applied at appropriate times during the growing cycle. Irrigation cycles are controlled by factors such as rainfall events, evapotranspiration rates, crop needs, and soil conditions. Both ground water and surface water resources supply water for irrigation. Readily available sources of water have eliminated much of the risk in farming and have boosted agricultural productivity in arid regions.

Sprinkler irrigation and flood irrigation are currently the most common types of irrigation in the United States, but interest in drip-irrigation methods seems to be growing. Drip irrigation delivers water directly to the plant and requires less water.

Harvesting

Harvesting methods vary depending upon the crop and the size of the farm operation. Row crops such as corn and soybeans are normally harvested mechanically. Other crops such as strawberries are much more delicate and are picked by hand. Advantages

Some corn harvesters are so sophisticated that they can cut down the stalk, remove the ears, shuck the corn and remove the kernels from ears while still in the field.

of mechanical harvesting include faster harvesting and reduced labor costs. Therefore, larger fields can be planted, and less food will spoil in the field. However some types of produce (e.g., tomatoes) must be picked before they are fully ripe to avoid damaging the fruit. Mechanization also requires larger energy inputs and may increase soil compaction.

Animal Husbandry

All agricultural operations that produce meat and animal products involve **animal husbandry** activities (i.e., the breeding, care, and feeding of animals). Common examples of animal production activities include beef cattle production, dairy operations, swine production, poultry and egg production, and sheep operations that produce food and fiber. Important aspects of animal husbandry are reviewed in the following sections.

Animal Breeding and Growth

Commercial livestock is carefully bred for various characteristics (e.g., larger breasts on poultry, more milk production in dairy cattle, higher wool production in sheep). Male animals not intended for breeding purposes are generally castrated. Most large animals are bred via artificial insemination. This reduces the risk of animals being injured during the breeding process, and allows farmers to choose from a wider array of bloodlines. Most poultry operations obtain young birds from commercial hatcheries that carefully control genetic traits.

The length of time required to produce marketable livestock depends upon the type of animal being raised as well as on its intended market. For example, most animals raised for meat are sold when

> Livestock production relies on a few select breeds. There is some concern that this lack of genetic diversity may make agricultural animals more susceptible to disease.

they reach the targeted marketable weight; this usually occurs before the animal reaches maturity. This meat tends to be tender, but it lacks the flavor of meat from more mature animals. Advances in animal nutrition as well as various growth-promoting hormones have allowed farmers to bring younger animals to market than in the past. This maximizes agricultural efficiency by reducing the time and feed necessary to produce meat, but it also leads to a blander-tasting product.

Land Use and Animal Confinement

Animal production facilities may be loosely grouped according to the degree of confinement used to control the movement of the animals. Animals such as hogs and poultry – which are **omnivores** – are normally produced in concentrated animal production facilities. Grazing animals such as cattle and sheep generally spend at least part of their life under less confined conditions.

Concentrated animal production facilities (also referred to as factory farms) house large numbers of animals in relatively small areas (e.g., poultry barns and livestock feedlots). The number of animals per unit area varies from one operation to another, but the animals live in very close proximity to each other. Many operations are mechanized, so labor requirements are usually minimal. This allows for a high level of productivity and a high economic return from agricultural land, but it also requires strict control mechanisms. When animals are housed indoors, the environmental conditions must be carefully controlled. Temperatures are controlled with heating and ventilation systems. Interior lighting is provided. Animal wastes can produce dangerous gases, so ventilation systems must operate continuously.

On the other end of the spectrum, range cattle may be free to wander and graze over vast tracks of land. The more efficient digestive systems of **ruminants** (e.g., cattle and sheep) allow them to convert foods with low nutritive value for humans (e.g., grasses and **forage**) into high quality protein. Through this process, additional land is brought into agricultural production and crop residues can be utilized. Depending upon the degree of confinement, grazing lands are referred to as either **pasture** or **range**, with pasture being the more confined area.

Feeding

Hogs and poultry are normally produced in concentrated animal production facilities where they obtain most of their nutrition from concentrated feeds. Automatic feeders dispense food and water; they are usually activated when the amount of food or water in the trough drops below a specific level. The concentrated feeds promote rapid growth and generally result in low residues (i.e., low volumes of feces). Various growth promoters and antibiotics are typically mixed with the feed.

Grazing is normally a very cost-effective feeding method, but growth rates are not as rapid as those obtained when concentrated feeds are used. Beef cattle spend 80-100 percent of their life on grazing lands. The remaining time is spent in a feedlot where they are fattened up for slaughter. As in other concentrated production facilities, feedlots use nutrient-dense foods and supplements to produce rapid weight gain.

Many of the things fed to agricultural animals would never be found in the animal's natural diet. For example, cattle are **herbivores**, but beef cattle are commonly fed supplements derived from animal products (e.g., blood meal and bone meal) to promote growth. This practice has recently come under scrutiny because this type of supplementation may have been responsible for the spread of Mad Cow Disease in Great Britain. Livestock are also fed surplus or expired consumer products (e.g., breakfast cereals and candy).

Animal Health

Livestock operations must maintain the health of their herds and flocks in order to produce high quality animals and reduce mortality losses. Animals are frequently inoculated against transmissible diseases. External parasites are usually controlled by dipping or spraying animals with pesticides. Internal parasites are controlled with antibiotics and dewormers. Disease vectors such as flies and cockroaches that infest barnyards and spread disease must also be controlled. Proper waste management helps stem the transmission of disease due to unsanitary conditions.

> The potential for disease transmission is very high among herd animals and animal raised in concentrated production facilities. Farmers use a number of control strategies to combat livestock diseases and parasites.

The potential for disease transmission is very high among animals in factory farms; therefore animals are routinely fed antibiotics with their feed. The stress associated with close confinement also fosters cannibalism in poultry and tail biting in swine. To combat these behaviors, poultry are often debeaked (i.e., the tip of the beak is cut off), and pigs are de-tailed. Mortality is often high under confined conditions.

Although disease transmission is not as rapid in pastures, cows and sheep are herd animals; therefore, diseases can still be transferred from one animal to another. Other sources of mortality in these open systems include predation, exposure, and accidents. Various means are used to control predators including poisons, guard animals, and fencing.

Waste Handling

All animals produce wastes. Feces contain high concentrations of nutrients, BOD, and bacteria; urine contains high concentrations of salts and nitrogen compounds. **Manure** consists of animal wastes mixed with bedding materials and stall litter. The consistency of manure ranges from solid to liquid and depends upon the mixture of wastes. Properly handled, manure can be used as a fertilizer or soil conditioner; improperly handled it can spread disease and contribute to nonpoint source pollution. Manure is usually stored for future use or disposal. The type of storage will be determined by a number of factors including the composition of the waste, the volume of the waste, the availability of land, and the ultimate use or disposal method.

General Farm Operations

Chemical Storage and Handling

Farms commonly use and store a variety of chemicals including petroleum products, fertilizers, soil conditioners, herbicides, insecticides, fungicides, and other pesticides. Many of these chemicals are stored in concentrated form and must be mixed prior to application. Because many of them are hazardous, workers should be thoroughly trained in proper handling techniques.

Energy

Modern agriculture is heavily mechanized. It requires large inputs of energy, both in the form of petroleum products to drive farm machinery and in the form of electricity to power pumps, lighting, heating, ventilators, milking machines, automated feeders, etc.

Checking Your Understanding

1. What are five of the basic commodities produced by the agriculture industry?

2. After what event did the switch to large scale farming became a national trend?

3. What factor in crop production is beyond the farmer's control?

4. What are five types of pesticides and their intended uses?

5. What are five uses for heavy farm equipment?

6. Give three examples of integrated pest management control methods.

22-3 Minimizing Environmental Impacts of Agriculture

Examples of environmentally sensitive areas include:

— Riparian Areas – Areas that border water sources.

— Wetlands – Soil areas saturated or covered with water all or part of the year that support certain species of vegetation.

— Critical Habitats – Areas that are essential to the survival of a wildlife population (e.g., winter forage areas, migration routes, calving or nesting areas, habitat used by endangered species, etc.).

— Unique Habitats – Areas that harbor unusual assemblages of plants and animals.

Agriculture is one of the most important sources of environmental pollution, particularly water pollution. This is largely because nonpoint sources have not been regulated as heavily as "end-of-pipe" pollution from industry.

While the industrialization of farming has increased productivity it has also created new environmental problems. Agriculture has traditionally been a major source of **nonpoint source pollution** (e.g., sediment, fertilizers, pesticides, animal wastes, and other organic matter). It has also been a major source of ecosystem disruption. In addition, the increase in concentrated production facilities, such as feedlots and poultry barns, has also created serious localized air and water pollution problems.

Ecosystem Effects

This section examines several of the environmental problems associated with agricultural activities and offers some suggestions about how these events can be modified to minimize environmental effects.

Problem: All agriculture disrupts natural ecosystems by displacing or destroying native species and impacting soil, water, and air quality, but some environments are more sensitive than others.

Potential Solutions: Restrict livestock access to sensitive ecosystems. Do not convert sensitive ecosystems to agricultural production. Leave native vegetation in place along field breaks and agricultural ditches.

Problem: Monoculture agriculture and concentrated animal production facilities foster the proliferation of diseases and pests.

Potential Solutions: Use crop rotation to suppress weeds, **pathogens**, and insect pests. Use disease and pest resistant crop varieties.

Problem: Broad-spectrum pesticides may kill nontarget species including beneficial organisms such as earthworms and honeybees. Some types of pesticides accumulate in the ecosystem and damage wildlife populations.

Potential Solutions: Use IPM to control insects.

Soil Conservation

Problem: **Soil erosion** is the mobilization of soil particles by either wind or water. It results in the loss of productive topsoil from agricultural areas and contributes to air and water pollution problems. Agricultural activities such as tillage, harvesting, grazing, and herbicide application, contribute to soil erosion by removing vegetative cover and disturbing the top layer of soil.

Potential Solutions: Use conservation tillage methods (e.g., leave crop residues in place to stabi-

Healthy soils are essential for successful agriculture. Unfortunately a number of agricultural activities contribute to soil erosion, soil depletion, and salinization.

lize the soil). Plant row crops following the contour of the land to minimize slope. Retire highly erodible fields from production. Use soil additives to improve soil texture and aggregation. Use cover crops while resting the soil.

Problem: Continuous cropping leads to soil depletion by removing essential nutrients and soil organic matter. Removing crop residues contributes to the problem.

Potential Solutions: Use crop rotation and **green manure** (i.e., cover crops that can be plowed under) to improve soil fertility and workability. Leave crop residue in the field. Use fertilizers, soil conditioners, and bulking agents (i.e., compounds that add organic matter to the soil.)

Animal Wastes

Problem: Animal wastes contain high concentrations of nutrients, BOD, and pathogens. If not properly managed, animal wastes pose a threat to human and environmental health as well as to the health of livestock and wildlife.

Potential Solutions: Maintain adequate manure storage facilities. Do not exceed the capacity of the storage facility. Do not house animals near riparian areas or near groundwater recharge areas. To avoid the spread of pathogens, do not apply fresh manure to fields.

Waste Chemicals

Problem: Agricultural chemicals may become waste products due to spills, inadvertent mixing, contamination, storage beyond their shelf life, cleanout of applicators, changes in farm practices, and excess diluted product after application.

Potential Solutions: Buy no more than is needed. Keep an inventory of products on hand. Use older products first. Use separate storage areas for different types of agricultural chemicals to minimize the chance of contamination. Do not mix more than what you expect to apply. Do not mix incompatible chemicals. Use designated mixing areas with adequate spill containment. Minimize the number of chemicals used. Sell or trade excess chemicals to farm cooperatives.

Health and Safety Issues

Problem: Farm workers are exposed to a number of hazardous chemicals, especially pesticides. Many farm workers are poorly trained and do not realize the risk to themselves or others. Their lack of knowledge increases the likelihood of spills, accidents, and inappropriate use.

Potential Solutions: Train all workers that handle hazardous chemicals or work in areas where pesticides are stored or applied.

Problem: Many people rely on groundwater as their source of drinking water. Agricultural chemicals including nitrate and pesticides are commonly found in the groundwater in agricultural areas. There are health risks associated with ingestion of these chemicals; therefore rural populations are at risk.

Potential Solutions: Avoid storing or applying agricultural chemicals near groundwater recharge areas. Use tile drains to collect excess water from agricultural fields. Avoid using pesticides that do not break down quickly into harmless byproducts. Do not apply more fertilizer and pesticide than necessary.

Water Quality

Sediments

Problem: Sediment clouds water, chokes watercourses, interferes with aquatic life cycles, damages pumps, and acts as a carrier for other pollutants (e.g., pesticides, nutrients, and pathogens).

Potential Solutions: Use soil conservation methods to minimize soil erosion. Use wetlands and native vegetation to slow runoff and filter sediments out of water resources.

> Agriculture is a major source of water quality problems. The primary pollutants generated by agriculture are sediments, nutrients, pesticides, organic matter, and pathogens.

Nutrients

Problem: Chemical fertilizers and animal wastes introduce nutrients into water supplies. These nutrients accelerate the growth of aquatic plants and microorganisms; the proliferation of these organisms may lead to eutrophication problems, such as low dissolved oxygen levels, fish kills, or destruction of submerged aquatic vegetation.

Potential Solutions: Do not overfertilize. Time fertilizer application to minimize runoff losses. Use manure storage facilities to contain animal wastes. Use water storage facilities to impound runoff onsite. Exclude livestock from riparian areas.

Pesticides

Problem: Pesticides are inherently toxic, so it is not surprising that they are toxic to some aquatic species. Sometimes this toxicity is apparent, as when aquatic grasses are killed by influxes of agricultural wastewater containing herbicides. However, sometimes the effects are more insidious. Pesticides are often hydrophobic and lipophilic (readily soluble in fats and oils); therefore some pesticides with long half lives can accumulate in the fatty tissues of aquatic organisms. This process is known as **bioaccumulation**. At the low end of the food chain, toxic effects are minimal, but animals at the higher end of the food chain may be significantly affected. Effects include reproductive problems, birth defects, impaired immune systems, and nervous system disorders.

Potential Solutions: Minimize pesticide use whenever possible. Use IPM rather than broad-spectrum pesticides. Use pesticides that rapidly break down to nontoxic compounds. Time pesticide applications to minimize losses due to wind drift and runoff.

Pathogens

Problem: Animal feces are loaded with bacteria and sometimes contain parasites. Some of these microorganisms are pathogenic. Luckily most pathogens are species specific, so with a few notable exceptions (see sidebar), humans are not

Cryptosporidia, a parasite found in the feces of some animals, causes gastrointestinal illness and has been implicated in gastroenteritis outbreaks in Milwaukee, Wisconsin (400,000 cases and 100 deaths in 1993) and in Carrollton, Georgia (13,000 cases in 1987). While the source of the organism in these outbreaks was never determined, the fact that Cryptosporidia has been found in many dairy herds has brought some attention to this sector, especially given the proximity of dairies to population centers.

especially vulnerable to animal pathogens. Unfortunately, wildlife that is closely related to domestic livestock is not so lucky. Populations of bighorn sheep, for example, have been decimated by diseases spread by domestic sheep being grazed on public lands.

Potential Solutions: Clean animal enclosures frequently. Use appropriate manure storage facilities to contain wastes onsite. Do not exceed the storage capacity of manure facilities. Do not graze livestock on public lands where they may come in contact with closely related wildlife (e.g., domestic sheep with bighorn sheep, cattle with buffalo). Minimize direct fecal contamination of water supplies by excluding livestock from riparian areas.

Salts

Problem: The salinity of irrigation return flows is generally higher than that of the source water because the irrigation water dissolves salts that exist in arid soils. The salts are then concentrated as the water evaporates. Dissolved salts and other minerals can have significant impacts on surface water and groundwater quality. Increased concentrations of naturally occurring toxic minerals – such as selenium and boron – can harm aquatic wildlife and degrade recreation opportunities. Increased levels of dissolved solids in public drinking water supplies can increase water treatment costs, force the development of alternative water supplies, and reduce the life spans of water-using household appliances. Increased salinity levels in irrigation water can reduce crop yields or damage soils so that some crops can no longer be grown.

Potential Solutions: Carefully control irrigation rates to avoid irrigation return flows. Use flushing flows to periodically remove salts from the crop root zone; do not allow the salt laden water to return to the water source.

Water Quantity

Problem: Agriculture consumes vast quantities of water, especially in the arid Southwest of the United States where agricultural needs far exceed the annual rainfall and groundwater recharge rates. Irrigation water is supplied by either vast water supply projects or by groundwater withdrawals. Both types of water supply have environmental impacts associated with them. Large-scale water projects destroy riparian habitats, disrupt natural sediment transport mechanisms, and often submerge productive agricultural land. Excessive groundwater use causes water tables to drop and may disrupt surface water sources (e.g., springs and perennial streams) and shallow wells. In coastal areas excessive groundwater withdrawal can lead to salt water intrusion into aquifers.

Potential Solutions: Use irrigation methods that conserve water. In areas with high evapotranspiration rates, choose crops with low water requirements. Reduce water needs by using mulch to minimize water loss from soil.

Air Quality

Nuisance Odors

Problem: Animal wastes and fertilizers have odors that many people find offensive. Although these odors are generally not health threatening, they can affect quality of life and cause tension with neighbors.

Potential Solutions: Minimize use of chemical fertilizers. Locate manure storage areas away from neighbors. Carefully manage animal wastes to provide healthful conditions for the animals and adequate manure storage facilities.

Particulates

Problem: Dust from agricultural fields exposed to wind erosion or disturbed by tillage, traffic, or harvesting creates air quality problems. These airborne particulates may contain toxic inorganic compounds (e.g., arsenic, selenium, boron, etc.), adsorbed pesticides, fungal spores, and high concentrations of nutrients. Breathing these contaminants can cause health problems in sensitive individuals.

Potential Solutions: Use water trucks to wet surfaces and reduce dust. Try to avoid activities that stir up dust when very windy conditions exist. Use windbreaks to shelter fields.

Smoke

Problem: Burning fields to remove crop residues and return nutrients to the soil creates localized air pollution problems. There are health risks associated with smoke inhalation.

Potential Solutions: Incorporate crop residues into the soil instead of burning; microbes in the soil will break down the wastes thus making nutrients available and improving soil texture.

Toxics

Problem: Many pesticides are applied by spraying; because of their toxicity, wind drift to non-target areas poses serious risks to human health and to ecosystems.

Potential Solutions: Do not spray pesticides under adverse climatic conditions. Spray in close proximity to the target to minimize wind drift. Minimize pesticide use.

Methane and Other Gases

Problem: Gases produced as manure is decomposed can 1) be toxic (hydrogen sulfide), 2) act as an asphyxiant (carbon dioxide and carbon monoxide), 3) be corrosive (ammonia), and 4) be explosive (methane). Roofed or enclosed storage areas may be especially hazardous since gas concentrations can reach dangerous levels.

Potential Solutions: Carefully monitor gas levels and provide adequate ventilation, especially during manure agitation and pump-out. Use filters on ventilation systems to prevent environmental releases. Use anaerobic digesters to treat wastes and collect methane for fuel use.

Checking Your Understanding

1. Name five nonpoint source agricultural pollutants.

2. What are two ways to reduce soil erosion?

3. What two nitrogen-containing materials can add nutrients to groundwater supplies?

4. Define bioaccumulation.

5. List four gases produced by decomposing manure.

22-4 Point and Nonpoint Source Pollution

Point Source Pollution

The proliferation of concentrated animal production facilities and chemically dependent monoculture farms has created serious localized air and water pollution problems. These pollution sources are best characterized as point sources. Examples of these sources include 1) animal waste storage facilities at factory farms, 2) animal feedlots, 3) fertilizer storage and mixing facilities, 4) pesticide storage and mixing facilities, and 5) pesticide dips used to control external parasites on livestock. Agricultural point sources do not typically resemble the "end-of-pipe" sources that are generally associated with the term point source pollution. However, they are non-mobile sources of relatively concentrated wastes, and as such they are easily identified sources of pollution.

Nonpoint Source Pollution

On the other hand, the origin of nonpoint source pollution is difficult to identify. Nonpoint source pollution originates over a wide geographical area. Nonpoint sources include urban runoff; agricultural soil erosion and runoff of chemicals from agricultural fields; leaching of agricultural chemicals into groundwater; and atmospheric deposition of contaminants from air pollution. Nonpoint sources often change temporally and spatially, making it difficult to predict when and where products will be used; for example, a specific pesticide will have to be applied whenever an outbreak of pests occurs, which is not predictable. Because nonpoint sources of pollution are usually widespread, intermittent, and undefined, solving a water quality problem caused by nonpoint source pollution is often difficult.

Checking Your Understanding

1. Name four point source pollutants that are associated with agriculture.

2. What are some of the likely sources of nonpoint sources of pollutants?

22-5 Regulations, Permits, and Programs

Clean Water Act (CWA)

Animal waste discharges from large confined livestock operations are regulated under Section 404 of the Clean Water Act. Section 319 requires the states to submit assessments of water pollution from nonpoint sources including agricultural sources. Lack of progress in controlling nonpoint-source pollution has prompted calls for revisions to the program.

Coastal Zone Management Act

The Coastal Zone Management Act Reauthorization Amendments of 1990 required states to develop strategies for protecting coastal water quality. These strategies include measures relating to nonpoint source pollution from agriculture.

Federal Insecticide, Fungicide, and Rodenticide Act (FIFRA)

FIFRA regulates pesticide manufacture, use, and disposal. Under FIFRA, EPA may restrict, cancel, or temporarily suspend pesticide uses that pose unreasonable risks to the environment or human health. Pesticide benefits are considered in these decisions. Each state is required to design and implement a plan to protect groundwater from pesticide contamination. These state plans may include severe local restrictions or bans on the use of those pesticides shown to be problem contaminants. Additional provisions in FIFRA that impact agricultural activities include:

—Registration Sunset – All pesticide registrations must be renewed every 15 years.

—Pesticide Phase Out/ Phase Down – A new restriction process to kick in if EPA concludes that a pesticide is "reasonably likely to pose a significant risk to the environment or humans."

—Reduction of Pesticide Risks and Pesticide Use – An effort by EPA to reduce pesticide risks by reducing use and promoting IPM, the latter requiring training to be included in certification training programs.

—Pesticide Use Record Keeping – Require record keeping for all pesticide uses, not just restricted use products.

—Enforcement Authorities – More inspections, record-keeping audits, addition of "whistle blower" and citizen suit provisions, as well as significant increases in penalties for improper use.

Endangered Species Act

The Endangered Species Act is designed to protect the habitat and encourage reproductive success of threatened or endangered species listed by The Fish and Wildlife Service. It expressly prohibits "taking" a listed species, meaning to "harass, harm, hunt, shoot, wound, kill, capture the species or modify/ degrade their habitat in such a way as to alter their behavior or impair their lives." Agricultural activities (including pesticide use) that could potentially harm an endangered species or its habitat are restricted by this regulation. As the list of endangered species grows and efforts to protect habitat expand (e.g., clean water for aquatic species, riparian areas, and wetlands for breeding, etc.), the impact of the law and regulations on pesticide use will most likely increase.

State and Local Regulations

States use a variety of approaches for addressing water quality problems: controls on inputs or practices, controls on land use, economic incentives, and education programs. Most states have passed laws or instituted programs that protect water quality by affecting some aspect of agricultural production associated with nonpoint source pollution.

Many states regulate pesticide use, and most require certification of pesticide applicators. Some states restrict where particular chemicals can be used, usually in response to observed groundwater problems.

Land-use laws are used by municipalities, counties, and other local governments to control the environmental impacts of agricultural activities. Land-use controls include zoning, land acquisition, and easements targeted to areas deemed critical for protecting water resources. Zoning ordinances are used in many areas – especially around the rural-urban fringe – to ban confined animal operations.

Farm Policies and Programs

Many farm activities are driven not by regulations but by government incentive programs. The preference for voluntary approaches is based on the inherent difficulty in regulating nonpoint sources of pollution, and on the belief that when educated about the problems and provided with technical and financial assistance, farmers will make improvements in production practices to achieve conservation and environmental goals. Some examples of programs that affect environmental quality are given below.

The Federal Agriculture Improvement and Reform Act of 1996 (FAIR) created several programs to address environmental protection. FAIR promotes environmental protection in the following ways:

— It provides federal matching funds to state and local farmland protection programs;

— It reauthorized the wetlands programs established by the 1990 Farm Bill;

— It established an agricultural air quality task force;

— It authorized funding to preserve farmland; and

— It integrated several environmental programs. The Environmental Conservation Acreage Reserve Program (ECARP) now encompasses the Conservation Reserve Program, the Environmental Qualities Incentives Program, and the Wetland Reserve Program. These programs provide incentives to farmers 1) to implement multi-year farm management plans that protect water quality by reducing the use of fertilizer, manure, and pesticides; 2) to retire land with highly erodible soil from production; 3) to use IPM, crop rotations, and other methods that promote efficient chemical use; and 4) to return prior converted or farmed wetlands to wetland condition.

Checking Your Understanding

1. What are five provisions of FIFRA that impact agricultural activities?

2. What are five ways the Federal Agriculture Improvement and Reform Act of 1996 promotes environmental protection?

Summary

The agricultural industries produce basic foods including meats, grains, fruits, vegetables, mushrooms, dairy products, and eggs as well as agricultural products that are used for animal feeds, fertilizers, fibers, and raw materials for other industries. In the last fifty years in the United States, there has been a dramatic shift away from the family farm that produced small amounts of many products to large factory farms that produce one single commodity.

Crop production continues to be heavily dependent on the weather and soil conditions as well as on crop varieties, fertilizers, and the availability of a large arsenal of pesticides to control weed and insect infestations. The heavy use of pesticides in the past has resulted in widespread environmental contamination today. One alternative that shows great promise is the use of integrated pest management, which relies on the use of natural predators, rather than chemicals, to provide the control. FIFRA regulates pesticide manufacture, use, and disposal.

Animal husbandry is the branch of agriculture that is responsible for the production of meat, dairy products, and eggs as well as fibers such as wool. The move toward factory farms or concentrated animal production facilities poses many new health and environmental problems. For example, it is more difficult to maintain healthy animals when they are living within close quarters; also, the handling and storage of large amounts of animal wastes give rise to nuisance odors, ammonia, and flammable methane gas.

Both crop and animal production are major sources of point and nonpoint runoff. Two known point sources in crop production are the storage and mixing facilities for both fertilizers and pesticides. Known point sources from concentrated animal production facilities include animal waste storage facilities, feedlots, and pesticide dip areas used to control parasites. Nonpoint sources are widespread, varied, and intermittent, but most agricultural sources are the result of occasional runoff. Soil erosion of cropland can result in sediments that cloud the water, choke watercourses, and interfere with aquatic life cycles. Runoff from animal production facilities provides nutrients that accelerate the growth of aquatic plants and microorganisms and may lead to eutrophication problems in addition to polluting groundwater. The CWA regulates waste discharges from large confined livestock operations and requires states to submit assessments of water pollution from nonpoint sources, including agricultural sources.

Critical Thinking Questions

1. What sort of programs, policies, techniques, incentives, or penalties do you think should be used to reduce nonpoint source pollution from agriculture?

2. What are the advantages and disadvantages of concentrated animal production facilities? Please consider economic, environmental, and societal issues in your answer. Name at least six advantages and six disadvantages.

23

The Food and Beverage Industries

by Walter Cordell, III

Chapter Objectives

Upon completing this chapter, the student will be able to:

1. **Describe** the processing steps used in each of these industries.

2. **Identify** common waste streams generated by each of these industries.

3. **Describe** opportunities to minimize and reduce waste in these industries.

4. **Evaluate** the economics of source reduction, recycling, and treatment technologies in these industries.

5. **Cite** specific regulations and permits that pertain to the food and beverage industries.

23-1 Introduction

The food industry and the beverage industry produce similar waste streams – both wastewater and solid waste – and similar environmental concerns. Each industry discharges wastewater high in Biochemical Oxygen Demand (BOD), which is an indication of the organic waste present. Each industry also discharges a wastewater stream that is acidic, which may corrode processing equipment and damage sewer pipes that carry the wastewater to the treatment plant. If not neutralized, the acidity can also kill the microorganisms used at the wastewater treatment plant to remove the various organic wastes. Regardless of where the wastewater is discharged, it must undergo some pretreatment to lower its organic content and raise its pH. For most facilities, pretreatment is required to meet regulation or permit requirements prior to discharge.

In the following sections, three examples are presented: tomato canning as an example of fruit processing and canning; soft drink bottling; and red meat processing. The tomato canning and red meat processing industries produce wastewaters high in total suspended solids (TSS), which are solids that remain suspended and do not easily settle out or float. Total suspended solids result in turbid water, which reduces sunlight penetration, thus adversely affecting aquatic life. The tomato canning and red meat processing industries also produce a solid waste stream. In tomato canning, dirt, stems, and leaves are added to the wastewater stream from washing the fruit. In addition, peels and seeds enter the waste stream during processing. These are screened out prior to discharging the wastewater from the plant site, but become part of the solid waste stream produced. In red meat processing, skinning and washing also generate TSS. These also become part of the solid waste stream when they are screened out.

23-2 Fruit Processing and Canning

Tomato Canning

The tomato canning industry produces many different products. Among them are ketchup, tomato paste, tomato sauce, canned tomatoes, salsa, barbecue sauce, and pizza sauce, to name just a few. The initial step in preparing tomatoes for canning, like most fruit, consists of washing the fruit to remove inedible materials such as dirt, twigs, and leaves. Once clean, the fruit is peeled to remove its tough outer skin, placed into a clean container, and sealed.

Container Preparation and Filling

Preparing the containers involves the rinsing of the inverted cans with hot water or blasting them with compressed air. If rinsing is used, a small clean wastewater stream is generated that can be reused elsewhere in the process. Some manufacturers also perform an empty container inspection prior to filling to eliminate defective containers. After cleaning, the containers are filled with fruit, sealed, and placed in a **retort** to be cooked. The retort is pressurized with steam to cook the fruit in the container, and the pressure is controlled to prevent the cans from rupturing.

Major Waste Streams

The tomato canning industry has two major types of waste streams: wastewater and solid wastes. The first wastewater is generated while unloading the tomatoes. Water is used to wash the tomatoes from trucks into water-filled **conveyance flumes**. The flumes are stainless steel or cement-lined trenches in which tomatoes are transported to elevators and delivered to the processing building. The fruit is processed as whole peeled tomatoes, using large quantities of water to do the peeling. In the whole peel process, the tomato is first heated slightly to split the peel, then quickly cooled with a jet of water, which also removes the peel in one piece.

Wastewaters high in BOD are generated by the unloading, washing, and peeling processes, partly because water-soluble sugars are dissolved from the fruit. The wash water also contains suspended or-

ganic solids such as seeds, leaves, broken fruit pieces, and dirt. When screened out of the wastewater, these materials become a part of the solid waste stream generated. Figure 23-1 indicates the steps in which the tomatoes are washed and sorted as well as those in which the wastewater and solid wastes are separated. A large amount of wastewater is also produced by performing routine cleanup of the processing facilities and equipment.

A typical cannery – depending on its water efficiency – will process 3,000-7,000 tons of tomatoes per day and discharge 3-7 million gallons of wastewater. Closer analysis reveals that for each ton of tomatoes processed, nearly 1,330 gallons of wastewater, six pounds of TSS, and eight pounds of BOD wastes are also produced.

Since U.S. tomato canneries process approximately 12,000,000 tons of fresh tomatoes per season and have an average discharge rate of six pounds of TSS per ton, about 72,000,000 pounds of solid material are discharged per season. Off-season production of tomato paste, which had been stored during the fresh season, may result in yet another 72,000,000 pounds of solid material. This means that tomato processing for just one season produces nearly 145,000,000 pounds of solid wastes to be disposed of or used in other ways. A large percentage of these solid wastes are recycled into useful by-products, such as animal feed. Some are applied to land as a soil supplement. Others go to a landfill.

Waste Stream Pollutants

There are several types of pollutants contained within the two major waste streams: sewer screenings, wet wastes, and pomace in the solid waste stream, and suspended and dissolved organic matter (TSS/BOD) in the wastewater. **Sewer screenings** are the materials removed from the wastewater by rotary or stationary screens, prior to discharge to the local sewage system. Rotary screens are more efficient, but stationary screens offer lower maintenance and operational costs. Flocculants may be added to enhance the efficiency of the screening process. Sewer screenings are not suitable as animal feed because they contain treatment chemicals and possibly metal contaminants, so they are disposed of as a soil supplement by land application. Where such applications

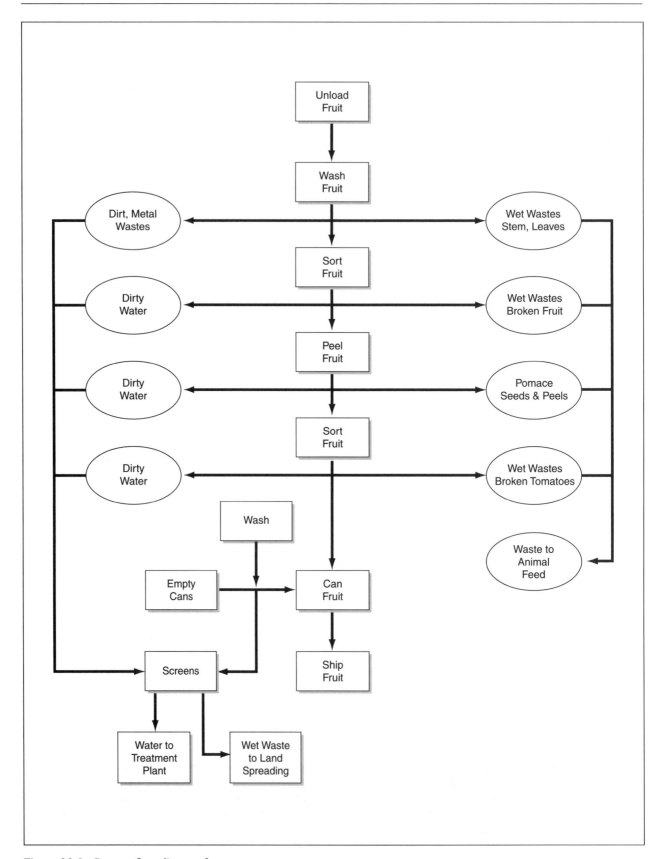

Figure 23-1: Process flow diagram for a tomato cannery.

are either not available or not economical, the sewer screenings are sent to a landfill.

Wet wastes consist of stems, leaves, broken tomatoes, and some inorganic contaminants. These are generated during the initial washing, the various rinsing stages, the sorting, and screening processes throughout the cannery. These wastes have some nutritional value and are typically sold as animal feed. **Pomace** is a waste stream composed of tomato seeds and peels that is produced during the middle processing stages and collected by screening process. Because pomace is high in nutritional value, it is used as a high-grade supplement for animal feed.

Knowing the percentage of soluble BOD in the waste stream is critical in determining the treatment needed to remove it. Much of the waste from a tomato cannery is water-soluble; highly soluble wastes are difficult to remove and usually require biological treatment methods. Approximately 35 percent of the wastewater discharged from a cannery is generated during unloading and conveyance of the tomatoes through processing. Broken and otherwise damaged tomatoes release water-soluble substances that contribute to a high BOD wastewater. An additional 35 percent is generated while peeling the tomatoes. Both waste streams are typically discharged directly to the sewer. The remaining 30 percent derives from miscellaneous sources within the manufacturing plant, such as equipment and container washing. The trucks delivering the tomatoes are also a major source of high BOD waste. **Liquid serum** – the fluid from damaged tomatoes – collects in the bottom of the trucks and is washed into the conveyance flumes.

The TSS in the wastewater stream come from dirt, stems, leaves, peels, and seeds, as well as some portions of the tomato body itself. Once screened from the wastewater, these materials become components of the solid waste stream. The wet wastes and pomace that make their way through the process and sewer screens also become a part of the TSS.

Wastewater Pretreatment Methods

Canneries use several wastewater pretreatment methods. Pretreatment is usually a requirement by the municipality, the state regulatory agency, or the NPDES permit. Such pretreatment methods as pH adjustment, mechanical screening, chemical treatment, aerobic and anaerobic digestion are common.

Adjustment of pH is critical to avoid damage to equipment, underground sewer pipes, and microorganisms. Damage to microorganisms is a particular problem created when large volumes of acidic wastewater are released near a treatment plant and do not have time for sufficient dilution or neutralization to occur. Adjustment of the pH prior to discharge is also important to prevent crop damage, if the cannery wastewater is to be used for agricultural irrigation purposes.

Mechanical screens are used primarily to remove suspended solids, but in some cases they are also used to reduce high BOD waste. Stationary screens and rotary screens with mesh openings of 0.01-0.05 inches are common. Screens are often used prior to all other forms of pretreatment to remove larger solids that may hinder other treatment processes. Flocculants are frequently added prior to screening in order to agglomerate smaller particles so they can be removed. The flocculants used are either cationic (positively charged) or anionic (negatively charged) polymers. Since overall cannery wastewater tends to have a slight negative charge, a positively charged polymer would be used in a single polymer system. A positively charged polymer followed by a negatively charged polymer would be used in a dual polymer system. Since polymers are expensive, this form of pretreatment is used only when maximum solids or BOD waste removal is necessary.

Biological treatment is used to remove BOD from the wastewater stream by converting soluble materials to insoluble materials, which can then be removed. Aerobic treatment processes use bacteria to convert the soluble organics to carbon dioxide and bacterial floc, which can subsequently be removed from the solution. Anaerobic treatment methods use anaerobic bacteria to convert the soluble organic compounds to methane and bacterial floc.

Source Conservation and Reduction Techniques

The tomato canning industry has been actively conserving resources – primarily water – for many years. In 1988, for example, tomato canneries used approximately 1,700 gallons of water per ton of tomatoes processed. By 1992, water usage had been reduced to about 900 gallons per ton. If the industry processes 12,000,000 tons of tomatoes per year, this water reduction represents an annual saving of 9.6 billion gallons of water for just this industry.

It is quite expensive to use biological treatment methods to reduce the BOD of a waste stream. The preferred alternative is, therefore, source reduction or other pollution prevention techniques. In an area where a cannery discharges its wastewater into a city sewer system, treatment charges for removal of organic wastes and total dissolved solids can be as high as $8 per pound. Therefore, to treat 1,000 pounds of high BOD waste or TSS can cost as much as $8,000! Conversely, for every 1,000 pounds of BOD waste and TSS removed by source reduction, $8,000 less is spent on treatment costs. For a typical 100-day fresh tomato season, this represents a saving of $800,000.

The first step in source reduction is to determine the processing areas that contribute pollutants to the wastewater stream. Reducing fruit breakage is the key step to reducing the amount of organic material in the wastewater when processing tomatoes. The more carefully tomatoes are handled, the less likely they are to break. Total suspended solids are introduced during tomato unloading, so a gentler and more efficient unloading process can help prevent solids that will later need to be removed.

Source reduction can also include modifying existing process equipment or adding new machinery. Peelers can be modified to use less water. Improved water collection systems under the processing equipment can collect the heavily BOD-laden water for product recovery, so it does not enter the waste stream. Concrete tomato flumes can be lined with smooth materials, like plastic or stainless steel, to reduce abrasion on the tomatoes during unloading and transport. Any modifications that result in less damage to the fruit will lower the BOD of the wastewater.

The addition of screens to remove tomato pieces at various points in the process presents opportunities to recover some product and reduce organic waste in the wastewater. Tomato pieces and heavily laden streams can be directed to other processes in the cannery and used as a raw material in such things as juice making. The use of centrifugal separators can remove dense particles from the water of the tomato conveyance systems, reducing the number of water changes needed to keep the flume water relatively clean. Several of the waste recovery techniques used to reduce the amount of organic wastes in the wastewater also concentrate the solid wastes. Once removed from the waste stream, the solid wastes become valuable by-products that can be disposed of by land application or, as a last resort, by landfilling.

Checking Your Understanding

1. What are the waste stream similarities between the tomato canning and the red meat processing industries?

2. How are tomatoes whole peeled?

3. What is pomace?

4. What is the source of liquid serum?

5. What is the largest waste stream generated by tomato processing?

6. What is the first step in wastewater source reduction?

23-3 Soft Drinks

There are nearly 450 different types of **soft drinks** on the market, and they account for 27 percent of all beverages consumed in the United States. In 1995, more than 13 billion gallons of soft drinks were consumed, bringing their retail market share to $52 billion. Not bad, for a product that is typically composed of 90 percent water, 9 percent flavoring, and one percent carbon dioxide.

While the purified water and carbon dioxide are well known ingredients, the flavoring composition is a safely and jealously guarded secret among the various manufacturers. The flavorings are composed of various percentages of the following ingredients: colors, caffeine, **acidulants**, preservatives, potassium, sodium, and sweeteners. The flavorings and carbon dioxide cause the soft drink to be slightly acidic, which is desirable as it adds a pleasant tartness.

Resource Conservation

Since soft drinks are composed primarily of water – the industry uses more than 10 billion gallons annually –there is no direct way to reduce this water usage. As in most industries, however, soft drink bottlers are reducing water and product losses through improved container filling technology. Waste audits and material balances are most concerned with product losses after the mixing tank, where the expensive ingredients are added. Any equipment that leaks more than a few gallons of product per day represents a considerable cost both in product loss and in wastewater treatment cost.

An area of greatest concern in recent years has been the increasing number of beverage containers entering the solid waste stream. Americans generated about 196 million tons of solid waste in 1990; one percent of that – corresponding to almost 2 million tons – was soft drink containers. Soft drink containers are made from aluminum, plastic, and glass, which makes them prime candidates for recycling efforts. Of the 82.8 billion containers shipped in 1995, 64.6 billion were aluminum, 16 billion were plastic, and 2.2 billion were glass. Of these, 57.5 billion, or nearly 70 percent were recycled. Aluminum and glass containers can be melted and made into new containers. PET (polyethylene terephthalate) plastic containers are typically shredded, melted, and remanufactured into carpet fiber and fiberfill.

In recent years technology has been able to reduce both the amount of material required and the weight of soft drink containers. Since 1972, the amount of aluminum needed to manufacture a 12 fluid ounce beverage can has been reduced by 35 percent; the amount of glass needed to produce a 16 fluid ounce beverage bottle has been reduced by 44 percent; and since 1977, the average weight of a PET soft drink bottle has been reduced by 25 percent. Such efforts have resulted in lowering not only the amount of material used, but also the amount of energy required to produce and deliver the product.

Major Waste Stream and Pollutants

The major waste stream from the soft drink and beverage industry generally contains sugar and has therefore a high BOD. Soft drink effluents are also acidic, making pH control necessary. The primary sources of BOD and low pH are spills from the mixing and bottling areas. Floor washes, filter backwashes, can and bottle rinses as well as cooling tower discharges also contribute to the wastewater produced. Figure 23-2 shows the basic soft drink production and bottling processes.

Pretreatment Methods

There are three methods of pretreatment suitable for wastewater streams produced by the soft drink industry: anaerobic and aerobic biological treatment and membrane technology. Since the biological treatment methods require a large capital expenditure, are expensive to operate, and require large areas for installation, most bottlers use membrane technology – especially if their wastewater volumes are less than 200,000 gallons per day. Membrane technology, however, may require additional systems for wastewater conditioning to meet the membrane manufacturers' performance guarantees. Such conditions as temperature, oil and grease concentrations, pH, and chlorine content of the wastewater may need to be modified to meet the performance guarantees.

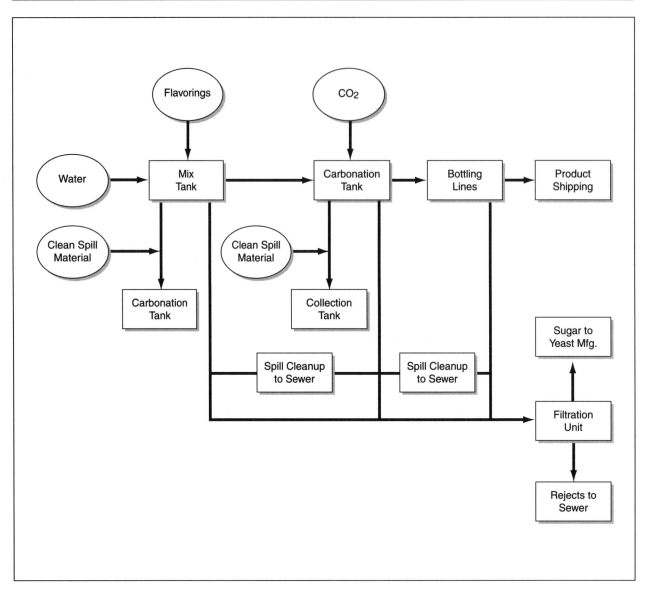

Figure 23-2: A flow diagram of soft drink bottling processes.

Membrane Technology – The choices for membrane technology include low pressure reverse osmosis, high pressure reverse osmosis, or nanofiltration depending on the characteristics of the wastewater. Membrane technology is typically used on the wastestreams with the highest organic concentrations because they produce a concentrated sugar by-product, which can be sold. All types of membrane technology rely on synthetic membranes of various pore sizes, and each one operates under a slightly different set of conditions. Low pressure reverse osmosis units typically operate at pressures lower than 10 psi, while high pressure units may operate at pressures as high as 30 psi. Nanofiltration units use membranes with a much smaller pore size and operate at pressures lower than 20 psi.

Source Conservation and Reduction Techniques

A typical soft drink bottling plant discharges approximately 100,000 gallons per day of wastewater containing an average BOD of 4,000 ppm. This translates into about 3,500 pounds of BOD per day. Most of this wastewater results from rinsing and sanitizing the equipment, since many bottlers produce several different products using the same bottling lines. The sugar concentrated through membrane technology can be turned into a valuable product by, for example, selling it to a manufacturer who can use it as a medium for growing yeast. In addition to yielding a profit, this operation has also eliminated a wastestream. With an average treatment cost of $2-$4 per pound of BOD, up to $13,000 per day can be saved in expensive biological pretreatment costs, while also saving resources by generating a marketable by-product.

Checking Your Understanding

1. What is the composition of the average soft drink, given in percentage values?

2. What is the largest wastestream generated by the beverage industry?

3. What type of technology is typically used to treat wastewater in the beverage industry?

23-4 Red Meat Processing

The **red meat** processing industry in the United States is a $60 billion business that includes more than just the traditional prime beef products and ground beef patties. It also includes top grade pork (except for sausage) and lamb products as well as the associated by-products such as **organ meats**. To simplify the understanding of the environmental concerns for these industries, the beef industry will be used as an example.

A typical beef processing plant will slaughter about 4,000 heads of cattle per day. Nearly every part of the animal is used. Red meat for human consumption accounts for about 400 pounds or 62 percent of the animal's live weight. The fat, hides, and **paunch wastes** are the primary nonedible by-products with fat comprising 100 pounds or about 18 percent of the live weight, while hides and paunch wastes make up 50 pounds or eight percent each of the live weight. The remaining four percent is hooves and bone pieces, which are ground into **bone meal**, and the organ meats such as the kidneys, liver, and heart. The hides are sold to tanners who convert them into leather. The paunch wastes and bone meal are sold as cattle feed supplements and fertilizer.

Major Waste Streams and Pollutants

As suggested earlier, red meat processing also has two major waste streams: solid wastes and wastewater. Figure 23-3 shows the major processing steps and the wastes associated with each of these steps.

The solid waste stream is composed primarily of manure, meat and bone fragments, and paunch wastes. Paunch wastes are the undigested stomach contents of the animal at the time of slaughter and it is removed either mechanically or by using high pressure, low volume water. Some paunch wastes are fed back to the livestock, since it still has some nutritional value: 65 percent water, 25 percent protein, and 10 percent carbohydrate.

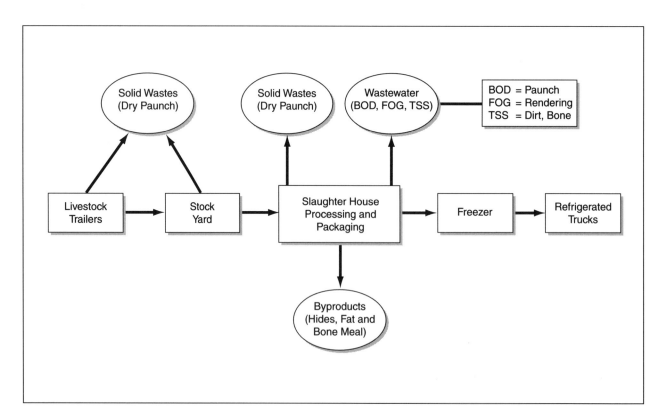

Figure 23-3: Flow diagram of red meat processing.

The primary pollutants in the wastewater stream are BOD, TSS, and **FOG** (fats, oil, and grease). The BOD consists of paunch wastes as well as blood and bits of animal flesh. Regardless of the paunch waste removal method used, some water is generated in the cleaning step. TSS are generated by washing the dirt from the animal prior to slaughter, as well as bits of hair, hide, and bone. As previously discussed, BOD wastes in the wastewater are removed by biological means, while TSS can be removed by mechanical screening. FOG entering the wastewater stream, however, can be removed by either chemical and or mechanical means.

Both the fat associated with the various by-products and the bone meal are removed by a process known as **rendering**. Steam is used as the heat source for melting the fat and the process is therefore known as wet rendering. Through the use of large kettles, this method can render up to 200 tons of material per day. The FOG portion is separated and sold to a soap manufacturer to be used as raw material.

Pretreatment Methods

There are three methods of pretreatment suitable for wastewater streams from slaughterhouses: mechanical screening, biological treatment, and dissolved air flotation. Mechanical screening is used to remove solid materials such as dirt, bone, and hide from the wastewater stream. For wastewater streams high in BOD wastes, either aerobic or anaerobic biological treatment is needed to remove the organic material.

If large amounts of FOG exist, then a dissolved air flotation separation technique may be used. The FOG are typically agglomerated through the use of polymers and then passed into a tank containing air bubbles. The air bubbles attach to the agglomerated fat globules and cause them to float to the top, where they are skimmed into a fat collection tank. It is important to control bubble size, since smaller bubbles provide more surface area and improve efficiency. Any method used either to reduce the BOD waste or to remove FOG will create a solid waste stream. Disposal of these solid wastes can occur either by simply selling them or by using them as soil supplement.

Source Conservation and Reduction Techniques

Little can be done to reduce the manure solid waste stream generated by this industry. However, water consumption during animal processing has been reduced by 50 percent since 1987: it now takes about 600 gallons of water per animal instead of 1,200. Due to the reduction of water usage the waste from this operation is mostly solid, which is desirable, especially when an inexpensive way of disposing of solid wastes is available, as is the case with land spreading.

It is obviously most cost effective to remove from wastewater those pollutants that require the most expensive treatment methods. While TSS wastes are easily and inexpensively removed, BOD and FOG wastes are not. In the slaughterhouse most BOD wastes are generated from paunch removal and cleaning, blood, and by-product wastes. Using a dry method of paunch content removal would reduce the BOD wastes from that source. Reduction in the amount of FOG in the wastewater could be accomplished by using a dry rendering process, where burners replace steam as the heat source. This would result in the FOG containing less water and, if collected separately, it would keep it from entering the wastewater stream.

Checking Your Understanding

1. What products are considered to be red meat?

2. What are four types of wastes that are present in the solid waste stream?

3. What are the three types of wastes present in the wastewater stream?

4. What is the purpose of the rendering process?

5. What type of waste can be removed through the use of dissolved air flotation?

23-5 Environmental Permits

Several federal environmental regulations as well as many state and regional regulations are applicable to the food and beverage industries. This section will focus on the federal regulations found in 40 CFR. Some reference to local enforcement of the federal regulations, however, may be necessary to understand the compliance scheme. The majority of the environmental regulations that impact the food and beverage industries are set forth in three important laws and their amendments: the Clean Air Act (CAA), the Clean Water Act (CWA), and the Comprehensive Environmental Response, Compensation, and Liability Act of 1980 (CERCLA).

Clean Air Act

The most recent amendment to the Clean Air Act is Title 5, which had an immediate impact on the industry and undoubtedly will continue to impact it for years to come. In the past the CAA, through local air agencies, permitted and regulated primary pollutants from combustion sources, such as boilers. Initially, adding new emission sources only required obtaining a new permit. Typically, there were no operating restrictions, only maximum annual amounts of pollutant emissions. These regulations eventually evolved to limiting these amounts to 30 ppm for nitrogen oxides, NOx, and 200 ppm for carbon monoxide, CO.

Title 5 has now expanded the scope of the law to include other emission sources, including hazardous air pollutants (HAPs), to the list of pollutants considered. Under the new provisions, a facility will be regulated on the basis of its potential to emit, rather than its actual emissions as was done in the past. The addition of new pollution sources will require that BACT (Best Available Control Technology) pollution control devices be installed. In many areas, continuous monitoring equipment is required, and the data are sent hourly to the regulating agency via a data management system. While the old permits were one or two pages long, Title 5 permits are likely to be several pages long. These permits closely regulate emission sources and in some areas these permits have limited boiler stack emissions to 25 ppm SOx, 15 ppm NOx, and 200 ppm for carbon monoxide. The regulating agency is also requiring monthly monitoring reports.

Clean Water Act

The primary effect of the CWA on the food and beverage industry is through the administration of the National Pollutant Discharge Elimination System (NPDES). The NPDES program regulates point sources, such as individual industries and publicly owned treatment works (POTW), that discharge to the surface waters of the United States. Permit conditions can vary by location, but typically wastewater parameters such as BOD, COD, pH, residual chlorine, turbidity, and temperature, as well as oil and grease are regulated. Facilities that discharge directly to surface waters usually have very stringent limits. In almost all cases, these limits will require some type of pretreatment. Typical limits may be as follows: BOD: 10 ppm; COD: 10 ppm; pH: 6.5-7.5; residual chlorine less than 0.5 ppm; and temperature within five degrees of the normal temperature of the water body.

Facilities that discharge indirectly to surface waters through a POTW will have a wastewater discharge permit that specifies the conditions allowing it to meet its NPDES permit conditions. If the wastewater treatment plant has sufficient treatment capacity, the industries discharging to it may have no pretreatment requirements; however, discharge fees are likely to be so expensive that it is more economical for companies to install their own pretreatment facilities. The POTW permits are likely to contain limits on many of the same pollutants as listed above. In addition, the NPDES program requires periodic testing for heavy metals, such as lead and copper, and for total toxic organics, such as pesticides and PCBs. Monthly reports on test results are typically required by the regulating agency, showing results of the testing in tabular format.

The CWA also regulates the storage of both petroleum and vegetable-based oils through the Spill Control and Countermeasure Plan (SPCC). Since many food processors store fuel oil for boilers and vegetable oil for food ingredients, they are typically covered by this regulation. The SPCC regulation requires preparing a spill plan if the facility stores more than a threshold quantity of oil, which is currently 1,320 gallons above ground or 45,000 gallons underground. The spill plan contains those policies and procedures to prevent spills as well as the actions to be taken in the event of a spill.

Comprehensive Environmental Response, Compensation, and Liability Act

Recent amendments to CERCLA (commonly known as Superfund) have created new hazardous material reporting requirements applicable to the food and beverage industries. In 1984, the Superfund Amendments and Reauthorization Act (SARA) Title 3 required the submittal of two forms: Tier II for the storage and use of listed chemicals and Form R to report the release of listed hazardous materials into the environment. The food industry typically uses large amounts of chlorine for water disinfection and ammonia for refrigeration systems. Both of these chemicals are listed in Title 3 and require annual submittal of the use (Tier II) and emissions (Form R) report forms.

Checking Your Understanding

1. Which amendment to the CAA has an effect on the food and beverage industries?

2. What wastewater limits are regulated under an NPDES permit?

3. What reporting forms are required under SARA Title 3?

Summary

Fruit processing and canning is comprised of two preparatory steps: food preparation and container preparation. Food preparation involves washing and pitting/peeling the fruit. Container preparation involves inverting the containers and washing or blowing them clean.

Fruit canning generates two major waste streams: wastewater and solid wastes. The wastewater is contaminated with organic molecules that cause Biochemical Oxygen Demand (BOD) and Total Suspended Solids (TSS). The BOD is caused by large amounts of soluble organic material, while the solid waste stream consists of dirt, stems, leaves, peels, and seeds. Pretreatment methods to reduce end-of-pipe pollution include pH control, mechanical screening, chemical treatment, and aerobic and anaerobic digestion.

Source reduction techniques include process modifications that cause less fruit breakage, collect heavily BOD-laden water for product recovery, use centrifugal separators to remove heavy materials, and screens to remove lighter pollutants.

Soft drink bottling generates wastewater that is acidic and high in BOD. The BOD comes from the high amounts of sugar in the drinks. Floor washes, filter backwashes, and product losses contribute to the high BOD loading. Pretreatment methods usually involve membrane technology as well as aerobic and anaerobic digestion. Source reduction techniques involve separating out the high sugar streams and selling the recovered sugar to other industries to be used as raw material.

Red meat processing generates a wastewater stream that is high in BOD and FOG, along with some TSS. The BOD comes from wet methods of paunch removal, while the FOG comes from rendering methods used to remove fats from the materials to be spread on land. TSS is from dirt, bone, and hide and can easily be removed by screening. A solid waste stream is also generated, which is made up of manure from the stockyard and paunch wastes. Pretreatment methods include mechanical screening, dissolved air flotation, and biological treatment. Source reduction techniques include dry paunch removal and dry rendering methods.

Critical Thinking Questions

1. What other methods of BOD/TSS reduction might be incorporated into a tomato processing operation?

2. What percentage of the total U.S. solid waste stream is composed of beverage containers? Does this percentage warrant further legislative pressure on the soft drink industry for more recycling efforts? Why or why not?

3. List the two types of rendering processes and how they are similar to or different from each other.

4. How does bubble size affect the operation of a dissolved air flotation unit?

24

The Furniture Finishing Industries

by Margaret Lee

Chapter Objectives

Upon completing this chapter, the student will be able to:

1. **Explain** the five basic steps in the furniture manufacturing process.

2. **Identify** and explain the regulation that drives the wood furniture industry.

3. **Identify** the air pollutants and their sources within the furniture manufacturing industry.

4. **Describe** the different waste reduction techniques available in the furniture manufacturing industry.

Chapter Sections

24-1 Introduction

The steps involved in the making of wood furniture consist of 1) obtaining the desired wood, 2) shaping it into a part, 3) finishing the parts for assembly, 4) assembling the parts, and 5) applying the surface finish. The finished furniture is then packaged for shipment to a retail store. Some of the basic steps in making furniture can be varied; for example, some furniture manufacturers elect to apply the surface finish to the parts prior to assembly, yet others ship the finished parts to the consumer for home assembly. The steps in a typical wood furniture manufacturing and finishing process and their associated wastes are shown in Figure 24-1.

There are about 10,000 furniture manufacturers in the United States, making mostly household and office furniture. Furniture manufacturing has traditionally been located in geographic areas near forests and wood sources.

The finish coating processes in wood furniture manufacturing involve a series of steps depending on the quality of furniture being produced. Typically, finishing processes for inexpensive furniture require only 6-12 steps, while the ones for high quality and expensive pieces require 30 or more steps. Wood finishing processes involve the application of different stains and coatings on the furniture. Each application must be allowed to dry, cool, and must also be sanded. The timeframe for each of these finishing phases varies from 15 minutes to 20 hours. Many different types of coatings can be applied: **stains**, **washcoats**, **sealers**, **toners**, and **topcoats**.

The purpose of a **coating** is to protect the furniture from deterioration and rot. It also highlights wood characteristics, gives color, good looks, and durability to the wood. Liquid coatings typically

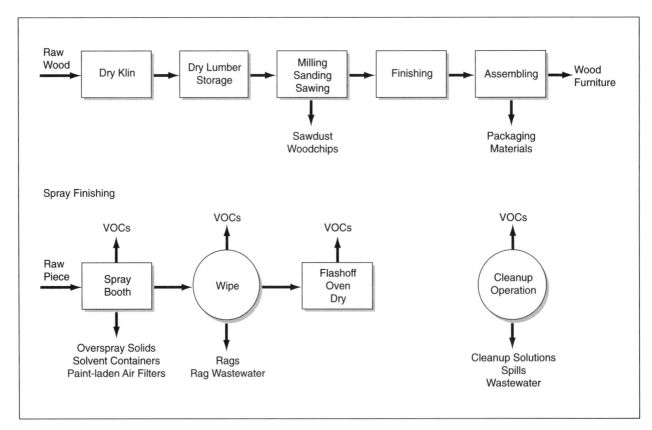

Figure 24-1: Typical steps in wood furniture manufacturing process.

consist of a nonvolatile solute and a volatile solvent. When the coating is applied, the **solvent** allows for the uniform spread of the nonvolatile resins onto the surface. The solvent then evaporates, leaving a dry film of resin on the surface of the wood. Some of the coatings used to protect wood are **polyurethanes**, **polyesters**, **acrylics**, and **nitrocellulose resins**. These resins are typically suspended in either organic or water-base$_d$ solvents. The organic solvents used are alcohols, ketones, esters, glycol ethers, aliphatic solvents, and aromatics. The solvents for water-based resins include alcohols, glycols, and glycol ethers. Regardless of the type of resin used, the coatings are applied to the furniture by brushing, wiping, spraying, rolling, or dipping.

Organic solvents that contain nitrocellulose resins, pigments, and additives have traditionally been used in the furniture finishing industry. The pigment gives the coating a color, while the additives – such as **plasticizers** – increase the workability of the coating. The most common method for applying the finishes is by spraying.

24-2 Major Pollutants and Sources

The major pollutants in the coating process are volatile organic compounds (VOCs) from the evaporation of organic solvents. These volatile organic compounds may be toluene, xylene, methyl ethyl ketone, or methyl isobutyl ketone. They react with nitrogen oxides (NOx) in the air and in the presence of sunlight to form ground-level ozone. The VOCs also have occupational and environmental health impacts. Prolonged inhalation of toluene can cause headache, nausea, impairment of coordination, and loss of memory. High concentrations may cause drunkenness, unconsciousness as well as hepatic and renal damage. Exposure to low levels of xylene produces mucous membrane irritation, nausea, vomiting, dizziness, and loss of coordination.

Emission Sources

Many potential emission sources exist in the finishing processes. Since spray coating is the most common technique to apply the coatings, the major solvent emission sources in a manufacturing plant consist of spray booths, **flashoff areas**, and ovens.

Cleanup operations may also generate VOC emissions. For a typical wood furniture manufacturer, 70 percent of the total VOC emissions occur at the spray booth. The VOC and solids content of various spray finishing emissions are shown in Table 24-1.

Formulation	VOC Content (lbs./gallon less water)	Solids Content (Percent by Volume)
Nitrocellulose	6.0	16
Acid Catalyzed	5.1 - 5.8	18 - 26
Waterborne	1.3 - 2.3	26 - 30
Polyurethane	3.4	30 - 60
Polyester (Acrylic)	3.0	30 - 50
Polyester (Styrene)	negligible	100
UV or EB cured	0 - 3.1	56 - 100

Table 24-1: VOC and solids content of several coating formulations.

Spray booths are manual – the items to be finished are manually placed in and removed from the booth – or automatic – a conveyor system is used to transport the furniture parts. The air emissions from spray booths may be purified by dry or waterwash methods. In the latter method, mostly used in the past, overspray particulates present in the spray booth emissions were removed by being drawn through a series of water curtains. Waterwash spray booths, however, are more costly to install, and the disposal costs of wastewater – considered a hazardous waste – is rising. The dry spray booth is now popular within the industry and requires only the removal of a dry filter containing the solids. Spray booths are also available in various sizes to accommodate the type and size of furniture being coated. Spray booths for the manufacture of residential furniture are usually loaded and unloaded manually and are eight feet high, 19 feet wide, and nine feet deep. Spray booths designed to accommodate office furniture tend to be a bit smaller, but use an automatic conveyor system to move the furniture.

Flashoff areas are set up for the solvent to volatilize before final cure in the oven. These areas are located either between spray booths or between the last spray booth and the oven. Some flashoff areas have forced-air circulation; most, however, do not

have a separate exhaust system. Ovens are used to cure the coatings before the next step in the finishing process. Ovens can be steam-heated using wood, coal, or gas. Infrared or ultraviolet ovens are also used.

During cleanup operations, volatile organic solvents are typically used to clean the application equipment, such as spray guns, feed lines, and coating reservoirs. Cleanup solvents are also used to clean out piping, spray booths, or machinery. VOCs are emitted into the atmosphere during these operations.

Checking Your Understanding

1. What are the five basic steps in the furniture manufacturing process?

2. What are five different types of coatings?

3. What was the typical composition of surface coatings used in the finishing industry?

4. What are the potential emissions sources in the furniture finishing process?

5. What is a flashoff area?

24-3 Regulations and Source Reduction Techniques

The law that is encouraging reformulation of coating products and alternative application technologies is Title III of the Clean Air Act Amendments (CAAA) of 1990. The Amendment was passed by Congress on October 27, 1990 and signed into law by President Bush on November 15, 1990. EPA has been regulating air toxics identified as hazardous air pollutants (HAPs) under Section 112 of the 1970 CAA. However, the 1990 CAAA greatly expanded EPA's rulemaking authority over HAPs. The law lists 189 chemicals that would be subject to control. Major sources that emit 10 tons/year or more of any one pollutant, or 25 tons/year or more of any combination, must apply maximum acheivable control technologies (MACT). Area sources that emit less than 10 tons/year of a single pollutant, or less than 25 tons/year of any combination, will eventually be regulated under the discretion of EPA.

The 1990 CAAA directs EPA to set emissions standards requiring the furniture industry to sharply reduce routine emissions of toxic air pollutants under the National Emissions Standards for Hazardous Air Pollutants (NESHAPs) program. The VOCs, listed as HAPs, released from solvents in the furniture finishing process will be regulated. Several states and localities have already implemented finishing regulations. They are Illinois, Indiana, New Jersey, New York City, Pennsylvania, and Southern California. The regulations set emissions standards based on the mass of VOCs emitted per gallon of applied solids or mass of VOCs emitted per surface area coated. Most of the regulations subtract water and exempt solvents from the standard limits. Exempt solvents include methylene chloride, ethane, methane, chlorofluorocarbons, and hydrochlorofluorocarbons. Regulations are set assuming that all of the VOCs in the coating evaporate at the same time.

Source Reduction Techniques

Several methods can be used to reduce the amount of air emissions generated by the application of coatings. These methods include coating reformulations, spray equipment modifications, and good housekeeping practices. Traditionally, coatings in the industry consisted primarily of nitrocellulose, pigments, additives, and organic solvents. This combination of coating materials emits the most VOCs. By reformulating the coating to a water-based type, the VOC emissions can be cut by more than 60 percent. Table 24-1 shows the approximate VOC and solids content of nitrocellulose and alternative coating types. Besides reformulating the coatings to reduce the VOC emissions, there are also many other advantages: reduction in odor, toxicity, and fire hazard.

Reformulation

By substituting polymeric resins, such as polyurethanes or polyesters, the solids content of coatings can be increased to 20-100 percent and their VOC emissions correspondingly reduced. The polymeric coatings are highly durable; however they require high temperatures – often too high for wood surfaces – to cure. Other disadvantages of these polymeric resin coatings are that they are difficult to spot repair and that dirt cannot be rubbed out of the finish after it has cured.

Ultraviolet (UV), electron beam (EB), and infrared (IR) technologies use various wavelengths of electromagnetic radiation to alter the physical and/or chemical properties of a **radiation-curable coating** by forming cross-linked polymer networks, thus changing it from a liquid to a solid. Radiation-cured coatings are made of organic materials such as polyesters, acrylics, and epoxies. These formulations contain little or no organic solvents; therefore, they produce few VOC emissions upon curing. This type of coating is also difficult to spot repair and cure because of the three-dimensional nature of furniture. Recessed surfaces that do not receive direct exposure to the radiation do not cure properly. Radiation-curable coatings, therefore, are usually applied to panels on flat line operations.

New coating reformulations are under development within the industry. The Unicarb™ coating technology developed by Union Carbide Chemicals and Plastics Company was introduced in July 1991 by Pennsylvania House, a wood furniture manufacturer in White Deer, Pennsylvania. This coating replaces part of the solvent with supercritical carbon dioxide. The supercritical carbon dioxide is compatible with nitrocellulose; it reduces its **viscosity** and minimizes by as much as 80 percent the amount of VOCs released during finishing operations.

VOC emissions can also be lowered by reformulation of cleanup solvents. When possible, cleanup solvents should be substituted with aqueous detergent-type cleaners; however, most aqueous cleaners are not compatible with solventborne coatings.

Process Changes

Another source reduction method is to modify the spray equipment. Many different types of spray systems are available for coating applications. Spray guns are characterized and evaluated based on their transfer efficiencies. **Transfer efficiency** – expressed as a percentage – is defined as the ratio of the mass or volume of solid coating deposited on the part compared to the total mass or volume of coating used during an application step. As spray transfer efficiency increases, the amount of VOCs decrease. Conventional air spray guns utilize compressed air to atomize the coatings, but have low transfer efficiencies and produce high quantities of VOC emissions. By using airless (hydraulic) or combined air/hydraulic types of spray equipment, the transfer efficiencies can also go up and the amount of VOCs go down. Use of a High Volume/Low Pressure (HVLP) sprayer, which utilizes high volumes of low pressure air and has a high speed turbine to atomize the coatings, can also provide transfer efficiencies of up to 70 percent.

Electrostatic spray guns have the highest transfer efficiency: up to 90 percent. These spray guns use electricity to place a negative charge on the coating particles. The charged coating particles are then attracted to the grounded positive furniture part. The wood surface is made conductive by the application of a conducting compound known as a **sensitizer**. Table 24-2 shows the transfer efficiencies of different types of spray guns.

Good Housekeeping Practices

Another source reduction technique is maintaining good housekeeping practices. Emissions of VOCs from cleanup materials and rags can be reduced by limiting their exposure to the air. Solvent-laden rags, for example, should be placed in a closed container so the solvents cannot evaporate. Small parts can be cleaned in a closed container, thereby minimizing emissions. The Gun Washer/Recycler can be utilized, which cleans the gun internally and externally in an enclosed container. The solvent is collected in a container after cleaning; the solids settle and can then be removed, making the solvent reusable. Solvent emissions can also be reduced by limiting the quantity of cleanup solvent available to each worker and providing him/her with proper cleanup training.

The use of a strippable coating on spray booth walls can assist in removing dry solid buildups from overspray. This makes the use of solvents for the removal of overspray solids unnecessary. To prevent the need for excessive washing and wiping with solvent, the furniture pieces should be finished in a clean environment. Dirt deposited on the workpiece requires repair and additional solvent usage.

Spray Gun Type	Coatings	Transfer Efficiency (Percent)
Conventional (Low Volume/ High Pressure)	Solventborne Waterborne	25 - 50
Airless	Solventborne Waterborne	35 - 65
Air-Assisted Airless	Solventborne Waterborne High Solids	40 - 70
Electrostatic	Solventborne Waterborne High Solids Powder Coat UV curable	35 - 90
High Volume/ Low Pressure	Solventborne Waterborne High Solids UV curable	40 - 70

Table 24-2: Transfer efficiencies of different types of spray guns.

Checking Your Understanding

1. What federal law is driving the coatings industry toward source reduction?

2. What are three major areas in which the coatings industry can accomplish source reduction?

3. What is transfer efficiency and how can it be used to rate different types of spray guns?

4. Why is the high-volume low-pressure spray gun better than the low-volume high-pressure spray gun in terms or overspray?

5. What is a Unicarb™ coating?

6. What is an alternative coating to conventional nitrocellulose coating?

Summary

There are more than 10,000 furniture manufacturers in the United States, mostly concentrated in areas near forests and wood sources.The steps involved in the manufacture of wood furniture are: obtaining the desired wood, shaping the parts, finishing the parts for assembly, assembling the parts, and applying the surface coatings. The application of a finish on wood furniture requires from six to more than thirty steps, depending of the quality of the final product. The different types of coatings are stains, washcoats, sealers, toners, and topcoats. Liquid coatings, consisting of nonvolatile resins and volatile solvents, are most often used and applied by air spraying.

The major sources of pollution are VOCs that are generated by solvent evaporation in the spray booth and flashoff areas. The spray booths used vary in size and shape; they may be loaded and unloaded either manually or with a conveyor system, and may use either a waterwash or dry filter system to capture the overspray particles.

The major legislation that is applicable to and driving change in the coating industries is Title III of the Clean Air Act Amendments of 1990, which regulates 189 chemicals, many of which are used in the furniture finishing industry. As a result, the industry is moving to reduce the amount of air emissions generated through coating reformulation, redesign of coating equipment, and implementation of good housekeeping practices. The use of polymeric resins and UV-, EB-, and IR-cured coatings have greatly reduced solvent VOC emissions. The switch to spray systems with higher transfer efficiencies has also reduced the VOC emissions. The highest transfer emissions – up to 90 percent – have been obtained using electrostatic spray guns that put a charge on the coating particles, which are therefore attracted to the grounded target piece. The implementation of good housekeeping practices includes reducing the amount of solvent used during cleanups; keeping solvent-laden rags in conatiners with lids, and using small parts washing systems that are enclosed and allow the reuse of the solvent.

Critical Thinking Questions

1. What is nitrocellulose and how is it produced? What are its advantages and disadvantages as a coating for residential furniture?

2. If the transfer efficiency of a spray system is 30 percent and the total mass of solids used during coating is 320 g, then what is the mass of solid actually deposited onto the furniture part?

3. Smog is formed when photoreactive VOCs are combined with nitrogen oxides in the presence of sunlight. Why are certain solvents, such as methylene chloride, exempted from the 1990 CAAA regulation?

4. Devise an economical way to quickly measure the transfer efficiency of a spray coating system. How do you obtain the necessary measurements?

5. Group Activity – Find out what a state's air regulations are pertaining to the furniture industry. In what ways will the regulations impact a typical wood furniture manufacturing process that utilizes conventional spray finishing process? Discuss possible process modifications, material substitution, or equipment modifications.

25

The Wood Preservation Industries

by J. Simonsen and J. J. Morrell

Chapter Objectives

Upon completing this chapter, the student will be able to:

1. **Describe** the various types of wood treatment processes.

2. **List** the common chemicals used in the wood treatment industries.

3. **Describe** the processes that generate wastes in the wood treatment industries.

4. **Explain** the environmental concerns and governmental regulations that apply to the wood preservation industries.

25-1 Introduction

Wood decays. This is a naturally occurring process caused primarily by fungi and is vital to the maintenance of an ecological balance on the planet. In certain environments, insects and marine borers are also organisms that cause wood degradation. The speed at which the wood decays is a function of moisture content, temperature, species, and availability of atmospheric oxygen. In general, fungal growth is insignificant in wood with a moisture content of less than 20 percent, at temperatures below freezing, or above 43°C. While wood decay is a natural process in the forest, it poses problems for consumers of wood products. It has been estimated, for example, that 10 percent of the wood harvested each year in the United States is used to replace wood that has failed in service due to fungal, insect, or marine borer attack. Attempts to prolong the useful life of wood products through the application of pesticides gave rise to the wood preservation industries.

Wood preservation is an ancient industry; tars and pitches have been used since ancient times to prolong the useful life of wood products. The beginning of the modern pressurized wood preservation industry was brought about by the development of the railway system in England. Decay and termites were destroying the wooden crossties (called "sleepers" in England) faster than they could be replaced. The English engineer John Bethell received a patent in 1838 for the pressure treatment of crossties with creosote. The creosote slowed the destruction of the crossties allowing for the expansion of the English railway system. A similar need for treating crossties with creosote was also found in the building of the North America transcontinental railway system in the 1800s. This chapter will discuss the use of creosote and a variety of other wood preservation chemicals as well as their application.

25-2 Factors Affecting Wood Treatment

A host of factors affects the transport of chemical pesticides deep into the structure of wood. For most treatment processes, the wood species, amount of heartwood, and moisture content at the time of treatment are the major variables. These variables, however, do not greatly affect anti-stain treatments, since these treatments are applied primarily to the surface of wood.

Characteristics of Wood

Treatment procedures and the resulting penetration patterns vary widely based on wood types. Tree species are broadly divided into **hardwoods** and **softwoods**, based upon their leaf morphology and wood structure. Softwoods tend to have fewer cell types and treat more uniformly, while hardwoods have more cell types that often give variable treatment results.

Sapwood and **heartwood** can also affect the treatment outcome. Sapwood is the outer living part of the trunk or stems that conducts fluids between the roots and foliage in the living tree. As these cells mature, they gradually die and form heartwood. The connections between individual cells, called **pits**, generally become **aspirated**, that is they become occluded or blocked, as the cells die. The heartwood of most species, therefore, is extremely resistant to the penetration of the treatment liquid. Blockage of the pits by either surface debris or aspiration is a major problem in all wood species that have relatively thin bands of sapwood, since sapwood tends to block the flow of the preservative into the wood.

Moisture Content (MC)

The **moisture content (MC)** at the time of treatment can also greatly affect treatment results. Since treatment processes are largely designed to coat or fill the cell **lumens** with preservative, the presence of excess moisture at the time of treatment can either block the flow or result in uneven deposition. Wood from a freshly fallen tree may have an MC between 40 and 200 percent (weight basis). Failure to remove this fluid will result in poor penetration patterns. In the case of oil-borne treatment solutions, for example, interaction between the water and the treatment solution can increase sludging and the amount of contamination produced. Most wood, therefore, is dried prior to treatment either by air seasoning or kiln drying to the point (approximately 30 percent MC) where little or no free water is present. The remaining moisture is bound to the wood structure and does not interact substantially with the treatment solution.

Moisture may be removed by several different pressurized processes. The oldest of these is the **Boulton seasoning process** in which the wood is placed in a **retort** along with an oil-borne treatment solution. The solution is heated to 70-100°C, and a vacuum is drawn to lower the boiling point of the water causing it to vaporize from the wood. In this process, a freshly harvested pole may be taken from 80-100 percent to 12-25 percent MC in 24-48 hours. The Boulton seasoning process, however, is energy intensive and the vacuum condensate results in large quantities of oil-contaminated wastewater. Moisture can also be removed by **steam conditioning**. In this process live steam is used to heat the wood for 17-20 hours. A vacuum is drawn, which lowers the boiling point of the water in the wood below the wood temperature, thus **flashing** off the water, which is recovered in the vacuum condensate. Regardless of the method used, once a majority of the water has been removed, the wood is pressure treated using the appropriate **impregnation** process.

As environmental regulations have become more restrictive, more wood preservation plants are electing to dry the wood prior to treatment. This practice avoids the need to process considerable amounts of preservative-contaminated wastewater that has been removed by the Boultonizing or steaming processes.

Checking Your Understanding

1. What organism is primarily responsible for the decay of wood?

2. Above and below what temperatures is there likely to be little decomposition of wood?

3. Who developed the first pressure treatment process using creosote to treat wood?

4. What are three things that have an effect on the transport of chemical solutions into wood?

5. How do softwoods and hardwoods differ?

6. What are the three steps involved in the Boulton seasoning process?

25-3 Wood Treatment Processes and Chemicals

Wood Treatment Chemicals

Since the use of creosote as a wood preservative became widespread, a variety of other substances have been developed, blended, and tested. By the very nature of the task at hand – being an effective pesticide – the substances found that were successful in prolonging the useful life of wood have also been substances that are generally harmful to the surrounding environment. Table 25-1 lists some of the more common wood preservatives and indicates their chemical composition and typical uses. In reviewing the table, it should be noted that such substances as arsenic, hexavalent chrome, and toxic metals are often present.

Sap Stain Control Processes

Freshly sawn or peeled wood requires short-term protection from molds and stain fungi. Although these fungi do not cause appreciable strength loss, their pigmented spores and hyphae do cause cosmetic blemishes that reduce the value of the wood. Short-term protection is generally required until the moisture content of the wood falls below 20 percent (weight basis). These treatments are generally applied to the cut surface and are designed to be prophylactic in nature, with the goal of protecting the wood for a period of 90-180 days.

The **sap stain** treatment chemicals are usually water-soluble or water-based emulsions that can be

Name	Abbreviations	Chemical Formulas	Application/Use
Ammoniacal copper zinc arsenate	ACZA	A mixture of CuO, H_3AsO_4, ZnO, and NH_3	Pressure treatment
Chromated copper arsenate	CCA	A mixture of CrO_3, CuO, and H_3AsO_4	Pressure treatment
Creosote	Creosote	A mixture of organic compounds and polynuclear aromatic hydrocarbons (PAHs)	Pressure treatment
Pentachlorophenol	PCP or Penta	$C_6Cl_5 - OH$	Pressure treatment
Sodium pentachlorophenate	NaPCP or Sodium penta	$C_6Cl_5O^- Na^+$	Sap stain control
3-Iodo-2-propynyl butyl carbamate	IPBC		Millwork
Didecyldimethyl ammonium chloride	Quat, DDAC, or Alkyl ammonium compounds	$(C_{10}H_{21})_2(CH_3)_2N^+ Cl^-$	Millwork and Sap stain control
Tributyl tin oxide	TBTO	$(C_4H_9)_3SnO$	Millwork
Copper naphthenate	Cupronol		Pressure and Brush treatments
Copper 8-quinolinolate	Copper 8, Oxine or Copper	$(C_9H_6ON)_2 Cu$ Copper-8-hydroxyquinoline	Pressure treatment
Acid copper chromate	ACC	A mixture of Cr^{6+} (CrO_3 or $Na_2Cr_2O_7 \cdot 2H_2O$) and copper Cu^{2+}(CuO or $CuSO_4 \cdot 5H_2O$) compounds	Pressure treatment

Table 25-1: Wood preservation chemicals and their uses.

applied by either spraying or quickly – 30 seconds – dipping the wood to coat all of its exposed surfaces. Waste generation during these processes can come from a number of sources, but the most significant source is the generation of contaminated sawdust and soil that was present on the wood prior to treatment. These materials tend to sink to the bottom of the treatment tank or collect on the spray system filters and must therefore be periodically removed. Air contamination also occurs due to losses of Volatile Organic Compounds (VOCs) from the treatment solution. Since surface water and groundwater contamination can result from rain washing over the freshly treated wood, it is typically either covered with sheets of plastic or stored in a covered area.

Sap Stain Control Chemicals

In the past, the most commonly used chemical for sap stain control was **sodium pentachlorophenol**, $C_6Cl_5O^- Na^+$, also called **sodium penta** or **NaPCP**. Worldwide, the water-soluble compound was applied by either spraying or dipping freshly cut lumber, **timbers**, and poles. Because NaPCP is easily converted into **pentachlorophenol** (PCP), C_6Cl_5-OH, it is now one of the most ubiquitous pollutants on earth. PCP has been detected in a wide variety of foodstuffs, including soft drinks, bread, rice, wheat, and fish. Human exposures range from 1-100 parts per billion (ppb); these levels are 1,000-10,000 times lower than the lethal concentration in the environment. Some studies suggest that PCP is a human mutagen, and there are authors who believe that human exposure to PCP currently poses a significant health hazard.

NaPCP is no longer in general use for sap stain control in North America and has been replaced by a variety of other less toxic – but also generally less effective – chemicals. Currently, the most widely used sap stain chemicals in the United States are 3-iodo-2-propynyl butyl carbamate (**IPBC**) in combination with didecyldimethylammonium chloride. This chemical combination can be used either in an organic solvent or as a liquid emulsion in water.

Millwork Processes

Intermediate protection of wood is generally performed where the wood is expected to provide long-service life under conditions where decay can occur, but at a fairly slow rate. The best examples of this

application are windows and door frames. Previous experience has shown that dipping in preservative for 30 to 180 seconds can produce a treatment sufficient to protect these materials in a properly designed and maintained structure. In most cases, whole units or parts are dipped in a preservative solution, drained, and then oven-dried to drive off excess solvent prior to application of the final finish. Alternatively, individual pieces move through a bath, then proceed through the drying and finishing stages. More recently, **millwork** producers have shifted to in-line spray systems to speed up production.

Millwork Treatment Chemicals

Traditionally millwork treatment solutions have used organic solvents. For many years, a five percent pentachlorophenol solution – made by dissolving PCP into a petroleum solvent – was the most popular mixture used for this application. Another millwork treatment chemical used for many years was **tributyl tin oxide (TBTO)**, $(C_4H_9)_3SnO$, also dissolved in a petroleum solvent. The use of organic solvents resulted in VOC emissions into the environment. This challenge is being addressed through the development of water-based systems. Water, however, causes swelling and unsightly wood grain raising that necessitates an additional sanding step. As a result, water-based systems are not widely used. The largest volume chemical currently in use for the treatment of millwork items is IPBC.

Checking Your Understanding

1. What is responsible for sap stain color?

2. What used to be the most commonly used chemical for sap stain control? What has it been replaced with today?

3. What are three different ways that preservation solutions are applied to millwork items?

4. What is the problem associated with using water-based treatment compounds on wood?

25-4 Wood Preservation Treatment Processes

Vacuum Treatment Processes

Although their use is limited in North America, vacuum treatment processes are used extensively in Australia and Europe for the protection of millwork items. In these processes, dry wood (<30 percent MC) is placed in a retort and a vacuum is applied to remove as much air as possible from the wood; then the preservative is transferred into the retort and the vacuum released. As a result of releasing the vacuum, the ambient air pressure helps force the treatment solution into the wood, producing a thin shell of treated wood.

Vacuum processes are normally used with oil-based preservatives and generate sludges as a result of the contaminants from the wood interacting with the preservative solution as well as with the wood sugars and extractives, which can be dissolved in – then later precipitated from – the treatment solution. The use of oil-borne solvents is also a source of VOCs.

Thermal Treatment Processes

Thermal processes are usually employed for treating **roundstock** (cut and untreated tree trunks with the bark removed) for utility poles. This use is largely confined to the Western red cedar, which has a highly durable heartwood, but is surrounded by a thin shell (<25 mm) of sapwood that requires treatment. The thermal treatment process immerses the dried wood – usually by air seasoning – in an oil-based treatment solution. The solution is heated to 90°C for 8-20 hours; subsequently, it is pumped out of the tank and then reintroduced, which results in a slight cooling effect. Because the wood is cooled by the reintroduced solution, the air inside the wood contracts – forming a partial vacuum – resulting in a greater uptake of treatment solution than would have occurred from just soaking in the hot solution.

The waste generated during this process includes sludges resulting from interactions between wood, dirt, and treatment solution. The use of high temperatures increases both VOC and energy losses, so most thermal processes are now carried out in closed vessels.

Pressure Treatment Processes

Pressure treatments are generally used to protect wood that will be exposed to high decay hazards including contact with the soil, or continuous wetting, such as in cooling towers or marine environments. Pressure treatment of wood is a large-scale industrial process. At the center is a large retort capable of withstanding both pressure and vacuum cycles. Treatment vessels vary in size from 7.5 m (25 ft) to more than 60 m (197 ft) in length and 1.8-3.6 m (6-12 ft) in diameter. They are usually constructed of steel and installed on a pedestal over a sealed concrete surface so that any leakage can be detected. The retort is connected to a series of tanks containing the various preservative solutions as well as to vacuum and pressure pumps. In some systems, another vessel is placed on top of the retort called a **Rueping tank**. The Rueping tank serves as a reservoir for the solution placed into the retort. In addition, some vessels are equipped with live steam or have heating coils to maintain solution temperature, while others have vapor condensation systems for recovering volatile oil and water vapors as they exit the retort under vacuum. The two pressure treatment processes most commonly used to transport fluids into wood are the full cell and empty cell processes.

Full Cell Process

The **full cell (Bethell) process** was developed in the 1830s by the British engineer John Bethell. The treatment cycle or charge begins by placing the wood to be treated into a container, variously known as a treatment vessel, retort, or pressure cylinder. The size of the container can range from 1.2 to 2.4 m (4-8 ft) in diameter and 6 to 48 m (20-160 ft) in length. The maximum rated pressure is typically 200 psi, although there are exceptions. As shown in Figure 25-1, an initial vacuum is drawn to remove air from the wood, followed by introduction of the treating solution (fill). This again is followed by an increase in pressure to force the solution into the wood. After a period of time, the pressure is released, the container is drained, and a final vacuum is drawn. This final vacuum removes excess treating solution

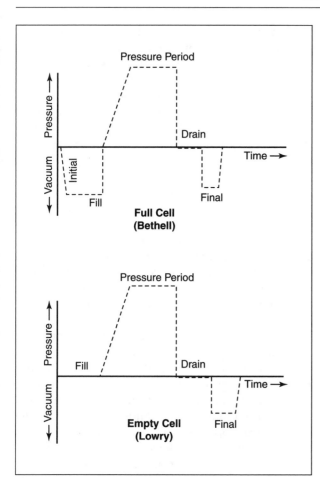

Figure 25-1: A comparison of the pressure vs. time diagrams for the full cell (Bethell) and the empty cell (Lowry) wood treating cycles.

from the surface and reduces exudation or **bleeding** of solution after treatment. The treated wood is then removed from the cylinder and stored. The full cell process introduces a maximum amount of treatment solution into a given amount of wood. It is generally used for creosote treatments for marine exposures where a high chemical loading is desirable, or with water-based systems where the solution concentrations can be manipulated to produce the desired chemical loading or retention.

Empty Cell Process

When a lower total absorption of treating solution will provide the desired protection, the **empty cell (Lowry) process** can be used. Empty cell process eliminates the initial vacuum and begins at either atmospheric pressure or with the injection of a small amount of air into the wood prior to introduction of the preservative solution (fill). The pressure is then raised until the desired chemical uptake has been achieved. In this process, the air present in the wood before treatment is compressed and as the pressure is released, this air expands out of the wood carrying with it excess treating solution. The recovered solution is called **kickback**. The result is a lower amount of treating solution in the wood – **retention** – than would be obtained with a full cell treatment under similar conditions.

The loading in the empty cell process can be controlled to some extent by the initial amount of air pressure used. Empty cell treatments are typically used with oil-borne treatments such as pentachlorophenol to reduce the amount of expensive petroleum solvent remaining in the treated wood. The kickback recovered from the wood lowers the overall treatment cost, but it increases the waste problems. Water from the wood, wood extractives, and sugars are dissolved in the kickback. These can later form sludges and emulsions that are difficult to separate from the treating solution and must be disposed of as hazardous wastes. Before EPA in the 1970s, these wood treatment sludges were typically disposed of in unlined sediment ponds on the treatment plant property. This practice resulted in the contamination of soil and groundwater at many locations, some of which are now Superfund cleanup sites.

In both the full and empty cell processes, a series of vacuums are applied after the pressure is released to recover excess treatment solution and to equalize pressure in the wood and its surroundings. This reduces the likelihood that preservative will bleed from the wood once it leaves the treatment facility. In addition, steam is sometimes applied after treatment to clean the wood surface and – in the case of some waterborne systems – to accelerate fixation of the chemicals to the wood. Both the recovered solutions and the steam are usually contaminated with the preservative and must be processed and handled as a hazardous waste.

Pressure treatment methods currently account for most of the chemical tonnage used in the wood treatment industry. As indicated by Table 25-1, there are several chemicals applied by pressure treatment methods; of these, creosote, pentachlorophenol, and inorganic waterborne preservatives are predominant. Each of these is classified as a restricted use pesticide by EPA.

Vacuum, Thermal, and Pressure Treatment Process Chemicals

Creosote – Creosote is a by-product of the petroleum coking process. It is a black, tarry substance with a distinctive odor and is not soluble in water. It is composed of a large number of different chemical compounds, some of them being polynuclear aromatic hydrocarbons (PAHs). Animal studies have shown creosote to be a carcinogen. Its uses are largely confined to railway crossties, marine pilings, and utility poles.

Pentachlorophenol – Industrial production of PCP started in the 1930s. Its use for treatment of utility poles and fence posts replaced creosote in the 1950s. One of the early advantages of PCP seemed to be that it is cleaner than creosote. Unfortunately PCP is contaminated with dioxins, although the infamous 2,3,7,8-tetrachlorodibenzo-p-dioxin has not been detected in PCP. Pure PCP is a solid at room temperature, so a 5-10 percent petroleum oil solution is typically used for wood treatment. Current regulations restrict the amount of dioxins permissible in PCP to reduce the risk of environmental or health effects.

Inorganic Waterborne Preservatives – While there are several products included in this classification, the treatment process today is predominantly that of **chromated copper arsenate (CCA)**. CCA is a mixture of copper oxide, arsenic acid, and hexavalent chrome, in the form of chromic acid. It is generally produced as a liquid concentrate with an active ingredient concentration of 50-55 percent. The concentrate is transported to the treatment plant and diluted with water to a 1-3 percent treating solution. Once in contact with the wood, the hexavalent chrome undergoes a chemical reaction with the wood components, especially the wood sugars, reducing it to the less toxic trivalent chrome and raising the pH from 2 to 5-6. The increase in pH precipitates the chrome, copper, and arsenic as a variety of mixed compounds. The precipitated and fixed CCA components are generally non-leachable, with water solubilities of only a few ppm. This treatment generally imparts a green color to the wood, although dyes and pigments can be added to produce a variety of colors. CCA is typically used for the treatment of lumber and timbers.

An alternative treatment to CCA is **ammoniacal copper zinc arsenate (ACZA)**, which uses ammonia to solubilize the copper. As the ammonia evaporates from the wood following treatment, the copper (along with the zinc and arsenic) precipitates and becomes resistant to leaching. ACZA is typically used in the Western United States where many of the wood species are difficult to treat with CCA.

Other Wood Treatment Chemicals – In addition to the previously described systems, other chemicals are employed for wood treatment. Copper naphthenate and copper-8-quinolinolate are oil-based preservatives that are less toxic to humans and animals than PCP or creosote. Acid copper chromate is a water-based preservative system used on a limited basis to pressure treat wood used in cooling towers.

Several other preservatives are either under development or have recently been put on the market. These include several ammoniacal copper solutions that have been introduced in recent years, such as ammoniacal copper quat, containing quaternary ammonium compounds as the co-biocide; ammoniacal copper citrate, containing ammonium citrate as the co-biocide; copper dimethyldithiocarbamate, containing sodium dimethyldithiocarbamate as the co-biocide; and copper azole systems, containing the agricultural pesticides tebuconazole and propaconazole as co-biocides.

Oil-based preservatives that are alternatives to the big three – CCA, creosote, and PCP – include chlorothalonil, chlorpyrifos, and isothiazolone. These agricultural pesticides have been shown to be effective wood preservatives in tests both in the laboratory and on the field. None of these compounds has achieved a significant market share, but their development is indicative of the continued interest in less toxic wood preservatives.

Checking Your Understanding

1. What are four types of exposure that will require pressure treated wood products?

2. In the full cell process, what is the purpose of the final vacuum that is drawn?

3. Which process, the full cell or empty cell, delivers a product with a higher loading of chemicals?

4. What is the name of the solution recovered after a vacuum has been placed on the wood?

5. What are three uses for creosote-treated wood?

6. What are the undesirable compounds found in pentachlorophenol treatment mixtures?

7. What substance is the primary waterborne preservative in used today?

25-5 Wood Preservation Wastes and Regulations

Prior to the 1970s, treatment plants were usually sited on cleared ground, pipes were run underground, and sometimes tanks were buried in the ground. As mentioned above, unlined ponds were used for the disposal of sludges and evaporation of contaminated water. Drippage of treating solution from the treated wood as it exited the retort and during storage in the yard – including runoff – was uncontrolled. VOCs from petroleum solvents and ammonia from the ammoniacal systems were not controlled or even monitored. Processes were put in place to minimize the bleeding of treating solutions from the wood after treatment, but not to the extent they are today.

As is also shown in Figure 25-2, modern treatment plants are typically designed with all tanks, lines, and retorts above ground and are contained in diked areas to prevent the escape of any materials. Monitoring wells are installed at strategic locations around the facility to detect any groundwater contamination as early as possible. Treated wood is held on a drip pad until drippage has stopped; the pad itself is highly engineered with leak detection systems (usually monitoring wells) at various locations around and under the pad. An increasing trend in wood preservation facilities is for the use of covered storage areas to prevent rainwater contamination from the freshly treated wood. Since most millwork operations are typically indoors, the most significant sources of waste generation remain VOC emissions and sludges from the treatment tanks and drip areas.

Applicable Wood Preservation Regulations

The wood preservation industries are a heavily regulated group, reflecting its substantial use of broad-spectrum pesticides. All pesticides, including those used for the preservation of wood, are regulated under the Federal Insecticide, Fungicide, and Ro-

Box 25-1 ■ The Toxicity Characteristic Leaching Procedure (TCLP)

The toxicity characteristic leaching procedure (TCLP) is used to determine if a waste is a hazardous waste. The toxicity procedure attempts to duplicate landfill disposal conditions and involves immersing the test sample in a pH 5 buffer solution and shaking. The solution is then filtered and analyzed for a variety of constituents. The headspace above the solution is analyzed for VOCs. Samples are classified as a hazardous waste if the concentrations of specific chemicals exceed the stated threshold concentrations.

Constituent	Threshold Concentration (mg/liter)
Arsenic	5.0
Barium	100.0
Cadmium	1.0
Chromium	5.0
Lead	5.0
Mercury	0.2
Selenium	1.0
Silver	5.0

denticide Act (FIFRA). This law governs the required labeling of wood preservatives. Without a registered label, a wood preservative product cannot be sold. Four wood preservatives are classified as restricted use pesticides: pentachlorophenol, creosote, chromated copper arsenate, and ammoniacal copper zinc arsenate. Users of these chemicals must pass a test on chemical handling and safety administered by their appropriate state agency. In addition, EPA periodically reviews biocides to determine if the benefits of use outweigh the risks. This process, called **Rebuttable Presumption Against Registration (RPAR)** has already occurred with creosote, PCP, CCA, and ACZA.

Operations within a treatment facility fall under RCRA, which regulates the disposal of industrial hazardous wastes. Sludges, contaminated treating solutions, and wastewater are likely to fall under RCRA regulations. In many cases, treatment plants endeavor to reduce waste generation through good industrial practices and recycling. For example, filtration of solutions between treatment cycles can remove wood sugars and debris that contribute to sludging. Sawdust, dirt, and other contaminants can be removed prior to treatment to reduce sludging. Water generated by steaming or Boulton seasoning can be used to make up waterborne preservative solutions in plants that use both oil and water-based systems, although care must be taken to ensure that

components of the wastewater do not alter the stability of the resulting treatment solution. All of these processes can markedly reduce waste generation at a treatment facility. Sludges from wood treatment plants are classified as hazardous wastes under RCRA. Thus, disposal is expensive and requires extensive documentation.

The Clean Water Act (CWA) regulates pollution of rainwater runoff from industrial plants through the non-point source discharge system (NPDS) permit program, which charges producers annual permit fees and specifies processes to reduce discharges. Wood treatment plants are under this regulation and are even more susceptible if treated wood is stored outdoors.

The Clean Air Act (CAA) regulates VOC and other pollutant emissions, such as ammonia. All industrial operations, including wood treatment plants, are affected by regulations issued under this law.

Treatment and Handling of Wood Preserving Wastes

As with any industry, there is a fine line between waste treatment and waste minimization. Waste treatment is an accepted industrial practice; how-

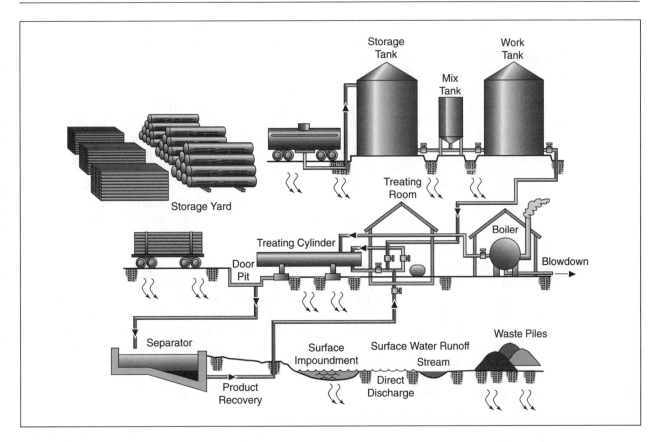

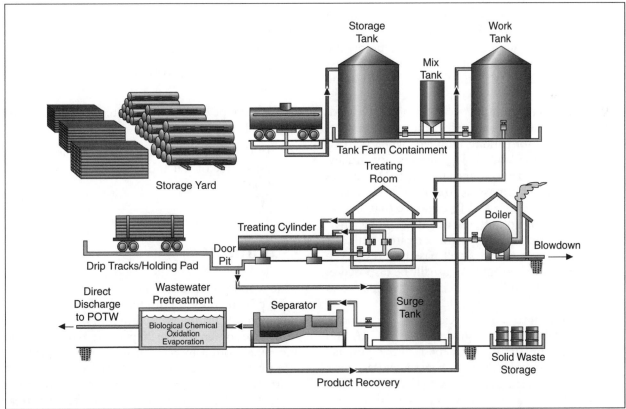

Figure 25-2: Schematic diagram of wood preservation industry with waste generation points identified (top) prior to 1970 and current (bottom).

ever, it results in an added bureaucratic layer, since the producer of the waste is now also viewed as a Treatment, Storage, and Disposal (TSD) facility and, therefore, falls under additional regulation.

Recycling and reuse are the most common forms of waste minimization in the industry. Water produced during Boultonizing, for example, may be used to dilute solutions of inorganic arsenicals or evaporated, taking care to limit the loss of any volatile compounds. The recovered residue may be reused, if its characteristics will not induce sludging of the treating solution.

Waste can also be minimized by careful attention to the quality of the treatment solution. Sawdust, dirt, wood sugars, and other contaminants can build up in the treating solution. Left untreated, these materials will sorb treatment chemicals and settle to the bottom of the retort or the storage tank, where they can pose problems. In addition, these sludges can produce unsightly blemishes on the treated wood. One solution to this problem is regular filtration of the treating solution to remove these materials before they induce solution instability. While the removed material must be handled as a waste, the volume of this material is much smaller than the volume resulting from the treatment solution being left unfiltered.

Ironically, the final waste produced by the wood preservation industries is the product itself: a wood product containing pesticides. At some point, the wood will reach the end of its useful life and must be discarded. Disposal of treated wood has become a major issue in some regions of the United States. EPA first recommends that treated wood be reused, wherever possible. Used utility poles, for example, are often used for parking bumpers, while railroad ties may become landscaping timbers. Eventually, when the wood is no longer useful, EPA requires that treated wood pass a Toxicity Characteristic Leaching Procedure (TCLP) to insure that any residual chemicals do not pose an unacceptable environmental risk. Generally, wood treated with the current labeled pesticides will pass this test, even when freshly treated, and can usually be disposed of in any lined landfill that accepts municipal solid waste.

In addition to activities related to current practices that fall under the guidelines of RCRA and FIFRA, a number of treatment facilities also fall under the Comprehensive Environmental Response, Compensation, and Liability Act (CERCLA). This act, also known as Superfund, is designed to iden-

tify and clean up contaminated facilities that pose the greatest risk to the environment. Nearly all older treatment facilities operated in a manner that allowed contamination of soil, surface water or groundwater to occur, as witnessed by more than 50 of the approximately 600 treatment plants now being designated as Superfund sites. This law also impacts the pesticide producers in the form of a tax they must pay to provide cleanup funds.

Perhaps the single greatest source of soil and groundwater contamination was the use of unlined ponds, which permitted treatment chemicals to percolate into the soil and eventually into the groundwater, where they still pose a risk and present a challenge to cleanup. Many other practices of older facilities also contributed to the contamination, but these practices reflected the perspective at the time that contamination did not pose a problem. Clearly, our attitudes have changed dramatically over the past two decades. These changes are reflected in both the efforts to remediate previous contamination as well as in efforts to ensure that releases from current facilities are minimized.

Checking Your Understanding

1. What changes have occurred since the 1970s in the way wood treatment plants are sited?

2. Why is FIFRA involved in the wood treatment industries?

3. What type of permit regulates non-point source discharges?

4. In the past, what practice was the single greatest source of soil and groundwater contamination?

Summary

The wood preservation industries have a long history of extending the useful life of products made from wood. Products used indoors, or those with moisture content below 20 percent, typically do not decay, although insect damage can be a problem. Products used outdoors, and especially in contact with the ground, require pressure treatment to achieve their maximum service time. Vacuum treatments are used primarily for millwork items and thermal treatments for poles. The two most frequently used pressure processes are the full cell and empty cell processes.

The full cell process places the maximum amount of treatment liquid in the wood. The empty cell process deposits less treatment liquid, but the solution expelled from the wood during treatment can create a waste problem. The three major wood preserving chemicals in use are creosote, PCP, and CCA (ACZA in the Western United States). Many factors affect the quality of the treatment, including moisture content, drying process, wood permeability, and sapwood/heartwood content.

The wood treatment industry is heavily regulated under FIFRA, RCRA, CWA, CAA, and CERCLA. Recycling and reuse are the major waste minimization processes. Minimizing contamination of the treatment solution is also important. Recent improvements in plant design help to reduce soil and groundwater contamination at the plant site. Remediation of old sites is prevalent in the wood treatment industry, with more than 50 Superfund sites listed. New, less toxic treatment chemicals and improved treatment processes are correcting past environmental abuses as the wood preservation industry enters the twenty-first century.

Critical Thinking Questions

1. As an exercise, calculate the potential drippage from a pressure treatment plant. Consider a plant in operation for 80 years that produces five charges per day in each of three retorts. After each charge, approximately 500 ml of 1.5 percent CCA preservative solution drips from the wood onto the soil in front of the retort. How much cumulative preservative has entered the soil? (Assume there are 200 work days in the year.)

2. FIFRA and RCRA are two of the laws that apply to the wood treatment industry. Briefly describe each law in general terms and discuss its effect on the wood treatment industry.

3. Calculate the amount of contaminated water produced annually by a medium-sized pole treatment plant using the Boultonizing process. Data: Typical plant production volume: 1,000 poles/month; 40 ft^3/pole; average initial MC: 85 percent; average MC after treatment: 30 percent; typical wood density: 30 pounds per cubic foot (pcf); water density: 8.3 lbs/gallon.

4. What would be the required flow of a water treatment system to handle the waste stream of question 4? (Assume a single shift per day.)

26

The Medical Industries

by Luis Nuñez

Chapter Objectives

Upon completing this chapter, the student will be able to:

1. **Identify** the major generators and major categories of medical waste.

2. **Explain** the three basic medical waste treatment technologies.

3. **Identify** the factors involved in the selection of a medical waste treatment technology.

4. **Explain** the issues associated with the final disposal of treated medical waste.

5. **Describe** the need for medical waste minimization and source reduction.

6. **Explain** the critical elements in and uses of a medical waste audit.

Chapter Sections

26-1 Introduction

In the brief survey – in Chapter 6 – about the microorganisms that surround us, it was noted that the vast majority of these are either harmless or beneficial to humans. If it were otherwise, we would be under continual assault from the microbial world. Although it is generally true that microorganisms pose little threat to humans, the microorganisms in medical waste are an important and obvious exception. These microorganisms have considerable potential to cause infection in humans and other creatures. Since hospitals and other medical facilities exist to provide services to sick people, includ-

Waste Category	Treatment Method
Infectious Wastes	
Contaminated Animal Carcasses	Incineration
Cultures and Stocks	Steam Sterilization
Contaminated Bedding/Patient Care Waste	Steam Sterilization or Incineration
Contaminated Small Equipment	Steam Sterilization or Incineration
Contaminated Large Equipment	Formaldehyde Decontamination
Waste Biologicals	Steam Sterilization or Incineration
Surgery Wastes	Steam Sterilization or Incineration
Human Blood	Steam Sterilization or Incineration
Autopsy Wastes	Incineration
Human Blood Products	Steam Sterilization or Incineration
Contaminated Laboratory Wastes	Steam Sterilization
Pathological Wastes	Steam Sterilization or Incineration/Grinding
Dialysis Unit Wastes	Steam Sterilization
Contaminated and Unused Sharps	Steam Sterilization and Incineration/Grinding
Other Medical Wastes	
Chemical Waste	See Chapter 4
Antineoplastic Drug Waste	Incineration
Low Level Radioactive Waste	See Chapter 27
Mixed Waste	See Chapter 4
Multiple Hazard Wastes	See Text

Table 26-1: Categories of medical waste and usual treatment methods.

ing many suffering from microbial infections, it is not surprising that their waste should require special consideration.

Though infectious waste is an important category of medical waste, it is only one of a number of categories under the umbrella term **medical waste**. Many other materials are classified as medical waste. These include materials that have either come into contact with infectious waste or are likely to cause infection or physical harm. Table 26-1 lists some categories of infectious and other medical wastes, as identified by EPA and other specialists. More than three million tons of infectious and medical waste are generated each year by a variety of establishments ranging from hospitals and doctor's offices to clinical and research laboratories. As of 1991, the estimated annual cost of treating this waste was more than ten billion dollars. As the medical industry expands, and the definition of medical waste broadens, the amount of waste produced is expected to grow to about ten million tons by 2005 – and the cost could approach fifty billion dollars a year.

Sheer volume of waste is not the only problem. A general lack of agreement on terminology also causes problems along the route from generation to final disposal. Medical waste management must respond to continually changing regulations at the federal, state, and local levels. Commercial waste services may be unavailable, unreliable, or of poor quality. And concern over medical waste does not end at the site of generation. For example, the State of Illinois Environmental Protection Agency provides legal enforcement for the safe and responsible handling of potentially infectious medical waste by hospitals, livestock producers, small generators, transfer-storage-treatment facilities, transporters, and landfill operators.

The problems associated with medical waste management are likely to increase as the United States population ages. Despite the problems mentioned above – which are largely administrative rather than technological – effective treatment for most medical waste is available. This chapter will focus on the treatment of medical waste and examine the generation of medical waste, the typical waste streams produced, the opportunities for source reduction, the available treatment technologies, and the critical elements of a medical waste audit.

26-2 The Medical Industry and Medical Wastes

In terms of volume, the largest generators of medical waste are the health care industry, the academic and industrial research laboratories supporting the health care industry, the pharmaceutical industry, the veterinary industry, the food industry, and the cosmetics industries. The wastes associated with the medical industry are presented in Table 26-1. It should be noted that this list is not complete, since there is disagreement among the authorities as to what constitutes a true medical waste. Most people agree that the routine paperwork generated in the accounting department of a hospital, for example, should be treated as a general or household waste. But what about old magazines that have been discarded from a hospital wing housing patients with infectious diseases? There are no clear-cut answers to questions such as these, but for the sake of prudence, the level of risk associated with the waste should always be considered. If there is any potential for harm, then the waste should be treated as if it were infectious and dangerous.

Standard hospital practice is to collect waste in color-coded plastic bags. Infectious waste is so commonly collected in red bags that it has become known as **red bag waste**. In Table 26-1 most of the categories and their origin are self-explanatory. Three categories that require further discussion are **pathological wastes**, **sharps**, and multiple hazard wastes. Pathological wastes are body tissue removed in testing, surgery, or autopsy and can vary from a few drops of cellular fluid to amputated limbs. Though steam sterilization will suffice to destroy the infectious characteristics of the tissue, for aesthetic reasons incineration or grinding is also required. Sharps are any metal or glass objects such as needles, scalpels, or sampling pipettes that can cause a puncture, cut, or abrasion. Again, steam sterilization is usually sufficient to treat the infectious characteristic of sharps; however, in addition to that sharps are ground because their ultimate disposal is in a landfill and the presence of intact sharps in such locations is generally considered unacceptable. **Multiple hazard wastes** are wastes falling into more than one of the following hazard categories: infectious, regulated chemical waste, and regulated radioactive waste. It is common practice that collected medical waste is placed in receptacles displaying the biohazard marker shown in Figure 26-1.

Figure 26-1: Biohazard warning sign.

Medical Waste Treatments

Figure 26-2 is a flow diagram for medical wastes from the point of generation to disposal. Three basic treatment options – described below – exist for treatment of medical waste: 1) incineration, 2) steam sterilization, and 3) chemical disinfection. The selection among these options involves far more than technical considerations.

Steam Sterilization

Steam sterilization rests on two principles. First, exposure to elevated temperatures for an appropriate period is an effective and time-proven way to destroy infectious microorganisms. If time and temperature were the only two critical variables in steam sterilization, then the use of dry air techniques should work as well. However, it has been shown that saturated steam kills microorganisms about 100 times faster than heating in dry air. This efficiency is due partly to the greater heat transfer in steam as compared to that in dry air, and partly to the greater ability of steam to disrupt the membranes surrounding the microorganisms.

Maintaining a saturated steam atmosphere is critical to the success of steam sterilization, which is

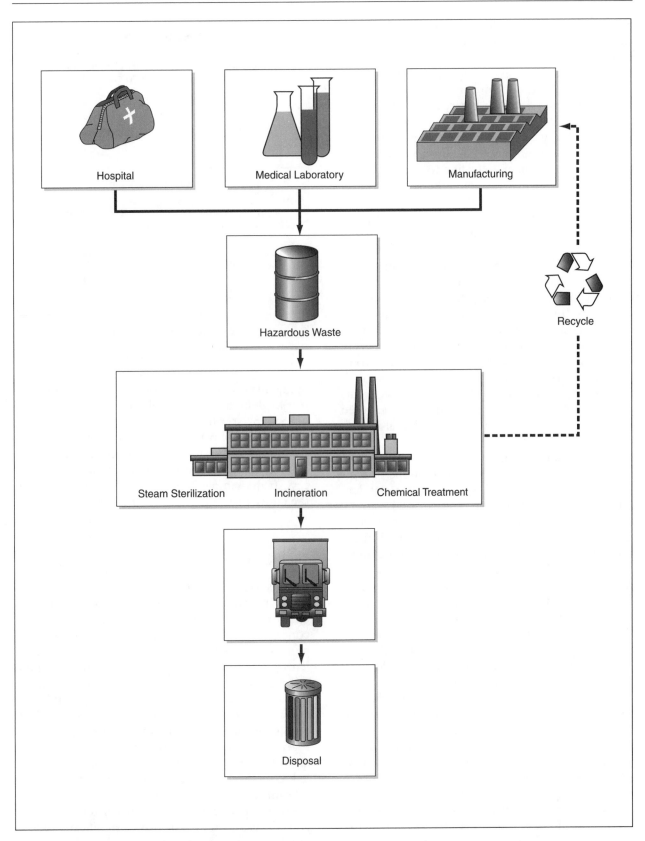

Figure 26-2: Medical waste flow diagram.

most commonly carried out in an **autoclave**. An autoclave is a piece of equipment composed of thick metal designed to withstand the temperatures (up to 165°C) and pressures (up to 30 psi) of the process. Its inner sterilization chamber is surrounded by a steam jacket as is shown in Figure 26-3. The advantage of this jacket is that it allows the sterilizing chamber to stay dry during the preheating period.

To operate an autoclave, the waste is loaded into the sterilizing chamber according to standard operating procedures that optimize the steam contact and heating. The lid is closed, and all safety precautions, such as latch-locking and safety-shielding, are followed. The outer steam jacket is first pressurized with steam and brought to the desired temperature. Once the walls of the inner sterilizing chamber have been preheated, steam is allowed to enter the sterilizing chamber. As the steam flows from the steam jacket into the sterilizing chamber, air is forced out and a pressure of 15-30 psi is reached and maintained. Temperatures ranging from 100-165°C are used for a predetermined period of time to accomplish the sterilization, after which the pressure is released by exhausting the steam, and the contents are removed from the chamber. Autoclaves are used to sterilize not only biological wastes, but also reusable medical equipment.

Though steam sterilization is far more effective than **dry heat** under the same conditions of temperature and pressure (see Table 26-2), dry heat does find considerable application. The ease of use as well as the elimination of the need for steam production facilities can often outweigh the combined disadvantages of higher temperatures and exposure times.

Steam sterilization can be applied to any waste that can be placed in an autoclave and is the preferred method of treatment for most infectious agents. A major exception is that of sharps. Although sharps can be sterilized in an autoclave, they are not structurally disrupted and can therefore still cause serious puncture wounds and other occupational safety problems. Further, a sterilized sharp tends to look the same as a contaminated one. This fact can also be a problem for autoclaved red bag waste. At the time of final disposal, there is no visible difference between treated and untreated waste. These problems can be addressed by the use of packaging that undergoes some clearly indicative transformation upon exposure to the sterilization process.

Routine inspection and servicing of steam sterilization equipment is necessary to ensure effective treatment. The United States Pharmacopoeia rec-

Steam Sterilization	
Temperature (° C)	Time (minutes)
116	30
118	18
121	12
125	8
132	2
138	0.8
Dry Heat Sterilization	
Temperature (° C)	Time (minutes)
121	360
140	180
150	150
160	120
170	60
180	30

Table 26-2: Exposure times and temperatures for steam and dry sterilization.

ommends the use of the bacterium *Bacillus stearothermophilus* as a biological indicator for monitoring steam sterilization procedures. Waterborne samples may be placed with the waste to be treated, submitted to the sterilization process, and later examined to determine the effectiveness of the process. If the sample is still alive, the equipment and the entire procedure must be reexamined.

Incineration

Various types of incinerators were discussed in Chapter 5. Incinerators marketed as medical, infectious, or pathological waste incinerators are usually of the controlled air type. Though incineration may appear to be the perfect solution to medical waste problems (in terms of destruction of potentially harmful microorganisms and volume reduction), medical incinerators face the same problems of public acceptance that arise with incinerators used for hazardous chemical waste. Increasingly strict permitting regulations and emissions standards are working against the introduction of new incinerators and have led to medical incinerators being identified as a significant source of air pollution. Nevertheless, many hospitals operate incinerators installed in the 1960s and 1970s, so they still play a significant part in medical waste management.

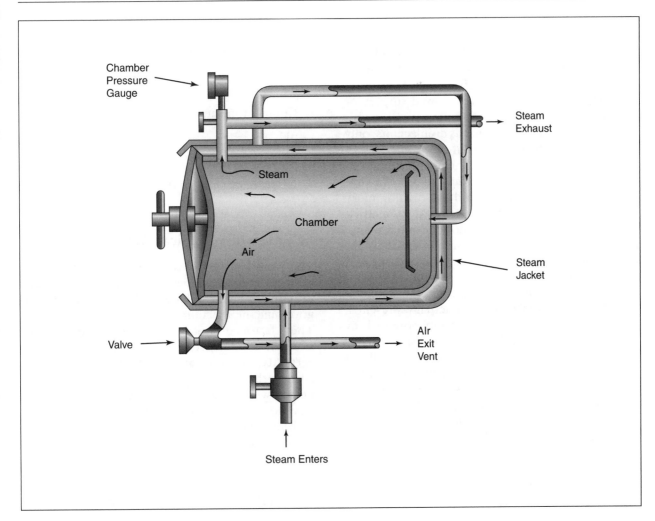

Figure 26-3: An autoclave for sterilization of medical waste.

The major problem today for most operating medical waste incinerators is their inability to meet particulate emission and pollution control standards. These incinerators were designed and installed to eliminate the hazards of infectious waste, a goal they accomplish so well that little concern exits regarding their ability to destroy biohazards. But emissions standards have tightened dramatically in the last three decades, and the performance of controlled air incinerators is now a matter of debate. It has been estimated that even among newly installed systems 25 percent fail to operate properly. Concern over reliability has also placed considerable importance on the correct type of wastes to be submitted for incineration. Today, medical wastes are about 20 percent plastics – much of which contain chlorine – that have a corrosive effect on the incinerator and exhaust acid gases. Glass is also a large waste component that can melt, fouling the ash removal system and eventually clogging the unit. An additional problem with incineration is the presence of noncombustible toxics in the waste such as heavy metals that can cause the ash to be disposed of as a hazardous waste.

So the question remains, why not use the best available commercial chemical waste incinerator? Two difficulties exist: first, even the best incinerators have many problems in overcoming their poor public image. Second, they are prohibitively expensive, even for the largest hospitals. The situation is difficult. The options are: 1) to retrofit pollution control devices onto existing incinerators, with no guarantee of success; 2) to seek regulatory relief in the form of an exemption; 3) to lessen the large cost of an effective incinerator by forming a coalition of hospitals; or 4) to relinquish incineration as a medical treatment option.

Chemical Treatment

In 1868 the British physician Joseph Lister (1827-1912) demonstrated that 2-5 percent solutions of phenol in water were highly effective in preventing the often fatal infections that followed surgery or compound fractures. Since that time, several chemicals have temporarily found use as disinfectants; only to be put aside when problems arose or when better disinfectants became known. The phenol employed by Lister had to be used in such high concentrations to be effective that it could cause major injury to tissue. It also had a peculiar odor. Many methyl- and chloro-derivatives of phenol and phenol analogues have been tried at one time or another, but they are now rarely used because of the toxicity problems they pose.

The case described above is only an example of how chemical sterilization can actually present a major problem to humans due to its potential for toxicity. For a chemical to act as a disinfectant it must obviously possess some **biocidal** (life-killing) **properties**, which always have the potential of being harmful to humans. The present climate of great caution toward the use, storage, and disposal of chemicals also works against their use. Even so, there are certain instances in which **chemical disinfection** is the recommended course of action. The disinfection of liquid wastes is one such category; here chemical treatment is superior to all other alternatives. Indeed, disinfection of large pieces of equipment can only be achieved with a chemical disinfectant wipedown; and disinfection of recyclable containers used to transport infectious waste can only conveniently be performed with chemicals.

Chlorine-containing compounds such as hypochlorites, iodine-containing compounds such as iodophor, formaldehyde, glutaraldehyde, simple alcohols, and ethylene oxide are among the chemicals that are most often used for chemical disinfection. Chlorine-containing compounds have the most widespread use, since they are the only compounds effective against the whole range of infectious organisms. Ethylene oxide and formaldehyde are often used as gas phase disinfectants. However, since they are regarded as human carcinogens, great care in their use is always exercised. When a chemical disinfectant is an option, the following factors must always be considered:

—Type of microorganism present;

—Degree of contamination;

—Amount of protein present;

—Properties of the disinfectant;

—Concentration and amount of disinfectant needed;

—Contact time and temperature required;

—pH;

—Mixing time; and

—Ultimate disposal of the treated waste.

Selection of a Medical Waste Treatment

The treatment methods outlined in the three preceding sections each have application under differing circumstances. Table 26-3 lists the major factors governing the selection of a waste treatment method. These factors do not operate in isolation from one another, and final selection involves a level of uncertainty. As rules and regulations continue to change, the criteria for selection of a treatment technology will also change. The best treatment will always combine maximum potential for lowering risk and cost while achieving compliance with the applicable regulations.

Final Disposal of Treated Medical Waste

The whole purpose of treating medical waste is to render it harmless from infection so it can be disposed of with the same level of risk as household waste. The three treatment technologies described above all perform the task of disinfection well, but two types of medical waste require special consideration: 1) sharps must undergo treatment that destroys their structural characteristics; and 2) materials with hazardous characteristics that are not removed by steam sterilization, chemical disinfection, or incineration – for example, radioactivity, or toxic metal content – must be disposed of with special care or only after further treatment.

Sanitary landfills are the normal disposal site for solid wastes. Most treated medical wastes are solid wastes, including incinerator ash, ground sharps, and steam sterilized waste. Though steam sterilization and incineration adequately address the infectious nature of medical waste, toxicity concerns remain, particularly with some incinerator ash. In these cases, disposal at a sanitary landfill is unacceptable, and a

Factor	Considerations
Onsite/Off site Tradeoffs	Availability Onsite Potential to be Installed Onsite Need for Control Over Waste Cost Control Relevance to Primary Function of Institute
Regulations	Full Understanding if Treatment Performed Onsite Regulations Rapidly Changing/Need to be Current
Complexity of Operation	Incineration Requires Extremely Skilled Operation
Suitability	Many Factors, e.g. –Liquids Best by Chemical Treatment –Volume Reduction Occurs with Incineration –Steam Sterilization is Preferred Technology
Volume Reduction	Affects Costs Since Landfill is Usually End-point
Environment, Safety & Health Concerns	Secondary Wastes Need Disposal Risks to Workers from Chemicals
Disposition of Secondary Wastes	Waste Chemicals to Sewer Chemically Treated Waste to Landfill Incinerator Ash to Landfill Steam Sterilized Waste to Landfill
Economics	Incinerator Most Expensive Chemicals Least Expensive

Table 26-3: Factors pertaining to the selection of a treatment.

hazardous waste landfill must be used instead. Liquids must not be disposed of at any landfill; sanitary sewers are appropriate for disposal of liquid wastes if local requirements are met.

Radioactive wastes are a common component of diagnostic and research activities and require special disposal consideration. Some radioactive wastes are so low in activity that they may be exempted from special regulation, while other waste may require indefinite storage. Other radioactive wastes may require storage for only an appropriate decay time. Still other radioactive wastes may be incinerated, disposed of in a sanitary sewer or appropriate landfill, or be discarded as regular trash. In general, the disposal of low-level radioactive medical waste is not a difficult problem, but the special nature of this type of waste must be recognized and handled appropriately.

Checking Your Understanding

1. What is the most common treatment method for infectious medical wastes?

2. What type of waste is placed in a red bag?

3. What are at least three problems associated with incinerating medical wastes?

4. Which chemical disinfectant is most effective against the whole range of infectious organisms?

5. Which options are used for the disposal of radioactive wastes?

26-3 Medical Waste Management

Source Reduction

The reduction of medical waste is particularly important, since the use of disposable equipment and supplies has produced a considerable waste increase in the past two decades. This trend has been driven by reduced costs in manufacturing medical supplies and also by the knowledge that the risk of infection can be reduced by the single-use-and-discard approach. However, the growing costs of disposal and the need to reduce risks to personnel who handle this waste is encouraging the industry to carefully examine alternatives. Experience has shown that by attacking the problem with a range of approaches, significant waste minimization can be achieved.

Substituting disposable supplies with reusable and recyclable supplies is an obvious approach to waste minimization. When adopting this approach, however, it is important to have accurate data on all of the costs – the total life cycle costs – associated with reusable supplies. Reusable supplies tend to be more expensive, require more sophisticated inventory control, must be sterilized, and must have a quality control system to address the problems of aging, wear, and adequate sterilization. These are the very same reasons that allowed disposable supplies to find a market.

Two treatment techniques discussed earlier can significantly reduce waste volume. Incineration can achieve volume reductions of more than 95 percent and can make use of the combustion energy of the waste. Steam sterilization tends to shrink plastics – especially the red bags – and achieves a volume reduction of about 20 percent. The use of compaction techniques after steam sterilization can also reduce the volume by an additional 50 percent.

Institutional changes and employee training, such as separating wastes at the point of generation, can reduce the amount of waste classified as medical waste; redesign of procedures so that fewer supplies are consumed can also offer significant savings.

Medical Waste Audit

The spectrum of medical waste management and minimization of waste activities includes 1) the iden-tification of wastes and the activities that generate them, 2) their safe collection and interim storage, 3) the treatment of selected wastes, 4) the transportation of treated and untreated wastes, and 5) final disposal. Since medical waste management is a rapidly changing and complex field, any tool that can help an institution manage these activities is very helpful. A medical waste audit is such a tool. The audit collects information on current practices, and subsequently allows them to be judged against alternatives that may reduce risk and/or cost. As such, it is a procedure rather than a hard-and-fast list of factors to be examined.

Table 26-4 lists some of the critical elements to be considered in an audit. The questions that need to be asked can be relevant to more than one element in the audit and will vary depending on the institution producing the waste. The three general questions that must be asked are:

1. Are genuine efforts being made to adequately treat waste in light of state-of-the-art procedures and legal requirements?

2. Is training being provided to all responsible personnel?

3. Are adequate records from both a legal and cost analysis perspective being kept? A series of well conducted audits can point to potential cost saving and can be a catalyst for employee creativity in finding new solutions to medical and waste problems.

Medical Waste Regulations

Since medical waste is a broadly defined term, it is not surprising to discover that a number of rules and regulations can be applied to the waste or its components. The most important of these are the Resource Conservation and Recovery Act (RCRA) and the Medical Waste Tracking Act (MWTA).

EPA has the power to regulate infectious wastes as a type of hazardous waste under RCRA; however, it has not chosen to do so, since the problem is not currently perceived as urgent. This perception is the result of generally good practices – and

Element	Typical Questions
Management and Policy	Are written procedures available? Has a capable individual been designated for oversight? Are emergency procedures written down? Are records and costs maintained? Do procedures comply with rapidly changing laws?
Personnel	Are personnel adequately trained? Are lessons from training being implemented? Are responsible personnel involved in policy and planning? Are records of training and problems maintained?
Collection and Storage	Are wastes properly categorized? Are proper containers and procedures used? Are storage facilities adequate for needs? Are inspections of system made? Are problems encountered and solved?
Treatment	Are wastes properly treated? Is an oversight/review system for treatment in place? Is an testing and verification plan available? Are costs and risks monitored and recorded? Are records of treatment maintained?
Transportation and Disposal	Is untreated waste transported for treatment off-site? Has the off-site treatment been adequately inspected? Are transporters aware of the nature of the waste? Is treated wasted landfilled? Is treated waste disposed of by sanitary sewer?

Table 26-4: Critical elements in a medical waste audit.

more pressing needs elsewhere! EPA has listed guidelines, but these are not legally enforced at this time.

In the summer of 1988, medical waste washed up on East Coast beaches. Concern about the risk of infection from AIDS-contaminated material fueled a media furor, brought medical waste into the public eye, and resulted in passage of the **Medical Waste Tracking Act (MWTA)** of 1988. This Act directed EPA to develop and test the effectiveness of regulations that would track medical waste generated in Rhode Island, Connecticut, New York,

New Jersey, and Puerto Rico. In brief, the regulations include a definition of regulated waste; the requirement of a tracking form for waste shipments; and a definition of scope as applying to generators, transporters, treatment operators, and disposal facilities. Sections of the regulations cover the requirements for separation, packaging, labeling, storage, decontamination, use of the tracking form, recordkeeping, reporting, and notification.

OSHA requires employers to operate workplaces free from hazards or from potential causes of harm. This general requirement can apply to medical wastes, and specific action has been taken concerning healthcare workers' risk of contracting AIDS.

Regulations regarding the radioactive material used in diagnosis and radiation therapy are set by the Nuclear Regulatory Commission (NRC). In general, these components of medical waste have low levels of radioactivity and do not pose the same threat as commercial or defense related waste. As a result, disposal and regulations corresponding to disposal are less burdensome than for high levels of radioactivity. But this situation could change. Incidents involving contamination resulting from the inadvertent recycling of radioactive material are growing more frequent and new regulations may be promulgated to address this problem.

Most states have regulations concerning medical waste. The regulations vary quite widely, but are generally at least as strict as federal regulations. Individual state environmental protection agencies should be consulted for specific details. Similarly, most local governments set out rules for the use of incinerators and sanitary sewers. Appropriate local authorities should be consulted for details.

Checking Your Understanding

1. The use of what type of equipment has lead to an increase in the amount of medical waste?

2. What are five problems associated with reusable medical equipment?

3. What are five critical elements to be considered when conducting a medical audit?

4. What are the major provisions of the Medical Waste Tracking Act of 1988?

Summary

We live in a world surrounded by microorganisms, the vast majority of which is beneficial and harmless to humans. Some, however, do have the potential to infect humans. Hospitals and other medical facilities that are exposed to infected individuals generate more than three million tons of infectious and medical (red bag) wastes per year. The potential for generating even more waste in the future is likely, as the U.S. population ages. A second problem is that there is still no clear definition on what exactly constitutes a medical waste.

There are three generally recognized categories of waste: pathological wastes (waste body tissue), sharps (disposable needles), and multiple hazard wastes. Multiple hazard wastes fall into one or more of the following categories: regulated chemical or radioactive wastes and/or infectious wastes. A variety of waste treatment methods has been developed, but the three treatment options most often used are steam sterilization, incineration, and chemical disinfection. Steam sterilization relies on the use of an autoclave to produce steam at elevated temperatures and pressures. Although incineration may appear to be the perfect solution for the disposal of medical wastes, it suffers from public acceptance problems and is subject to increasingly strict air emission standards. Many of the plastics disposed of today, for example, contain chlorine and their incineration leads to corrosive effects on the incinerator as well as acid exhaust gases. Chemical treatment relies on a variety of biocides including iodine-containing compounds or hypochlorites. Simple alcohols, formaldehyde, and ethylene oxide are also frequently used.

The whole purpose of treating medical waste is to render it harmless from infection so it can be disposed of with the same level of risk as household waste. All three treatment methods are effective in disinfecting the waste, but sharps must also be ground to destroy their structural characteristics. When attempting to reduce the volume of medical wastes, all aspects of the problem must be investigated. Substituting reusable and recyclable supplies for disposable items is an obvious approach. Reusable supplies, however, present some disadvantages: higher costs, more sophisticated inventory control, and sterilization between uses; they also require some type of quality control system to address aging, wear, and adequate sterilization. The volume of solid wastes can obviously be greatly reduced by incineration, which also presents an opportunity for co-generation. Up to a 20 percent volume reduction can be realized by steam sterilization of plastics and, when followed by compaction, it can result in an additional 50 percent reduction.

One of the valuable waste management tools available is the medical waste audit. The information collected by the audit includes 1) the types of wastes and the activities they were generated by, 2) the methods used for their safe collection and interim storage, 3) the treatment option used, 4) the method of transportation, and 5) their final disposal. Conducting the audit allows for a comparison of current practices against possible alternatives and may result in reducing both risks and costs. Medical wastes are governed by a number of rules and regulations. The most important of these are RCRA, the Medical Waste Tracking Act of 1988 (MWTA), OSHA, and the NRC.

Critical Thinking Questions

1. Bacteria are able to thrive in deep-sea thermal vents at elevated temperatures. Consider the following scenario: as part of an investigation into advanced waste treatment technology a deep sea thermal vent bacterium is genetically engineered in a research laboratory to be able to survive at atmospheric pressure and still withstand high temperatures in saturated steam. The bacterium is found to be potentially infectious. How would this affect current medical waste treatment procedures?

2. Review the sections of this book on incinerators and clean air regulations. In light of this information discuss the pros and cons of incineration for medical waste.

3. Imagine that you are the manager of a commercial sanitary landfill. You are asked to receive waste and you come to believe that it may con-

tain medical waste. What issues should come to mind in deciding whether to accept the waste?

4. You are a businessperson who is approached by an inventor to invest in production of a new chemical for the disinfection of medical waste. What questions would you ask in order to help you decide on whether or not to invest?

5. Outside Assignment – Arrange a visit to a local hospital or medical waste generating facility and examine the waste management system. Discuss what you see in light of what you have learned in this chapter.

6. Group Activity – Divide into four subgroups. The first group will contact (via phone calls, Internet) the State Environmental Protection Agency and explore state rules on medical waste. The second group will contact local authorities and explore rules concerning sanitary sewer and landfill waste. The third group will contact the local hospital or local medical waste generating facilities and explore management practices. The fourth group will collect the results of the first three groups and summarize the local medical waste situation. The class should then identify and decide if any waste minimization activities are legally possible.

27

The Nuclear Industries

by William "Pat" Miller

Chapter Objectives

Upon completing this chapter, the student will be able to:

1. **Define** the nature of radioactivity and radiation.

2. **Describe** at least four uses of radioactive materials.

3. **Describe** at least two processes that generate radioactive wastes.

4. **Diagram** how radioactive materials are used to generate electrical power.

5. **Describe** the key provisions of regulations pertaining to radioactive waste management.

6. **Explain** three methods used to reduce the volume of radioactive waste.

27-1 Introduction

An isotope is one of two or more species of atoms of the same chemical element that vary in the number of neutrons in their nuclei and, therefore, in their atomic masses. The various isotopes of each element are designated by including their atomic mass as either a part of their symbol (^{12}C and ^{14}C) or as a part of their name (carbon-12 and carbon-14). **Radioactive isotopes, radioisotopes,** or **radionuclides** are the naturally occurring or manmade isotopes of an element that have an unstable nucleus. When an atom of a radioactive isotope decays, it emits one or more subatomic particles and/or energy.

The particles and/or energy that are released during decay are referred to as **radiation**. The most common forms of radiation that originate from the nucleus of a radioisotope are **alpha particles, beta particles, positrons,** and **gamma rays**. Alpha particles are helium nuclei containing two protons and two neutrons, and are therefore positively charged; beta particles are fast-moving electrons, and therefore negatively charged; positrons are positively charged fast-moving electrons; and gamma rays are a form of **electromagnetic radiation** composed of **photons** of various wavelengths. An **x-ray** is also an electromagnetic radiation, but of lower energy, and is produced by the rearrangement of the electrons in the energy levels. All of these are forms of **ionizing radiation**; this means they possess enough energy to cause an electron to be removed from either an atom or a molecule, resulting in the formation of a positively charged ion.

The **neutron** is a neutrally charged particle that is very slightly heavier than a proton. Neutrons are found, along with protons, in the nuclei of all atoms except hydrogen atoms and can be ejected from specific nuclei, such as beryllium-9 or lithium-7, by alpha particle bombardment. **Fissile** atoms, such as ^{235}U or ^{239}Pu, also release neutrons when they undergo **fission**. When a released neutron is absorbed by another uranium-235 atom, it can result in the splitting of that nucleus into two approximately equally sized smaller nuclei, called **fission fragments**, along with the release of one or more free neutrons. The majority of the energy released by this fission – 83 percent – is carried by the fission fragments.

The degree of **radioactivity** or activity unit for radioisotopes is determined by the number of nuclei decaying per second. The standard activity unit is the **becquerel (Bq)**, defined as one disintegration per second, or the **curie (Ci)**, which is 3.73×10^{10} disintegrations/second. The **half-life** of a radioisotope is the time required for one half the atoms in a sample to undergo disintegration. The half-life of radioisotopes varies from fractions of a second to many millions of years. Once a radioisotope sample has undergone a period of time representing seven half-lives, the remaining radioactivity would be only $1/2^7$ or 0.78 percent of its initial activity.

27-2 Uses of Radioisotopes

Since the discovery of radioactivity by Becquerel in 1895, many varied uses for radioactive substances have been found. In the following sections, the timeline for a few of these uses will be noted. Perhaps one of the most significant commercial applications is that of nuclear power generation, a topic mentioned here, but that will be considered in more detail separately.

Defense

The production and first use of radioactive materials for defense purposes started with the Manhattan Project during World War II and resulted in the production and use of two atomic bombs. The Clinton Engineering Works in Tennessee was built as a part of that project to produce the necessary amounts of enriched uranium-235. The Hanford Engineering Works in Washington was the site of the first plutonium production reactor in June 1943. By 1955, the U.S. Navy had launched the Nautilus, its first nuclear powered submarine.

Nuclear Power Generation

In 1957 the first commercially operated pressurized water nuclear power plant started producing electricity at Shippingport, Pennsylvania. By 1995 the United States had 109 operating nuclear power plants producing 22.49 percent of our total electrical power. Worldwide, the number of nuclear power plants reached 437 during that same time frame, generating 17 percent of the electricity consumed globally.

Medicine

The first medical research, diagnosis, and therapy applications were performed in the 1930s using radioisotopes produced by cyclotrons at Berkeley and MIT. The Atomic Energy Act of 1946 created a radioisotope distribution program that allowed private use of radioisotopes, thus launching the field of nuclear medicine. In the intervening years, nuclear medicine has made important contributions to evaluation and therapy for patients with cancer, cardiac disease, pulmonary disease, cerebrovascular disease, and many other disorders. Patients with cancer, for example, are given radiation treatments using gamma rays from either ^{137}Cs or ^{60}Co sources.

It is estimated that currently more than 10 million nuclear medical procedures are performed annually in United States hospitals. Radioisotopes are also used extensively as tracers in medical and pharmacology research. In addition to diagnosis and treatment, sterilization of medical tools has been routinely done for many years using a radioisotope source.

Industrial

Various types of measurements are made with devices that use radioactive sources. Gages have been made and are being used to measure 1) the thickness of metal or paper as it is being manufactured; 2) liquid densities or levels in containers; 3) moisture contents of soils; and 4) flaws in metal objects. In the last application, known as **radiography**, a radioactive isotope is used as a portable high-energy source of gamma rays that is used to penetrate the metal object and produce the photographic image.

In addition to the medical field, radioisotopes have also found uses as tracers in the production of chemicals, paper and cellulose, cement, metallurgy, energy, electronics, in automotive manufacture, in environmental and sanitary engineering, and in mineral production. These tracers are used in a wide range of applications including flow measurements, leak detection, residence time studies, mixing and blending analysis, and volumetric measurements.

Agriculture

Tiny fruit flies lay their eggs in ripening fruit. Upon hatching, the larvae consume the fruit reducing it to an unmarketable product. In 1995, a Mediterranean fruit fly infestation was successfully eliminated in Chile with the help of sterile fruit flies. At the

Box 27-1 ■ Nuclear Power Plant

The present generation of reactors operating in the United States uses three percent enriched uranium-235. During operation the uranium-235 atoms are consumed as their nuclei undergo fission or splitting. The two fission fragments carry away, in the form of kinetic energy, about 83 percent of the energy released. Their kinetic energy is then absorbed by the circulating core coolant and converted into thermal energy. As illustrated in the figure below, in a Pressurized Water Reactor (PWR) the core coolant water is pumped through the steam generator where the heat is transferred to a second closed water system producing steam. This steam turns the turbine converting the thermal energy into mechanical energy, which in turn, results in the production of electrical energy by the generator. The exhaust steam from the generator is passed through a condenser, and the condensate is reused as feedwater. The cooling water for the condenser comes from either a cooling tower or a large body of water, i.e. a large lake, a river, or an ocean. The condenser cooling water removes the waste heat from the energy conversion process.

A typical nuclear power plant operating at 3,000 megawatts thermal with a duty cycle of 80 percent will consume approximately 1,100 kilograms of uranium-235 in a year. Assuming that the efficiency is 33 percent, the operation results in the production of 1,000 megawatts of electrical power, releasing the other two thirds of the thermal energy produced as waste heat.

Radioactive Wastes from Nuclear Power Reactors

During normal operations, nuclear power reactors generate low-level waste (LLW) primarily from contamination of the reactor coolant. The treatment of the contaminated water by various methods creates ion exchange resin, sludge, and filter wastes. The residues are wet solids and require treatment before final disposal. The gaseous wastes released from a reactor are treated by filtration, absorption, or hold-and-decay methods. Solid wastes include paper, protective clothing, filters, used tools, exchanged parts, and scrap. Decontamination of the facility and equipment contributes to the LLW produced.

In a typical operating cycle, one third of the fuel assemblies will be removed and replaced annually. The **spent fuel** will initially be stored in a water pool, which provides both for radiation shielding and cooling. The spent fuel, by definition, is a high-level waste. When the reactor reaches the end of its operational life it is decommissioned. Most of the waste generated during decommissioning is LLW; however, the amount of radioactive wastes resulting from the decommissioning of a 1,300 megawatt (electricity) pressurized water reactor is about 15,000 tons. This is about the same amount of waste produced during 40 years of its operation, including scheduled maintenance of the fuel assemblies.

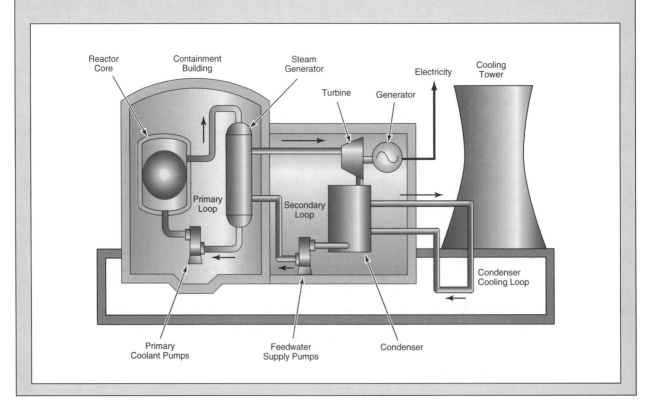

heart of the effort was the rearing of a vast number of male fruit flies in facilities where they could be sterilized by low doses of radiation. Once they were released into the wild, they mated, but failed to produce offspring, which broke the reproduction cycle and controlled the infestation.

A similar use of the sterile male technique was used to control the screwworm, a particularly damaging pest. The female screwworm fly lays its eggs in wounds, in the nose and/or navel of domesticated animals and humans. Once the eggs hatch, the larvae begin to consume the surrounding tissue. Through the use of sterilized male screwworm flies imported from Mexico, in 1988-1990, Libya was able to halt a screwworm infestation. This technique is also credited with the eradication of the screwworm fly in the Southern United States.

Radioactive isotopes have been used as tracers for investigating usage of fertilizers and to develop new plant varieties. For example, by exposing plants to carbon dioxide labeled with the radioactive carbon-14 isotope, it is possible to analyze photosynthesis and the movement of various metabolites throughout the plant. Also, a new strain of rice was developed by exposing its genetic structure to radiation. This strain produces rice with twice as much protein as other rice varieties.

Although its application remains controversial, it has been shown that the shelf life of many foods can be increased by irradiation with gamma rays from cesium-137 or cobalt-60. This method of preservation has been used on food eaten by astronauts on space flights and is routinely used to treat spices marketed in the United States. Many countries now allow irradiation of food to reduce spoilage and loss by insects.

Checking Your Understanding

1. How are the various isotopes of an element designated by the use of symbols?

2. What are the five most common forms of radiation that are emitted from the nucleus?

3. What is an ionizing radiation?

4. What would be the remaining radioactivity of a sample after a time period of two half-lives?

5. What are five uses of radioisotopes, other than for the production of electricity?

27-3 The Nuclear Fuel Cycle

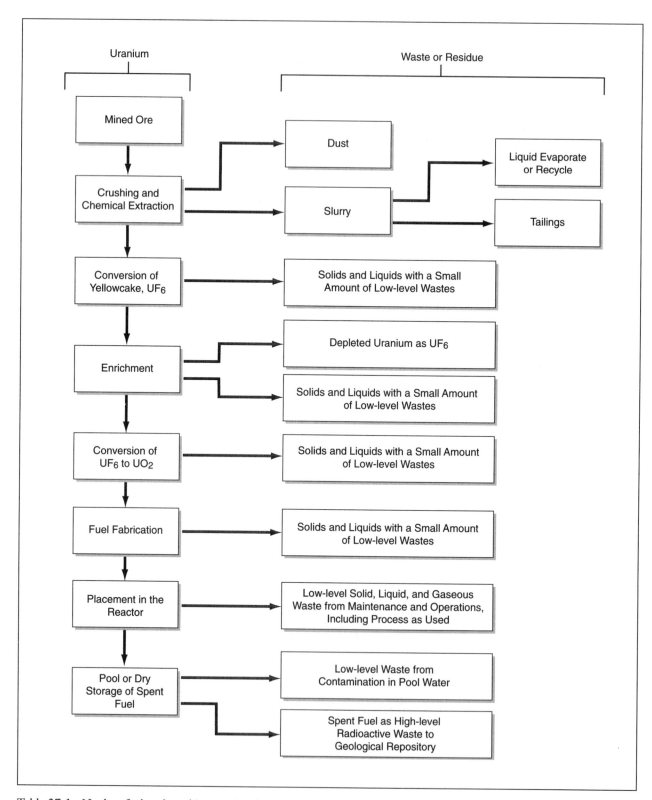

Table 27-1: Nuclear fuel cycle and its associated waste streams.

This section will discuss all the steps of the nuclear fuel cycle including supplying, using, and processing fuel for nuclear reactors, and disposal of wastes. Table 27-1 shows the various steps involved and associated waste streams generated by the nuclear fuel cycle.

Mining and Chemical Separation

The element **uranium**, U, is widely distributed in the earth and oceans. Most uranium ores being mined contain between a 0.02-0.5 percent mixture of uranium isotopes. The processing requires first that the ore be crushed and then that a chemical extraction process be used to remove the uranium. The concentrate produced by the series of chemical steps is called **yellowcake**. The mining and milling tailings typically still contain some small amounts of uranium, thorium, radium, and their radioactive daughter products. The volume of tailings produced per pound of uranium extracted will obviously vary according to the concentration of uranium in the ore.

Enrichment

To function in nuclear reactors that use graphite or heavy water as moderators to slow down fast neutrons, natural uranium with as low as 0.71 percent uranium-235 can be used for fuel. The so-called light water reactors, which use ordinary water as the moderator, require enrichment of the uranium-235 to about three percent. In the United States the enrichment of uranium-235 is accomplished by use of a uranium hexafluoride process. In this process, either yellowcake, UO_2, or triuranium octoxide, U_3O_8, is converted into gaseous uranium hexafluoride, UF_6, through a series of steps. The gaseous UF_6 is then separated and purified into its various isotopes by use of a fractional distillation method.

Fuel Element Fabrication and Storage

For use in light water reactors the enriched uranium is converted from the uranium hexafluoride back into uranium dioxide. The uranium dioxide powder is pressed into pellets that are fired to produce a ceramic-like material. The pellets are then placed in zirconium alloy tubing and sealed; these tubes are referred to as **fuel pins**. A fixed number of fuel pins is assembled in a grid pattern to form a fuel assembly. The fuel assemblies are shipped to the reactor and loaded into the core according to the refueling schedule. Depending upon the reactor type and refueling schedule, the fuel assembly typically remains in the core of the reactor for up to five years.

One of the nagging problems associated with the continued use of nuclear energy is that of disposal of the spent fuel assemblies. There are no high-level (spent fuel) commercial nuclear fuel reprocessing plants operating in the United States. This means that there is no reuse or recycling of the uranium and plutonium in the spent fuel. As a result the spent fuel is being placed in storage while awaiting final deposition. At this time most spent fuel is simply being stored at the reactor site, either in water pools or in dry storage containers. The construction of a monitored retrievable storage facility has been studied, but this, too, would only be an interim storage place until a final disposal repository becomes operational.

Checking Your Understanding

1. What is the typical concentration of uranium in ores being mined?

2. What is the concentration of uranium-235 that is present in the fuel for light water reactors?

3. What is the current status of U.S. spent nuclear reactor fuel storage and recycling?

27-4 Waste Regulations and Management

Waste Classification by Source

Radioactive wastes consist of many elements and chemical compounds and may also exist in the gaseous, the liquid, or the solid state. Radioactive waste is classified in the regulations according to the level of radioactivity and the half-life of the radioisotopes present. These classifications determine how the waste is to be handled. On the other hand, the origin of a radioactive waste – whether from defense activities or from a commercial generator – determines how and where its disposal will occur.

Since World War II, defense-related radioactive wastes have been generated primarily by the chemical separation techniques used to produce nuclear reactor fuel. These generator facilities are located at Hanford in the state of Washington; at Savannah River in Georgia; and in Idaho Falls, which is the nuclear fuel production site for the navy. The fabrication of weapons has also produced considerable waste and is responsible for the contamination at other sites, such as the Rocky Flats in Colorado. Commercial waste sources are primarily from reactors generating electrical power or from the facilities producing their fuel. Lesser amounts of waste are produced by industrial facilities, research, and medical institutions.

Waste Definitions

Atomic Energy Act

The goal of the Atomic Energy Act of 1954 was to provide for the control of source material, special nuclear material, and by-product materials. The Act defines each type of the radioactive wastes it regulates. High-level radioactive waste (HLW) is defined as 1) irradiated reactor fuel; 2) liquid waste resulting from the operation of the first cycle solvent extraction system, or equivalent, and the concentrated wastes from subsequent extraction cycles, or equivalent, in a facility for reprocessing irradiated reactor fuel; and 3) solids into which such liquid wastes have been converted.

Transuranic (TRU) waste is defined as waste material contaminated with alpha-emitting isotopes with atomic numbers above uranium (>92) with half-lives of more than 20 years and concentrations greater than 100 nanocuries/gram (a nanocurie is defined as being 37 disintegrations per second).

Low-level radioactive waste (LLW) is radioactive waste that is neither HLW, nor TRU, nor mill tailings. (**Mill tailings** are classified separately from low-level waste derived from nuclear power reactors and from radioisotope production and use.) Low-level wastes are further classified in 10 CFR 61 for near-surface disposal – as opposed to geological repositories – as Class A, Class B, and Class C. The determination of waste class is based on the radioisotope present, its half-life, and its radioactivity concentration, as measured in curies per cubic meter (Ci/m^3). Class A is the least restrictive class. Class B requires more stringent requirements on the stability of the waste after disposal. Class C must meet requirements to protect it against future inadvertent intrusion. If the waste does not meet Class C specification, then it must be disposed of as a high-level waste.

Mixed wastes are wastes that contain both radioactive materials and hazardous chemicals. Hazardous wastes are substances that are flammable, toxic, explosive, and/or corrosive. If pesticides, carcinogens, or heavy metals such as silver, cadmium, lead, or mercury, are part of the radioactive waste mixture, then they must be classified as a mixed waste. Because of the volume of mixed wastes generated and the problems associated with their disposal, methods are being developed to separate and treat their nonradioactive components separately.

Exempt wastes are items that have a relatively low radioactivity content and do not have to conform to radioactive waste regulations. One category, specified in 10 CFR 31, is manufactured items. This means that items such as static elimination devices, ion-generating tubes, luminous devices in aircraft, calibration sources, ice-detection devices, and prepackaged *in vitro*/clinical testing kits are exempt. In a second category, excreta from patients receiving radioactive material for treatment or diagnosis of medical conditions are exempted by 10 CFR 20.2003(b), which allows for their release into the sanitary sewer system. The provisions in 10 CFR 20.2005 allow a licensee to dispose of some licensed material as if it were not radioactive. This provision includes items such as liquid scintillation fluids and animal tissues that have less than 0.05 microcurie (1.85 kBq) per gram and exempts their disposal.

Low-level Waste Disposal

Commercial Sites

All of the regulations concerning the disposal of low-level radioactive waste at a commercial site are found in Title 10 of the CFR. For example, the regulations covering burial of low-level waste are set forth in 10 CFR 61; the regulations for disposal of waste by a licensee in 10 CFR 20; the Nuclear Regulatory Commission (NRC) rules for packaging and transportation of licensed materials are found in 10 CFR 71; and the standards for protection against radiation are contained in 10 CFR 20. However, the Department of Transportation (DOT) regulations covering transportation of radioactive materials, which include packaging, marking and labeling, placarding, monitoring, accident reporting, and filling out shipping papers are found in 49 CFR Parts 170 through 189.

According to 10 CFR 61.59, for a waste to be received at a near-surface commercial disposal site, it must meet the following general conditions:

1. The waste cannot be packed in cardboard or fiberboard boxes.

2. If it is a liquid waste, it must be solidified or packaged in sufficient absorbent material to absorb twice the volume of the liquid.

3. If a solid waste contains a liquid, the liquid must represent less than one percent of the volume.

4. The waste must not be readily capable of detonation, of explosive decomposition at normal pressures and temperatures, or of explosive reaction with water.

5. The waste must not contain or generate quantities of toxic gases, vapors, or fumes harmful to people transporting, handling, or disposing of the waste.

6. The waste must not be pyrophoric.

7. If in gaseous form, the waste must be packaged at pressures that do not exceed 1.5 atmospheres at 20°C. Total radioactivity must not exceed 100 curies per container.

8. The waste containing hazardous, biological, pathogenic, or infectious material must be treated to reduce, to the maximum extent practicable, the potential hazard from the non-radiological materials.

In addition, 10 CFR 61.59b states that the waste is to have structural stability, which is intended to ensure that the waste does not structurally degrade nor affect the overall stability of the site through slumping, collapse, or other failure that might lead to water infiltration. Most of the low-level waste, therefore, is shipped to the disposal site in steel drums or rectangular steel containers. To be acceptable for burial, the waste package must have less than five percent void space.

State Responsibility

The Low-level Radioactive Waste Policy Act of 1980 and its 1985 Amendments were enacted by Congress to make states responsible for disposal of the low-level wastes generated within their borders. However, the states were given the choice to form **compacts** and share a disposal site or to remain non-aligned and proceed individually. In 1997 only two low-level waste burial disposal sites were available in the whole country; one in Barnwell, South Carolina and a second at Richland, Washington. The state of California has selected Ward Valley, a site near Needles, but its operations are still pending.

The availability of a low-level waste disposal site depends not only on the state in which the generator is located, but also on whether that state is in a compact. At present, generators located in non-aligned states have to either continue to store their waste or give up using materials that require disposal at a licensed facility. For generators who do have a low-level waste disposal facility available, the cost becomes a second issue. A 300 pound, 55-gallon drum of low-level waste generated in Washington and disposed of at the Northwest compact's Richland site, for example, would cost nearly $1,000. It should also be noted that this figure does not include the cost of the container or other expenses associated with preparing the waste for shipment.

Due to the high cost associated with the disposal of low-level radioactive wastes, it is wise to use those P2 activities such as good housekeeping practices, reuse and recycling of liquids and equipment, and process changes to keep the volume of waste generated to a minimum. Yet another permitted way – 10 CFR 20.20001(2) – to reduce waste volume is to store short half-life (< 90 days) radioisotopes through 10 half-lives, which reduces their radioactivity to 0.1 percent of its initial activity. This reduction most likely puts the waste below regulatory concern, which in turn makes it possible for the material to be disposed of in the regular trash or incinerated as a nonradioactive substance.

There are several other techniques that can be used to reduce the volume of low-level radioactive waste. One of the more common methods used for solid wastes is supercompaction. Eliminating the voids within the packaging can reduce volumes by 3-6 times, depending on the density and springback of the uncompacted materials. Incineration is the preferred method for disposing of dry wastes and can result in volume reductions of 100-200 times. The resulting ash can then be mixed with cement and disposed of as a solid. Incineration is also the preferred method for disposing of materials containing toxic chemicals, organic solvents, and petroleum products. The incineration of liquid scintillation fluids and animal tissue is also allowed under 10 CFR 20.2004, if its radioactivity is less than 0.05 microcuries of ^{3}H or ^{14}C per gram.

As shown in Table 27-2, low-level waste generators are separated into the following categories for reporting purposes: 1) utility, 2) academic, 3) medical, 4) industrial, and 5) government. Commercial nuclear power plants are included in the utility category. The academic category includes university hospitals and university medical and non-medical research facilities. Medical generators include hospitals and clinics, research facilities, and private medical offices. The industrial category encompasses private entities such as research and development companies, manufacturers including radiopharmaceutical manufacturers, nondestructive testing, mining, and fuel fabrication facilities. The government category includes state and federal agencies.

If this waste (792,177 ft^3) were stacked on a football field (160 by 300 feet), it would make a pile just 16.5 feet high. Not only is this considered to be a small volume but – as shown below – the annual volume of low-level radioactive waste being shipped to commercial facilities appears to be decreasing.

Year	Volume, ft^3
1980	3,700,000
1991	1,369,303
1992	1,743,279
1993	792,177

Generator Category	Volume (cubic ft)	Percent of Volume	Activity (curies)	Percent of Activity
Academic	12,168.41	1.54	110.65	0.02
Government	95,492.90	12.05	28,665.18	4.46
Industrial	276,987.34	34.97	2,037.18	0.32
Medical	5,135.80	0.65	21.45	0.00
Utility	402,392.57	50.80	612,372.35	95.21
Total	792,177.02		643,206.81	
Distribution by Site				
Barnwell	605,442.67	76.43	611,784.73	95.11
Richland	186,734.35	23.57	31,422.08	4.89

Table 27-2: Distribution of low-level waste received at the Barnwell, SC and Richland, WA commercial disposal sites.

For a low-level waste to be deposited in a commercial site, it must have stable physical and chemical characteristics. Several different methods are used to provide the waste with the necessary mechanical, physical, chemical, and radiological stability for the subsequent stages of radioactive waste management. Cement, bitumens, thermoplastics, thermosets, and waste containers made from composite-reinforced plastics can be used to provide structural integrity and protect the material from chemical attack. If the waste contains carbonates, nitrates, silicates, asbestos, glass fibers, incinerator ash, concrete, or chemical process wastes, a vitrification process may be used to convert the waste into a glasslike material. Although this process is expensive and has been under consideration primarily for converting high-level radioactive waste into a glass, it could also be used for the treatment of low-level wastes.

Various other processes have been used on liquid and wet solid wastes. Ion exchange, filters, evaporation, centrifuging, and reverse osmosis have been used on liquid wastes to concentrate the low-level constituents. Wet solids have been dried, centrifuged, dewatered, filtered, and dehydrated to remove excess water. The net result of all these processes is to reduce the volume and produce waste that is physically and chemically more stable.

Unregulated Wastes

Disposal of Naturally Occurring or Accelerator Produced Radioactive Materials (**NARMs**) fall under state rather than federal regulations. Naturally Occurring Radioactive Materials (**NORMs**) include: radium needles and sources; radium/beryllium neutron sources; radium-containing compasses, aircraft instruments, and watch markers; radium daughters (^{210}Pb, ^{210}Bi, and ^{228}Th); and unlicensed uranium and/or thorium samples, if present in exempt concentrations. The need for proper disposal of material classified as NORM is illustrated by a leaking radium source purchased at a garage sale. The resulting cleanup of the contamination to a private residence in Tennessee, for example, was estimated to have been more than $200,000 in donated time, materials, equipment, and waste disposal.

Mill Tailings from Mining

EPA has published standards (40 CFR 192) for the emissions from uranium mill tailings for air and water exposures. Uranium ore tailings consist of the residue remaining from the ore grinding and chemical extraction processes. During these processes the uranium concentration will have been reduced to 5-10 percent of the initial (0.02-0.5 percent) concentration. The only decay product of concern remaining is the radioisotope radium-226. This isotope has a half-life of 3.82 days and decays to radon-222, a gas that escapes into the atmosphere and becomes a source of exposure. Congress passed the Uranium Mill Tailings Radiation Control Act in 1978, which required the Department of Energy to establish a program of assessment and remedial action at 24 inactive uranium mill sites. Included in this program were properties in the vicinity of the mill sites that were contaminated with residual radioactive material. This project is proceeding to stabilize, dispose, and control the residual radioactive materials to meet EPA standards.

High-level Wastes

The disposal of high-level radioactive waste is also controlled by the Nuclear Regulatory Commission (NRC), using EPA standards. The following are some of the 10 CFR 60 requirements:

1. The issuance of a license for operation must not constitute an unreasonable risk to health and safety of the public.

2. The high-level waste must be retrievable for up to 50 years after first emplacement operations start.

3. The high-level waste must remain contained within its package for at least 300 years.

4. The geologic setting of the repository must be located so that the fastest travel time of groundwater to the accessible environment is at least 1,000 years.

High-level and Transuranic Wastes

Several sites are being extensively tested to determine if they will meet the geological criteria for a high-level disposal site. Among these are Yucca Mountain site in Nevada and the Waste Isolation Pilot Plant (WIPP) outside Carlsbad, New Mexico. WIPP is a research and development facility designed to demonstrate the safe disposal of defense-related transuranic waste and cannot be used for commercial waste. As designed, the WIPP storage facility will bury the waste 2,150 feet below the surface in a thick salt deposit.

Checking Your Understanding

1. What factor determines the disposal site for a radioactive waste?

2. What Act defines high-level and low-level radioactive wastes?

3. In addition to radioisotopes, what else must be present for a waste to be considered as mixed?

4. What types of radioactive wastes are exempted from disposal restrictions?

5. What are the eight general conditions that must be met for a low-level radioactive waste to be buried in a near-surface commercial disposal site?

6. What treatment method produces the largest solid waste volume reduction?

7. What are two sites that are being developed for long-term deposit of high-level radioactive wastes?

Summary

All naturally occurring elements exist as a mixture of isotopes, with varying atomic masses. Of those isotopes, some have stable and others have unstable nuclei that tend to undergo a spontaneous release of subatomic particles and/or energy. Those isotopes, called radioisotopes, release radiation in the form of alpha particles, beta particles, positrons, and very powerful electromagnetic radiation, including gamma rays and x-rays. Some elements are fissile and release neutrons. When one of these neutrons is absorbed by the nucleus of a neighboring atom, it can cause it to split producing two smaller fission fragments and the release of more neutrons. The degree of radioactivity is measured by a unit called the becquerel and is determined by the number of nuclei decaying per second.

The energy resulting from fission reactions has been used for a variety of purposes including power generation, medicine, industrial measurements, agriculture, and military weapons. Uranium is the element mined for nuclear power generation fuel. The ore is mined, crushed, and through a series of chemical extraction steps, is converted into a concentrate called yellowcake. The yellowcake is enriched by converting it to gaseous uranium hexafluoride and then separating the gas into its various isotopic fractions. The enriched uranium hexafluoride is then converted into uranium dioxide, pressed into pellets, and fired into a ceramic-like material. These pellets are sealed in zirconium alloy tubes and become the fuel pins for the reactor refueling.

Radioactive wastes consist of many elements and chemical compounds; they may exist in the gaseous, the liquid, or the solid state. The wastes are classified and regulated according to the level of radioactivity and the half-life of the radioisotopes present. These classifications determine how the waste is to be handled, but their origin determines how and where it will be disposed of. The Atomic Energy Act of 1954 defined each type of radioactive waste it regulates. High-level radioactive waste is defined as: irradiated reactor fuel; liquid waste resulting from the operation of the first cycle solvent extraction system and the concentrated wastes from subsequent extraction cycles; and solids into which such liquid wastes have been converted. Transuranic waste is defined as waste contaminated with alpha-emitting isotopes of elements with atomic numbers above uranium, with half-lives longer than 20 years, and in concentrations greater than 100 nanocuries/gram. Low-level radioactive waste is waste that is not high-level waste, not transuranic waste, and not mill tailings. Mixed wastes are those wastes containing both radioactive materials and hazardous chemicals. Because of the volume of mixed wastes generated and the problems associated with their disposal, new methods are being developed to separate and treat the nonradioactive components.

Finally, there are those relatively low radioactive substances that are exempt wastes. This category includes ice-detection devices, prepackaged in vitro/ clinical testing kits and other manufactured items. Low-level wastes are regulated by Title 10 of the Code of Federal Regulations and NRC rules. Because of the regulations, they are shipped to commercial burial sites in steel drums or rectangular-shaped containers with less than five percent void space. The Low-level Radioactive Waste Policy Act of 1980 and its 1985 Amendments required states to provide low-level waste disposal sites within their borders or permit them to form compacts to share disposal sites. At this time, only two low-level disposal sites are operational. Disposal costs, therefore, are high, and alternative ways to reduce total volume are recommended.

High-level wastes are also regulated by NRC, but using EPA standards. Several sites are being extensively tested for the long-term entombment of high-level and transuranic wastes. WIPP is an example of a research and development site designed to demonstrate the safe disposal of defense-related transuranic wastes.

Critical Thinking Questions

1. Research the volume of LLW currently being sent to disposal sites. Based on the past and current trends, present arguments for the need to build ten more LLW disposal sites in the United States.

2. Determine local institutions (medical and educational) and companies that are licensed to use radioactive materials. How much material do they ship for disposal at a low-level radioactive site?

3. Do the regulations and facilities under study provide for the safe disposal of high-level radioactive wastes? Justify your answer.

4. Compare the generation of electrical power based on the use of nuclear, petroleum, and coal fuel sources. Which fuel type offers the least long-term damage to the environment and why?

28

Consumer-Related Waste Issues

by Melinda A. Trizinsky

Chapter Objectives

Upon completing this chapter, the student will be able to:

1. **Identify** wastes commonly generated in the home or office.

2. **Recognize** the hazards associated with various household chemicals and office supplies and be able to discuss their potential for environmental damage.

3. **Identify** the environmental costs associated with consumer-generated hazardous waste disposal.

4. **Identify** how consumer choices affect waste generation and energy consumption throughout the industrial sector.

5. **Discuss** changes that have occurred in waste generation and the use of resources as a result of the political consciousness of consumers.

Chapter Sections

28-1 Introduction

Chapter 7 of this book provided an overview of pollution prevention concepts and methods that can be used by industry. The industry-specific chapters described the operations of a particular type of industry – which the student may not necessarily be familiar with – and demonstrated how waste can be reduced through the use of various pollution prevention strategies. The present chapter deals with more familiar issues since it examines pollution prevention opportunities from the consumer's perspective. Consumers generate pollution directly and indirectly. This chapter examines ways for consumers to reduce their environmental impact and improve quality of life by reducing exposure to toxic substances in the home, on the job, and while commuting. It will also consider the impact consumers have on how large-scale industries select their materials, how they produce, and how they package their products.

Waste at Home

Everyone generates wastes. Do you know what sort of wastes you generate in your home? If your home is typical, you throw away nonhazardous wastes such as food wastes, junk mail, packaging materials, yard wastes, newspapers, old clothes, and furnishings. You probably also generate a fair amount of hazardous waste: batteries, paints, cleaning supplies, pesticides, drain cleaners, disinfectants, expired medicines, beauty products, fuels, aerosols, solvents, and automobile maintenance products.

Residential areas contribute substantial amounts of fertilizer, pesticides, petroleum products, animal wastes, and sediment to surface water supplies. New regulations may soon try to mitigate this non-point source pollution by restricting the use of fertilizers and pesticides near waterways and by requiring that storm water runoff be treated before being discharged to waterways.

In addition to solid wastes, you also dispose of wastewater contaminated with soaps, cleaning supplies, beauty products, drain cleaners, food wastes, etc. This type of wastewater is collected in the drain system in your home and may be treated by the local sewage treatment plant, or your home may have its own septic system. Your home may also contribute to non-point source pollution as water runs off your yard or driveway.

You probably also contribute to air pollution problems. Some sources of residential air pollution include using aerosols, using home heating fuels, leaks from your air conditioner, burning yard wastes, using solvent-based paints and cleaners, using gas-powered yard maintenance equipment, and – most importantly – owning and driving a car.

Waste at the Office

Any place inhabited by people for eight hours a day or longer will generate wastes. At the very least, people will generate wastewater and – although the office is not necessarily designed as an eatery – it is inevitable that workers will eat and drink in the office. Virtually all offices feature a coffeepot; some have full kitchens. Whether workers prepare food onsite or bring in their lunch, scrap food and packaging will be part of the office waste stream.

These types of waste, however, are minimal compared with the amount of paper waste that is generated. Paper wastes account for more than 75 percent of the solid waste generated in the average office. Box 28-1 describes the average composition of solid wastes generated in an office.

Additional solid wastes generated by offices include disposable products (i.e. pens, pencils, toner cartridges, and stamp pads). Most offices use disposables, although refillable versions are available. Businesses come and go, and when they do, office equipment and furnishings are frequently replaced. In addition to the transient nature of business, computer technology has advanced so rapidly that most offices find that their computer systems become obsolete in a few years. Eventually these items find their way into the office waste pile.

Box 28-1 ■ Solid Waste Generation in an Office

The average employee in an office building generates 2.9 lbs. of solid waste per eight-hour workday. The average composition of this solid waste stream (by weight) includes high grade paper (39.6%), low grade paper (20.2%), newsprint (7%), cardboard (2.8%), other paper (7.4%), glass (11.8%), plastic (2.6%), food waste (2.9%), metal (1.9%), and other miscellaneous waste (3.9%). These percentages reflect waste composition before recycling.

Generally, offices do not contribute significantly to outdoor air pollution. However, modern office buildings often have hermetically sealed windows as part of their climate control system. While this system helps save energy, it can also trap indoor air pollutants. New furnishings, construction materials, paints, and carpeting can release hazardous chemicals into indoor air. Copiers and printers can also release airborne pollutants, and cigarette smoke – if permitted – is full of irritants and carcinogens. Thus, indoor air in modern office buildings can be more polluted than outdoor air. This condition is known as sick building syndrome.

Checking Your Understanding

1. Name ten hazardous wastes generated in homes.

2. What is the largest volume of waste generated in offices?

3. What does the term sick building syndrome refer to?

28-2 Service Industries

The U.S. economy is dominated by **service industries**. As indicated by their title, these industries do not manufacture products. Rather, service industries may 1) research consumer needs, 2) design products, 3) finance production, 4) test products, 5) advertise products and services, 6) deliver products and advertising, 7) sell products to consumers, 8) maintain and clean products and facilities, 9) repair products, 10) insure products and facilities, and 11) eventually dispose of wastes generated. In other words, service industries support industrial manufacturing. Large industries may have in-house divisions that perform some of these functions but, at some point, even these large manufacturers rely on outside services to move their products around in the economy. Service industries also include professional services, communication industries, live entertainment, recreational facilities, health and beauty services, food service, and the medical profession. The medical industries have been covered separately in chapter 26. To illustrate the importance of service industries in the United States, Table 28-1 lists the Standard Industrial Codes (SIC) used by the government to categorize various segments of the economy. SIC 40 to 97 are service industries.

Service industries considered in this chapter are end-consumers of products and energy produced by other industries. By thinking of service industries as consumers, parallels can be drawn between activities and wastes in these industries and individual consumers. Much like in the home, the wastes generated onsite by service industries generally include broken or obsolete equipment and furnishings, cleaning and maintenance products, waste paper, packaging materials, construction materials, food scraps, wastewater, air pollutants, and spent fuels and lubricants. As Box 28-1 and 28-2 show, the mix of wastes will depend on the nature of the business. A fast food restaurant will generate a much different mix of wastes, for example, than an office. A waste audit will help identify the wastes being generated by a particular business.

Transportation

The personal vehicle is the preferred mode of travel in the United States, accounting for approximately 85 percent of the passenger miles traveled. Air travel is a distant second and accounts for about 12 percent of the total. This means that only about three percent of passenger travel occurs via public transportation (bus and rail) or by human locomotion (walking and bicycling). There were approximately 141 million automobiles and 26 million light trucks on American roads in 1990. Combined, these vehicles travel approximately two trillion miles per year and consume about 16 million barrels of oil per day. Despite fuel efficiency increases over the last two decades, fuel consumption has been increasing about 2.6 percent per year since 1980. Energy costs associated with fuel production, road construction, road maintenance, motor vehicle manufacturing, and vehicle maintenance account for the energy equivalent of an additional four million barrels of oil per day.

Box 28-2 ■ Solid Waste Generation at a Fast Food Restaurant

A typical fast food restaurant serving hamburgers to 2,000 customers per day generates an average of 238 pounds of MSW a day. The average composition of this waste stream (by weight) includes corrugated boxes (34%), food waste (34%), coated paper (7%), customer wastes including diapers, etc. (6%), polystyrene cups, lids, etc. (4%), uncoated paper napkins (4%), plastics (3%), and miscellaneous waste (8%). Items such as fryer grease are typically recycled.

01	Agricultural production crops	36	Electronic and other electric equipment	65	Real estate
02	Agricultural production livestock	37	Transportation equipment	67	Holding and other investment offices
07	Agricultural services	38	Instruments and related products	70	Hotels and other lodging places
08	Forestry	39	Miscellaneous manufacturing industries	72	Personal services
09	Fishing, hunting, and trapping	40	Railroad transportation	73	Business services
10	Metal mining	41	Local and interurban passenger transit	75	Auto repair, services, and parking
12	Coal mining	42	Trucking and warehousing	76	Miscellaneous repair services
13	Oil and gas extraction	43	U.S. postal service	78	Motion pictures
14	Nonmetallic minerals, except fuels	44	Water transportation	79	Amusement and recreation services
15	General building contractors	45	Transportation by air	80	Health services
16	Heavy construction, ex. building	46	Pipelines, except natural gas	81	Legal services
17	Special trade contractors	47	Transportation services	82	Educational services
20	Food and kindred products	48	Communications	83	Social services
21	Tobacco products	49	Electric, gas, and sanitary services	84	Museums, botanical, and zoological gardens
22	Textile mill products	50	Wholesale trade and durable goods	86	Membership organizations
23	Apparel and other textile products	51	Wholesale trade non-durable goods	87	Engineering and management services
24	Lumber and wood products	52	Building materials and garden supplies	88	Private households
25	Furniture and fixtures	53	General merchandise stores	89	Services, not elsewhere classified
26	Paper and allied products	54	Food stores	91	Executive, legislative, and general
27	Printing and publishing	56	Apparel and accessory stores	92	Justice, public order, and safety
28	Chemicals and allied products	57	Furniture and homefurnishings stores	93	Finance, taxation, and monetary policy
29	Petroleum and coal products	58	Eating and drinking places	94	Administration of human resources
30	Rubber and misc. plastics products	59	Miscellaneous retail	95	Environmental quality and housing
31	Leather and leather products	60	Depository institutions	96	Administration of economic programs
32	Stone, clay, and glass products	61	Nondepository institutions	97	National security and international affairs
33	Primary metal industries	62	Security and commodity brokers	99	Nonclassifiable establishments
34	Fabricated metal products	63	Insurance carriers		
35	Industrial machinery and equipment	64	Insurance agents, brokers, and service		

Table 28-1: Standard Industrial Codes.

The production, maintenance, and use of personal automobiles as the basis of our transportation system consume about 25 percent of the gross national product annually.

In addition to the air pollution caused by burning the equivalent of 20 million barrels of oil per day, personal automobiles impact the environment and our quality of life in other ways. Roads occupy 60,000 square miles of U.S. land; this is about two percent of our total land area. The proportion of land devoted to the automobile is especially high in urban areas where about one third of the land area is occupied by roads and parking areas. Car use contributes to urban sprawl, congestion, noise pollution, and decreases in pedestrian and bicyclist safety. The disposal of waste oil, tires, roadside litter, batteries, abandoned vehicles, and auto parts adds to the solid waste burden. Gasoline is a hazardous substance and contains benzene, a known carcinogen; therefore, driving and fueling our cars accounts for one of our largest exposures to toxic chemicals. Leaking underground storage tanks at gas stations across the country have contaminated soil and groundwater with these same chemicals. Luckily, leaded gasoline has finally been phased out in the United States, so we no longer spew toxic lead into the environment. However, the residual lead deposited near roadways prior to the ban still impacts the mental abilities of people, particularly children, living near major roads.

Checking Your Understanding

1. What types of services are provided by the service industries?

2. What aspect of service industries makes their waste generation patterns similar to the average consumer's?

3. What is the SIC Code for private households?

4. What percent of the total United States land is occupied by roads?

28-3 Consumer-generated Wastes

> The average person in the United States generates more than 3/4 of a ton of solid waste per year.

Solid Wastes

Homes, offices, and small businesses are the primary sources of **municipal solid wastes (MSW)**. These are the mixed solid wastes generated within a municipal jurisdiction that are collected by city or county waste haulers. After collection, they may be landfilled, recycled, composted, or incinerated by municipal government programs. This category of solid waste does not include industrial wastes unless the industry is disposing of its wastes through agreements with the municipality. The United States produces more MSW than any other country. In 1995,

approximately 208 million tons of MSW were generated; this corresponds to approximately 4.3 lbs. of solid waste per person per day. Less than 27 percent of this solid waste is being recycled; therefore, more than 150 million tons of solid wastes per year are disposed of in landfills or incinerated. Figure 28-1 shows the breakdown of the waste by type of product discarded while Figure 28-2 shows the composition of these wastes by type of material, i.e., paper, glass, and metal.

Paper

The United States is the world's largest user of paper and the largest producer of paper waste. Paper and cardboard account for almost 40 percent of the solid MSW generated in the United States. Paper products in MSW include corrugated boxes, paper packaging, newspapers, books and magazines, other

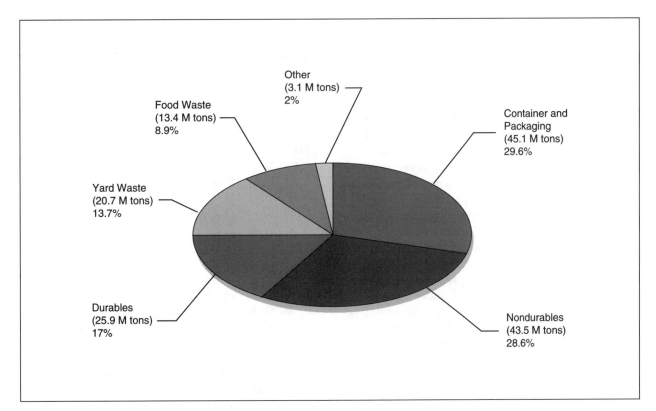

Figure 28-1: Waste types discarded in MSW in 1995 (after recycling).

Paper Product	Waste Generated*	Waste Recycled*	Percentage Recycled	Percentage Disposed
Newspapers	13.13	6.96	53%	47%
Books & Magazines	3.54	0.89	25.1%	74.9%
Other Printed Material	19.02	4.89	25.7%	74.3%
Corrugated Boxes	28.8	18.49	64.2%	45.8%
Other Paper Packaging	9.26	1.41	15.2%	84.8%
Paper Goods	7.79	0	negligible	100%
Totals	**81.54**	**32.64**	**40%**	**60%**
* Millions of tons in MSW				

Table 28-2: U.S. paper waste and recycling (1995)

printed materials, and disposable paper products such as tissues, plates, cups, napkins, etc. Table 28-2 shows how many millions of tons of waste paper were generated in the United States in 1995. Most of this paper is recyclable; the table also shows how much was recycled in each of these categories in 1995.

Although recycling is better than disposal, reducing the amount of waste paper generated is a better approach. The paperless office promised by the computer industry has tremendous appeal, but the reality is that computers have made it easier than before to generate large volumes of waste paper. Paper usage has been increasing for at least 25 years and – as Table 28-2 shows – only about 40 percent of this paper is currently being recycled. The average office worker in the United States uses about 350 pounds of paper per year. Of course, not all of this paper ends up as waste paper, but there are other environmental costs associated with this paper use (i.e., increased logging with its associated environmental costs, increased air pollution from paper production and transport, etc.).

Glass

Glass is very durable, non-porous, non-reactive, and non-toxic. Therefore, it is frequently used for food and beverage containers. Approximately 13 million tons of glass are discarded annually; 90 percent of this glass is in the form of food and beverage containers that are fully recyclable. Only about 25 percent of the glass discarded annually is recycled; therefore almost 10 million tons of glass are buried in landfills annually.

Plastics

The amount of plastic waste generated in the United States has been increasing at a rate of about 10 percent per year for the last 20 years. Plastics are used primarily as packaging materials, toys, furniture, components in appliances, housings for electronics, utensils, bags, film, and coatings. Approximately 18 million tons of plastics were incinerated or landfilled

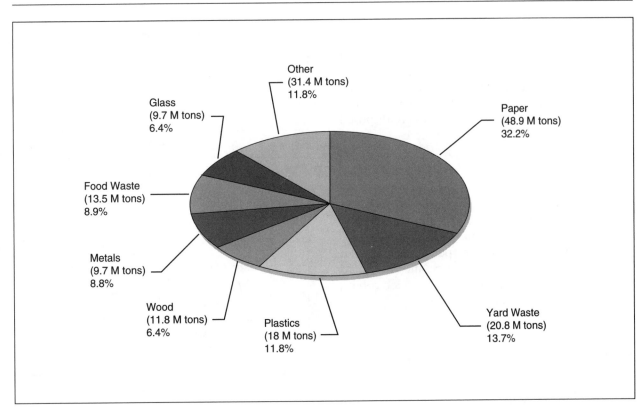

Figure 28-2: 1995 U.S. MSW waste composition (after recycling).

in 1995. Although plastics account for less than 12 percent of MSW, they are produced from nonrenewable petroleum sources, so we are losing valuable resources when we throw plastics away. Plastic trash that doesn't make it into MSW presents another problem. Most plastics do not biodegrade, so when plastics are disposed of improperly (e.g. roadside trash, disposal at sea) they accumulate in the environment and harm wildlife. Plastic refuse is particularly noticeable along roads, beaches, and waterways. Although about half of U.S. households have access to plastics recycling programs, only 20-25 percent of recyclable plastics is being recovered.

Metals

Almost 16 million tons of metals are discarded annually as part of MSW in the United States. Food and beverage containers account for a large percentage of these waste metals. Because of the large energy costs associated with the mining and production of metal containers, these packaging materials should be recycled. A 95 percent energy savings is achieved when an aluminum can is produced from recycled materials instead of virgin ore. Aluminum and steel cans are recyclable through most municipal recycling programs. Other scrap metals can frequently be recycled through scrap metal dealers. About 40 percent of MSW metals are currently being recycled in the United States.

Food Waste

Food wastes include food scraps, spoiled food, and discarded food. These wastes account for about 14 million tons of MSW annually. This value does not include agricultural and manufacturing wastes associated with food production.

Yard Waste

Yard wastes are plant materials associated with yard maintenance (grass clippings, leaves, branches, dead plants, trimmings, etc.). Most plant material can be composted or mulched for reuse in the yard, but almost 21 million tons of yard wastes are disposed of annually as MSW in the United States.

Other Non-hazardous Solid Wastes

It is difficult to generalize about this category of solid waste. It can include construction debris (gypsum wallboard, insulation, carpeting, etc.), dead animals, used clothing and textiles, and unwanted consumer products (appliances, toys, electronics, home decor, furniture, etc.).

Hazardous Wastes

Many people don't realize how many products used in the home and office are hazardous. In fact, the two most likely places to be exposed to hazardous chemicals are the home and the workplace. Usually, OSHA tries to ensure a safe workplace, but ultimately everyone is responsible for their own safety, especially in the home. Consumers should learn to read product labels before they buy the product. They should look for the words Caution, Flammable, Warning, Combustible, Danger, Explosive, Poison, Volatile, Toxic, Corrosive; these terms indicate that the product requires special handling or disposal. Table 28-3 lists some common consumer products that contain hazardous components.

Most product labels tend to advertise the virtues of the product rather than emphasize information on product safety. The consumer must know what to look for and how to read the fine print on a label. It is the consumer's responsibility to know how to use a product safely. Federal law requires that most hazardous products include specific types of information about the product on their labels. The Federal Insecticide, Fungicide and Rodenticide Act (FIFRA), for example, regulates labeling of products that contain pesticides. The Federal Hazardous Substances Act (FHSA) regulates labeling of all other hazardous products.

Hazardous consumer products should not be disposed of with regular non-hazardous trash because municipal landfills are not designed for haz-

Product
Automotive Fluids
Fuels
Household Cleaners
Disinfectants
Polishes
Pesticides
Pet Care Products
Paint Products
Photographic Chemicals
Swimming Pool Chemicals
Personal Hygiene Products
Batteries
Toner Cartridges
Solvents & Degreasers
Drain Openers
Gas Cylinders

Table 28-3: Hazardous consumer products.

ardous wastes. Contact your local waste disposal agency for the proper handling of hazardous waste. Do not dispose of excess product by pouring it down the drain; this may disrupt your septic system or else contaminate treatment plant sludge. If disposed of improperly, these hazardous chemicals in consumer products may contaminate air, soil, groundwater, and surface water.

Air Pollution

Most people blame factories and power plants for air pollution and don't really consider their own contribution to the problem. True, the amount of air pollution an individual produces is fairly small compared to a power plant, but collectively we are responsible for polluting the air far more than industrial polluters. The motor vehicles we drive are the greatest source of air pollution in the world. Passenger vehicles and trucks consume nearly 50 percent of the petroleum used in the United States; this accounts for about 21 percent of the total energy used in this country. The end products from internal combustion engines are discharged to the air. These products include carbon dioxide, carbon monoxide, NO_x compounds, particulates, and partial combustion products. Several of these compounds contribute to the smog and ground level ozone that degrade air quality in urban areas; they also contribute greenhouse gases to our atmosphere. Carbon dioxide is the most desirable end product of fossil fuel combustion because it is non-toxic and is a natural component of the atmosphere, but even this product has negative consequences when produced at current rates. Carbon dioxide levels appear to be increasing in the atmosphere; many scientists believe that this build-up may lead to global warming.

Other ways of contributing to air pollution include smoking tobacco products; using products that contain or release volatile organic compounds (VOCs); using gas-powered yard maintenance equipment; burning yard wastes; using wood-fired stoves and fireplaces to heat homes; using aerosols and other products that contain ozone-depleting chemicals (ODCs); using solvent-based paints and cleaning supplies; and using too much energy. Because most electricity is produced from fossil fuels, energy consumption is probably the second most significant way in which individuals harm air quality.

The choices we make as consumers can impact our health and the environment. Most people spend about 80 to 90 percent of their time indoors. Thus, their greatest exposure to airborne pollutants may occur indoors. In an effort to reduce energy needs, many homes and businesses have replaced leaky doors and windows and increased the insulation of exterior surfaces. While this helps save energy, it can also trap indoor air pollutants, making it even more important to be aware of hazardous air pollutants in indoor air. New furnishings, construction materials, paints, carpeting, and cleaning products can release hazardous chemicals into indoor air. Faulty heaters can produce deadly carbon monoxide gas. Radon gas released by geologic decay can accumulate indoors and cause dangerous exposure to radiation. In areas where people smoke the air quality is greatly diminished because tobacco smoke is full of irritants and carcinogens.

Water Pollution

There are two basic types of wastewater produced by homes and businesses: wastewater that is collected by interior plumbing and runoff that collects sediment and pollutants from exterior surfaces, pavement, and the yard.

Wastewater generated inside homes and businesses is usually treated either at a publicly owned treatment works (POTW) or by a septic system onsite. This wastewater usually contains a mixture of **sewage** and **graywater**, unless the plumbing system separates the two. In plumbing systems that separate the two, sewage refers solely to wastewater generated in toilets; this waste stream may contain human pathogens and cannot be reused without treatment. Graywater includes water from sinks, tubs, showers, dishwashers, and laundry facilities; this water does not contain human wastes and can potentially be reused in the yard, if it doesn't contain too many harmful chemicals. Graywater may be contaminated with cleaning products, grease, disinfectants, food wastes, drain cleaners, personal hygiene products, beauty aids, and anything else people pour down the drain.

Runoff from residential and business areas contributes to non-point source pollution. Most of this water is discharged to surface waters without treatment. Pollutants in this water may include pesticides, fertilizers, motor oil, sediment, and animal wastes.

Checking Your Understanding

1. What is the approximate amount of solid waste generated per person per day in the United States?

2. Has the use of plastics increased or decreased in the last ten years, and by what percentage?

3. What are ten words that warn you that a product may need special handling or disposal?

4. What is the largest source of air pollution in the world?

5. How does graywater differ from sewage?

28-4 As a Consumer

As consumers we make many kinds of choices that can impact the environment, our health, the economy, and our quality of life. Because of this, we should consider the environmental impact of the products we purchase. Perhaps if all the environmental costs associated with a product were obvious, then consumers would make different choices than they currently make. However, many of the costs are not obvious, even to careful observers. This text has taught you a little about the wastes that are produced by various industries, so you are now in a slightly better position than most consumers to make intelligent choices about the products you buy.

U.S. citizens have the highest per capita consumption rates in the world. The United States also produces more waste per capita than all other nations. The rates at which we consume resources are not sustainable, and we rely heavily on imports to supply our appetites. In essence, we are borrowing resources from the rest of the world and from the future to maintain our lifestyles. Unfortunately, we are also exporting our consumer culture to the rest of the world, and consumption rates are rising globally.

Although we seem to live in a disposable society, it doesn't have to be this way. As a consumer, you should learn to avoid buying disposable goods. Cheaply made products may seem like a bargain, but they may be unreliable and need to be replaced after a relatively short lifetime. Buying well-made, durable goods encourages craftsmanship and reduces environmental impact. For example, buying a well-made piece of furniture may cost you more today, but it may last longer than you do. How else do you explain all of those antiques? Buying quality products will save you money in the long run, and because these products cost a little more, you may think more carefully about your purchases and consume less. Try to avoid impulse purchases and fads. You should also buy products that can be repaired rather than replaced when they malfunction.

As a consumer, try to purchase **green products** if they are available. Green products are produced to be more environmentally friendly than most consumer products. Some characteristics that distinguish green products include:

—Minimal packaging;

—Significant recycled material content in product or packaging;

—Non-toxic and non-polluting;

—Energy efficient;

—Not animal tested;

—Organically grown;

—Reusable;

—Recyclable;

—Minimally processed (e.g., unbleached, free of artificial flavors, colors, and fragrances);

—Produced from renewable sources;

—Other environmental benefits.

Each of us must take responsibility for the impact caused by the products we buy. These effects extend beyond the impact these products may have in our home and when they become wastes. As consumers, we drive the demand for various products. If we do not buy products with high environmental costs, then alternatives will be developed.

Pollution Prevention Options for Consumers

The pollution prevention options presented in this text to reduce industrial pollution can also be applied at the level of the consumer. The general approach is to determine what sorts of waste are being generated, then find ways to reduce the amount of waste generated, reuse products, recycle what you can't reuse, and substitute environmentally friendly products for hazardous products or products with high environmental costs. These options can be used at home, in the office, and by service industries.

Reduce

The best way to reduce the amount of waste we generate is to change our habits as consumers. Part of the reason we produce so much waste is that we buy more than we need. This over-consumption occurs on several levels. Many of the products we buy are over-processed, over-packaged, over-sized, or under-utilized. We buy things on impulse or in response to external stimuli, then later we find we don't need or want these items. The proliferation of inexpensive consumer goods has fueled over-consumption and has encouraged our throwaway lifestyles.

Changing the way we do business, for example, can have major impacts on the amount of waste paper generated in the United States. There are many easy, low-tech ways to reduce the amount of paper used in offices. Consider the following:

— Minimize paperwork by eliminating unnecessary reports, memos, and forms.

— Redesign necessary forms to use less paper or use electronic forms.

— Reduce report size.

— Print or copy on both sides of the paper, and make fewer copies.

— Use lightweight paper for drafts and internal documents.

— Update mailing lists to remove duplicate names and out-of-date entries.

— Don't use cover sheets on faxes that you send.

— Make fewer copies. Share copies and don't make more copies than you need.

— Use reusable envelopes for interoffice mail.

— Use bulletin boards or circulate copies of announcements, memos, documents, reports, and publications.

Many paper reduction options have been made possible by new technologies. Computers, faxes, the Internet, voice mail, and e-mail can reduce the amount of paper wasted in our day-to-day lives. For example :

— Proof and edit documents electronically and preview before printing.

— Use software that can help you identify typos and grammatical errors.

— Share documents electronically.

— Send and receive faxes directly through your computer.

— Use electronic mail and voice mail.

— Use electronic forms and reporting.

— Use the Internet to advertise and sell your product or services.

— Read news on the Internet instead of buying a daily newspaper.

— Do not download and print unnecessary information. If you are only interested in a portion of a document, print only the sections you need.

You can reduce the volume of wastewater you discharge to your home septic system or local sewage treatment plant by conserving water. If you have a septic system, you should be aware that 75 percent of drainfield failures are due to hydraulic overloading. By conserving water, you can help prevent your system from overloading and contaminating ground water and surface water. You can conserve water by using low flow plumbing fixtures, repairing leaks, only running washing machines and dish washers when fully loaded, taking short showers instead of baths, and not letting faucets runs unnecessarily. Do not over-water your lawn or garden; over-watering may increase leaching of fertilizers and pesticides to groundwater. Your wastewater will be easier to treat if you avoid using hazardous household products and minimize your use of phosphate detergents.

Sitting down to leisurely meals with family and friends is one of the great joys in life. However, most Americans find it difficult to enjoy three leisurely meals a day. Our busy schedules, long workdays, and convenience-oriented lifestyles seem to encourage the consumption of overly processed and packaged foods or takeout meals from fast food restaurants. Besides being unhealthy, these eating habits create a tremendous amount of packaging waste and promote the use of disposable cups, plates, and silverware. Consider the following options to reduce food-related wastes:

— Employees should keep a mug, plate, bowl, and silverware at their workplace and bring their meals in reusable containers.

— Work with food vendors to reduce food and packaging waste by providing condiments in bulk dispensers, giving customers a discount when they use their own cups, etc.

— When purchasing and using perishable food, buy and prepare only what you can use in a reasonable amount of time; this will reduce waste associated with spoilage.

—Buy nonperishable items in bulk containers to reduce packaging wastes.

Conserving energy is a good way to reduce the amount of air pollution you generate. Some energy conservation ideas include:

— Conserve electricity by using energy-efficient appliances and lighting.

— Turn off lights and electric appliances when not in use.

— Adjust your thermostat according to the season. Set it a few degrees higher in warm weather and a few degrees cooler in cold weather.

— Insulate your home and office, and use energy efficient windows and doors.

— If you use motorized equipment, choose fuel-efficient models with emission control equipment. Better yet, use non-motorized equipment; it will reduce noise and air pollution and might provide healthy exercise as well.

VOCs evaporate easily and contribute to air pollution. Therefore, avoid consumer products that contain VOCs such as solvent-based paints and cleaners. Be careful while refueling motorized equipment to avoid spills and do not "top off" your tank. If you do use solvent-based products, you can extend the life of the product and reduce air pollution if you seal containers tightly to prevent VOCs from evaporating,

The best way to reduce the pollution associated with driving a personal vehicle is to drive less. Consider using alternate modes of transportation such as public transportation, walking, or biking. Don't use your car when you don't need it. Maintaining your vehicle in good working order will extend its useful life and reduce pollution. Good driving hab-

its and car maintenance can reduce the amount of air pollution produced. Some suggestions that can reduce air pollution and save money by reducing the number of miles driven, improving gasoline mileage, and reducing the amount of time spent in traffic include:

—Plan your trips to accomplish as many tasks as possible.

—Choose your route to drive the fewest miles.

—Plan your route and your travel time to avoid traffic delays.

—Use ride sharing to reduce the number of vehicles on the road.

—Choose a vehicle with good gas mileage to maximize the number of miles per gallon of gasoline burned.

—Use clean fuels and an energy-conserving grade of oil.

—Drive at a steady, moderate speed; it is more fuel-efficient.

—Stop and start evenly; it reduces gas consumption.

—Don't idle an engine unnecessarily. Letting an engine idle for 30 seconds or longer uses more fuel than stopping and restarting the engine.

—Travel light; the less weight you carry, the less fuel is needed.

—Keep you tires inflated properly and check your car's alignment.

—Don't remove or tamper with pollution control devices.

—Keep air filters and catalytic converters clean to improve gas mileage.

—Don't use your air conditioner unnecessarily; it can reduce gas mileage by as much as 20 percent. Have any leaks in your air conditioning system repaired. The CFCs used in air conditioners destroy the earth's protective ozone layer when released to the environment.

Use low maintenance landscaping (i.e., plants that grow slowly, that are pest and drought resistant, and that don't need to be trimmed) around your yard and business. These low-maintenance plants generate less waste, use less water, usually use

If every American commuter car carried just one more passenger, we could save more than 600,000 gallons of gasoline and prevent 12 million pounds of "greenhouse gases" from making their way into the atmosphere everyday.

less fertilizer, and don't require pesticides. Consider replacing your lawn with a low maintenance groundcover. Low maintenance plants can usually be maintained without gas-powered yard maintenance equipment and the attendant noise and air pollution. Your landscaping choices can also affect your energy costs in other ways. You can landscape your yard to reduce heating and cooling costs in your home. Use trees and hedges as windbreaks to protect your home from cold winter winds that rob heat from your home. Plant **deciduous trees** on the southern side of your home. Deciduous trees lose their leaves in the winter and allow the winter sun to help warm your home, but their summer leaves will shade your home and reduce your air conditioning needs.

You can reduce the offsite impact of runoff from your yard if you make sure that all areas prone to erosion are properly landscaped and graded to minimize soil erosion. Grass swales (low areas in the lawn), gravel trenches along driveways or patios, and porous walkways increase infiltration and decrease runoff. You can also decrease runoff by installing wood decking, bricks, or interlocking stones instead of impervious cement walkways and driveways. Use pesticide and fertilizers sparingly to avoid contaminating runoff with these compounds. You should also compost or mulch your yard wastes to convert waste products into useful products that will reduce landscape maintenance costs.

Reuse

Many products are discarded before their useful life is over. Community groups, schools, and charities are often happy to get donations of furniture and old but serviceable appliances, computers, and electronics. There are also many organizations that take donations or clothing and household goods. By buying used goods, you keep useful products out of landfills, save resources and reduce pollution by reducing the demand for new products.

Reuse is not limited to durable goods. Even wastepaper can be reused. For example, one-sided copies or outdated letterhead can be reused as draft paper in fax machines and copiers, or this paper can be used as scratch paper or for memos. Some types of paper goods are relatively durable, e.g., file folders, and can be relabeled and used for new projects. Other types of paper can find a second life serving a new function. For example, shredded wastepaper can

be used as packing material. Lastly, don't forget that other people may be interested in the printed material that you have already read; consider donating old trade journals and magazines to local libraries, schools, hospitals, nursing homes, reading programs, English as a Second Language programs, etc.

Recycle

Even if you do reduce the amount of waste you generate and reuse products when possible, inevitably you will still generate waste. You can reduce the amount of solid waste being incinerated or landfilled and reduce energy costs if you recycle. Glass, paper, cardboard, steel, aluminum, and plastic can be recycled. Check to see which materials are recycled in your neighborhood and begin saving your recyclable wastes. Some communities even recycle organic wastes through large scale composting programs. These programs reduce the amount of biodegradable material being entombed in landfills and provide useful material for community gardens and parks. If recycling programs aren't available in your neighborhood, ask your community leaders to initiate the appropriate programs. You can also encourage recycling by buying products with post-consumer recycled content; this creates demand for recyclable materials and helps make recycling economically viable.

If your office is not recycling, find out how you can set up a program. Estimate how much waste your office produces; then design an appropriate program. If your office is small, consider combining your recyclables with other offices nearby. Office recycling programs generally work best if they are convenient and unobtrusive. Provide each office with a container for recyclable materials. Provide clearly labeled recycling bins near copiers, shipping and receiving areas, and in employee eating areas to collect white paper, mixed paper, newspaper, magazines, cardboard as well as nonpaper products (glass, aluminum, plastic, etc). Enlist the help of your janitorial staff to collect recyclables, and consider an incentive program that rewards good recycling habits. It is easier to recycle properly if your office doesn't use products that are difficult to recycle, e.g., thermal fax paper, glossy/plastic coatings, plastic windows, bright colored papers, adhesive labels.

Even some hazardous wastes can be regenerated or recycled, such as used motor oils, solvents,

Product	Alternatives
Drain cleaner	Pour boiling water down the drain. Use a plunger or a plumber's "snake".
Chlorine bleach	Baking soda and water, Borax, or natural sunlight (you may use bleach as a disinfectant).
Paints and Solvents	Use water-based (latex, acrylic) paint if possible
Paint remover/stripper	Heat guns may be used for removing many paints, but only in well-ventilated areas. Avoid using them for lead-based paints.
Pesticides	Learn which insects are beneficial in managing "pests". Keep your lawn and garden weed-free. Remove and destroy infected plants. Refer to an organic gardening book.
Houseplant insecticide	Spray soapy water on leaves, then rinse.
Herbicides	Hand pull weeds or mulch generously. Cover garden with plastic in the fall to prevent weed germination. Also, use biological controls such as lady bugs or praying mantises.
Oven cleaners	Use baking soda for scouring. For baked-on grease, heat oven to 200 degrees, turn off, and leave 1/4 cup ammonia in a dish in the oven for several hours to loosen. Then scrub with baking soda. Save the ammonia to be used again.
Furniture polish	Make a non-toxic polish by melting 1 tbsp. Carnauba Wax into 2 cups mineral oil. For lemon oil polish: dissolve 1 tsp. Lemon oil into 1 pint mineral oil.
Spot remover	Immediately soak in water, lemon juice, club soda, or corn meal and water.
Silver cleaner	Soak silver in 1 quart warm water with 1 tsp. Baking soda, 1 tsp. salt, and a small piece of aluminum foil.
Window cleaner	Use a pump spray container filled with 2 tbsp. Vinegar in 1 quart water, or rub newspaper on the glass.
Toilet bowl cleaner	Use toilet brush and baking soda, mild detergent, or 1/2 cup bleach.
Mothballs	Use cedar chests or place cedar chips around clothes.
Roach spray	Boric acid (sold as a powder), or chopped bay leaves and cucumber skins. A 50:50 mix of boric acid and brown sugar in a dish may also be effective. Be sure to keep these materials away from children and pets.

Table 28-4: Alternatives to toxic products.

and car batteries. Other hazardous products, such as excess paint or pesticide can often be donated to community groups.

Product Substitution

When possible, purchase a non-hazardous product to minimize your exposure to hazardous chemicals and to protect the environment. Table 28-4 lists safer alternatives to some common hazardous consumer products.

Checking Your Understanding

1. What are the four general approaches that can be used to reduce the wastes generated by consumers?

2. What is the most common type of solid waste material disposed of in the United States?

3. If we doubled the recycling rate for the material identified in the previous question, how many fewer tons of solid waste would be disposed of as MSW per year?

Summary

Consumers generate nonhazardous and hazardous wastes both directly and indirectly. In the home nonhazardous wastes include such items as packaging materials, yard wastes, newspapers, old clothing, and junk mail. Batteries, paints, cleaning supplies, pesticides, and drain cleaners are some of the hazardous wastes discarded. In addition to these nonhazardous and hazardous solid and liquid wastes, there are also solvents and aerosol cans as well as cars that generate air pollution. We also produce wastes during work and during recreation.

The service industries are also considered consumers; the wastes they generate are broken or obsolete equipment and furnishings, cleaning products, waste paper, construction materials, wastewater, air pollutants, and spent fuels and lubricants. Also, ap-proximately 85 percent of passenger transport is done in personal vehicles, which significantly adds to the amount of noise, land, and air pollution.

Municipal solid wastes are mixed wastes that are collected and sometimes recycled, composted, incinerated, or landfilled. The volume is staggering: it averages more than four pounds per person per day, of which less than 27 percent is currently being recycled. The United States is the world's largest user of paper, with paper products such as tissues, plates, cups, and napkins accounting for 40 percent of municipal solid waste. Glass also makes up a significant part of the solid waste stream and only about 25 percent of it is recycled. For the past 20 years the amount of plastic discarded has been increasing by about 10 percent per year. Only 20-25 percent of the plastics are recycled, which is a real concern considering that plastic is made from nonrenewable petroleum resources. Nearly 16 million tons of metals are discarded in the solid waste stream annually and only about 40 percent is currently being recycled. Food, too, – nearly 14 millions tons of it – ends up as scraps, spoiled, or discarded waste.

The two places we are most likely to be exposed to hazardous substances are our workplace and our home. Also, because many of these hazardous substances are used at home, they end up in the municipal solid waste stream as hazardous waste. Read labels carefully, watching for words like Caution, Flammable, Warning, Combustible, Danger, Explosive, Poison, Volatile, Toxic, or Corrosive. Hazardous consumer products should not be disposed of in the regular nonhazardous trash because landfills are not designed to protect the environment from them.

Most people spend 80-90 percent of their time indoors. There are many airborne pollutants escaping into the air from building materials, paints, carpeting, and cleaning products. In some areas, radon gas can accumulate indoors reaching unhealthful levels and tobacco smoke is full of harmful irritants and carcinogens. The home also produces two wastewater streams: sewage and graywater. In most homes the two are mixed; however, graywater from showers, dishwashers, and laundry facilities, for example, does not contain human wastes and can potentially be used for landscape irrigation purposes.

As consumers, the kind of choices we make can impact the environment, our health, the economy, and our quality of life. Before we purchase a product, we need to pause and consider its environmental impact. People in the United States have the highest per capita consumption rate in the world. Remember, if we continue consuming at the current rate, we are borrowing resources from the rest of the world and from future generations. Undoubtedly, it would be a good idea to abandon our throwaway society in favor of buying well-made and repairable products that offer the possibility of a long and useful life. It is time to find ways to reduce the amount of waste generated, reuse products, recycle what we cannot reuse, and substitute environmentally friendly products for hazardous products or products with high environmental costs. There are many things we all can do as consumers to reduce environmentally harmful practices.

Critical Thinking Questions

1. Examine your transportation habits and name five realistic ways in which you could decrease the environmental impact associated with your transportation needs. Try to implement these changes for a week, then describe the pros and cons of these changes from your perspective.

2. Conduct the necessary research to determine the cost and environmental impact of producing aluminum cans from the ore as opposed to recycled aluminum.

3. Conduct a survey of your local recycling program. List the items that are currently being recycled, study the statistics supplied by the waste management company, and determine economics and the long-term impact of the program.

Glossary

33/50 Program – An EPA sponsored program intended to reduce emissions of the 17 TRI targeted pollutants. Its targets were a 33 percent reduction in TRI chemicals by 1992 and a 50 percent reduction by 1995.

Abietic Acid – A carboxylic acid, $C_{19}H_{29}COOH$, with a phenanthrene ring system that is derived from rosin.

Absorption – The penetration of one substance (sorbate) into the inner structure of another (sorbent).

Acceptor Ion – Ions such as those of boron, gallium, or indium that are implanted to create a p-type, nonconducting region on a semiconductor chip.

Acid – Acids are variously defined as substances that can ionize in water producing hydronium ions (Arrhenius), are proton donors (Brønstead-Lowry), or substances with a deficiency of electrons (Lewis).

Acid Cleaning – An aqueous cleaning process that uses acids to remove scale and oxides from metal surfaces.

Acid Mine Drainage (AMD) – A liquid high in metal and low in pH, produced by minerals exposed to oxygen and water in rock disturbed by mine operations.

Acid Pickling Solution – An acidic solution (usually, hydrochloric acid) used to clean and etch the surface of items in preparation for painting or coating.

Acidulants – Mixtures of organic acids occurring naturally in fruits and vegetable that are used as additives in food processing. They serve as bacteriostats in processed foods, aid in the sterilization of canned foods by lowering the pH, act as chelating agent for metal ions that catalyze rancidity, and as flavor enhancers by offsetting excessive sweetness with their tart taste.

Acrylics Resins – Resins that are the polymerization product of acrylic acid or acrylates; that result in a colorless, transparent plastic.

Activated Sludge – Sludge particles produced in raw or settled wastewater (primary effluent) by the growth of organisms (zoogleal) in aeration tanks in the presence of dissolved oxygen. The term activated comes from the fact that the particles are teeming with bacteria, fungi, and protozoa.

Activated Sludge Process – A biologically based wastewater treatment process. Activated sludge is added to wastewater and the mixture (mixed liquor) is aerated and agitated. After some time in the aeration tank, the activated sludge is allowed to settle out by sedimentation and is disposed of (wasted) or returned to the aeration tank, as needed.

ACZA – See Ammoniacal Copper Zinc Arsenate.

Additive Processing – Also known as sequential build, in this process each layer of a PWB is built on the previous one, therefore, eliminating the later need for lamination. The process starts with a bare catalyzed laminate with no metal surfaces. The laminate is then coated with a photodielectric material that is imaged and developed. The board is then selectively plated or metallized to form the desired interconnect pattern. These steps are repeated until the overall structure is completed.

Adhesive Bonding – A process that uses a non-metallic bond to join two surfaces.

Adit – An underground tunnel that intersects the earth's surface at one end and terminates in rock at the other.

Administrative Requirements – During a cleanup, the administrative requirements specify how the action is to be documented and approved.

Adsorption – The adherence of atoms, ions, or molecules of a gas, liquid, or solid to the surface of another substance by means of chemical and/or physical attraction. Adsorbents are substances that present large active surface areas. Activated charcoal, alumina, and silica gel are the most commonly used adsorbents.

AEA – See Atomic Energy Act.

Aerated Impoundment – Also known as aerated lagoons, they tend to be rectangular containments with a length-to-width ration of 2 to 1, with a fitted liner. The containment is 8 to 12 feet deep and by means of aerators sufficient oxygen is furnished to maintain dissolved oxygen at all depths.

Aerobes – Bacteria that require an environment containing dissolved oxygen, O_2, for respiration.

Aerobic Impoundment – Any impoundment that is kept aerobic by providing sufficient oxygen at all depths through an aeration method. See Aerated Impoundment.

Aerobic Respiration – Energy-yielding metabolism in which the terminal electron acceptor for the substrate oxidation is molecular oxygen.

Agglomerate – Combination or aggregation of colloidal particles suspended in a liquid into clusters or flocs of approximately spherical shape. It is usually achieved by neutralization of the electric charges that maintain the stability of the colloidal suspension.

Air Pollution Control System (APCS) – A variety of end-of-pipe devices (dry mechanical collectors, fabric collectors, wet & dry scrubbers and electrostatic precipitators) designed to remove both particulate and gaseous pollutants from a gaseous waste stream.

Air Sparging – A process that takes advantage of the different vapor pressures of liquids that have been commingled and is similar to air stripping, but is performed *in situ*. Through a series of openings in the soil, compressed air is pumped into the soil or groundwater causing the more volatile substances to join the air stream. By means of a vapor extraction system, the air-vapor mixture is captured and the volatile component removed.

Air Stripping – The separation of two or more liquids on the basis of their vapor pressures. Huge volumes of air are used to generate a foam that greatly increases the contact surface between liquid and gas. The liquid with the higher vapor pressure enters the gaseous phase as a foam.

Algae – One of the variety of Protista, most of which are unicellular or photosynthetic autotrophs. Microscopic plants, many of which live in water but may also be attached to rocks or other similar substances.

Alkaline – A substance having a pH greater than 7. See Base.

Alkaline Cleaning – An aqueous cleaning process that uses alkalis and other additives to clean metal surfaces.

Alkalinity – The condition of being alkaline.

Alkylation – See Catalytic Reforming.

Alloy – A metal comprised of two or more metallic elements.

Alpha Particles – A helium nucleus with a positive two charge. It is a relatively large, slow moving particle that can travel only a few centimeters in air or less than 0.005 mm in aluminum.

Alternative Synthetic – Pathways for Pollution Prevention An OPPT sponsored research grant program to insure the inclusion of P2, through environmentally-conscious design of chemical products and processes, as a part of all chemical synthesis. They are the central focus of the Green Chemistry Program.

AMD – See Acid Mine Drainage.

Ammoniacal Copper Zinc Arsenate (ACZA) – A waterborn wood preservative typically used in the western United States where many of the wood species are difficult to treat with CCA. As the ammonia evaporates, the metals form a non-leachable precipitate.

Anabolism – The phase of metabolism in which simple substances are synthesized into the complex cellular materials.

Anaerobic Bacteria – Bacteria that live and reproduce in an environment containing no free or dissolved oxygen. Anaerobic bacteria obtain their oxygen by breaking down chemical compounds that contain oxygen, such as sulfate ions, SO_4^{2-}.

Anaerobic Digester – A large air-tight tank in which anaerobic reactions take place. Used for the final treatment of sludge in a wastewater treatment operation, it relies on acetogens and methanogens to reduce the volume by 40-60 percent.

Anaerobic Filter – A fixed-film anaerobic treatment process, in which the anaerobic organisms are established on the filter's packing and the wastewater allowed to flow gently either upward or downward through it.

Anaerobic Impoundment – They are constructed in a manner similar to an aerated impoundment, except for a lower surface to volume ratio. Because they are not mixed or aerated and are typically 8-16 feet deep, principally anaerobic bacteria degrade the waste.

Anaerobic Respiration – Energy-yielding metabolism in which the terminal electron acceptor from substrate oxidation is an inorganic compound other than molecular oxygen, such as sulfate or nitrate ions. Also known as fermentation.

Anaerobic Suspended Growth – Similar to an activated sludge or lagoon process, except that a sealed container is used to exclude oxygen.

Anilox Roller – A screened roller in which cells of uniform size and depth are engraved; the size and depth of these cells controls ink film thickness during flexographic printing.

Animal Husbandry – Activities associated with the breeding, care, and feeding of animals.

Anions – A negatively charged ion. Any ion that migrates to an anode during electrolysis.

Annealing – A heat treatment process used to reduce hardness of metals and improve machinability.

Anodes – The positive electrodes of an electrolytic cell, to which negatively charged ions travel when an electric current is passed through the cell. NOTE: In a battery or fuel cell, the anode is the negatively charged electrode.

Anoxic – Insufficient oxygen; low oxygen levels.

Anoxic Denitrification – The anaerobic application of denitrifying bacteria to remove nitrogen, in the form of nitrate, by converting it to nitrogen gas. The denitrifying bacteria obtain energy from the conversion of nitrate to nitrogen gas, but since most nitrified effluents are low in carbon-containing matter, an external source of carbon may be required for cell synthesis.

Antimony – A metal with a silvery white metallic luster and having the chemical symbol Sb. It is used to alloy other metals.

AOC – See Area of Contamination.

APCS – See Air Pollution Control System.

°API – See API Gravity.

API Gravity (°API) – A scale of density measurement adopted by the American Petroleum Institute (API). It runs from 0.0 (equivalent to 1.076 g/cm^3) to 100.0 (equivalent to 0.6112 g/cm^3). The API values as used in the petroleum industry decrease as density increases. The formula for its calculation is 141.5 divided by the specific gravity, less 131.5 equals the °API value.

API Separator – A large settling tank in which a water-oil mixture is placed. The floating oil is removed and salvaged by skimming, the heavy sludges sink to the bottom for removal, and the remaining water is sent to a POTW for further treatment before release into the environment.

Applicable or Relevant and Appropriate Requirements (ARARs) – The ARARs is a threshold criteria for a CERCLA cleanup. The substantive cleanup standards in the ARARs are borrowed from other areas of the law. Waste type and contaminants determine the ARARs that pertain to storage, treatment, transportation, and disposal of the waste. The CERCLA applicable regulations apply directly to the cleanup situation, where the relevant and appropriate requirements are less strongly linked to the situation under consideration. To be considered (TBC) requirements, although neither relevant nor appropriate, are also binding on the cleanup operation.

Aqueous – Relating to, containing, or dissolved in water.

Aqueous Metalworking Fluids – Cutting fluids with greater than 90 percent water content, usually a mixture of water soluble oils and water.

Aqueous Vapor Degreasing (AVD) – A process that uses perfluorocarbon compounds in a modified vapor degreaser, fitted with 150 percent freeboard and a –20°C chiller coil to condense vapors.

ARARs – See Applicable or Relevant and Appropriate Requirements.

Architectural Paint – Paints, stains, and undercoats such as primers and sealers that are used on the outside of structures.

Area of Contamination (AOC) – A continuous and widespread area that may contain varying types and concentrations of contamination at a CERCLA site. Waste managed within the AOC can be handled much more simply than waste that is removed from the area.

Arsenic – A metalloid with a steel gray metallic luster and having the chemical symbol As. It is used as a poison in pesticides.

Ashing – The process of removing residual photoresist material using a high temperature plasma to prevent damage to the underlying layers.

Aspirated – The condition encountered in wood where the pits that allow the flow of liquids between cells are blocked or occluded by the adherence of the pit membranes to the pit walls.

Atmospheric Evaporators – Evaporators that spray wastewater over packed media, grids, or plates, while blowing ambient air across them.

Atom – The Greek Philosopher Democritus ~350 BCE proposed that all matter was composed of uncuttable parts that translated into the word atom. See Atomic Theory.

Atomic Energy Act – The 1954 revision of the AEA of 1946 established a separate regulatory pathway for civilian and governmental radioactive wastes activities; the Nuclear Regulatory Commission regulates civilian and the DOE regulates most government waste activities.

Atomic Theory – Modern atomic theory proposes that each element has its own kind of atom and that it has a small dense nucleus. Within the nucleus is where the positively charged protons and neutrally charged neutrons reside. Around the nucleus are energy levels that contain the negatively charged electrons. Since atoms do not have a net charge, the number of protons and electrons within an atom must always be equal.

Attachment Pad – The substance that a semiconductor chip is attached to before being assembled into an outer protective case.

Autoclave – A sealable, steam jacketed vessel, able to withstand elevated temperature and pressure; used for sterilization.

Autogenous Mill – Self-grinding; a mill in which the material being ground is also the grinding media.

Automatic Blanket Wash – A device on a printing press that meters out a specific amount of blanket wash to be used during the cleaning process

Autotrophs – Literally, self-feeders. They are organisms that make their own organic compounds from an inorganic carbon source such as CO_2. They get their catabolic energy by either photosynthesis or from redox reactions involving inorganic chemicals.

AVD – See Aqueous Vapor Degreasing.

Back – Refers to the ore (rock) above an underground opening such as a drift or stope.

Back Washing – Removal of the trapped solute particles on a filtering medium by reversing the flow of the filtrate.

BACT – See Best Available Control Technology.

Bacteria – Any of the unicellular, prokaryotic microorganisms of the kingdom Monera (sometimes separately listed as Eubacteria and Acheabacteria). They vary considerably in their oxygen and nutritional requirements. Most are chemohetrotrophs using organic matter for both their food and energy.

Baghouse – A series of fabric dust-collection bags, assembled on a metal frame, that particle-laden air is drawn through. Powered by large fans that draw the dirty air through fabric that may have pore sizes as small as 0.5 micron; a layer of suspended solids accumulate on the outside of the bags. Periodically, puffs of air or mechanical rapping is used to shake the collected particles from the sides of the bags.

Ball Mill – A mill in which the grinding media are iron or steel balls.

Bar Screen – The largest filtering medium in common usage. With spacings up to 1 inch (2.54 cm), bar screens or bar grates are used to prevent large objects from entering a water treatment system and damaging the equipment.

Barrel – A liquid volume unit used for the measurement of oil. It is equivalent to 5.61 ft^3, 42.00 gallons, or 158.98 liters.

Barrel Plating – Used for multiple small workpieces where minor nicks or abrasions are not a problem. The items are poured into a barrel that is then placed in the process bath and electrified. Using a slow rotation, each part comes into contact with the electrified barrel, causing a thin coating of the metal to form on its surface.

Barren Solutions – Process solutions that contain no dissolved metal.

Base – Bases are variously defined substances that can either ionize in water producing hydroxide ions (Arrhenius), are proton acceptors (Brønstead-Lowry), or substances that have an unshared pair of electorns (Lewis).

Bauxite – A mineral composed of hydrous aluminum oxide. The most common ore of aluminum.

BDAT – See Best Demonstrated Available Technology.

Becquerel (Bq) – A measure of radioactivity that is defined as one disintegration per second.

Belt Filter – Two filter belts, constructed of synthetic material with variable pore size, that are put on either side of a mixture of solids and liquids (sludges). By passing the belt-sludge-belt sandwich through an s-shaped series of rollers, called a shear zone, a filter cake can be produced with up to 75 percent of the extractable liquid removed.

Bending – A metal forming process that applies pressure to form an angle or curve in a part.

Benthic Zone – The transition zone between the bottom of a body of water and soil/rock, it usually contains a strong reducing and anaerobic environment.

Best Available Control Technology (BACT) – An air pollution emission limitation based on the maximum degree of emission reduction that, considering energy, environmental, and economic impacts and other costs, is achievable.

Best Demonstrated Available Technology (BDAT) – The most stringent demonstrated level of control technology for that source. Treatment technologies that have been shown to remove ignitable (D001), corrosive (D002), and reactive (D003) characteristics from wastes, therefore rendering them deactivated.

Beta Particles – A fast-moving negatively charged electron emitted from the nucleus of a radioisotope. Beta rays can cause skin burns and are harmful once within the body, but the skin can be protected by a thin sheet of metal.

Bethell Process – See Full Cell Process.

Billets – Small, nearly square pieces of iron or steel.

Bioaccumulation – The accumulation of a toxin within an organisms.

Biocells – See Biological Treatment Cells.

Biochemical Oxygen Demand – The dissolved oxygen required to decompose organic matter in water; a numerical estimate of contamination in water expressed in milligrams per liter of dissolved oxygen. BOD testing measures the rate of oxygen consumption under controlled conditions of time and temperature. Standard test conditions include dark incubation at 20°C for a specified time, usually five days.

Biocidal Properties – See Biocides

Biocides – Nonspecific pesticides with life killing properties that are often used in substances as disinfectants.

Bioextraction – A type of hydrometallurgical process wherein bacteria are used to free metals from their ores. The bacteria use part of the mineral as an energy source, thereby freeing the metal and leaving an ore pile or vat of pregnant solution from which the metal can be recovered.

Biological Reactor – See Fixed-Film Bioreactor.

Biological Treatment Cells – Also known as biotreatment cells. Typically hydrocarbon-laced soil is placed over slotted PVC pipes to a depth of 6-8 feet. By means of a vacuum pump, air is drawn down through the cell providing oxygen to the aerobes. Typically performed onsite.

Biological Treatment Technology – The use of naturally occurring microorganisms (plants and animals) to decompose waste organic chemicals into harmless by-products.

Bioreactors – See Fixed-Film Bioreactor.

Bioremediation – A managed or spontaneous process in which microorganisms are used to transform harmful substances into nontoxic or less harmful compounds; thereby remedying or eliminating environmental contamination.

Black Liquor – The liquid resulting from the digestion of wood chips in the presence of white liquor, heat and pressure, and the brown pulp washing process. Weak black liquor contains unused sodium sulfide and sodium hydroxide; acetic, formic, and sugar acids; xylans; and methyl alcohol. It is about 15 percent solids with the remaining 85 percent being water. When this liquor is evaporated its solids are concentrated from 15 to about 65 percent and it is called strong black liquor. The burning of strong black liquor consumes the organic components, but its inorganic components solidify into a substance called smelt.

Bleaching – A process used to improve the whiteness of a product.

Bleeding – The exudation of liquid treating solution from wood after treatment.

Block Caving – A method of mining involving under-cutting or blasting and withdrawing the lower portion of an ore body to create an underground opening. The ore above is allowed to slough or cave into the opening.

Blowout – The uncontrolled escape of oil, gas, or water from a well.

BOD – See Biochemical Oxygen Demand.

BOE – See Buffered Oxide Etch.

Boiling Point – The temperature of a liquid at which its vapor pressure is equal to or slightly greater than atmospheric pressure.

Boiling Point Fractions – The portions of a mixture that are characterized by having closely related boiling points.

Bonding – A process of joining two surfaces.

Bonding Wires – The wires used to connect the semiconductor chip to the lead frame.

Bone Meal – Ground hooves, horns, and bone from the red meat processing industries. Bone meal is used as a cattle feed supplement or as a fertilizer.

Boulton Seasoning Process – The oldest wood seasoning process. The wood and an oil-born treatment solution are placed in a retort, heated, and then a vacuum drawn to vaporize the water from the wood. Although it is a very effective process, it is energy intensive and produces a large quantity of oil contaminated wastewater.

Bound Water – Water molecules that are tightly held by various chemical groups, such as carboxyl, hydroxyl and amino groups. Hydrogen bonding is often the responsible forces attracting the opposing polar portions of the molecules.

Bq – See Becquerel.

Brazing – A bonding process that uses a non-ferrous metal filler with a melting point of greater than 800°F, but less than the melting point of the metals being joined.

British Thermal Unit (BTU) – A measurement of thermal energy. By definition it is the amount of heat required to change the temperature of one pound of water by 1°F.

Bronze Age – Approximately 3,500 BCE to 1,500 BCE A historical time frame between the Copper and Iron Ages when man began to alloy metal and produce bronze.

Brownfields – Abandoned, idled, or under-used industrial and commercial facilities where expansion or redevelopment is complicated by real or perceived environmental contamination. EPA launched the Brownfields Initiative in 1995.

BTU – See British Thermal Unit.

BTU Value – BTU/cubic foot is the accepted unit for comparison of the heating values of various fuels.

Bubbling Fluidized-bed Incinerator – Refractory-lined vessels containing a 3-6 foot deep bed of graded, inert granular material – usually silica sand. Large volumes of heated air are pushed through the bed, and the waste injected into or just above the swirling hot sand. Depending on the waste being incinerated, typical APCS equipment follows.

Buffered Oxide Etch (BOE) – A mixture composed of hydrochloric acid buffered with ammonium fluoride that is used to remove silicon dioxide without etching away any underlying silicon or polysilicon.

Buffering Agent – A substance capable of neutralizing small amounts of either acid or base added to a solution, thereby keeping the pH stable.

Byssinosis – Also known as brown lung disease, it is a debilitating lung disease resembling emphysema, resulting from inhalation of cotton dust.

Cable Tool Technique – A drilling process in which a heavy bit is repeatedly raised and dropped into the drill hole, with each impact punching a little deeper into the earth, until the desired depth is reached.

Calcination – A process in which inorganic materials are heated to high temperature but do not fuse, resulting in changes to the physical and/or chemical properties of the material being heated. A process often used in the benefication of ores.

Captives – Electroplating facilities that perform in-house metal plating for parts the company also manufactures.

Carbonate Precipitation – A chemical treatment method that makes use of the fact that a number of metal carbonates are insoluble, e.g. lead and cadmium.

Carbonitriding – A heat treatment process that adds carbon and nitrogen to metal surfaces to increase their hardenability.

Carburizing – A heat treatment process that adds carbon to metal surfaces to increase their hardenability.

Carcinogenic – A substance that is cancer producing.

Case Hardening – A two stage heat treatment process used to harden the surface of metal parts.

Casing – Large (7-51 cm) sections of steel pipe that are cemented into place behind a drill bit. Their function is to prevent leaks and borehole collapse during the drilling and production processes.

Catabolism – The metabolic process used by living organisms to degrade their food and extract energy.

Catalytic Cracking – The use of a catalyst in the petroleum cracking process to improve both the amount and quality of gasoline produced. In the fluid bed catalytic cracking process, the finely divided catalyst is whirled into contact with the entering liquid and vapor hydrocarbons. Fresh catalyst is continuously added and spent catalyst continuously removed from the process. Steam stripping is used to separate the cracked hydrocarbon vapors from the catalyst and also helps to maintain the fluidized state of the bed.

Catalytic Reforming – Also known as alkylation, a process that uses heat to crack low-octane petroleum fractions and reform them into branched-chain, cyclohexanes or aromatic hydrocarbons. The principal reaction involves the catalytic combination of an isoparaffin and an olefin. The process results in high-octane components that greatly improve the octane rating of gasoline.

Cathode – The negative electrode of an electrolytic cell, to which positively charged ions migrate when a current is passed through the cell. (NOTE: In a battery or fuel cell the cathode is the positive electrode.)

Cations – Positively charged ions. Any ion that moves toward the negative electrode during electrolysis. Cations are usually metal ions, such as Na^+, Ca^{2+}, Fe^{3+}, etc.

Caustic – See Caustic Soda.

Caustic Soda – Commonly known as lye or sodium hydroxide (NaOH). The most economical and often used strong base.

Causticizing – The process of reacting green liquor with lime, CaO, which after clarification results in

the regeneration of the white liquor and a precipitate called lime mud, $CaCO_3$. The causticizing reaction actually occurs in two steps, the first being the slaking of lime into $Ca(OH)_2$ and the second being the conversion of Na_2CO_3 into NaOH.

CCA – See Chromated Copper Arsenate.

Cellulose – The principal constituent of wood (about 45 percent by mass) and paper, is a white linear, partially crystalline polysaccharide containing about 10,000 anhydroglucose units.

Cementation – See Immobilization.

Centrate – The liquid passing through a filtering medium and removed by a centrifugal filtering operation.

Centrifugal Filtration – A filtering technique based on the application of centrifugal force by placing the mixture to be separated into a rapidly (1,500-2,000 rpm) rotating drum that has a porous outer filtering medium. Centrifugal force causes the filtrate to pass through the filtering medium and the solute particles to collect forming a filter cake.

Centrifugal Force – The force that propels an object outward from a center of rotation.

CERCLA – See Comprehensive Environmental Response, Compensation, and Liability Act.

CERCLA Off Site Rule – A rule cited in 58 FR 49204, Sept. 22, 1993, that specifies that any waste shipped off-site must be shipped to a CERCLA receiving TSD facility.

CESQG – See Conditionally Exempt Small Quantity Generator.

Char – The material remaining after the volatile components have been driven off by a pyrolytic heating process.

Characteristic Hazardous Waste – Any substance that is not on the hazardous waste lists, but exhibits one or more of the hazardous characteristics of ignitability, corrosivity, toxicity, and reactivity as defined in Title 40 of the Code of Federal Regulations.

Chelate – A type of coordination compound in which a central metal ion is attached by coordinate links to two or more nonmetal atoms in the same molecule. A common substance that will chelate calcium or magnesium ions is ethylenediaminetetraacetic acid (EDTA), for example.

Chelating Agents – Substances capable of forming coordination compounds with a central metal ion such as Ni^{2+}, Cu^{2+} or Zn^{2+}.

Chemical Disinfection – Exposure of wastes to certain chemicals that possess the ability to kill microorganisms; commonly used to treat liquid medical waste and large pieces of equipment.

Chemical Feedstocks – The raw materials used for the manufacture of petroleum-based consumer products, including plastics, synthetic fibers, and many household and industrial chemicals.

Chemical Oxygen Demand (COD) – Refers to the amount of oxygen consumed under specified conditions in the oxidation of the organic and oxidizable inorganic matter contained in a wastewater.

Chemical Precipitation – Any treatment process that causes a soluble substance to become less soluble or insoluble in that solvent.

Chemical Properties – The unique set of characteristics that describe how a substance reacts chemically with another substance, e.g. flammability.

Chemical Pulp – Fibers from wood or other plants that have been isolated by dissolving the lignin that holds the fibers together. Chemical wood pulps are dark brown if unbleached or nearly white if bleached to remove remaining lignin.

Chemical Treatment Technologies – Any treatment process that is used to change both the chemical and physical characteristics of a waste. Chemical treatment methods change the characteristics of a pollutant by converting it into a new substance.

Chemical Vapor Deposition (CVD) – The formation of a thin layer of a substance on a substrate by a chemical process. Vapor deposition is a low pressure process that combines appropriate gases in a reactant chamber at elevated temperatures to produce a uniform film thickness, e.g., formation of silicon dioxide layer using silane and oxygen.

Chemicals – All matter is composed of chemicals or pure substances. Pure substances are defined as always having a constant composition, and are further subdivided into either elements or compounds.

Chemophobic – A fear of chemicals.

Chemotrophs – Organisms that obtain catabolic energy from redox reactions on an external chemical source.

Chip – A tiny rectangular slice of semiconducting material cut from a larger wafer on which a transistor or an entire integrated circuit is formed; also called a die.

Chromated Copper Arsenate (CCA) – A waterborn preservative composed of a mixture of copper oxide, arsenic acid, and hexavalent chrome in the form of chromic acid. Once in contact with the wood the hexavalent chrome is reduced to trivalent chrome and due to a pH increase, precipitates the metals in a variety of non-leachable compounds.

Ci – See Curie.

Circulating Fluidized-bed Incinerator – Refractory-lined vessels containing graded, inert granular material – usually silica sand. Its depth, however, is not fixed because of the high velocity heated air introduced at the bottom of the chamber. This results in a suspension of the wastes in the sand along the entire height of the combustion chamber. Off-gases and sand next pass through a cyclone filter that captures and recycles the sand to the combustion chamber. Typical APCS equipment follows.

Clarifier – A tank of varying design, size, and shape in which gravity sedimentation is used to separate and remove solids from a liquid waste stream. See Lamella and Inclined Plate Clarifiers

Classification – The processes involved in the separation of waste from valuable minerals.

Classifier – A piece of equipment that uses the physical properties of minerals to separate them; e.g., cyclones and thickeners.

Claus Process – A process used for converting hydrogen sulfide gas (H_2S) removed from sour crude, into sulfur and water. The hydrogen sulfide gas is burned in the presence of a bauxite catalyst, but under starved oxygen conditions. Although some sulfur dioxide is formed and later treated to remove its sulfur, the equation for the idealized reaction would be: $2 H_2S + O_2 \longrightarrow 2 H_2O + 2 S$

Clean Levels – The predefined cleanup levels, which are established using a variety of factors including: 1) chemical specific ARARs; 2) acceptable risk-based levels; 3) meeting delisting criteria; or 4) meeting predetermined disposal levels.

Cleaner Technologies Substitutes Assessments – A DfE tool developed to detail human and environmental health, exposure and risk, economic and performance information on traditional and alternative manufacturing methods and technologies for a given industry.

Closed Loop – A process system in which effluents are recycled, that is, treated and returned for reuse.

Coagulant – A substance that induces coagulation. The coagulant neutralizes the charges (typically negative) on colloidal sized solute particles. Prior to the addition of a coagulant the particles refuse to settle due to their similar electrical charges.

Coagulate – The transformation of separate dissolved solute particles into larger clumps.

Coating – The application of material to a substrate.

COD – See Chemical Oxygen Demand.

Coining – A metal forming process in which a part is stamped to produce three-dimensional relief on one or both sides.

Coking – Tall chambers or drums into which the bottoms, slop oil, and oily wastes are placed for the coking process. During the heating process, large molecules are broken and some undergo isomerization into smaller fuel-sized products. The remaining materials undergo pyrolysis and form high quality petroleum coke. Once the pyrolysis is completed, water is used to cool the drum and its ends removed. A hole is drilled lengthwise through the coke and it is removed by high pressure water jets cutting it into pieces.

Cold Cleaning – A solvent cleaning method in which solvent is applied at or below ambient temperature.

Cold Pressure – A metal shaping operation that uses physical pressure to alter the form of cold metals.

Combustion – The oxidation of a substance occurring with the speed necessary to produce noticeable heat and light. Combustion of organic compounds usually results in the production of water, carbon dioxide, and/or carbon monoxide. Combustion of metallic and nonmetallic elements result in their oxides.

Co-metabolism – The fortuitous modification of one substance by an enzyme that routinely acts on another. The primary substrate supports the growth of the microorganism that produces one or more enzymes of low specificity that also act on the co-metabolized substrate. The co-metabolized substrate is usually altered only slightly and does not enter the catabolic or anabolic pathways of the host organism.

Comminution – The crushing and grinding processes involved in liberating valuable minerals from ore.

Compact – An agreement between two or more states to share a low-level radioactive waste disposal site. This is allowable under the provisions of the Low-level Radioactive Waste Policy Act of 1980 and its 1985 Amendments, which made states responsible for disposal of low-level wastes generated within the state.

Competent Rock – Rock with structural integrity that can handle high stress and is not easily deformed.

Compliance Risk – Risks that relate to the impact of a change on the manufacturer's ability to comply with laws and regulations.

Compound – A pure substance containing two or more elements bonded by either ionic or covalent bonds. Compounds can be broken down by chemical means into their elemental components. Compounds have different physical and chemical properties than the elements from which they are composed.

Compounded Metalworking Fluids – Fluid mixtures used in metal shaping operations that generally contain natural oils, mineral oils, or synthetic lubricants mixed with additives that enhance some property missing in the parent fluid

Comprehensive Environmental Response, Compensation, and Liability Act (CERCLA) – Commonly known as the superfund law, it requires the cleanup of releases of hazardous substances in air, water and groundwater, and on land. Both new spills and leaking or abandoned dumpsites are covered. It also established a trust fund to pay for cleaning up hazardous substances in the environment and gives EPA authority to collect the cost of cleanup from the parties responsible for the contamination.

Concentration – The relative content of a component. In chemistry, concentration is frequently described as the number of moles per liter, also known as molarity, M.

Concurrent Design – A practice that integrates product and process considerations by having all those responsible for the various stages or components of a product's life cycle participate in its design from the beginning.

Conditionally Exempt Small Quantity Generator – CESQGs generate less than 100 kg of hazardous waste and less than 1 kg of acutely hazardous waste per month and accumulate no more than 1,000 kg of hazardous and 1 kg of acutely hazardous wastes at any one time.

Conducting Salts – A term used in the electroplating industry to refer to those salts that are highly soluble and transfer electrical charges easily throughout process solution.

Continuous-flow Stirred Tank Reactor – A well mixed reactor to which reactants are added continuously and from which products are withdrawn continuously.

Conveyance Flume – A cement or steel lined trench filled with water that is used to transport fruit from the truck to the processing building.

Copper Age – Approximately 4,000 BCE to 3,500 BCE A historical time frame between the Stone Age and the Bronze Age when man began to use metals (copper) to make tools.

Coring – Cutting and bringing a section of rock to the surface for further examination.

Corona – A visible electrical discharge (bluish glow) caused by the ionization of a gas in a strong electrical field.

Corrective Action Management Units – EPA defined areas that were intended to reduce the burden of managing waste at related, but non-contiguous areas of contamination. At this time, however, CAMUs have proven to be of little value and management of waste outside their AOC remains burdensome.

Corrosion – The electrochemical deterioration of metals due to reaction with their environment, which is accelerated by the presence of acids or bases.

Corrosivity – One of the characteristics of a RCRA defined hazardous waste. A waste is considered to be corrosive if it has a pH ≤ 2 or ≥ 12.5, or if it is a liquid that corrodes steel at a rate greater than a designated corrosion rate.

Cottrell Precipitator – See Electrostatic Precipitator.

Countercurrent Rinses – Rinse systems in which the movement of the workpiece and the flow of rinse water are in opposite directions. As the workpiece leaves the process bath, it enters the first rinse bath. The workpiece is then moved into the second rinse bath, where the concentration of process chemicals is less. In the final rinse, the concentration of process chemicals is very low and fresh replacement water is continually being added.

Country Rock – Non-ore bearing rock adjacent to ore.

Covalent Bond – A chemical bond formed between atoms by the sharing of one or more pairs of electrons.

Covalent Compound – A stable molecule formed through the use of covalent bond(s) wherein two or more elements share one, two, or three pairs of electrons.

Cresote – A black tarry substance with a distinctive odor that is a byproduct of the petroleum coking process. It is a mixture of organic compounds, some of them being polynuclear aromatic hydrocarbons (PAHs).

Crib – A framework of timber used as support in underground mines.

Critical Temperature and Pressure – A set of conditions, defined by temperature and pressure, above which a gas can not be turned into a liquid. For water that point is 374°C and 3,200 psi.

Cross-media Transfer – The movement of a contaminate from one media (air, water, or soil) to another. For example, through cross-media contamination the air pollutant sulfur dioxide can become a water contaminate in the form of acid rain.

Crude Oil – A highly complex mixture, including straight- and branched-chained paraffins, cycloparaffins, and aromatic hydrocarbons, containing a low percentage of sulfur and trace amounts of nitrogen and oxygen compounds. It is the mixture from which various oil products such as petroleum gas, gasoline, kerosene, jet fuel, diesel, gas oil, fuel oil, lubricating oil, wax and asphalt are separated in the refining process.

Crusher – A machine used to reduce the ore particle size.

CSTR – See Continuous-flow Stirred Tank Reactor.

CTSA – See Cleaner Technologies Substitutes Assessments.

Cure – The transformation of a liquid on a substrate to a solid coating by a chemical change.

Curie (Ci) – A measure of radioactivity that is equal to 3.7×10^{10} disintegrations/second.

CVD – See Chemical Vapor Deposition.

Cyanide – Solutions containing cyanide ions, CN^-, are frequently referred to simply as cyanide. All cyanide compounds are highly toxic and readily absorbed through the skin. If they come in contact with acid, they will release HCN, the deadly gas that is used in a gas chamber.

Cyclone Filter – A conical filtering device that uses centrifugal force to remove more dense particles from gaseous waste streams.

Cyclone Separator – A type of classifier often used to de-water minerals before entering the extraction circuits and to separate small light particles from large heavy ones.

DC – See Direct Current.

Decantation – The process of removing the clarified or supernatant liquid so as to not disturb the sediment.

Dechlorination – Any process in which the element chlorine is chemically removed from a chlorinated organic compound.

Deciduous Tree – A tree that grows and sheds its leaves according to the season.

Decomposer – Organisms that obtain their energy from breaking down dead organic matter.

Dehalogenate – Any process in which a member of the halogen family is chemically removed from a haloorganic compound. See Dechlorination.

Deink – The process of removing ink such as during paper recycling.

Delisting – A RCRA listed waste that meets risk-based contamination levels and no longer presents a threat to the public and the environment may be delisted.

Denitrification – An anaerobic process performed by denitrifying bacteria where nitrite, NO_2^-, or nitrate, NO_3^-, ions are reduced to nitrogen gas, N_2, that escapes into the atmosphere.

Denitrifying Bacteria – Bacteria that convert nitrite, NO_2^- or nitrate NO_3^- ions into nitrogen gas, N_2.

Density – The physical property of a substance that is a measure of its mass per unit volume.

Depaint – A general term used to refer to the removal of paint from a surface by any method.

Design for the Environment (DfE) – A voluntary EPA program who's goal is to develop and distribute P2 and risk assessment information on alternative products, processes and technologies for small- and medium-sized industries. CTSAs are developed for the industry, and outreach occurs through use of fact sheets, bulletins, case studies, software, videos, and training materials.

Desmear – A process designed to remove epoxy-resin smear from inner layer copper interfaces.

Destruction and Removal Efficiency (DRE) – An EPA standard used to rate incinerator destruction and removal efficiency. Usually expressed in terms such as four nines (99.99%) or six nines (99.9999%).

Destructive Distillation – An operation in which a highly carbonaceous material is subjected to high temperature in the absence of oxygen, resulting in its separation into volatile and nonvolatile components and char.

DfE – See Design for the Environment.

DI Water – Water that has been deionized by passing it through an ion exchange unit where positive ions are replaced by an equivalent number of hydrogen ions and negative ions are replaced by an equivalent number of hydroxide ions.

Die – See Chip.

Dioxin – The commonly accepted name for a mixture of the 70 different members of the family of chlorinated 2,3,7,8-tetrachlorodibenzo-p-dioxin (TCDD).

Dip Coating – A paint or coating applied by dipping the item into the paint or coating.

Dipoles – Equal but opposite electric charges separated by a distance; for instance, at the hydrogen and oxygen ends of a water molecule.

Direct Current (DC) – An electric current the flows in one direction, as opposed to alternating current (AC) in which the direction changes every cycle.

Disk Refiner – A device used for making mechanical pulp or modifying chemical pulp for papermaking.

Dissolved Air Flotation – The use of finely dispersed gas bubbles that combine with suspended or emulsified waste particles to form a particle-gas composite that has a density less than water. When the composite reaches the surface it is skimmed off. See Electroflotation.

Distillation – The separation of a mixture of liquids on the basis of their different boiling points (vapor pressures). A distillation apparatus is typically composed of the distillation pot (still bottom) where the liquids are heated and evaporated, a cooling tower or condenser where the vapors are condensed, and a distillate collection container.

Donor Ion – An ion such as antimony, arsenic, phosphorus, or bismuth that creates an n-type conducting area on a semiconductor chip.

Dopants – An impurity intentionally added in very small, controlled amounts to change the electrical properties of a semiconductor, thereby creating a transistor.

Doping – The process of adding an impurity to a semiconductor to change its electrical conductivity.

Doré – A mixture of gold, silver, and impurities.

Dragline – An articulated excavating machine that removes overburden by casting a bucket and retrieving it. As the bucket is retrieved, it fills. The full bucket is lifted, swung to the side, and emptied.

Drag-out – The liquid clinging to a workpiece when it is removed from a process or rinse bath. The amount of drag-out is a function of the workpiece size and shape and the viscosity, concentration, surface tension, and temperature of the liquid.

Drag-out Rinses – The series of rinses in a countercurrent rinse system, with the first rinse following the process bath known as the drag-out rinse.

Drawing – A metal forming process in which metal sheet stock is forced into a die by a punch. The shape is formed in the space between the punch and die.

DRE – See Destruction and Removal Efficiency.

Drift – A horizontal opening excavated into rock underground. Used to refer to underground openings that parallel an ore body.

Drill Stem Testing – The lowering of an instrument into a bore hole to collect samples of fluids present and to record the fluid pressure in the well hole.

Drilling Mud – Also known as drilling fluid, it is an aqueous mixture of bentonite clay that serves as a thickener. Other additives may be included to change its pH or adapt it to a specific set of working conditions. Its primary purposes are to control the subsurface fluid and gas pressures and to cool, lubricate, and help carry away the drill bit cuttings.

Drip Tanks – A drag-out recovery tank that is used to collect process solution, but unlike a drag-out tank is not filled with water.

Dry Heat – Exposure of waste to an elevated temperature and pressure – in the absence of steam – for sufficient time to kill harmful microorganisms; exposure time is much greater than for steam sterilization.

Dry Ice – The solid state of carbon dioxide, which only exists under normal atmospheric pressures at temperatures below –78.5°C. Its density is 1.56 g/cc and it sublimes directly to the gas.

Dry Plasma Etch – The use of high electrical power and sulfur hexafluoride or a combination of tetrafluoromethane and oxygen to produce a dry plasma that cleans the wafer prior to chemical vapor deposition.

Drying – The separation of volatile liquids from solid or semisolid material by vaporization.

Dual Media Filters – A layered filtering medium, composed of an upper layer of activated charcoal or anthracite coal, followed by lower layers of progressively smaller rock or sand. Larger solute particles are trapped in and on the upper layers and smaller particles are captured by the rock and sand layers. The filter is periodically cleaned and regenerated by back washing the filter, collecting the trapped solids, and allowing the variously sized layers to reestablish their original position and density.

Dual-component Paints – Paints, including epoxy paints, 100 percent solids polyurethane paints, and similar coatings that require the mixing of two liquid components to produce the coating. In epoxy-based paints these components are a base and a catalyst, while in polyurethane systems it is an isocyanate and polyol-rich liquids. Once mixed these paints start to cure.

Ductility – The ability of a metal to be drawn into a wire.

Dyeing – A process used to add color to a product.

E-beam – A high-energy electron source with a wavelength shorter than ultraviolet light. Due to its shorter wavelengths, high-energy electron-beam lithography has been demonstrated to produce semiconductor features on silicon that are just 0.08 microns wide.

ECM – See Environmentally Conscious Manufacturing.

Ecosystem – A community of organisms interacting with each other and with the chemical and physical environment.

Electrochemistry – The branch of chemistry dealing with the relationship between electricity and chemical reactions.

Electrodeposition – The process used to deposit a metal by electrolysis.

Electrodes – Solid electric conductor through which an electric current enters or leaves a solution. The negatively charged electrode is called the cathode and the positively charged electrode the anode.

Electrodialysis – A form of dialysis in which an electric current aids in the removal of ions in the solution. The substance to be separated is placed between semipermeable membranes. By applying a potential to positively and negatively charged electrodes placed on either side, the ions tend to migrate into the chamber of the oppositely charged electrode. This results in salt-free water remaining in the middle chamber.

Electroflotation – A form of dissolved air flotation. Oxygen and hydrogen gas bubbles are generated by the electrolysis of water. The waste particles, which often carry a charge, migrate to the electrodes and coalesce on the emerging bubbles.

Electroless Copper – A protective coating of copper deposited in a bath without the application of an electric current, but rather by chemical reduction caused by an element higher on the activity series.

Electroless Nickel – A protective coating of nickel deposited in a bath without the application of an electric current, but rather by chemical reduction caused by an element higher on the activity series.

Electrolysis – The passage of electricity through an electrolyte.

Electrolyte – Any substance that contains ions and will conduct electricity when dissolved in water.

Electrolytic Recovery – The use of DC electricity to concentrate dissolved metals and possibly destroy cyanide in drag-out rinses.

Electromagnetic Radiation – Energy in the form of photons with various wavelengths. Starting with the longest wavelength and lowest energy, the complete electromagnetic spectrum is composed of radio, microwave, infrared, visible light, UV rays, X-rays, gamma rays, and cosmic rays, which have the shortest wavelengths and highest energies. All forms of electromagnetic radiation with wavelengths shorter than visible light are ionizing radiations.

Electromotive Series – Also known as the activity series, an arrangement of metals in the order of their strengths as reducing agents. For example, the arrangement for some of the more common metals is: K, Na, Mg, Al, Zn, Fe, Pb, H, Cu, Hg, Ag, Pt, Au.

Electronic Assembly – See Printed Wiring Assemblies.

Electrons – The negatively charged particles present in all atoms. The electrons circulate around the nucleus in energy shells or orbitals. In an atom the number of electrons is always equal to the number of protons. The electron has a mass of 9.1×10^{-28} g, but appears to occupy most of the space in an atom. Atoms may undergo oxidation (loss) or reduction (gain) reactions changing the number of electrons. The resulting charged particles are ions of that element; e.g., a sodium atom loses an electron to become a sodium ion.

Electroplating – The deposition of a thin layer of metal on an object (cathode) by passing an electric current through an aqueous solution of a salt containing ions of the element being deposited; e.g., Cr^{6+}.

Electrostatic Paint Guns – A spray gun that places an electrostatic charge on the powder or paint droplets so that it is deposited on the grounded workpiece.

Electrostatic Precipitator – Also known as the Cottrell precipitator after its inventor Fredreck G. Cottrell, it uses a high electric charge between conducting plates to create an electrical field. When particles enter this field, they become charged and are attracted to the oppositely charged plate. The accumulated particles are periodically freed by a mechanical rapping of the metal plates. The collected particles drop into a collection hopper below.

Electrowinning – An electrochemical method of recovering metals from a solution by converting it from its ionic form to the elemental form.

Element – A pure substance that can not be further broken down by chemical means. Examples include carbon, magnesium, gold, and chlorine. All known naturally occurring and human-made elements are listed on the Periodic Table of the Elements.

Empty Cell Process – A treatment process used when a lower total absorption of treating solution is required than provided by the full cell method. The wood is placed in a retort along with the preservative solution. The pressure is raised compressing the air in the wood and forcing the solution into the wood. After the desired amount of preservative has been absorbed, the pressure is reduced and the excess solution exuded.

EMS – See Environmental Management System.

Emulsifier – A chemical agent, usually with polar and nonpolar properties, that allows immiscible liquids, such as oil and water, to be mixed.

Emulsion Cleaning – A semi-aqueous cleaning method that uses common organic solvents, dispersed in water with the aid of emulsifying agents, to clean metal surfaces.

End-of-life Management – Decisions made following the product's useful life related to the product's disposal, disassembly, or other use of the product or its components.

End-of-pipe – Refers to treatment processes applied to hazardous waste after its generation, as opposed to front-of-pipe or pollution prevention technology.

Energy Levels – The areas around an atom's nucleus identified by K, L, M, N, O and P that are occupied by the electrons.

Energy Star Buildings Program – Part of the Green Lights Program, a comprehensive program with a focus on the development and use of energy efficient heating, cooling, and air handling equipment.

Energy Star Homes Program – Part of the Green Lights Program, designed to assist new home builders in developing energy-efficient homes and identifying financing opportunities for new home buyers.

Energy Star Loans Program – Developed in conjunction with financing companies and banks, EPA developed the loan program to increase retail sales and make the purchase of high efficiency Energy Star heating and cooling equipment more affordable.

Energy Star Office Equipment Program – Part of the Green Lights Program, in which energy-efficient computers, printers, fax machines, and copiers have been developed and identified by the Energy Star logo.

Energy Star Program – A cluster of energy saving programs designed to reduce the use of electrical energy. It is a part of the Green Lights Program and is managed by the EPA's Atmospheric Pollution Prevention Division in Washington, D.C.

Energy Star Residential HVAC Program – Part of the Green Lights Program, this program is designed to help homeowners retrofit their homes with heating, ventilation, and air conditioning equipment that is more energy efficient.

Environmental Compartment – Also referred to as environmental media. That part of the environment – water, air, soil, and biota (living things) – that contaminants are either carried by or transmitted through.

Environmental Management System (EMS) – A part of an effective evaluation, accounting, and auditing processes to measure the impact that an environmentally conscious manufacturing system is having on a manufacturing business. It is also a part of the ISO 14000 management system.

Environmental Protection Agency (EPA) – The federal agency founded in 1970 with the responsibility of developing and enforcing regulations pertaining to the protection of environmental quality.

Environmentally Conscious Manufacturing (ECM) – A strategy that has reduction, re-manufacture, and reuse as its main goals. ECM strategies, therefore, include changing input materials, processes, inventory control, to components that can

undergo repair, rework, or refurbishment, and can be recycled or reused.

EP Toxicity Test – A test designed to simulate the leaching actions that occur in a landfill, to identify wastes likely to leach hazardous concentrations of particular toxic constituents into the groundwater as a result of improper management.

EPA – See Environmental Protection Agency.

Epoxy Paint – A two-component paint in which a liquid base is mixed with a liquid catalyst. The mixture hardens as the base and catalyst react forming a solid coating. The two components can be mixed either before or during application.

Etch – Any action that cuts into a material, such as the chemical reaction between an acid and a base metal.

Etch Resist – A substance unaffected by an etchant that is selectively applied over copper to protect the copper from the etchant. Etch resists may be organic (photoresists) or metallic (tin, tin-lead, nickel-gold).

Etchant – The strong acids or bases used in etching to achieve controlled dissolution of metal surfaces.

Etchback – A process by which an amount of printed wiring board substrate material is dissolved or otherwise removed from the walls of drilled holes. The purpose of etchback is to expose a greater inner layer of copper surface area for interconnection with subsequently plated copper.

Etching – A surface treatment that uses chemical reagents to produce a specific design or surface appearance through the action of controlled dissolution.

Ether Process – A chemical reaction that results in the formation of an ether, such as methanol and isobutylene that produces methyl tertiary butyl ether.

Eukaryotic – One-celled microorganisms that have a true nucleus.

Evaporate – Mineral created when supersaturated solutions precipitate salts as they evaporate.

Evaporating Pond – Shallow ponds containing large surface areas that can economically evaporate large amounts of solvent, concentrating the solute to the point of dryness. Typically sited in areas where favorable climatic conditions and large amounts of solar energy are available.

Evaporation – The removal of volatile solvent from a solution or relatively dilute slurry by vaporizing the solvent.

Evaporative Recovery – The process of boiling water off the first rinse bath in an electroplating operation to recover valuable chemicals that can be returned to the plating bath. If the process is carried out at pressures below atmospheric (vacuum), the water can be removed at a lower temperature, requiring less energy.

Executive Order 12780 – See Federal Agency Recycling and the Council on Federal Recycling and Procurement Policy.

Executive Order 12856 – See Federal Compliance with Right-to-Know Laws and Pollution Prevention Requirements.

Executive Order 12873 – See Federal Acquisition, Recycling, and Waste Prevention.

Exempt Wastes – Specific items, with relatively low radioactivity content, that are exempt from waste disposal regulations.

Exothermic – Describes a process or chemical reaction that is accompanied by the evolution of heat, e.g., combustion reactions.

Extraction (Ores) – The process of removing metals from their minerals.

Extraction Circuits – The series of processes that extract metal from the ore.

Extractions (Solutions) – Separation processes based on solubility differences (solute – solvent affinities).

Extractive Metallurgy – The science of extracting metals from rock and producing a product used in industry.

Extractive (Wood) – A minor, non-integral component of wood removed by extracting with water and organic solvents containing hundreds of compounds including pine oil, turpentine, and rosin.

Extruding – A forming process in which a blank is forced through a die to form a specific shape.

Exxon *Valdez* – In 1989, the oil tanker, Exxon *Valdez*, ran aground on Bligh Reef in Prince William Sound, spilling an estimated 270,000 barrels (11 million gallons) of crude oil, which fouled 1,300 miles of Alaskan coastline.

Fabric – A cloth made by weaving, knitting, or felting fibers.

Fabricator – A business that uses metal to make consumer or industrial products.

Facultative – A microorganism that gains its energy by aerobic respiration if oxygen is present, but can switch to fermentation when oxygen is absent.

Facultative Lagoon – Constructed in a manner similar to aerated impoundments; however, they do not have forced aeration. They are typically 5-20 feet deep, allowing for aerobic degradation on the surface and anaerobic activities near the bottom.

Fatigue Cycles – The repeated application of stress on metal parts that cause them to loose their structural integrity, i.e., airplane landing gear.

Federal Acquisition, Recycling, and Waste Prevention – Also known as Executive Order 12873, signed on October 20, 1993. Requires federal agencies to promote cost-effective waste reduction and recycling of reusable materials from waste generated by federal government activities.

Federal Agency Recycling and the Council on Federal Recycling and Procurement Policy – Also known as Executive Order 12780, signed October 31, 1991. Among its several sections, it requires that federal agencies promote cost-effective waste reduction and recycling of reusable materials from wastes generated by federal government activities.

Federal Compliance with Right-to-know laws and Pollution Prevention Requirements – Also known as Executive Order 12856, signed on August 3, 1993. Establishes federal policy and requires federal agencies to incorporate the activities required by EPCRA and PPA into all aspects of agency programs.

Federal Government Environmental Challenge Program – A program aimed at fostering waste reduction efforts by federal agencies and contractors. It requires agencies like DOD and EPA to draft provisions for P2 and environmental compliance into governmental agreements with their contractors.

Felting – The process of making a non-woven fabric made of such materials as wool, fur, or hair, matted together by application of heat, moisture, and great pressure.

Fermentation – Also called anaerobic respiration. Any anaerobic process induced by living organisms that breaks complex organic compounds into more simple substances.

Fiber – A natural or synthetic filament capable of being spun into yarn.

Filter Cake – The solid that accumulates on the filter medium, often improving its ability to capture increasingly smaller particles. When the cake reaches a thickness that causes the flow rate to fall below process standards, it must be removed and the process repeated.

Filter Press – A filter composed of a meshed barrier supported on a stacked series of plates that have corrugated surfaces. The mixture to be separated is pumped, under high pressure, into the center of the filter and the filtrate allowed to escape through the mesh lined channels that capture the solids. Air/hydraulic pressure is then applied to complete the dewatering.

Filtering Medium – Almost any water-insoluble porous material having a reasonable degree of rigidity can serve as a filter medium. A substance as simple as a bed of sand can be used, where the voids between the grains provide the porosity. Cotton, woven wire, nylon, and glass cloth are sometimes used. Plastic membranes containing pore sizes down to 0.001 mm (micron) are used.

Filtrate – The liquid that passes through a filtering medium.

Filtration – The process of separating suspended solids from a fluid (liquid or gas) by passing the mixture through a filtering medium. The pore size of the filtering medium determines the size of suspended solids removed.

Financial Risk – Risks that relate to whether or not the costs of making a change will result in an appropriate return on investment.

Fines – Small portions of fiber that detract from the papermaking process and do not contribute strength to the final paper or any small particles resulting from a process.

Finishing – The addition of a process or chemical to produce a desired effect in the product.

Fissile – A substance capable of undergoing fission induced by low-energy neutrons.

Fission – The splitting of a heavy nucleus into two approximately equal parts, accompanied by the release of a relatively large amount of energy and generally one or more neutrons.

Fission Fragments – The two nearly equally sized nuclei fragments produced as the result of the fission of a fissile nucleus.

Fixation – See Immobilization.

Fixed Hearth Incinerator – A small capacity incinerator that is used for the thermal destruction of both solid and liquid waste. It has both a primary and secondary combustion chamber. Due to its lower temperatures, it is not suitable for the destruction of chlorinated liquid wastes.

Fixed-film – Any of a number of processes, e.g., trickling filters, bioreactors anaerobic filters, RBCs, etc. that have a slime of biological growth on its treatment surface.

Fixed-film Bioreactor – Also known as a biological reactor, or bioreactor, it is an enclosed unit that allows for the controlled exposure of a wastewater to a biological film. Within the reactor are a series of metal honeycombs on which the aerobic biological slime grows.

Flame Hardening – A heat treatment process which hardens metal surfaces by using an oxyacetylene flame to rapidly heat the metal surface to the hardening temperature.

Flashing – A process whereby water is boiled at a temperature <100°C by applying a vacuum to the system.

Flashoff Area – An area subject to ambient drying.

Float Cell – A tank in which flotation occurs.

Floc – A floc is composed of clumps or tufts of solid resulting from coagulation of colloidal-sized suspended particles. A shortened term used to mean flocculation. In wastewater treatment applications, it refers to aqueous, gelatinous clumps of unattached microbial growth that feed on the organic waste. It contains a wide array of aerobic bacteria, fungi, animal-like protozoa that eat both bacteria and fungi, and a few insect larvae and worms. See Flocculation.

Flocculant – A substance that induces flocculation. Examples of inorganic flocculants or flocculating agents are lime, alum, and ferric chloride; polyelectrolytes are examples of organic flocculants.

Flocculation – The formation of clumps or tufts of solids following coagulation. It is typically initiated through the addition of a flocculating agent and enhanced by slow stirring, which improves the chances of the solute particles colliding.

Flotation – A method of mineral separation based on the selective wetting and attachment of air bubbles to the mineral particles.

Flow Diagram – A pictorial representation of a process showing the entry point of all raw materials, and the exit point of all product(s), by-product(s) and/or waste stream(s).

Fluid – Any material or substance that changes shape or direction uniformly in response to an external force imposed upon it. Liquids, gases, and even solids composed of finely divided particles (quick sand) can behave like a fluid.

Fluidized-bed Incinerator – Refractory-lined vessels containing a bed of graded, inert granular material – usually silica sand. Large volumes of heated air are pushed through the bed, and the waste injected into or just above the swirling hot sand. Two variations on the fluidized-bed technology exist, they differ in separation methods of the off-gases produced. See Bubbling Fluidized-bed and Circulating Fluid-bed Incinerators.

Flux (Soldering) – A substance that promotes the fusing of metals or prevents the formation of oxides.

Fluxes (Metallurgical) – Materials added to a melt to facilitate the separation of metals from waste.

Fly Ash – The very fine ash produced by combustion and carried off in the flue gases. It is generally a mixture of alumina, silica, unburned carbon, and metallic oxides.

FOG – An acronym used in the meat processing industries for fats, oils, and grease; the primary pollutants in wastewater streams.

Forage – A mixed vegetation eaten by grazing animals.

Forging – A metal forming process that produces a specific shape by applying external pressure that either strikes or squeezes a heated blank into a die.

Formic Acid – A carboxylic acid, HCOOH, that has a strong odor and is a potent reducing agent.

Forming Processes – Metal shaping operations that alter the shape of a metal object without reducing the amount of metal in the piece.

Fountain – The reservoir on a printing press for materials such as ink or fountain solutions.

Fractional Distillation – A distillation process in which a mixture is separated into a series of components with similar boiling points. The mixture is introduced into a column as hot vapors. As they rise and cool, they encounter liquid that has condensed at that temperature. Any component remaining as a vapor then continues upward until it contacts more liquid at a cooler temperature. The process is repeated many times, resulting in separation of the components based on their boiling points. A process used to separate crude oil into its components.

Fractional Distillation Tower – The tower in which separation of hot crude oil vapors takes place. (See Fractional Distillation.)

Fractionating Column – See Fractional Distillation Tower.

Fractions – Also known as distillate, fractions are the separate components with similar properties obtained from the distillation of a mixture. The most important fractions resulting from the fractional distillation of crude oil are naphtha, gasoline, fuel oil, kerosene, and tarry residues.

Freeboard – The space between the liquid and the chiller coils in a vapor degreaser.

Fuel Pins – Sealed zirconium alloy metal tubes packed with enriched uranium dioxide pellets that are assembled in a grid pattern to form a fuel assembly in a light water reactor.

Fuels – Products used in combustion processes to produce energy.

Full Cell Process – A wood preservation method developed by John Bethell that impregnates the wood with the most preservative. The wood is placed in a retort, a vacuum drawn to remove air from the wood, followed by the introduction of the treating solution and an increase in pressure to force the solution into the wood. Later the pressure is released and a final vacuum drawn to remove excess treating solution.

Fungi – A heterotropic eukaryote that digests its food externally and absorbs the resulting small nutrient molecules. Molds, mushrooms, toadstools, and yeasts are examples of fungi.

Fungicides – Pesticides used to kill or control fungi, including molds, yeast, and rot.

Furans – Also known as dibenzofuran. Actually a group of chlorinated compounds with properties and structure very similar to the dibenzo-p-dioxins.

Fusible – A material capable of being melted.

Galena – A common ore of lead composed primarily of lead sulfide.

Gamma Rays – Electromagnetic radiation (photons) of extremely short wavelength and intensely high energy. Gamma rays originate in the nucleus and are frequently accompanied by either alpha or beta particle emission. Gamma rays are extremely penetrating and lethal, but may be absorbed by dense materials like lead.

Gangue – The waste fraction of ore that does not contain economically recoverable metals.

Gas Chromatograph – Used for the separation and identification of volatile substances.

Gas Diffusers – Devices used to generate small gas bubbles in a liquid.

Gasoline – A mixture of volatile hydrocarbons suitable for use in an internal combustion engine that uses spark plug ignition and has an octane number of at least 60.

Gasoline Blending – A physical process in which measured quantities of components are combined to form a homogeneous mixture. Blending a specific grade of gasoline involves combining the appropriate petroleum fractions and adding proprietary additives.

Glassy Slag – An obsidian-like slag resulting from the fusion of non-combustible minerals and metals that have been heated to high temperatures.

Grade – A measure of the amount of valuable metal in ore. Usually express in terms of units of metal per unit of ore. Example: 0.1 oz gold/ton ore

Granulated Activated Carbon – Prepared by the destructive distillation of wood, animal bones, coconut shells, or other carbonaceous materials. After the charcoal is prepared, it is activated by heating it to temperatures of 800-900°C in an atmosphere of steam or carbon dioxide, which results in it becoming more porous. It is estimated that one gram of activated carbon can contain 10,000 ft³ (930 m³) of surface area.

Gravimeter – An instrument for measuring variations in the gravitational field of the earth by detecting differences in weight of an object of constant mass at different points on the earth's surface.

Gravure Printing – A printing process that uses an image area set below the plate's surface. The image is transferred when the plate is flooded with ink and the excess ink wiped away with a metal blade.

Graywater – Wastewater collected in a building's drain system that does not contain human urine or feces.

Green Chemistry Challenge – A EPA Design for the Environment partnership with the chemical industry. Designed to recognize and promote fundamental breakthroughs in green chemistry, i.e. chemistry that accomplishes pollution prevention goals and is useful to industry.

Green Chemistry Program – A program designed by EPA's Office of Pollution Prevention and Toxics to promote the development of products and processes that reduce or eliminate the use or generation of toxic substances associated with the design, manufacture, and use of chemicals.

Green Lights Program – A voluntary, non-regulatory EPA program aimed at promoting energy efficiency through investment in energy-saving lighting. Its goals are to save business money and create a cleaner environment by reducing pollutants released into the environment.

Green Liquor – An aqueous solution made by dissolving smelt, composed of sodium carbonate, Na_2CO_3, and sodium sulfide, Na_2S, into water.

Green Manure – Cover crops that can be plowed under to improve soil fertility and workability.

Green Products – Products designed and produced to be less harmful to the environment and to promote better utilization of resources. Some characteristics that distinguish green products include minimal packaging, significant recycled material content, non-toxic and non-polluting, energy efficient, organically grown, reusable, recyclable, minimal processing (e.g. unbleached, free of artificial flavors, colors, and fragrances), and produced from renewable sources.

Greenhouse – An enclosure in which plants are grown under artificial conditions.

Grinding – A process of size-reducing ore to an extremely fine powder.

Ground Level Ozone – Ozone that is produced as the result of a chemical reaction between sunlight and one of the EPA-designated volatile organic compounds (VOC) and other pollutants.

Gypsum – An evaporate mineral composed of hydrated calcium sulfate ($CaSO_4 \cdot 2H_2O$). The principal ingredient of plaster and sheet rock.

Half-life – The time for one-half of the atoms in a radioisotope sample to disintegrate.

Halogen – The electronegative members of group VII on the Periodic Table. They are strong oxidizing agents and include fluorine, chlorine, bromine, iodine, and radioactive astatine.

HAPs – See Hazardous Air Pollutants.

Hardening – A heat treatment process which increases the hardness and brittleness of metals.

Hardwood – Wood from angiosperms (the deciduous or broad leafed trees).

Haulage Drifts – In an underground mine, a drift used to transport ore from the working area to the shaft.

Hazardous Air Pollutants (HAPs) – All other air toxics, not included as primary or secondary air pollutants, that may be reasonably anticipated to result in health problems. Include such things as arsenic, asbestos, benzene, beryllium, mercury, radionuclides, radon-222, and viny chloride.

HCFCs – See Hydrochlorofluorocarbons.

Heap Leaching – The process of sprinkling the top of the pile with a solvent to leach the metal from the mineral.

Hearth – A frame of metal bars to hold fuel when burning.

Heartwood – That part, in the center of a live tree, which contains dead wood. The heartwood generally is less permeable than the sapwood and therefore more difficult to treat with preservative solutions.

Heat Exchanger – A device used to transfer heat from one substance, usually exiting a process, to a substance entering a process.

Heat Set Inks – Inks that require heat to cure.

Heat Treating – A process using controlled heating and cooling cycles to alter the physical properties of alloys.

Hemicellulose – A major component of wood (20-35 percent by mass), it is a white polysaccharide containing 50-200 sugar units.

Herbicides – Pesticides used to kill or control plants, especially weeds.

Herbivores – Animals that obtain their growth requirements by consuming plants.

Heterotrophs – Literally, other-feeders. They are organisms that obtain their carbon from organic compounds. Most also get their catabolic energy from the organic compounds, but some heterotrops can also obtain their energy by photosynthesis.

Hexavalent Chrome – Chromium metal can exist in various ionic forms. Hexavalent chrome refers to chromium atoms that have lost six electrons and have an oxidation number of plus six, Cr^{6+}.

HHW – See Household Hazardous Waste.

Hierarchy of Hazardous Waste Management – A 1976 EPA policy statement that established source reduction as the highest priority, followed by recycling, treatment, and disposal to the land. The policy was intended to encourage industry to reduce hazardous waste at its source, rather than use end-of-pipe treatment and/or disposal.

High-level Radioactive Waste (HLW) – Defined as irradiated reactor fuel; liquid waste resulting from the operation of the first cycle solvent extraction system, or equivalent, and the concentrated wastes from subsequent extraction cycles, or equivalent, in a facility for reprocessing irradiated reactor fuel; and solids into which such liquid waste has been converted.

High Volume/Low Pressure (HVLP) – These spray systems use high volumes of low pressure air to atomize the paint. This improves transfer efficiency, because low pressure reduces turbulence and bounce-back of paint.

HLW – See High-level Radioactive Waste.

Hole Barrels – The interior surface of the cylindrical holes in a printed wiring board (PWB).

Household Hazardous Waste (HHW) – Hazardous items used in the household that have served their intended purpose and have become waste. Unusable paints and cleaning products are examples.

HVLP – See High Volume/Low Pressure.

Hydraulic Gradient – A difference in pressure that results in the movement of water from the point of higher pressure to lower pressure.

Hydrocarbons – Organic compounds that consist exclusively of the elements carbon and hydrogen. They are commonly subdivided into the following

organic compounds that consist exclusively of the elements carbon and hydrogen. They are commonly subdivided into the following organic families: alkanes, alkenes, alkynes, and aromatics.

Hydrochlorofluorocarbons (HCFCs) – A group of compounds that have a carbon backbone and varying numbers of hydrogen, chlorine, and fluorine atoms attached, e.g., $CHClF–CClF_2$.

Hydrocracking – Developed in the early 1930s, hydrocracking involves the use of small amounts of hydrogen, in combination with a specialized catalyst. It greatly reduces the amount of coke formed and increases the amount of high quality gasoline produced. The catalyst is removed from the resulting alkylate, recycled and reused. The hydrocarbon portion is sent on for further blending.

Hydrofluoric Acid – A colorless, fuming liquid that will attack glass and any other silicon-containing material. It is toxic by ingestion and inhalation, highly corrosive to skin and mucous membranes. TLV: ceiling 3 ppm. Used for etching glass, processing uranium, and as a catalyst in the alkylation (reforming) process.

Hydrogen Bonding – A weak, but effective electrical attraction between polar covalent molecules. Hydrogen bonding affects a substance's melting and boiling points as well as its water solubility.

Hydrogen Ion – Also called a proton. A hydrogen ion, H^+, is the smallest positive ion, consisting of only a single proton and no electrons.

Hydrolysis – A chemical reaction in which water reacts with another substance to form two or more new substances.

Hydrometallurgy – The process of extracting metals from ore by dissolving the ore in a solution.

Hydronium Ion – An ion, H_3O^+, formed by the attachment of a proton, or hydrogen ion, to the unshared pair of electrons on water's oxygen atom.

Hydrophobic – Literally translated it means a fear of water. Substances that are hydrophobic cannot dissolve in and are antagonistic to water. Most petroleum products, fats, waxes, and resins fall into this category.

Hydrothermal Fluids – Underground solutions of water and dissolved minerals heated by rising magma.

Hydrotreating – A process used to remove sulfur compounds from petroleum products by converting them, in the presence of hydrogen gas and a catalyst, into hydrogen sulfide gas (H_2S).

Hydroxide Ion – An ion, OH^-, typically formed by the dissociation of soluble metal hydroxides, e.g., LiOH, NaOH, KOH, etc. when placed in water.

Hydroxide Precipitation – A chemical precipitation reaction where the cation is a metal and the anion is the hydroxide ion, OH^-.

Hyperaccumulators – Plants that concentrate metals to levels that would ordinarily be toxic to most other plants.

Hyperfiltration – Hyperfiltration is one of the new membrane separation technologies that is due to the continued improvements being made in synthetic semipermeable membranes. Other processes also relying on these membranes are microfiltration, ultrafiltration, nanofiltration, reverse osmosis, and electrodialysis, each of which permits for the separation of a contaminate (solute) from a liquid phase (typically water).

IBPC – 3-Iodo-2-propynyl butyl carbamate in combination with didecyldimethylammonium chloride is currently the most widely used sap stain control in America. It can be applied in either an organic solvent or as a liquid emulsion in water.

IBSIN – See Innovations in Building Sustainable Industries.

IC – See Integrated Circuits.

IDW – See Investigation Derived Waste.

Image Area – The area on a printing plate that prints.

Immobilization – A general term that refers to the processes used to convert liquid and semi-solid wastes into a solid that is stable and exhibits physical properties that make the waste safe for disposal.

Impregnation – The process of transporting preservative chemicals into wood by a pressure treatment process.

In Situ – The term indicates that a given procedure is conducted not just in the field or on the site, but in the natural or original place or position in the environment exactly where it is found.

Incineration – The complete thermal destruction of a substance's chemical properties; generally the result of burning at temperatures of 1,200-2,400°C.

Inclined Plate Clarifier – Similar to the slant tube clarifier, except that plates are used in place of the tubes. Both types of clarifiers have the advantage of being small in size for their throughput capability.

Indicator Paper – A test paper impregnated with an organic substance that changes color in the presence of, or based on the concentration of, some other substance, e.g. litmus paper.

Induction Hardening – A heat treatment process which hardens metal surfaces by using a high frequency electric current to heat the metal surface to the hardening temperature.

Ingot – Masses of metal in the form of bars or other convenient shapes. Such as purified silicon from which wafers are cut to make semiconductors.

Injector Nozzle – A nozzle that, when coupled with high pressure air or steam stream, causes the atomization of a liquid waste stream. This results in an increase in surface area, which improves its chances for complete incineration.

Innovations in Building Sustainable Industries (IBSIN) – An EPA program that is designed to provide incentives for industry, government, and communities to work together to promote sustainable development goals, such as reducing TRI toxics and implementing product stewardship programs.

Inorganic Compounds – Chemical compounds that are not hydrocarbons or their derivatives. In general any chemical compounds that do not contain the element carbon, except for CO, CO_2, CS_2, HCO_3^-, CO_3^{2-} and CN^-.

Insecticides – Pesticides used to kill insects or interfere with their reproductive cycle.

Insoluble – A substance that does not or only partially dissolves in a solvent, e.g. 1.0 g/100 ml.

Integrated Circuits (IC) – A circuit of transistors, resistors, and capacitors constructed on a single semiconductor wafer or chip, in which the components are interconnected to perform a given function.

Integrated Pest Management (IPM) – A system of methods used to control crop pests without heavy reliance on commercial pesticides. IPM techniques include choosing pest resistant crops, rotating crops to avoid recurring outbreaks of host specific pests, and using tillage practices to control pests and weeds. It also stresses the use of biological controls such as introducing host specific parasites to control pests, using traps baited with insect attractants, and creating environmental conditions that favor proliferation of beneficial species.

Internal Combustion Engine – There are two types of internal combustion engines: those using reciprocating pistons and continuous combustion turbines. Piston engines use either spark plug or compression ignition. They use four basic strokes; fuel-air intake into the cylinder, followed by compression and ignition. The large volumes of hot, expanding gasses produce the power-stroke, followed by the exhaust stroke to remove the combustion products.

International Organization for Standardization (ISO) – Based in Geneva, Switzerland, ISO is a specialized international agency whose members are the national standards bodies of 111 countries. The American National Standards Institute (ANSI) is the U.S. member body. The goals of ISO standards are to facilitate the efficient exchange of goods and services.

Intrinsic Bacteria – Bacteria that are native or natural to an area.

Investigation Derived Waste (IDW) – Wastes generated as the result of characterizing a site, which may include such things as drill cuttings, decontamination fluids, waste from well installation and monitoring activities, samples and analytical residues, and PPE.

Ion – An atom or group of atoms (polyatomic) that has either gained or lost electron(s), resulting in it

having an electrical charge. Positively charged ions are called cations and negatively charged ions are called anions.

Ion Exchange – In its typical industrial application, wastewater containing unwanted positive and/or negative ions are brought into contact with a polymeric resin that is "loaded" with exchange ions. Once the unwanted ions contact the resin, the exchange occurs, releasing the exchange ions and retaining the unwanted ions. Once the resin's supply of exchange ions is exhausted, the resin can be regenerated by using a recycle-reuse back-flushing process.

Ion Implant – A process that changes the electrical characteristics of a semiconductor by placing impurities such as boron, gallium, indium, antimony, arsenic, phosphorous, or bismuth ions below the surface.

Ionic Bond – The electrical attraction occurring between a positively charged ion (cation) and a negatively charged ion (anion). Considered to be one of the major types of chemical bonding. Ionic solids tend to have hard crystals, high melting and boiling points, are typically nonflammable, and may readily dissolve in water e.g., NaCl.

Ionic Compound – A chemical compound formed by the aggregation of two or more different kinds of ions such as sodium chloride, NaCl.

Ionization – The process of creating a charge particle or ion. Light, X-rays, gamma rays or electrical energy can be used directly to ionized atoms into ions by removing electrons. When placed in water, molecules of many substances can be pulled apart forming ions. These ions result from having an unequal number of electrons and protons in each of the parts.

Ionizing Radiation – Any radiation capable of displacing one or more electrons from an atom or molecule, thereby producing ions.

Irrigation – A system that supplies water to crops by artificial means.

ISO – See International Organization for Standardization.

ISO 14000 – A set of industrial management standards that will help any organization address environmental issues in a systematic way and thereby improve its environmental performance.

Isoelectric – Two or more ions or an ion and an atom that have the same electron configurations.

Isomerization Process – A process that results in the rearrangement of a straight-chained to a branched-chain hydrocarbon. It can also result in the fusing of many smaller into one larger hydrocarbon.

Isooctane – Also known as 2,2,4-trimethylpentane, it is a branched-chain hydrocarbon with a boiling point of 99.2°C and an autoignition temperature of 417°C. It is used in organic synthesis, and fuels, and is mixed with n-heptane to prepare standard fuel mixes to determine the anti-knock properties of gasoline.

Isotope – Atoms of the same element that vary in the number of neutrons in the nucleus. For example, all carbon atoms have 6 protons, but some have 6 neutrons, known as carbon-12 and others have 8 neutrons called carbon-14. The ratio of these two isotopes is the basis for the radiocarbon dating of artifacts.

JIT – See Just-in-time.

Job Shops – Small businesses generally employing less than 50 employees and with annual sales less than $5 million, that provide electroplating services for other manufacturers.

Just-in-time – One of the techniques used as a part of a P2 and waste minimization program. Raw materials are purchased as needed, when they are needed, which results in reducing both inventories and waste.

K Listed Waste – As specified in Subtitle C of RCRA, K listed wastes are from a specific source, e.g., K051, AIP separator sludge from the petroleum refining industry or K053, tank bottoms (leaded) from the petroleum refining industry.

Kerosene – A clear, nearly colorless, oily liquid with a strong odor. It has a moderate fire risk, with explosive limits in air ranging from 0.7-5.0 percent and an autoignition temperature of 228°C. Primary uses include fuel for rockets and jet engines.

Kickback – The treating solution recovered during a pressure treating cycle when a pressure differential exists between the interior and exterior of the wood with lower pressure on the exterior. The pressure differential causes liquids to flow to the surface and leave the wood. The pressure differential can be created either by applying air pressure to the wood before treatment, applying a vacuum after treatment, or both.

Knitting – The interlacing of fibers or yarn in loops to make a fabric.

Kraft Pulping Process – The process of making strong paper or paper board from wood pulp, produced from wood chips boiled in an alkaline solution containing sodium sulfate.

LAER – See Lowest Achievable Emission Rate.

Lagoon – Also known as ponds. See Surface Impoundment.

Lamella Clarifier – A lamella clarifier is a variation on the inclined plate clarifier. By use of lamella plates, the effective surface area for sedimentation to take place is greatly increased, thus the clarifier requires less space.

Laminate Stage – The assembling in a press of the layers of copper foil and substrate into a multilayer panel.

Land Disposal Restrictions (LDRs) – A 1984 amendment of RCRA regulations that banned land disposal of listed hazardous wastes unless they had met treatment standards.

Landfarming – Variously known as land cultivation and land treatment, it involves mixing excavated contaminated soils to be bioremediated with clean topsoil. By occasionally tilling, fertilizing, and keeping it moist, intrinsic bacteria are stimulated to accomplish the remediation process. Advantages include that it is a simple surface process that requires no special engineering skills or exotic tools, and is often done on site.

Lapping – A polishing process using a rotating wheel or disk holding an abrasive or polishing powder on its surface.

Large Quantity Generator – Hazardous waste generators that exceed SQGs and CESQGs limits fall into this classification.

Latex Paint – Water-based paints that cure and harden through the evaporation of water and polymerization.

Layering – The addition of a conducting, semiconducting, or insulating deposit over the surface of a wafer prior to forming a new layer.

LDRs – See Land Disposal Restrictions.

Leachate – Solutions containing ions dissolved from minerals present in the soils or rock.

Leaching – A hydrometallurgical process that uses aqueous solutions to dissolve metals.

Lead Frame – The legs of an integrated circuit are used to connect the chip to the printed wiring board.

Letterpress – A printing process in which the image area is raised above the non-image area on the printing plate. The image area is inked by rollers and the image transferred to the substrate.

Life Cycle Design – A proactive approach that takes into consideration the impact of all design decisions on the manufacture, distribution, use, service, recovery, and disposal of a product.

Lignin – A major component of wood (18-35 percent by mass), lignin is a highly branched, high molecular weight polymer of phenylpropane units connected by C–C, or C–O bonds.

Lime – Calcium Oxide (CaO) made from heating and oxidizing limestone ($CaCO_3$). In water, lime undergoes a reaction and forms Ca^{2+} and OH^-.

Liquid Injection Incinerator – A liquid injection incinerator consists of a refractory-lined combustion chamber and a series of atomizing nozzles. The combustion chamber may be either horizontal or vertical. Following combustion, flue gases are cooled and treated by an APCS.

Liquid Serum – A high BOD liquid waste that accumulates on the bottom of transport trucks due to fruit crushing during transport.

Liquid Petroleum Gas (LPG) – A compressed or liquefied gas obtained as a by-product in petroleum refining or natural gasoline manufacture, e.g., butane, isobutane, propane, propylene, butylenes, and their mixtures.

Listed Wastes – Defined in four separate lists in 40 CFR, each type of waste is coded to identify its source. Letter "F" are from nonspecific sources; "K" from specific sources; "P" for commercial chemical products that are intended for discard and considered to be acutely hazardous, and "U" for those commercial chemical wastes that are simply hazardous.

Lithography – A printing process in which the image is on a smooth surface that has been chemically treated so that the image area will accept only ink and the non-image area will accept only water.

Lithotrophs – Organisms whose biosynthesis rely on only inorganic carbon sources like carbon dioxide (CO_2), carbonate ions (CO_3^{2-}) and bicarbonate ions (HCO_3^-).

Litmus – An organic substance derived from lichens that is red in solutions with pH $\leq$ 4.5 and blue in solutions pH $\geq$ 8.3.

LLW – See Low-level Radioactive Waste.

Locatable Mineral – Minerals containing valuable constituents (primarily metals) in which individuals can acquire an ownership interest by staking a valid mining claim in accordance with provisions of the Mining Law of 1872.

Logging – The use of electrical measuring, sounding, and radiation instruments to analyze the walls of a bore hole. The data provides information about the porosity, conductivity, and composition of the rock formation.

Loom – A device used to construct fabric by interlocking fibers or yarn at right angles.

Lowest Achievable Emission Rate (LAER) – Under the Clean Air Act, this is the rate of emissions that reflects the most stringent emission limit that is contained in the implementation plan of any state for the source unless the owner or operator of the proposed source demonstrates the limits are not achievable. It also means the most stringent emissions limit achieved in practice, whichever is more stringent.

Low-level Radioactive Waste (LLW) – Radioactive waste that is not a high-level waste (HLW), transuranic (TRU) waste or mill tailings. Low-level waste typically relatively large in volume, but containing only a small amount of radioactivity.

Lowry Process – See Empty Cell Process.

LPG – See Liquid Petroleum Gas.

LQG – See Large Quantity Generator.

Lubricants – Substances of any composition that have the ability to reduce friction between two moving solids.

Lumens – The hollow interior portion of wood cells.

Machine Shop – An area in which metal shaping operations are carried out.

Machining Processes – Metal shaping processes that use machine tools to refine the shape of a workpiece by removing small amounts of material from it.

MACT – See Maximum Achievable Control Technology.

Magma – Molten rock present in the earth's mantel that rises and cools forming igneous rock.

Magnesite – A mineral composed of magnesium carbonate.

Magnetic Separation – The process of sorting minerals based on their magnetic properties.

Magnetometers – An instrument used to measure the intensity of magnetic fields.

Malleability – The ability of a metal to be hammered out into a thin sheet.

Manure – Animal wastes mixed with bedding materials and stall refuse, frequently applied to enrich agricultural fields.

Masking – The shielding of areas from paint, prior to the application of the paint.

Material Handling – The transport and storage of raw materials, products and wastes.

Material Safety Data Sheet (MSDS) – Chemical information sheets provided by the chemical manufacturer that include information such as: chemical and physical characteristics; long and short term health hazards; spill control procedures; personal protective equipment to be used when handling the chemical; reactivity with other chemicals; incompatibility with other chemicals; and manufacturer's name, address and phone number. Employee access to and understanding of MSDSs are important parts of the HazCom Program.

Materials Balance – A comparison of the quantities of all materials entering, to the amount of product(s), by-product(s), and waste stream(s) generated by a process. When these quantities are added to the block flow diagram, it creates a powerful tool for analyzing process efficiency.

Matte – A mixture of metal sulfides that form a homogenous solution during smelting.

Matter – Anything that occupies space and has mass.

Maximum Achievable Control Technologies – MACT is the control technology that can be used effectively on a specific piece of equipment to reduce the maximum amount of emissions.

Maximum Contaminant Levels (MCLs) – The maximum levels of certain contaminants allowed by the SDWA in drinking water from public systems. Under the 1986 amendments, EPA has set numerical standards or treatment techniques for an expanded number of contaminants.

MC – See Moisture Content.

MCLs – See Maximum Contaminant Levels.

Mechanical Pulp – Fibers produced by grinding wood (or other materials) against a rough surface or metal plate with raised bars to liberate individual fibers. These pulps are light gray in color but turn yellow with age as the lignin discolors.

Medical Waste – An umbrella term for wastes generated by medical and related (veterinary, cosmetic, research, etc.) establishments that require special treatment.

Medical Waste Tracking Act (MWTA) – Passed in 1988, this Act directed the EPA to develop and pilot test the effectiveness of regulations that would track medical waste along the eastern seaboard of the United States. The regulations define the waste; require use of a tracking form for waste shipments; and define the scope as applying to generators, transporters, treatment operators, and disposal facilities. The regulations also include the requirements for separation, packaging, labeling, storage, decontamination, use of the tracking form, record keeping, reporting, and notification for medical wastes.

Metal Ions – Metals, by definition, lose electrons during oxidation-reduction reactions. Depending on the number of electrons lost, the resulting metal ion may carry a charge of +, 2+, etc. e.g., Na^+, Cu^{2+}, Zn^{2+}.

Metallurgical Properties – The inherent properties of a metal or alloy.

Metallurgy – The study of the chemical and physical properties of metals.

Metalworking Fluids – The fluids used in metal shaping operations to reduce heat and friction, wash away chips and metal debris, inhibit corrosion, and improve machined surface characteristics.

Methyl Tertiary Butyl Ether (MTBE) – $CH_3-O-C_4H_9$. An oxygenate being used in the formulation of clean burning fuel (gasoline).

Microbes – A non-technical term used to refer to the whole range of microscopic or sub-microscopic sized, aerobic and anaerobic organisms. For example, the major types of aerobic microbes in the soil include bacteria, actinomycetes, fungi, algae and protozoa.

Microfiltration – A pressure driven separation process that uses a microporous membrane to remove solute particles larger than 0.2 microns, such as bacteria and some viruses, from the filtrate.

Microorganisms – Microscopic and sub-microscopic organisms, especially bacteria and protozoa.

Mill Tailings – The residue from processing ore to extract uranium. Tailings contain the decay products from uranium along with some uranium.

Millwork – A general term used to describe windows, doors, and other assembled wooden structures that are used in buildings. Synonym: joinery.

Mineral – A naturally occurring non-organic homogenous solid with a specific atomic arrangement and composition resulting in a predictable crystalline structure and possessing recognizable physical characteristics.

Mineral Oils – Oils derived from crude oil.

Minimization of Waste – An umbrella term that includes source reduction, recycling, and waste treatment processes. It results in reducing the weight, volume, or toxicity – to the extent feasible – of a waste prior to any off site treatment, storage, or disposal.

Mixed Waste – Wastes that contain both radioactive materials and hazardous chemicals.

Mixing and Blending Processes – Physical processes that disperse solids or gases in liquids, blend liquids, or produce fluid motion. The distinction between mixing and blending is vague. Both processes physically mix reactants, but blending results in a more homogenous product that cannot be separated by physical processes.

Mixture – A heterogeneous association of substances, with a variable composition, which cannot be represented by a chemical formula.

Moisture Content (MC) –

$$MC = \frac{\text{weight of wood tested} - \text{weight of dry wood}}{\text{weight of dry wood}} \times 100\%$$

Molecules – Any neutrally charged combination of two or more atoms that derive mutual benefit from the sharing of one or more electron pairs.

Monera – One of the five kingdoms, based on cell type, of prokaryotic organisms (sometimes separately listed as Eubacteria and Archaebacteria) that have a nutritional mode of absorption, photosynthesis or chemosynthesis, comprising the bacteria, blue-green algae, and various primitive pathogens.

Monoculture – A large assemblage of plants or animals of a single species (e.g. a corn field).

MSDS – See Material Safety Data Sheet.

MSW – See Municipal Solid Wastes.

MTBE – See Methyl Tertiary Butyl Ether.

Multiphase Reactor – A reaction vessel in which the reactants are present in more than one phase, for example solid/liquid, liquid/gas, or two or more immiscible liquids.

Multiple Hazard Wastes – Wastes falling into more than one of the following hazard categories: infectious, regulated chemical waste, and regulated radioactive waste.

Multiple Level Hearth Incinerator – A refractory-lined steel shell containing a series of vertically stacked flat hearths. A series of rabble arms plow the waste material successively across the hearths, causing it to work its way from top to bottom. Typically used for treatment of sludges and municipal wastes. However, it can also be used to incinerate gases and liquids.

Municipal Solid Wastes – The mixed solid wastes generated within a municipal jurisdiction that are collected by city or county waste haulers. After collection, they may be landfilled, recycled, composted, or incinerated by municipal government programs.

Mushrooming – Having the metal being plated onto a PWB exceed the thickness of the resist.

MWTA – See Medical Waste Tracking Act.

NAAQS – See National Ambient Air Quality Standards.

Nanofiltration – A pressure filtering process that uses a synthetic membrane with pore sizes down to 0.001 μm (micron).

NaPCP – See Sodium Pentachlorophenol.

Naphtha – A term applied to the flammable refined or unrefined petroleum fractions that are used for gasoline blending. It has a boiling point range between 30-60°C, an autoignition temperature of 287°C, and with an explosive limit between one and six percent in air.

NARM – A radioactive isotope produced by an accelerator.

National Ambient Air Quality Standards (NAAQS) – A part of the Clean Air Act that sets standards for the criteria pollutants in outside air throughout the country.

National Emissions Standards for Hazardous Air Pollutants (NESHAPs) – Emissions standards set by EPA for an air pollutant not covered by NAAQS that may cause an increase in deaths, or in serious, irreversible, or incapacitating illness.

National Pollutant Discharge Elimination System (NPDES) – A requirement of the CWA that discharges meet certain requirements prior to discharging waste to any water body. It sets the highest permissible effluent limits, by permit, prior to making any discharge.

National Priority List (NPL) – A list identifying those sites – identified through the use of the Hazard Ranking System model – that pose a relative risk to public health and the environment from hazardous substances in groundwater, surface water, air, and/or soil. To be eligible for a CERCLA cleanup, the site must be on the NPL.

Natural Oils, Fats, and Derivatives – Water insoluble substances of animal or vegetable origin.

Negative Photoresist – A photoresist where unexposed areas are developed and removed, resulting in a negative image.

NESHAPs – See National Emissions Standards for Hazardous Air Pollutants.

Netting – A process where yarns are coarsely interlaced, resulting in such fabrics as mosquito netting.

Neutralization – A term generally used to mean any chemical reaction in which water is formed by the interaction of an equal number of H^+ from an acid and OH^- from a base.

Neutron – A neutral particle found within the nuclei of all atoms except hydrogen. Like the proton, the neutron also has a mass of 1.67×10^{-24} g. Although it does not contribute to the atom's charge it does to its mass. Neutrons may be ejected from the nucleus by fission; called fast neutrons. Neutrons have tremendous penetrating power and a highly damaging effect on living tissue. Fast neutrons are responsible for propagating the chain reaction in atomic bombs, but can be slowed by a moderator for controlled use in a nuclear power reactor.

Newly Identified Waste – Waste materials identified or listed as hazardous by the EPA after November 8, 1984 and not subject to the Land Disposal Restrictions until new rules for those wastes are promulgated.

Nitriding – A heat treatment process that adds nitrogen to metal surfaces to increase their hardenability.

Nitrification – An aerobic process in which bacteria change ammonia and organic nitrogen into oxidized nitrogen, usually nitrate.

Nitrifying Bacteria – Bacteria that can change ammonia and organic nitrogen into oxidized nitrogen, usually nitrate.

Nitrocellulose Resin – A protective or decorative coating that dries by evaporation of an organic solvent, rather than by polymerization. The solvents used include ethanol, toluene, xylene, and butyl acetate together with nitrocellulose and alkyd resins to improve durability. Typically used for coatings on metal, paper products, textiles, plastics, furniture, and nail polish.

Nitrogen Fixation – The bacterial transform of atmospheric nitrogen, N_2, into ammonia, NH_3.

Noble Gases – The elements helium, neon, argon, krypton, xenon and radon found in group VIII on the Periodic Table. Nobel, in chemistry terminology describes elements that are either completely unreactive or react only to a limited extent.

Non-heat Set Inks – Inks that cure by absorption of the ink's carrier solvent (either water or volatile organic solvent) into the substrate.

Non-image Area – The area on a printing plate that does not print.

Non-locatable Mineral – Valuable inorganic substances such as coal, oil, oil shale, and gas; and rock such as building stone, phosphate, and pot ash that are not governed by the Mining Law of 1872 and, when occurring on public lands, are leased to individuals with the United States government retaining an ownership right in the form of a royalty.

Nonmetals – Any of the elements found above and to the right of the zig zag line that cuts across the Periodic Table. Common examples would include oxygen, nitrogen, phosphorous, carbon and the halogens.

Nonpoint Source Pollution – Pollution originating from widespread, intermittent, and poorly defined sources.

Nonpolar Covalent – Covalent bonds in which the centers of positive and negative coincide exactly.

Non-stoichiometric – A condition where the amount of another chemical present is not in the exact proportion required for a reaction. This deficiency may range from only slightly below the stiochiometric equivalent to nearly a complete absence of the chemical.

NORM – Natural occurring radioactive material. A NORM is a subset of NARM and refers to material not covered under the Atomic Energy Act whose radioactivity has been enhanced by mineral extraction or processing activities, e.g., production wastes from the oil and natural gas industries.

Normalizing – A heat treatment process which reduces the hardness of steel and improves its machinability.

NPDES – See National Pollutant Discharge Elimination System.

NPL – See National Priority List.

NRC – See Nuclear Regulatory Commission.

n-Type – Conducting areas on a semiconductor produced by implanting antimony, arsenic, phosphorus, or bismuth ions.

Nuclear Fuel Cycle – The cycle that includes all the steps in supplying, using, and processing fuel for nuclear reactors, including the disposal of wastes.

Nuclear Power Plant – Any device, machine, or assembly that converts nuclear energy into some other form of useful energy, such as mechanical or electrical power.

Nuclear Regulatory Commission (NRC) – In conjunction with the Atomic Energy Act of 1954, the NRC draft and enforce laws that govern all uses of nuclear power – military and civilian. The NRC is primarily responsible for civilian activities, while the DOE oversees most military activities.

Nucleus – The small (diameter = 1×10^{-13} cm) dense area in the center of the atom, where the protons and neutrons are located.

Obligate Aerobes – Also known as strict aerobes, they are bacteria that must have molecular oxygen, O_2, to survive.

Obligate Anaerobes – Also known as strict anaerobes, they are organisms that cannot survive in the presence of oxygen

Obsidian – A glass-like, silica rich, igneous rock formed on or at the earth's surface. The glass-like appearance results from rapid cooling of magma resulting in formation of microscopic sized mineral crystals.

Octet Theory – A theory proposed by Albrecht Kossell and advanced by G. N. Lewis, that except for elements close to hydrogen, all other elements attempt to gain eight electrons in their outer energy levels by either transferring or sharing electrons with another element.

Oil Spill Prevention and Liability Act of 1990 – Passed in response to the 1989 Exxon *Valdez* oil spill. Among its several provisions is the requirement that future tankers be double hulled to provide an extra margin of safety against punctures and spills. In addition, a scale was developed to determine a ship's retirement age called its OPA life. Once a tanker exceeds its OPA life, the Act states that it can no longer visit U.S. ports or be used to transport oil in U.S. waters.

Omnivores – Animals that obtain their growth requirements from both plant and animal sources.

Open Pit Mine – A large hole excavated into the earth's surface for the purpose of recovering deep ore. The waste rock above the ore is placed in piles on the surface.

Ore Chutes – Inclined or vertical tunnels bored in rock designed to transport, by means of gravity, broken ore into haulage vehicles or skips.

Ore Dressing – The preparation of ore for metal extraction by crushing and grinding processes.

Organ Meats – Meat other than red meat, including liver, kidneys, brain, heart, and intestines.

Organic Compounds – All hydrocarbons and their derivatives that contain the element carbon. Excludes certain carbon-containing compounds. See Inorganic compounds.

Organohalogen – Any aliphatic or aromatic organic compound that contains one or more halogens.

Organotrophs – Organisms whose biosynthesis requires an external source of organic carbon, such as other dead organisms.

Orphan Waste Streams – A waste stream generated during a cleanup activity that does not have a treatment method and, therefore, may have to be stored in perpetuity.

Osmotic Pressure – The pressure associated with the passage of water through a selectively permeable membrane, from the side with lower solute concentration to the side with the higher solute concentration. The pressure exerted by osmosis is considerable and accounts, in part, for the elevation of sap from the root hairs to the top of very tall trees.

Outcrops – The part of a rock formation that is exposed at the earth's surface.

Out-of-process Recycling – As opposed to in-process recycling where the waste is recycled back into the process, thereby reducing the amount of raw material purchased, out-of-process recycling reclaims the waste through onsite or off site treatment units.

Overburden – Waste rock that overlies valuable mineral deposits and must be removed prior to recovering ore. Generally refers to waste rock from surface mines.

Overspray – The paint spray that does not reach its intended target.

Oxidation – Originally, used only to describe the combination of an element with oxygen. Today, it is more broadly defined as any process in which one or more electrons is lost to another substance.

Oxidizing Agents – Substances that gain electrons during oxidation-reduction or redox reactions.

Oxygenated Fuels – Hydrocarbon fuel blends that are a requirement of the 1990 Clean Air Act Amendments and are intended to reduce carbon monoxide emissions from the automobile engine by incorporating either alcohols or ethers.

Ozonation – The process of applying ozone to a substance.

Ozone – Sometimes referred to as triatomic oxygen, O_3. It is an allotropic form of oxygen that is a strong oxidizing agent.

Ozonolysis – The oxidation of an organic material by ozone. See Ozonation.

P2 – See Pollution Prevention.

Pantone – A series of standard colors used in the printing industry.

Paper – A mat or web of cellulosic fibers used for printing, sanitary products, packaging, and specialty products.

Paper Additives – Materials used to control the papermaking process and improve the quality of the final sheet including biocides, retention aids, sizing agents, and strength agents.

Paper Fillers – Pigments such as clay, calcium carbonate, and titanium dioxide used to enhance printing properties of paper and replace (more expensive) pulp.

Parasites – Organisms that obtain their nutrition from another living organism. They do not generally kill their host, but they do weaken the host by robbing it of vital nutrition

Passivator – A substance that causes a loss of normal chemical activity.

Passive – The loss of normal chemical activity, such as what happens in an electrochemical cell when electrolysis causes an oxide coating to be formed on the electrode surface.

Pasture – A grazing area with good vegetative cover, usually enclosed by fences, sometimes irrigated.

Pathogens – Agents of disease or infection. They are usually microscopic organisms, such as bacteria, protozoa, or viruses. They multiply within the tissues of the host.

Pathological Wastes – Wastes consisting of body tissue removed in testing, surgery, or autopsy and can vary from a few drops of cellular fluid to amputated limbs.

Paunch Waste – The undigested stomach contents of a ruminant animal at the time of slaughter.

PCB – See Printed Circuit Board.

PCBs – See Polychloronated Biphenyls.

PCP – See Pentachlorophenol.

PEL – See Permissible Exposure Limits.

Pellicle – A thin plastic sheet that is mounted a short distance from the surface of a reticle to keep its surface clean.

Pentachlorophenol (PCP) – A water insoluble white solid at room temperature, that is typically dissolved in oil for use as a wood treatment. PCP solutions contain dioxins, although the most feared 2,3,7,8-tetrachlorodibenzo-p-dioxin has not been found.

Perfluorocarbon – A fluorocarbon compound in which all hydrogen atoms that are directly attached to the carbon atoms have been completely replaced by fluorine, e.g., tetrafluoromethane, CF_4.

Periodic Table – An arrangement of the known and human-made elements by symbol and organized in such a fashion that vertical rows form chemical families with similar chemical properties.

Permeability – The ability of rock to transmit water.

Permeate – The liquid (water) that has passed through a synthetic membrane.

Permissible Exposure Limits (PEL) – A time weighted average concentration of an airborne contaminant that a healthy worker may be exposed to 8 hours per day or 40 hours per week without suffering any adverse health effects. It is established by legal means and is enforceable by OSHA.

Pesticides – Lethal chemicals used to control pest populations.

Petrochemicals – Those substances derived from crude oil that have a high degree of purity and unique properties. Examples of petrochemicals would include the olefins: ethylene, propylene, and isobutylene and the aromatics: benzene, toluene and styrene.

Petroleum Coke – The relatively pure carbonaceous residue resulting from the pyrolysis of petroleum refining wastes. Petroleum coke is used in the production of aluminum and other substances such as silicon and calcium carbides.

pH – A logarithmic scale developed by Søren Sørensen in 1909, for expressing the degree of acidity/alkalinity of aqueous solutions. By definition, $pH = -\log_{10}[H_3O^+]$. An acid solution contains an excess of H_3O^+ and has a pH number that is less than seven. A basic solution contains an excess of OH^- and has a pH number that is more than seven. A pH = 7.00 only occurs in neutral solutions.

pH Meter – An instrument with either one or two electrodes design to measure a solution's H_3O^+ concentration and read directly as its pH.

pH Scale – An acid-base scale defined by $pH = -\log[H_3O^+]$. Solutions with pH < 7 are acid and pH > 7 are basic. Neutral solutions have pH = 7.

Photodegrading – The breakdown of a more complex substance into simpler substances as the result of exposure to light energy, *only*.

Photodehalogenation – The removal of a halogen (fluorine, chlorine, bromine or iodine) from an organic compound by the action of light.

Photolithography – The technique for making integrated and printed circuits by photographing the circuit pattern on a photosensitive substrate and chemically etching away the background.

Photolysis – The exposure of a substance to UV radiation, with the intent of causing it to photodegrade into simpler substances.

Photons – A unit of electromagnetic radiation. All forms of electromagnetic radiation are composed of photons with varying wavelengths. Photons move at the speed of light and have no rest mass.

Photoresist – A photosensitive liquid polymer that also serves as the etch resist after exposure to light and development.

Photo-tool – A mask used to determine which areas of the photoresist are struck by light and which are not.

Phototrophs – Organisms that obtain catabolic energy from sunlight.

Physical Properties – The unique set of properties possessed by each pure substance, like color, odor, hardness, malleability, ductility, electrical conductivity, density, etc. No two pure substances have the same exact set of physical properties.

Physical Treatment Technologies – Any of a number treatment processes that are designed to separate, concentrate, or remove a wastes based on their physical properties, but do not destroy its characteristics.

Phytoremediation – The use of selected varieties of plants to extract metals from contaminated soil or water. See Hyperaccumulators.

Pickling – A chemical treatment that removes oxide or scale from the surface of a metal, usually through the use of hydrochloric acid.

PICs – See Products of Incomplete Combustion.

Pigments – Coloring agents that are added to paint.

Pinenes – Any of a class of isomeric terpene hydrocarbons, $C_{10}H_{16}$, derived from sulfate wood turpentine and used as a solvent for protective coatings, polishes, and waxes.

Pipeline – Long tubular conduits, often placed underground, with pumps and valves for flow control. About one-third of all petroleum products are transported from refineries to consumers through a national network of over 220,000 miles (352,000 km) of pipelines.

Pits – The open connections between individual cells in wood.

Placer – A deposit formed by the transport and concentration of rock or minerals from their place of origin to another place by means of moving water, air, or ice.

Plasma – A neutral mixture of positively and negatively charged particles interacting within an electromagnetic field and reaching temperatures upward to 15,000°C (27,000°F).

Plasma Arc Torch – One of the electrodes and source of the plasma arc in a plasma arc incinerator.

Plasma Etching – Used to remove epoxy resin and/ or fiberglass from hole bores. The plasma is formed through the combination of an etch gas, such as oxygen and carbon tetrafluoride, and 200 to 2,000 watts of electrical power.

Plasma Thermal Treatment Unit – A copper-lined steel shell where the tip of the natural gas burner serves as one electrode and the copper cladding at the bottom of the reactor serves as the other. Boosted by the electrical discharge, the core of the burning gas-plasma may reach temperatures up to 15,000°C. It can be used to destroy mixed wastes such as PCBs and soils contaminated with large amounts of metals. Nonvolatile compounds are reduced to a non-leachable glassy product.

Plastic Media – The use of plastic media is known as plastic media blasting (PMB) and is considered to be a soft media blasting technique. Plastic media in three different hardness grades and six grain-sizes are available. Typically it is applied using conventional sand blasting equipment.

Plasticizers – A substance added to increase flexibility of a coating and facilitate its application.

Plate and Frame Filter Press – See Filter Press.

Plug Flow Reactor – A vessel through which pre-mixed reactants flow and in which a reaction takes place. Progress of the reaction corresponds with the position in the reactor.

PMN – See Premanufacture Notifications.

POCs – See Products of Complete Combustion.

POHCs – See Principal Organic Hazardous Constituents.

Polar Covalent – Covalent bonds in which the center of the positive charge is not in the center of the negative charge. This results in a molecule that has a partially positive and partially negative charged portion. Water is probably the most important polar covalent liquid.

Polar Covalent Molecule – A molecule formed by sharing one, two, or three pairs of electrons, in which the center of the shared electron cloud does not coincide with the center of the positive charge. This lopsidedness results in the electron rich end of the molecule becoming partially negative (δ-) and the electron deficient end of the molecule becoming partially positive (δ+).

Polish (Water) – A physical pretreatment method used to remove dissolved substances from water prior to use in such applications as electroplating.

Polishing – A surface treatment process which uses abrasives to remove or smooth out surface defects (scratches, pits, or tool marks) that affect the appearance or function of a part.

Pollution Prevention (P2) – Essentially synonymous with source reduction – reducing the generation of wastes or contaminants at the source, and thereby reducing releases to the environment. As defined by the PPA, it does not include out-of-process recycling, waste treatment, or combustion of wastes for energy recovery.

Pollution Prevention Act of 1990 (PPA) – The PPA focused industry, government, and public attention on reducing the amount of pollution through cost-effective changes in production, operation, and raw materials use. It also includes other practices that increase efficiency in the use of energy, water, or other natural recourses, and protect our resource base through conservation, and sustainable agriculture.

Pollution Prevention Opportunity Assessment – A tool used in a company's pollution prevention program to provide a baseline of waste and energy usage and to identify and evaluate opportunities for pollution and energy conservation.

Polychloronated Biphenyls (PCBs) – One of several aromatic compounds containing two benzene nuclei with two or more substituted chlorine atoms. They are toxic and persistent in nature; their manufacture was discontinued in the United States in 1976.

Polyelectrolyte – A coagulant composed of high molecular weight polymers of either natural (protein, gum arabic) or synthetic origin.

Polyesters Resins – The polycondensation product of dicarboxylic acids with dihydroxy alcohols.

Polyimide – Any of a group of polymers that have an imide group (–CONHCO–) in the polymer chain.

Polymerization – A process usually carried out with a catalyst, heat, or light, and often under high pressure, in which a large number of small molecules combine to form a chain-like large molecule.

Polyol-rich Liquids – Liquids containing alcohols that have three or more hydroxyl groups, e.g., glycerin.

Polyurethane Coatings – Two component paints in which a liquid isocyanate is combined with a polyol-rich liquid. The mixture forms a solid coating as the two constituents undergo a chemical reaction. Components can be mixed either before or during application of the coating.

Polyurethane Resins – A polymer composed of ethyl carbamate, $CO(NH_2)OC_2H_5$, as the typical repeating unit in a polyurethane resin.

Pomace – The fruit canning waste stream produced during middle processing stages, which is composed of seeds and peels and used as a high-grade animal feed supplement.

Porosity – A measurement of the pore size or interstices of a substance.

Porous Medium – Any substance that contains pores or interstices. When used as a filtering media, almost any water-insoluble porous material having a reasonable degree of rigidity. See Filtering medium.

Porous Pot – A device composed of a rectifier, a ceramic pot that houses a cathode (protecting it from direct contact with the process solution), and an anode that surrounds the pot and that is in direct contact with the bath. When placed in a spent permanganate desmear bath, it results in the anodic re-oxidation of the manganate ion, MnO_4^{2-}, back to the permanganate ion, MnO_4^-.

Positive Photoresist – A photoresist where exposed areas are developed and removed, resulting in a positive image.

Positrons – A fast-moving positively charged electron emitted from the nucleus of a radioisotope. Positrons are similar to beta particles, except for their charge.

Pot Life – The length of time that a paint remains a usable liquid in a spray system.

POTW – See Publicly Owned Treatment Works.

Powder Coatings – Finely-divided particles of epoxy, aromatic urethane, aliphatic urethane, polyester, or similar materials that are applied to an item and then melted in an oven to produce an impermeable coating adhering to the item.

PPA – See Pollution Prevention Act of 1990.

PPOA – See Pollution Prevention Opportunity Assessment.

Precipitants – Any chemical in combination with another chemical that results in the formation of an insoluble or slightly soluble substance.

Precipitate – Small particles that have settled out of a liquid or gaseous suspension by gravity. An insoluble salt that forms in a solution.

Pregnant Solution – Solutions containing dissolved metals.

Premanufacture Notifications (PMNs) – PMNs are required under Toxic Substance Control Act prior to the production of a new chemical substance or the review of a previously approved chemical for a significant new use to insure that they do not pose a risk.

Primary Pollutant – Substances that are air pollutants, which enter the atmosphere directly from a source.

Principal Organic Hazardous Constituents (POHCs) – Difficult-to-burn compounds that are easily detected. They are selected by a regulatory agency for each waste feed during a trial burn.

Print-and-Etch – A process for making the circuitry on inner layers of rigid multilayer PWBs. A copper-clad laminate is coated with a photoresist and exposed to light through a photo-tool. The exposed/developed photoresist is then removed and the copper below removed (subtracted) with an etchant. The remaining etch-resist is then remove, leaving only the circuitry that was beneath.

Printed Circuit Boards (PCB) – See Printed Wiring Board.

Printed Wiring Board (PWB) – A sheet or board of non-conducting substrate onto which the interconnecting conductors and some of the circuit components have been printed, electroplated, or etched. Also known as printed circuit board (PCB).

Printed Wiring Assemblies (PWA) – The combination of a printed wiring board (PWB) and its components: also known as an electronic assembly.

Process Bath – Tanks that contain chemical solutions that make up one part of a process, e.g., the electroplating bath.

Process Circuits – A group of processes, on a flow diagram, that perform a major function.

Product Life Cycle – The system that starts with acquisition of raw materials, includes all refinement and manufacturing processes, use and service, retirement, and disposal of residuals produced in each stage.

Product Yield – The amount of product formed from reactants. These yields are often expressed in terms of percentage yields relative to the maximum product that could be stoichiometrically formed from the reactants.

Production Risk – Risks that relate to the ability to maintain production rates, or minimize production losses from making a change.

Products of Complete Combustion – Substances that have reached their maximum combining capacity with oxygen, e.g. H_2O, CO_2, SO_3.

Products of Incomplete Combustion – Substances that have not reached their maximum combining capacity with oxygen, e.g. C, CO. When additional oxygen is supplied in a secondary combustion chamber, for example, these PICs both become CO_2.

Prokaryotic – A type of cell lacking a membrane-enclosed nucleus and other membrane enclosed organelles. Found only in the kingdom Monera, which includes all bacteria.

Protista – Members of the kingdom Protista.

Proton – The positively charged subatomic particle present in the nucleus of all atoms. No two elements can have the same number of protons. The number of protons determines the element's atomic number and name. The mass of the proton 1.67×10^{-24} g is the basis for the atomic mass unit (a.m.u.).

Protozoa – Protista that live primarily by ingesting food. A heterotropic, animal-like Protista.

p-Type – Nonconducting areas on a semiconductor produced by implanting boron, gallium, or indium ions.

Publicly Owned Treatment Works (POTW) – A waste treatment works owned by a state, local government unit, or Indian tribe, usually designed to treat domestic wastewaters.

Pug Mill – A machine in which materials are mixed, blended, or kneaded into a desired consistency.

Pulp – Fibers from vegetable matter, especially wood, that have been separated from each other by chemical or physical means or a combination of chemical and physical means.

Pump-and-Treat Technology – Groundwater treatment technologies that rely on pumping the contaminated water to the surface, followed by a chemical or biological treatment, e.g., precipitation or bioremediation.

Pure Substance – Matter that has a constant composition. A pure substance may be further subdivided into either elements or compounds.

Purification – A process that removes impurities that account for less than 3-5 percent of a process stream.

PWA – See Printed Wiring Assemblies.

PWB – See Printed Wiring Board.

Pyrite – A commonly occurring iron sulfide mineral (FeS_2). Commonly known as fool's gold.

Pyrolysis – A nonoxidative transformation of a substance into one or more other substances by heat alone.

Pyrolytic – Transformation of a substance into one or more other substances by heat alone. Similar to destructive distillation.

Pyrolytic Thermal Treatment Unit – A thermal treatment unit that is used to transform a substance into one or more other substances by heat alone, i.e., without oxidation. Pyrolytic treatment methods employ heating of a substance in the absence of oxygen or in starved amounts of oxygen.

Quarry – A shallow surface excavation where building materials and gravel are mined.

Quartz – A common silicon rich mineral with the formula SiO_2. Quartz is a major ingredient in glass.

Rabble Arm – A mechanical arm used to plow the waste material successively across the hearths while they are being heated.

Rack Plating – In the plating industry, rack plating refers to the use of a frame, called a rack, to hold a number of parts as they are automatically moved from the process bath to a series of rinse baths.

Radiation – The particles and/or energy that are released as the result of a radioisotope undergoing decay.

Radiation Curable Coating – Coatings composed of polyesters, acrylics, and epoxies that undergo a chemical cross-linking, when they are exposed to ultraviolet (UV), electron beam (EB), or infrared (IR) forms of electromagnetic radiation.

Radioactive Isotopes – Those isotopes of an element that have an unstable nucleus and emit subatomic particles and/or energy.

Radioactive Management Areas (RMAs) – A small area that has been designated for the temporary storage or processing of a radioactive waste. Free-release criteria apply to civilian RMAs, but DOE has not adopted free-release criteria and any addition of radiation, no matter how small, is grounds for treating the waste as a LLW.

Radioactivity – Atomic emission resulting from natural or artificial nuclear transformation. The energy of the process is emitted in the form of alpha particles, beta particles, or gamma rays. The amount of radioactivity or activity unit is the curie, which represents 3.73×10^{10} disintegrations per second.

Radiography – The production of radiographs or photographic images by the use of X-rays or nuclear radiation.

Radioisotopes – Also known as a radionuclides, they are the isotopic form of an element, either naturally occurring or artificially produced, that exhibits radioactivity. Radioisotopes are frequently used as diagnostic and therapeutic agents in medicine, in biological tracer studies, and for many industrial purposes.

Radionuclides – See Radioisotopes.

Range – An open grazing area usually covered with natural vegetation.

Rankine Temperature Scale – An absolute temperature scale related to the Fahrenheit scale like the Kelvin scale is related to the centigrade scale. Rankine temperatures are obtained by adding 459.7° to the Fahrenheit temperature.

RBC – See Rotating Biological Contactor.

RCRA – See Resource Conservation Recovery Act.

RCRA Subtitle C Exemption – A 1978 EPA proposal to exempt gas and oil muds and oil brines from the hazardous waste management standards. After further study, it was amended in 1980 to include drilling fluids, produced water, and any other wastes associated with exploration, development, or production (E & P) of crude oil or natural gas.

Reactor – A vessel in which a chemical reaction takes place.

Rebuttable Presumption Against Registration – A periodic review by EPA to determine of the continued use of a biocide presents benefits that outweigh its risks.

Reciprocating Grate Incinerator – A refractory-lined steel shell containing a series of hearth grate steps. Materials to be incinerated are introduced at the top step and as the steps slowly reciprocate, the burning materials are moved toward the bottom step. This type of incineration is typically used when the waste stream components vary greatly in size, such as is found in municipal trash.

Record of Decision – A public document that explains which cleanup alternative(s) will be used at an NPL site. The record of decision is based on information and technical analysis generated during the remedial investigation/feasibility study and consideration of public comments and community concerns.

Recycling – The series of activities by which products or other materials are diverted from a waste stream for use as raw materials in the manufacture of new products.

Red Bag Waste – Infectious waste commonly collected in red plastic bags used in hospitals.

Red Meat – Beef, pork, and lamb meat and their many associated byproducts such as liver, brains, kidneys, tongue, heart, and intestine, but not sausage.

Redox – A shortened combination of the terms oxidation-reduction. A type of chemical reaction in which the oxidizing agent gains one or more electrons from the reducing agent.

Reducing Agents – Substances that lose electrons during oxidation-reduction or redox reactions.

Reduction – Originally, only used to describe the combination of an element with hydrogen. Today, it is more broadly defined as any process in which one or more electrons are gained from another substance.

Refined Products – Those substances resulting from the refining of crude oil that are not considered petrochemicals due to a lower degree of purity and a wider range of boiling points. They are typically marketed for fuel blending and other applications.

Refinement – A manufacturing process that involves removing unwanted compounds from a material to adapt it for use.

Refinery – The facility where petroleum refining occurs. The processes centering around the fractional distillation of crude oil to produce naphtha, low-octane gasoline, kerosene, fuel oil, and asphaltic residues. It also includes the processes involved in catalytic and hydrocracking for the production of high-octane gasoline.

Refining (Smelting) – A smelting process that removes impurities from the metal or separates one metal from another.

Refining (Papermaking) – The separation of fibers in mechanical pulping or the surface modification of chemical pulp fibers for papermaking.

Refractory – An earthy, ceramic material of low thermal conductivity and capable of withstanding extremely high temperatures without changing its shape.

Regenerant – The solution containing concentrated ions resulting from the regeneration of an ion exchange resin.

Remanufacture – Refers to repair, rework, or refurbishment of a component, part, or product.

Remediation Wastes – Wastes composed of the investigation derived remedy and site cleanup stages of a corrective action.

Rendering – The process of removing the fat from meat and by-products by using steam or heat prior to land disposal.

Residue – The large molecules resulting from petroleum refining and used for tar and asphalt surfacing and in the production of petroleum coke.

Resin – A semisolid or solid complex amorphous mix of organic compounds. It is used in ion-exchange units to hold the exchange ions until contact is made with the unwanted ions.

Resource Conservation Recovery Act (RCRA) – A federal law enacted in 1976 to deal with both municipal and hazardous waste problems and to encourage resource recovery and recycling.

Respiration – The aerobic or anaerobic harvest of energy from food within a cell.

Responsible Care® – A voluntary Chemical Manufacturers Association program intended to promote continuous improvement in health, safety, and environmental quality within the chemical industry.

Restricted Waste – Wastes that may not be disposed of to the land, without meeting the current treatment standards.

Retention – The amount of preservative left in wood after treatment. Usually expressed in units of weight per volume, i.e., pounds per cubic foot (pcf) or kilograms per cubic meter (kg/m^3).

Retention Time – The average time that reactants remain in a reactor.

Reticle – A mask used in photolithography, which is one to ten times the actual size of the pattern they produce.

Retort (Distillation) – A vessel in which ore is heated during a distillation extraction process.

Retort (Wood Treatment) – A pressure cylinder used for the pressure treatment of wood. Synonyms: treatment cylinder or pressure vessel.

Retort (Food) – A pressurized container used to heat canned food after sealing.

Reuse – The additional use of a component, part, or product after it has been retired from its original use.

Reverse Osmosis – The application of pressure to reverse the normal flow of water across a semipermeable membrane from an area of higher solute concentration into an area with less solute concentration. The process can be used to remove molecules and ions that are larger than water molecules (>0.0003 μm), e.g., sodium and chloride ions. Pressures required may range from 800-1,000 psi.

Rhodochrosite – A mineral composed of manganese carbonate.

RMAs – See Radioactive Management Areas.

Roasting – A thermal process commonly used in metal extraction; it is typically used to change the oxidation state of elements in the mineral.

Rock – A solid composed of minerals. Rocks are classified based on the way they are formed and the minerals that compose them.

Rock Bolts – Rods inserted and cemented with resins into holes drilled in rock. They increasing the stability of the back and sides of underground mine openings.

ROD – See Record of Decision.

Rod Mill – A mill in which the grinding media are iron or steel rods.

Rodenticides – Pesticides used to kill rodents (e.g., rats, mice, gophers).

Rolling – A metal forming process in which metal is passed through a set or series of rollers that bend and form the part into the desired shape.

Room-and-Pillar – An underground mining method that involves excavating and remove ore from large areas called rooms while leaving blocks of ore called pillars between the rooms to support the roof.

Rosin – Obtained form pine trees and composed of resin acids of the abietic and pimaric types. See Abietic Acid.

Rotary Bit – A diamond encrusted, tempered steel bit used to bore a well

Rotary Drilling Rig – The major components consist of a tower or derrick, sections of threaded pipe and a rotating drill bit. The drill bit is cooled and lubricated by drilling mud that is pumped through a Kelly into the top of the drill string. The drill mud, containing the cuttings, returns up the annulus between the wall of the well and the drill string where it is physically separated and the drilling mud recycled.

Rotary Kiln – Considered to be one of the most versatile types of incinerators. A long cylindrical, refractory-lined shell mounted at a slight incline from horizontal. Solid wastes are introduced into the upper end and/or liquids may be injected through auxiliary nozzles. Rotation of the outer shell enhances mixing with air and promotes combustion. Typically, kilns are fitted with secondary combustion chambers to insure more complete combustion.

Rotary-drum Vacuum Filtration – A rotary-drum vacuum filtration unit is composed of a compartmentalized, porous cylinder covered by a filtering medium. By application of a vacuum to the rotating drum, a layer of sludge forms on the surface and is dewatered prior to its removal.

Rotating Biological Contactor – Also known as an RBC, it is an aerobic fixed-film biological treatment method. It is composed of a series of discs that are coated with a microbial slime and are rotated through a trough containing the waste. Due to the rotation, the disks spend approximately one-half of their time exposed to the air.

Roundstock – Cut but untreated tree trunks that have the bark removed and will be used for such things as utility poles.

RPAR – See Rebuttable Presumption Against Registration.

Rubblize – To make into rubble.

Rueping Tank – A vessel, usually placed above the retort, that is used to store treating solution or, when Boultonizing, to prevent vapor locks in the retort.

Ruminants – Grazing animals with multi-chambered stomachs that are able to digest grasses and other plants with low nutritive value to humans (e.g., cows and sheep).

Runoff – That portion of a precipitation event that flows on the earth's surface.

Run-of-mine Rock – The rock that comes from the mine before its size is reduced.

Sacrificial Layer – A layer that causes preferential corrosion of a metal coating for the sake of protecting the substrate metal, such as a zinc coating used to protect steel. Also known as sacrificial protection.

Salt – Any ionic substance that dissociates in water producing a positive ion other than a hydrogen ion, and a negative ion other than a hydroxide ion, e.g. NaCl.

Sandblasting – The process of using air-driven grains of sand, metal shot, slag, pumice, or other media to remove the surface coating and prepare the workpiece surface.

Sap Stain – Fungi that attack and often discolor wood, but do not cause loss of strength or decay. Typically, they are limited in extent to the sapwood, leaving the heartwood unaffected.

Saponifiers – Chemical substances such as NaOH and KOH that can convert fats, esters, and long-chained carboxylic into water soluble salts.

Sapwood – The outer portion of a tree which contains living cells. It is typically more permeable and, therefore, more easily treated than heartwood.

SARA – See Superfund Amendment and Reauthorization Act.

Save Money and Reduce Toxics Program – Chevron's SMART program was initiated in 1987 with the focus of industrial source reduction, toxic chemical use substitution, and recycling for hazardous and nonhazardous solid wastes.

SBR – See Sequencing Batch Reactor.

SC1 Solution – A solution composed of ammonia, hydrogen peroxide, and DI water that was developed by RCA in 1970 to remove organic impurities and particulate matter from a silicon wafer surface.

SC2 Solution – A solution composed of hydrochloric acid and hydrogen peroxide that was developed by RCA in 1970 to cause a new natural oxide layer to form on a silicon wafer surface.

Scouring – A wet cleaning process using chemical, mechanical, or both to remove impurities from fiber.

Screen – A rectangular structure containing openings of consistent size. Typically used in series to separate broken rock or gravel into fractions of same or similar sized materials.

Screen Printing – A printing process wherein mesh is stretched over a frame. The non-image area of the mesh is blocked, so that as a rubber squeegee is drawn across the screen, ink is forced through the image area, which is not blocked, onto the substrate.

Scrubber – A device used for the purification of a dirty gaseous waste stream.

Sealers – An organic substance that is soft enough to spread on the substrate and hardens to form a permanent bond with the substrate.

Secondary Pollutants – Substances, such as ozone, that are air pollutants but were not released directly into the air. They are the product(s) of chemical reaction(s) occurring on some other substance that was released to the atmosphere.

Sedimentation – Sedimentation is the settling of solids from a liquid suspension driven by the force of gravity.

Seed Crystal – The minute crystal used as the nucleating body when growing crystals from saturated liquors or vapors. The seed crystal provides the organizational structure around which more of the substance adds.

Seed Layer – A conductive material coating (usually electroless copper) that is deposited in the hole barrels before the electrolytic copper plating process.

Seismograph An instrument for measuring and recording the vibrations of the earth.

Semibatch – A process characterized by the continuous addition of some reactant in addition to chemicals initially added to a batch reactor.

Semiconductor – Any of various solid crystalline substances, such as silicon or germanium, having more electrical conductivity than an insulator, but less than a good conductor such as copper or gold.

Semipermeable Membrane – A microporous structure, either natural or synthetic, which acts as a highly efficient filter in the range of molecular dimensions, allowing passage of ions, water and other solvents, and very small molecules, but impermeable to colloidal or larger particles.

Sensitizer – An electrically conducting substance that is applied to wood when using an electrostatic spray gun system.

Sequencing Batch Reactor (SBR) – This process is a modification of the activated sludge process. The reactor vessel, containing idle microorganisms, is filled with wastewater and allowed to react while being aerated and mixed. After a specified time the batch is allowed to settle and the clear liquid removed. Part of the sludge is wasted, and a part left for seed to the next batch.

Service Industry – A business that provides services instead of manufacturing a product.

Sewage – Human or animal waste products collected for waste treatment and disposal.

Sewer Screenings – A fruit processing waste stream composed of residual peels, leaves, stems, dirt, and metals removed from the fruit in the washing process.

Sewing – A process where a threaded needle is passed through and used to stitch the yarn into a solid fabric.

Shaft – A vertical hole in the earth's surface constructed to access underground ore deposits.

Shallow Land Burial – Disposal of low-level radioactive waste in trenches just below ground level.

Shapemaking – A metalworking or manufacturing process that forms materials into the shape of the product. Shapemaking processes can preserve all the material, as in extrusion, or remove material such as in machining or grinding.

Shaping – See Shapemaking.

Sharps – Any metal or glass objects such as needles, scalpels, or sampling pipettes, that can cause a puncture, cut, or abrasion through which infection can pass.

Shear Zone – The area on a belt filter where the belt-sludge-belt sandwich pass through an S-shaped series of rollers that applies progressively increasing pressure.

Shearing Processes – A metal shaping processes that cut away excess material from the workpiece.

Shrinkage Stoping – An underground mining method involving the upward excavation of a stope utilizing previously shot material as a platform for continued excavation of the back.

SIC Codes – Standard Industrial Classification Codes used to classify industries for statistical and regulatory purposes based on their primary functions and products.

Silicide – A compound of two elements, one of which is silicon.

Silicon – A semiconducting nonmetallic element. It is soluble in a mixture of nitric and hydrofluoric acids and alkalis. It combines with oxygen to form SiO_2, which is the major component of silica and silicate rocks, such as quartz and sand.

Six Nines – Refers to the 99.9999% removal and destruction efficiency of a hazardous material. Set as the EPA (TSCA) standard for incinerators permitted to burn chlorinated hydrocarbons like PCBs.

Sizing – The application of a compound such as a polymer or oil to a yarn or fiber to improve or increase its stiffness, strength, smoothness, or weight.

Skips – Bucket-like containers used to elevate ore from the lower depths of an underground mine, up a shaft to the surface.

Slag – A layer of non-metallic elements formed during smelting; a waste.

Slagging – The formation of a non-combustible residue during incineration. It is usually composed of metal oxides and silicates.

Slaked – The process of adding calcium oxide, CaO, to water to produce calcium hydroxide, $Ca(OH)_2$.

Slant Tube Clarifier – A clarifier that has a series of two inch square parallel tubes that are set at an angle of between 45° and 60° to the surface of the liquid. The liquid enters at the bottom and flows upward through the clarifier. Solids removal takes place within the tube section.

Sludge Wasting – Refers to the elimination of a part of the sludge-solids form a suspended growth treatment method. What biomass is not returned for seed is wasted.

Sluice Box – A classifying device that uses running water and a series of riffles to separate gold from sand and gravel.

Slurry – A mixture of solids and liquid, in which the solids are suspended in the liquid medium.

Small Quantity Generator – A category of hazardous waste generators established by RCRA based on production of more than 100 kg but less than 1,000 kg of hazardous waste per site per month, and accumulation of less than 6,000 kg at any one time.

SMART – See Save Money and Reduce Toxics Program.

Smear – The nonconductive epoxy-resin that was softened by a hot drill bit and smeared across the inner copper layers in the hole barrel.

Smelt – The inorganic components present in strong black liquor and recovered as a solid from the bottom of the recovery boiler. Smelt is primarily composed of sodium carbonate, Na_2CO_3, and sodium sulfide, Na_2S. When smelt is dissolved into water, the resultant solution is known as green liquor.

Smelting – An extraction process that uses heat to melt and separate metal-containing minerals.

SMOBC – See Solder-mask-over-bare-copper.

SMT – See Surface Mount Technology.

Smut – A general term referring to any finely divided base metal particles, smudges, or sooty-looking matter on the surface of the workpiece.

Sodium Borohydride Precipitation – A mild reducing agent, $NaBH_4$ can cause the precipitation of certain soluble metal ions by converting them into their insoluble metal form. This process can be used for removal of lead, mercury, gold, platinum and nickel from waste solutions.

Sodium Hydrogen Sulfite – Also known as sodium bisulfite, $NaHSO_3$. An economical reducing agent.

Sodium Hypochlorite – The chemical substance responsible for the action of common bleach. The hypochlorite ion, ClO^-, is the active ingredient and is a strong oxidizing agent.

Sodium Penta – See Sodium Pentachlorophenol.

Sodium Pentachlorophenol (NaPCP) – A water soluble salt, $C_6Cl_5O^- Na^+$, that in the past was the most often applied sap stain control treatment. It was applied by either spraying or dipping fresh cut lumber, but its use and conversion to pentachlorophenol lead to wide-spread contamination.

Soft Drinks – Nonalcoholic drinks typically composed of 90 percent purified water, 9 percent flavoring, and 1 percent carbon dioxide. There are 450 different types of soft drinks sold and they account for 27 percent of all beverages consumed in the US.

Soft-baked – The thermal process for hardening the photoresist not removed by exposure and development.

Softwood – Wood from gymnosperms (the conifers or evergreen trees often with needle-shaped leaves).

Soil Depletion – A process whereby soil nutrients and organic matter are used up.

Soil Erosion – Mobilization of soil particles by either wind or water.

Solder – A low-melting alloy usually of the lead-tin type used for joining metals at temperatures below 425°C. The solder acts as an adhesive and does not form an intermetallic solution with the metals being joined.

Solderability – Possessing the characteristics that allow it to be soldered.

Soldering – A metal bonding process that fastens metals together with a nonferrous metal that has a low melting point (<800°F).

Solder-mask-over-bare-copper (SMOBC) – A PWB surface finishing method where tin-lead is used as both the etch-resist and finish on a bare copper surface.

Solidification – See Immobilization.

Soluble – The ability of one substance to dissolve into another. For general purposes, salts that exceed 1 g/100 ml of solution are said to be soluble.

Solute – The substance, typically water, doing the dissolving or present in the greater amount.

Solution Mining – An *in situ* mining method that involves pumping reagents into an ore deposit and recovering solutions containing dissolved ore.

Solvated – The accumulation of up to six water molecules around a cation or anion. Typically represented by $Na^+_{(aq)}$ or $Cl^-_{(aq)}$.

Solvating Agents – Substance used to dissolve and put another substance into a solution.

Solvent – The volatile portion of a liquid coating. When the liquid coating is applied to the furniture, the solvent dissolves and dilutes the resins onto the substrate and evaporates.

Solvent Cleaning – A surface preparation process that uses organic solvents to remove unwanted grease, oil, or other organic films from metal surfaces.

Solvent-based Paint – Paints in which the pigment is carried by an organic solvent.

Sorbate – The substance that is absorbed, adsorbed, or which entraps another substance.

Sorbent – A substance that absorbs, adsorbs, or entraps another substance.

Sorption – The process of absorption, adsorption or entrapment of one substance by another.

Sour Crude – As used in the petroleum industry, sour crude refers to oil that has greater than 0.5 percent by mass sulfur present.

Source Reduction – Considered to be synonymous with pollution prevention, it is any practice that reduces the amount of hazardous substance or pollutant generated by making equipment or technology modifications, process or procedure modifications, reformulation or redesign of products, substitution of raw materials, and improvements in housekeeping, maintenance, training or inventory control.

SPC – See Statistical Process Control.

SPCC – See Spill Prevention Control and Countermeasures Plan.

Species – A metal impurity found in a melt.

Spent Fuel – Nuclear fuel elements (rods) that have been removed from a power reactor after having been used to produce power. Spent fuels are highly radioactive and contained in fission products.

Spill Prevention Control and Counter-measure Plan – Required under the Clean Water Act for any industry that owns or operates large above-ground oil storage tanks to establish an allowable oil discharge limit.

Spin, Rinse, and Dryer (SRD) – A machine designed to spin, rinse, and dry, with nitrogen gas, the wafers between IC production steps.

Spinning (Shapemaking) – A metal forming process in which pressure is applied to a shape while it spins on a rotating form, forcing the metal to acquire the form's shape.

Spinning (Textile) – The process of converting short lengths of fibers into a continuous thread or yarn.

Spoil – A term that refers to waste rock from a mine. Commonly used to refer to waste rock from coal mines.

Spoil Pile – An accumulation of waste rock from a mine.

Spray Booth – An enclosed, ventilated area with an air velocity of 100 cfm at the face of the booth and fitted with filters to prevent paint overspray from traveling out the exhaust vent. Waterwash booths provide a continuous sheet of water down the face of the rear booth panel. The sheeting water collects the overspray from the painting operation and the particulates can then be collected as a paint sludge.

SQG – See Small Quantity Generator.

Square Set Stoping – An underground mining method that uses timbers to support the back.

SRD – See Spin, Rinse, and Dryer.

Stabilization – See Immobilization.

Stains – An organic protective coating similar to a paint, but with much lower solids content.

Standard Industrial Classification Codes – See SIC Codes.

Starved Oxygen – An atmosphere where the amount of oxygen present is not in the exact proportion required for the reaction. This deficiency may range from only slightly below the stiochiometric equivalent to nearly a complete absence of oxygen.

Statistical Process Control (SPC) – A pollution prevention technique that involves performing analytical work and record keeping on process baths as a potential cost reduction measure.

Steam Conditioning – A moisture removal process that uses live steam to heat the wood, then a vacuum to flash the water into steam.

Steam Sterilization – Exposure of waste to a water saturated atmosphere at elevated temperatures and pressures for sufficient time to kill harmful microorganisms.

Still – A device for the conversation of a substance from the solid or liquid state into the vapor state. A term derived from distillation.

Stoichiometric Conditions – The exact proportion of reacting chemical substances required for a given reaction.

Stope – A room or large open area underground from which ore is excavated.

Strainer – A device for straining, sifting, or filtering. May be constructed of metal mesh with various hole size.

Stratospheric Ozone Layer – Ozone in the upper atmosphere (stratosphere) that protects the earth from much of the ultraviolet light emitted by the sun.

Strike – To form a thin preliminary deposit on an article in an electroplating bath at a rapid rate preliminary to a longer and slower deposition.

Stringfellow Acid Pits – Located in Riverside County, California, it was placed on the NPL listing on 9/1/83. The 17 acre site served as a hazardous waste disposal facility from 1956-1972. Over 34,000,000 gallons of industrial waste, primarily from metal finishing, electroplating and pesticide production were deposited in its evaporating ponds. Heavy rain storms in 1969 and again in 1978 resulted in an overflow of these ponds and heavy contamination of nearby streams.

Strip Mines – Shallow surface mining method that involves removing waste rock from one area of the mine and disposing of it in an area that has been previously mined.

Stripping – Processes that use either mechanical or chemical action to remove surface coatings or surface layers.

Structural Treatment – A manufacturing process used to alter the product materials to enhance one or more of its characteristics or extend its useful life.

Sub-level Stoping – An underground mining method involving the excavation of a large vertical stope from several levels. Ore is drilled and shot at each level and gravity fed to ore chutes beneath the stope.

Sublimation – The direct change of a substance from the solid to the gaseous phase without passing through the liquid phase.

Substantive Requirements – These requirements specify actions that must be performed and are often borrowed from other applicable federal legislation.

Substrate – A specific substance on which an enzyme acts; a food.

Subtractive Plating Process – See Print-and-Etch.

Sulfide Ore – Ore formed in a reducing environment that contains sulfur anions instead of oxygen. Sulfide ores usually contain high levels of several metals.

Sulfide Precipitation – The use of sodium sulfide, Na_2S, ferrous sulfide, FeS or other soluble sulfide, at a pH ≥ 8, to form highly insoluble metal sulfides.

Sulfuric Acid – CAS: 7664-93-9. The most widely used industrial chemical. A dense, oily liquid that is strongly corrosive. Now frequently used as the acid

catalyst in the alkylation (reforming) process in place of the more toxic hydrofluoric acid.

Supercritical Water Oxidation – A process resembling the wet air oxidative method, except it uses water above its critical temperature (374°C) and pressure (3,200 psi) to oxidize the waste. This technology has been demonstrated, but it is still considered to be a new treatment method.

Superfund – See Superfund Amendment and Reauthorization Act.

Superfund Amendment and Reauthorization Act (SARA) – A 1986 amendment to CERCLA that added strict cleanup standards strongly favoring permanent remedies, stronger EPA control over the process of reaching settlement with responsible parties, a mandatory schedule for initiation of cleanup work and studies, individual assessments of the potential threat to human health posed by each waste site, and increased state and public involvement in the cleanup decision-making process. A separate Title III of SARA provides a framework for emergency planning and preparedness and requires chemical manufacturers to provide community groups with information on their inventories of hazardous chemicals and releases of chemicals to the environment.

Superfund Site – See National Priority List.

Super Tankers – A class of very large ships (>70,000 dead weight tons) that are capable of carrying more than 4 million barrels of oil.

Supported Stoping – An underground mining method in which rock bolts are used to support the development of a stope.

Surface Impoundments – Also known as lagoons or ponds, they are systems in which microbial oxidation, photosynthesis and sometimes anaerobic digestion is combined to breakdown organic compounds. They are similar to activated sludge units, without sludge recycling.

Surface Mount Technology (SMT) – The attachment of ICs directly to the PWB, without the aid of holes or connector leads. It allows components to be more densely packed on the board and eliminates the need for drilling holes.

Surfacing – A manufacturing process used to alter the surface of the part for a specific function such as corrosion protection, wear resistance, or improved appearance.

Surfactant – Any compound that reduces surface tension when dissolved in water. They are also known as detergents, wetting agents, and emulsifiers.

Suspended Growth – Those aerobic and anaerobic biological treatment processes that rely on unattached zoogleal masses to consume the organic wastes. Treatment methods relying on suspended growths include activated sludge, surface impoundments, SBRs, anaerobic digestion, fluidized bed reactors, etc.

Sustainable Development – An economic activity that increases prosperity without the destruction of the environment.

Synthetic Lubricants – Chemical compounds that are used in metal shaping operations. They are manufactured to provide a slippery medium that reduces friction as surfaces move past each other.

TAAs – See Temporary Accumulation Areas.

Tailings – Non metal bearing minerals left after separation processes; a waste.

Tailings Pond – Large evaporation pond where waste minerals are pumped.

T$_B$ – See Watson Boiling Point.

TBC – See To Be Considered.

TBTO – See Tributyl Tin Oxide.

TCE – See Trichloroethylene.

Toxicity Characteristic Leaching Procedure – An EPA leaching test used to determine if heavy metals are sufficiently immobilized.

TCLP – See Toxicity Characteristic Leaching Procedure.

Technical Risk – Risks that relate to the characteristics of technology being used to solve a problem.

Tempering – A heat treatment process used after hardening to relieve the internal stain in hardened steel and increase its toughness.

Temporaty Accumulation Areas (TAAs) – A TSD facility, falling under the 90 day accumulation rule, as designated by the CERCLA off site policy.

Terpenes – Unsaturated hydrocarbons occurring in most essential oils and oleoresins of plants. The terpenes are based on the isoprene unit, C_5H_8, and may be either acyclic or cyclic with one or more benzenoid groups, e.g., pinene.

Tertiary Treatment – Used after secondary treatment to remove remaining unwanted substances, e.g., phosphates, nitrates, and heavy metals. Depending on the substances to be removed, activated carbon, ion exchange, reverse osmosis, electrodialysis, ultrafiltration, molecular sieves, etc. may be used.

Thermal Cracking – See Coking.

Thermal Treatment Technology – Synonymous with thermal destruction. Any of a variety of treatment technologies that use heat (combustion) or high temperatures (wet air oxidation) to destroy the waste.

Thermal Treatment Unit – Any of a number of different pieces of equipment that use heat to destroy or makes a waste less hazardous. For hazardous wastes, incineration is the most commonly used treatment, however, it may also include pyrolysis and wet oxidative treatment processes.

Thickener – A type of bowl classifier, often used in the treatment of sewage.

Three T's – Time, Temperature and Turbulence: In combustion technology, sufficient time coupled with high temperatures and enough turbulence to promote adequate mixing of the fuel and oxygen are considered to be the three elements essential to the complete destruction of a substance.

Threshold Planning Quantities (TPQ) – The quantity of a substance that is required to have a Material Safety Data Sheet under the OSHA Hazard Communication Standard and that was reported under SARA Title III amendment during the previous calendar year.

Tillage – Process that moves soil, especially cultivation.

Timbering – The use of lumber to support the back and sides of underground mines.

Timbers – Sawn lumber products equal to or greater than 4 inches in the smaller dimension.

To Be Considered (TBC) – Requirements that are neither applicable nor relevant and appropriate, but are still binding on a cleanup operation.

Tolling – The process in which a recycler picks up a waste from a generator, treats it, and returns it to the generator for a fee.

Toner – An organic pigment that does not contain an inorganic pigment or carrying base.

Topcoats – The last coat of paint or other substance that is applied to a surface.

Total Quality Management (TQM) – A broad management approach used to ensure quality in an operation by focusing on long-term goals. The key elements of TQM include a focus on customer, multi-disciplinary teamwork, and working in cooperation with suppliers.

Toxic Release Inventory – A list of seventeen pollutants that serve as the reduction targets for the 33/50 Program.

Toxic Substance Control Act (TSCA) – A 1976 federal act enacted to control the risks of chemical substances entering commerce. The act deals with both newly created chemicals and chemicals entering into commerce that may do serious damage to humans and the environment before their potential danger is known and with existing chemicals that may require more stringent control.

Toxicity Characteristic Leaching Procedure – A toxicity characteristic test based on a leaching procedure that replaced the similar EP toxicity test after June 1, 1990. All waste generators must now use the TCLP to determine whether a waste is hazardous, due to the toxicity characteristic.

TPQ – See Threshold Planning Quantities.

TQM – See Total Quality Management.

Tractor-Trailers – A combination of a 18-wheeled truck and trailer, equipped with large tanks.

Tramp – An unwanted metal that threatens to contaminate a process or product.

Transfer Efficiency – The ratio of the amount of material used in a painting operation to the amount actually sticking to the part being painted.

Transuranic Waste – Waste material contaminated with alpha-emitting isotopes having atomic numbers above uranium (>92) that have half-lives greater than 20 years and are in concentrations greater than 100 nanocuries/gram.

Treatment, Storage, or Disposal Facility (TSD) – Site where a hazardous waste is treated, stored, or disposed. TSDs are regulated by EPA and states under RCRA.

TRI – See Toxic Release Inventory.

Trial Burn – A test of an incinerator's ability to meet all applicable performance standards when burning a hazardous waste under specified worst case conditions. The parameters demonstrated in the trial burn become the parameters specified in its permit.

Tributyl Tin Oxide (TBTO) – A compound that provides intermediate wood protection and for years was used to treat millwork items. For years TBTO, $(C_4H_9)_3SnO$, was dissolved in petroleum solvents, but its application is now being studied because of the resulting VOC emissions.

Trichloroethylene (TCE) – A stable, colorless, photoreactive liquid commonly used as a de-greaser, $CHCl=CCl_2$.

Trickling Filter – A fixed-film secondary treatment method, where wastewater is allowed to trickle over a solid media that is covered with a zoogleal slime. Organic wastes are removed from the wastewater by aerobic and anaerobic bacteria and other microorganisms.

TRU – See Transuranic Waste.

TSCA – See Toxic Substances Control Act.

TSD – See Treatment, Storage, and Disposal Facility.

TTU – See Thermal Treatment Unit.

Tuft – A fluffy cluster of particles, usually as the result of the addition of a coagulant.

Ultrafiltration – A pressure filtering process that uses a synthetic membrane with a pore size small enough (< 0.1 μm) to remove bacteria, microscopic solids, and oils.

Uniform Hazardous Waste Manifest – A document required by RCRA when shipping hazardous wastes. It lists EPA identification numbers of the shipper, carrier, and the designated treatment, storage, and disposal facility, in addition to the standard information required by DOT.

Unit Operations – A term used in process analysis to designate processes that have identifiable inputs, outputs, and functions and can thus be analyzed independently.

Uranium – A metallic element that has three naturally occurring radioactive isotopes: ^{234}U (0.006%), ^{235}U (0.7%) and ^{238}U (99%). It is removed from its ore under oxidizing conditions and processed into uranium dioxide, UO_2, also known as yellowcake (see Yellowcake).

Use Cluster – A new approach being used within the DfE's CTSAs to compare the risk, performance, and cost trade-offs of alternative chemicals, processes, and technologies.

Useful Life – The period during which a product operates safely and meets performance standards when maintained and used properly.

UTS – See Universal Treatment Standards.

UV-curable Inks – Inks that cure upon exposure to ultraviolet light.

Vacuum Evaporator – Evaporators that make use of the principle that water will boil at temperatures below 212°F, when placed under varying degrees of vacuum.

Vadose Zone – The layer(s) of rock and soils that lie between the earth's surface and an aquifer.

Vapor Degreasing – A solvent cleaning method in which the part to be cleaned is suspended in a vapor phase solvent (usually a nonflammable, chlorinated hydrocarbon). The part is cleaned as the solvent condenses on the cooler metal surface.

Vapor Pressure – The pressure generated by the vapor above a liquid at a given temperature.

Vapor Recompression Evaporators – Evaporators that use steam to vaporize a liquid, then the vapors are compressed and heated. These vapors are then used as the heat source for vaporizing additional liquid.

Venturi – A narrow central portion, or throat, in a tube-like area. As a gas or fluid moves through the constricted area its rate of flow is increased and its pressure decreased.

Vias – Holes drilled through a PWB for making layer-to-layer electrical interconnections. The vias are plated with copper to complete the conductive pathways between layers.

Viscosity – The internal resistance of flow exhibited by a fluid. Liquids with high viscosity are thick, slow pouring, honey-like substances.

VOC – See Volatile Organic Compound.

Volatile Organic Compound – As defined by EPA in the Clean Air Act, VOCs are certain carbon-containing compounds that, when released into the atmosphere, interact with sunlight and other pollutants to produce ground level ozone.

WAC – See Waste Acceptance Criteria.

Wafer Fabrication Processes – The steps involved in these processes are oxidation to form a surface film of silicon dioxide, pattern transfer and imprinting using photolithography, etching, doping, and layering.

Wafers – The thin slices of semiconductor, such as silicon, used as a base material on which a single transistor or integrated circuit is formed.

Warp Yarns – The set of yarns placed lengthwise in the loom, crossed by and interlaced with the weft, forming the lengthwise threads in a woven fabric.

Washcoats – One of several finish coats, such as diluted sealer or shellac, that allows for proper stain penetration and a good finish.

Waste – A term that refers to a substance that is no longer useful, such as waste rock from a mine.

Waste Acceptance Criteria – The criteria that has been established to ensure that a waste sent to a facility meets the RCRA permit requirements for treatment and disposal.

Waste Audit (Medical) – A procedure that collects information on waste management practices to allow for evaluation of the management system.

Waste Dumps – Piles of worthless rock resulting from mining activities located on the earth's surface.

Waste Exchange – An exchange program based on the idea that waste generated by one industry may be a feedstock for another. Both informational and materials exchanges exist.

Waste Minimization – When process change fails to eliminate a waste stream, other methods need to be found to reduce its environmental impact. Waste minimization audits, good housekeeping practices, inventory control, recycling, reuse, waste exchanges and employee training are examples of other waste minimization strategies.

Waste Reduction Always Pays – Dow's WRAP program was started in 1986, with a focus on industrial source reduction and on-site recycling of SARA Title 313 reportable compounds.

Wastewater Treatment Methods – The vast majority of wastewater treatment technologies are biological in nature. They are employed after pretreatment and primary treatment methods have removed the larger debris. The goal of secondary treatment is to remove the dissolved and suspended organic matter. In some instances, this is followed by tertiary treatment methods to remove remaining nitrates, phosphates, and heavy metals.

Watson Boiling Point (T_B) – The average of five boiling points taken after 10 percent, 30 percent, 50 percent, 70 percent, and 90 percent of the sample of crude oil has been removed during a distillation process of a crude oil sample. The average is converted and reported in Rankine degrees ($°R = °F + 459.7$).

Watson K Factor – A value calculated using the Watson boiling point (T_B) that is used to evaluate the relative value of crude oil. Crude that yield low values for the Watson K factor are considered to be more valuable, since they have a higher proportion of the lower boiling fractions that are used for gasoline production. The value is determined by taking the cube root of T_B and dividing this value by the specific gravity of the sample.

Weather – The process of braking down rock both physically and chemically as it is exposed to moisture, air, organisms and the forces of nature on or near the earth's surface.

Weaving – The process of interlacing fibers or yarn to make a fabric.

Web – A continuous roll of paper or other substrate used in a printing process.

Weft Yarns – The set of yarns running the width of a woven fabric and that cross the warp yarns.

Weldability – Possessing the characteristics that allow it to be welded.

Welding – A bonding process in which the metal pieces to be joined are melted and fused together.

Wet Air Oxidative – A method used to destroy organic materials contained in a liquid phase by bringing them into contact with oxygen under conditions of both high pressure (up to 3,000 psi) and temperature (350-750°C).

Wet Scrubber – A device used for the purification of a dirty gaseous waste stream that employs a liquid such as water as an integral part of the cleansing process.

Wet Wastes – A fruit canning waste stream composed of seeds, peels, leaves, and stems, and used as a low-grade animal feed supplement.

White Liquor – A strongly alkaline aqueous mixture composed chiefly of caustic soda, $NaOH$, and sodium sulfide, Na_2S, used in the digester to delignify wood. In the kraft liquor recovery cycle, white liquor is regenerated from green liquor by the addition of lime, CaO, in a process called recausticizing.

White Water – Water recovered from the paper machine and used to make additional paper.

WRAP – See Waste Reduction Always Pays.

Xenobiotic – Compounds that are human-made and considered to be foreign to the environment.

X-ray – Short wavelength electromagnetic radiation, falling between UV and gamma rays, that can be produced by electron transitions in the inner energy levels of heavy atoms. X-rays possess the ability to damage or destroy tissue, but can be absorbed by dense materials such as lead.

Yarn – A thread made of natural or synthetic fibers that is formed from a twisted aggregate of fibers.

Yellowcake – Produced by digesting powdered uranium ore in a hot nitric-sulfuric acid mixture and filtering to remove the insoluble portion. The sulfate ions are removed by barium carbonate precipitation and the uranyl nitrate is extracted with ether. After re-extraction into water, the mixture is heated to drive off nitric acid, leaving uranium trioxide, UO_3, which must be reduced with hydrogen to produce the yellowcake or uranium dioxide, UO_2.

Zimpro – The trade name for the wet air oxidation method perfected by F. J. Zimmerman. Brings about the oxidation of organic materials by exposure to air pressure (150-3000 psi) and high temperatures.

Zincated – A pretreatment necessary before aluminum and magnesium can be electroplated. After cleaning, the part is etched in phosphoric acid to remove the oxide coating, dipped in nitric acid to activate the surface, and then dipped in sodium zincate, Na_2ZnO_2, solution. After rinsing, the part is immediately plated.

Zooglea – A jelly-like film or free-floating mass composed of a complex population of microorganisms.

Acknowledgments

Chapter 1: Introduction to Waste Streams

Table 1-1: The decade of environmental legislation. A. J. Silva.

Figure 1-1: Number of federal laws enacted each year regulating the use, manufacture, transportation, and disposal of hazardous substances. James T. Dufour, *Hazardous Waste Management*, California Chamber of Commerce, Sacramento, CA, 1989.

Figure 1-2: EPA's hierarchy of hazardous waste management. EPA.

Figure 1-3: Simple flow diagram for Kola's® bottling and return operations. A. J. Silva.

Figure 1-4: Expanded flow diagram for Kola's® bottling and return operation. A. J. Silva.

Table 1-2: Daily inventory for Kola® bottling plant. A. J. Silva.

Table 1-3: Daily materials balance tabulation. A. J. Silva.

Figure 1-5: Quantities of materials entering and leaving the bottling plant. A. J. Silva.

Figure 1-6: Liquids flow diagram. A. J. Silva.

Table 1-4: Daily quantities of liquids in and out of the plant. A. J. Silva.

Figure 1-7: Material balance for receiving/sorting station. A. J. Silva.

Table 1-5: Daily materials balance for receiving/sorting station. A. J. Silva.

Table 1-6: Daily materials balance for solids at the receiving/sorting station. A. J. Silva.

Table 1-7: Liquid mass balance calculation. A. J. Silva.

Chapter 2: Waste Stream Regulations

Table 2-4: Chemicals listed for the semiconductor manufacturing industry that are scheduled for Maximum Achievable Control Technology (MACT) Standards. From *Federal Environmental Regulations Affecting the Electronics Industry*, Design for the Environment Printed Wiring Board Project, EPA 744-B-95-001 September 1995, p.6.

Table 2-5: Reportable Quantities (RQs) that may apply to the semiconductor manufacturing industries. From *Federal Environmental Regulations Affecting the Electronics Industry*, Design for the Environment Printed Wiring Board Project, EPA 744-B-95-001 September 1995, p.13.

Table 2-6: Priority pollutants used in semiconductor packaging that may be present in discharge. From *Federal Environmental Regulations Affecting the Electronics Industry*, Design for the Environment Printed Wiring Board Project, EPA 744-B-95-001 September 1995, p.16.

Table 2-7: Hazardous and non-conventional chemicals used in semiconductor manufacturing. From *Federal Environmental Regulations Affecting the Electronics Industry*, Design for the Environment Printed Wiring Board Project, EPA 744-B-95-001 September 1995, p.17.

Table 2-8: Semiconductor Best Practicable Control Technology Currently Available (BPT) effluent limitations. From *Federal Environmental Regulations Affecting the Electronics Industry*, Design for the Environment Printed Wiring Board Project, EPA 744-B-95-001 September 1995.

Table 2-9: Semiconductor Best Available Control Technology Economically Available (BAT) effluent limitations. From *Federal Environmental Regulations Affecting the Electronics Industry*, Design for the Environment Printed Wiring Board Project, EPA 744-B-95-001 September 1995.

Table 2-10: Semiconductor New Source Performance Standards (NSPS) effluent limitations[1]. From *Federal Environmental Regulations Affecting the Electronics Industry*, Design for the Environment Printed Wiring Board Project, EPA 744-B-95-001 September 1995.

Table 2-11: Semiconductor Best Conventional Pollution Control Technology (BCT) effluent limitations. From *Federal Environmental Regulations Affecting the Electronics Industry*, Design for the Environment Printed Wiring Board Project, EPA 744-B-95-001 September 1995.

Table 2-12: Semiconductor pretreatment standards for common metals, chemical etching and milling, electroless plating, and electroplating facilities discharging 38,000 liters or more per day Pretreatment Standards for Existing Sources (PSES) limitations. From *Federal Environmental Regulations Affecting the Electronics Industry*, Design for the Environment Printed Wiring Board Project, EPA 744-B-95-001 September 1995.

Table 2-13: Printed wiring board pretreatment standards for facilities discharging 38,000 liters or more per day Pretreatment Standards for Existing Sources (PSES) limitations. From *Federal Environmental Regulations Affecting the Electronics Industry*, Design for the Environment Printed Wiring Board Project, EPA 744-B-95-001 September 1995, p. 32.

Table 2-14: Some examples of listed wastes found in semiconductor packaging. From *Federal Environmental Regulations Affecting the Electronics Industry*, Design for the Environment Printed Wiring Board Project, EPA 744-B-95-001 September 1995, p. 46.

Table 2-15: EPA toxic characteristic contaminants that may be found in semiconductor manufacturing waste. From *Federal Environmental Regulations Affecting the Electronics Industry*, Design for the Environment Printed Wiring Board Project, EPA 744-B-95-001 September 1995.

Table 2-16: Threshold planning and reporting quantities for some EPCRA-Designated Extremely Hazardous Chemicals used by the semiconductor packaging industry. From *Federal Environmental Regulations Affecting the Electronics Industry*, Design for the Environment Printed Wiring Board Project, EPA 744-B-95-001 September 1995.

Table 2-17: Chemicals used by the semiconductor industry that are listed in the toxic release inventory. From *Federal Environmental Regulations Affecting the Electronics Industry*, Design for the Environment Printed Wiring Board Project, EPA 744-B-95-001 September 1995.

Table 2-18: An excerpt from a hazardous materials regulations table. EPA.

Box 2-1: From *Federal Environmental Regulations Affecting the Electronics Industry*, Design for the Environment Printed Wiring Board Project, EPA 744-B-95-001 September 1995, p. 19.

Chapter 3: Physical Treatment Technologies

Table 3-1: A list of EPA's physical treatment reporting codes. EPA.

Figure 3-4: Cationic polymer forming a floc. Craig Baker.

Figure 3-5: Mixing and flocculation tanks used to remove suspended wastes. EPA.

Figure 3-7: Solids separate from the contaminated water and agglomerate on the plates of the lamella clarifier. EPA.

Figure 3-9: An air flotation system in which air bubbles are attached to tiny particles and move with them to the surface. Craig Baker.

Figure 3-10: Approximate particle size for various physical separation processes. EPA, *Physical/Chemical Treatment of Hazardous Wastes*, CERI-90-16 April 1990. p.5-15.

Figure 3-12: Cross-section of a belt filter. EPA.

Figure 3-14: Cross-section of a rotary-drum vacuum filter. EPA.

Figure 3-15: A filter press and expanded view of the filter plates. Modified from *A Compendium of Technology*, EPA/625/8-87/1014, September 1987, p. 13.

Figure 3-16: A basket centrifuge cross-section. EPA, A *Compendium of Technologies Used in the Treatment of Hazardous Wastes*, Center for Environmental Research Information Office of Research and Development Cincinnati, OH 45268, EPA/625/8-87/1014, September 1987, p. 4.

Figure 3-17: Cross-section of a dual media filter. EPA.

Figure 3-18: Interior of a bag house and bag shaking action. Adapted from California Air Resources board: Compliance Division, *Baghouses*, June 1991, p. 300-11.

Figure 3-19: A bank of cyclone filters. Adapted from EPA *Technology Transfer Handbook: Control Technology For Hazardous Air Pollutants*, EPA/625/6-86/014, June 1986.

Figure 3-20: A wet venturi scrubber. California Air Resources Board: Compliance Division, *Hot Mix Asphalt Facilities*, August 1990 p. 300-41.

Figure 3-21: DuPont/Oberlin microfiltration system. EPA *Physical/Chemical Treatment of Hazardous Wastes Speaker Slide Copies and Supporting Information* CERI-90-16 April 1990. p. 5-19.

Figure 3-22: A capillary ultrafiltration module. EPA *Physical/Chemical Treatment of Hazardous Wastes Speaker Slide Copies and Supporting Information* CERI-90-16 April 1990. p. 5-20.

Figure 3-23: Reverse osmosis membrane and membrane detail. Modified from EPA, *Physical/Chemical Treatment of Hazardous Wastes*, CERI-90-16 April 1990, p.5-21.

Figure 3-24: Separation of a NaCl solution in an electrodialysis process. EPA, *Physical/Chemical Treatment of Hazardous Wastes*, CERI-90-16 April 1990, p. 5-23.

Figure 3-25: Flow schematic for separation of a plating bath using electrodialysis. From Jim Potter, editor, *Alternative Technology for Recycling and Treatment of Hazardous Wastes, The Third Biennial Report*, The Department of Health Services, Toxic Substances Control Division, Alternative Technology and Policy Development Section, State of California, July, 1986, p. 32.

Figure 3-26: Mineral deposits at Mono Lake. Photo by Dr. Allan Schoenherr.

Figure 3-27: An *in situ* air sparging system. California Resouces Board's Compliance Division: California Air Resources Board, Soil Decontamination, July 1991, p. 300-5.

Figure 3-31: Electrostatic precipitator corona generation. Adapted from California Air Resources board: Compliance Division, *Electrostatic Precipitators*, June 1990, Figure 301.3.

Figure 3-32: An electron from the corona region converts a molecule into a positive ion, which becomes attached to a dust particle. The charged dust particle is removed by being drawn to the negative electrode. Adapted from California Air Resources board: Compliance Division, *Electrostatic Precipitators*, June 1990, Figure 301.3.

Figure 3-34: Adsorption vs. absorption. EPA, *Physical/Chemical Treatment of Hazardous Wastes*, CERI-90-16 April 1990. p. 3-18.

Figure 3-35: Example of an activated carbon sorption-regeneration process. EPA, *Physical/Chemical Treatment of Hazardous Wastes*, CERI-90-16 April 1990. p. 8-33.

Figure 3-36: In situ solidification/stabilization using soil-cement mixing wall method. EPA, *The Superfund Innovative Technology Evaluation Program: Technology Profiles*, EPA/540-5-90/006 Nov. 1990, p. 80.

Chapter 4: Chemical Treatment Technologies

Figure 4-8: Dr. Arnold Beckman, inventor of the pH meter. Photograph courtesy of Beckman Coulter, Inc.

Figure 4-9: Equipment used in a simple neutralization process. EPA.

Figure 4-10: Solubilities of metal hydroxides as a function of pH. EPA, *Development Document for Effluent Limitations Guidelines and Standards for the Metal Finishing Point Source Category*, EPA 440/1-83/091, June 1983.

Chapter 5: Thermal Treatment Technologies

Figure 5-1: Typical features of a liquid injection incinerator. Potter, Jim, editor, *Alternative Technology for Recycling and Treatment of Hazardous Wastes, The Third Biennial Report*, The Department of Health Services, Toxic Substances Control Division, Alternative Technology and Policy Development Section, State of California, July, 1986. p. 82.

Figure 5-2: Liquid injector nozzle atomizes the liquid. EPA, *A Compendium of Technologies Used in the Treatment of Hazardous Wastes*, Center for Environmental Research Information Office of Research and Development, Cincinnati, OH 45268 EPA/625/8-87/014.

Figure 5-3: Typical features of a rotary kiln incinerator. EPA.

Figure 5-4: Typical features of a fixed hearth incinerator. EPA.

Figure 5-5: Typical features of a multiple level hearth incinerator. EPA.

Figure 5-6: Typical features of a bubbling fluidized-bed incinerator. Modified from *A Compendium of Technologies Used in the Treatment of Hazardous Wastes*, Center for Environmental Research Information Office of Research and Development, EPA, Cincinnati, OH 45268, EPA/625/8-87/014 September 1987, p. 34.

Figure 5-7: Typical features of a circulating fluidized-bed incinerator. Modified from *The Superfund Innovative Technology Evaluation Program: Technology Profiles*, EPA/540-5-90/006 Nov. 1990. p. 64.

Figure 5-9: Features of Retech's plasma centrifugal furnace. EPA, *The Superfund Innovative Technology Evaluation Program: Technology Profiles*, EPA/540-5-90/006 Nov. 1990. p. 74.

Figure 5-10: Pyrolytic thermal treatment unit. Modified from *A Compendium of Technologies Used in the Treatment of Hazardous Wastes*, Center for Environmental Research Information Office of Research and Development, EPA, Cincinnati, OH 45268, EPA/625/8-87/014 September 1987, p. 38.

Figure 5-11: Typical features of the Zimpro wet air oxidative method. From Nemerow, Nelson Leonard and Dusgupta, Avijit, *Industrial and Hazardous Waste Treatment*, Van Nostrand Reinhold, New York, 1991, p. 198.

Figure 5-12: A cross-section of a wet scrubber. Modified from Sell, Nancy J., *Industrial Pollution Control: Issues and Techniques*, 2nd ed., Van Nostrand Reinhold, New York, 1992. p. 55.

Figure 5-13: Monitoring the control panel an oven incinerator. Photograph Courtesy of Laidlaw Environmental Services (Aragonite), Inc.

Chapter 6: Biological Treatment Technologies

Figure 6-6: The nitrogen cycle. From Boyce, Ann, *Introduction to Environmental Technology*, Van Nostrand Reinhold, 1997.

Figure 6-10: An activated sludge process schematic. Craig Baker.

Figure 6-11: Diagram for an oxidation ditch activated sludge system. EPA.

Figure 6-13: Operational steps in a sequencing batch reactor (SBR). EPA.

Figure 6-14: Cross-section of a trickling filter. Craig Baker.

Figure 6-15: Cross-section of microbial film on rock (trickling filter medium). Craig Baker.

Figure 6-16: The basics of a rotating biological contactor. Craig Baker.

Figure 6-17: Aerating the soil in a landfarming operation. Photograph Courtesy of Republic Environmental Technologies of Nevada, Inc.

Figure 6-19: Cross-section of a typical air sparging system. EPA.

Figure 6-20: Honeycomb support lattice in a fixed-film bioreactor. Modified from EPA, *The Superfund Innovative Technology Evaluation Program: Technology Profiles*, EPA/540-5-90/006 Nov. 1990. p. 18.

Chapter 7: Minimization of Waste and Pollution Prevention

Figure 7-1: EPA waste priorities summary. EPA.

Figure 7-2: A product recycling label. Recycling label.

Figure 7-5: Schematic showing differences between reuse and in-process recycling. EPA.

Box 7-7: Carol Browner photo. Photograph Courtesy of Steve Delaney/EPA

Box 7-8: Hammer Award photo. Photograph Courtesy of EPA/GLTS.

Figure 7-7: Green Lights logo. EPA.

Figure 7-8: Energy Star program logo. EPA.

Figure 7-9: DfE logo. DfE.

Figure 7-10: Green Chemistry logo. DfE.

Figure 7-11: Responsible Care® logo. CMA's Responsible Care®.

Chapter 10: The Petroleum Industry

Figure 10-1: Rotary drill rig. Photo Courtesy of ARCO.

Chapter 12: The Mining Industries

Figure 12-3: Room-and-pillar mining. Source: ASARCO EIS, Published by United States Forest Service, p. 22-23.

Figure 12-7: Typical open pit mining operation. Borax mine, Boron, CA photograph courtesy of U.S. Borax Inc.

Chapter 15: The Electroplating and Metal Finishing Industries

Figure 15-4: Closed loop (A) and open loop (B) evaporative recovery systems. From EPA, *Guidelines to Pollution Prevention: The Metal Finishing Industry*, EPA/625/R-92/011, October 1992.

Figure 15-5: Electrolytic recovery. From EPA, *Guidelines to Pollution Prevention: The Metal Finishing Industry*, EPA/625/R-92/011, October 1992.

Figure 15-6: Conventional ion exchange system. From EPA, *Guidelines to Pollution Prevention: The Metal Finishing Industry*, EPA/625/R-92/011, October 1992.

Chapter 17: Integrated Circuits and Electronics Assembly Industries

Figure 17-2: Steps in a semiconductor production process. EPA.

Figure 17-3: Cross-section of a reticle. EPA.

Table 17-1: Solder types. EPA.

Table 17-2: Typical wastes generated in the electronics industry. EPA.

Figure 17-7: Semiaqueous cleaning process. EPA.

Chapter 18: The Printed Wiring Board Industries

Figure 18-1: Overview of rigid PWB manufacturing process sequences. From EPA, 744-R-95-005, September 1995.

Figure 18-2: Use of photo-tool for inner layer exposure. From Printed Wiring Board Industry and Use Cluster Profile, EPA 74-R-95-005, September 1995.

Chapter 19: The Pulp and Paper Industries

Figure 19-1: Scanning electron micrograph of a paper sample. Christopher J. Biermann.

Figure 19-2: Flow diagram for the papermaking process. Christopher J. Biermann.

Figure 19-3: The kraft chemical pulping process. Christopher J. Biermann.

Chapter 20: The Graphics and Printing Industries

Table 20-5: Typical printing and allied industries operations: materials used and hazardous wastes that may be generated. EPA.

Table 20-6: Printers compliance checklist. EPA.

Chapter 21: The Textile Industries

Figure 21-1: Knitting is the construction of fabric by interlocking loops of one or more yarns. From *Fairchild's Dictionary of Textiles*, 7th Edition, by Tortora & Merkel, 1996. Used with permission, Fairchild Publications, a division of ABC Media, Inc.

Figure 21-3: Diagram of a weaving loom. From *Fairchild's Dictionary of Textiles*, 7th Edition, by Tortora & Merkel, 1996. Used with permission, Fairchild Publications, a division of ABC Media, Inc.

Chapter 24: The Furniture Finishing Industries

Table 24-1: VOC and solids content of several coating formulations. From *Pollution Prevention in the Finishing of Wood Furniture: A Resource Manual and Guide*, Virginia Department of Environment Quality, Waste Reduction Assistance Program, October 1993.

Chapter 27: The Nuclear Industries

Table 27-1: Nuclear fuel cycle and its associated waste streams. From Ann Boyce, *Introduction to Environmental Technology*, Van Nostrand Reinhold, New York, 1996.

Table 27-2: Distribution of low-level waste received at the Barnwell, SC and Richland, WA commercial disposal sites. Source Department of Energy, LLW-205 1993.

Chapter 28: Consumer-Related Waste Issues

Table 28-1: Standard Industrial Codes.

Figure 28-1: Waste types discarded in MSW in 1995 (after recycling). EPA.

Figure 28-2: 1995 U.S. MSW waste composition (after recycling). EPA.

THE LEGACY

Bibliography

Chapter 1: Introduction to Waste Streams

Dufour, M.S., Editor. *Hazardous Waste Management*. California Chamber of Commerce, 1989.

Hazardous Materials Training and Research Institute. *Basic Industrial Processes*. Raleigh, NC: Martini Print Media, Inc., 1992.

Chapter 2: Waste Stream Regulations

U.S. Environmental Protection Agency. *Federal Environmental Regulations Affecting the Electronics Industry*. Design for the Environment Program Economics, Exposure and Technology Division, Office of Pollution Prevention and Toxics, Environmental Protection Agency, EPA 744-B-95-001, September 1995.

Chapter 3: Physical Treatment Technologies

Dawson, Gaynor W. and Basil W. Mercer. *Hazardous Waste Management*. New York, NY: John Wiley & Sons, Inc., 1986.

Muralidhara, H.S. *Solid/Liquid Separation: Waste Management and Productivity Enhancement*. Columbus, OH: Battelle Press 1990.

Sell, Nancy J. *Industrial Pollution Control: Issues and Techniques*, 2nd Edition. New York, NY: Van Nostrand Reinhold, 1992.

Chapter 4: Chemical Treatment Technologies

Baum, Stuart J. and Charles W. J. Scaife. *Chemistry, A Life Science Approach*, 2nd Edition. New York, NY: Macmillan Publishing Company, Inc., 1980.

Holum, John R. *Elements of General and Biological Chemistry*, 6th Edition. New York, NY: John Wiley & Sons, 1983.

Lewis, Sr., Richard J. *Hawley's Condensed Chemical Dictionary*, 12th Edition. New York, NY: Van Nostrand Reinhold, 1993.

Sell, Nancy J. *Industrial Pollution Control: Issues and Techniques,* 2nd Edition. New York, NY: Van Nostrand Reinhold, 1992.

Chapter 5: Thermal Treatment Technologies

Brunner, Calvin R. *Handbook of Hazardous Waste Incinerator.* BlueRidge Summit, PA: TAB Professional and Reference Books, Division of TAB BOOKS Inc., 1989.

Freeman, Harry M. *Innovative Thermal Hazardous Organic Waste Treatment Processes.* Park Riley, NJ: Noyes Publication, 1985.

Gill, James H. and John M. Quiel. *Incineration of Hazardous, Toxic, and Mixed Wastes.* Cleveland, OH: North American Mfg. Co., 1993.

Potter, Jim, Editor. *Alternative Technology for Recycling and Treatment of Hazardous Wastes, The Third Biennial Report.* State of California Department of Health Services, Toxic Substances Control Division, Alternative Technology and Policy Development Section, July, 1986.

Sell, Nancy J. *Industrial Pollution Control: Issues and Techniques,* 2nd Edition. New York, NY: Van Nostrand Reinhold, 1992.

U.S. Environmental Protection Agency. *Superfund Treatment Technologies: A Vendor Inventory.* Environmental Protection Agency, EPA 540/2-86/004(F), September 1986.

Chapter 6: Biological Treatment Technologies

Adams, Michael W. and Robert M. Kelly. "Enzymes from Microorganisms in Extreme Environments," *Chemical & Engineering News.* Washington, D.C.: American Chemical Society, pp. 32-41, December 18, 1995.

Andrews, William A., Editor. A *Guide to the Study of Soil Ecology.* Englewood Cliffs, NJ: Prentice-Hall, Inc., 1973.

Atlas, Ronald M. "Bioremediation," *Chemical & Engineering News.* Washington, D.C.: American Chemical Society, pp. 32-42, April 3, 1995.

Boyd, Vickey. "This Bug's for You," *Environmental Protection.* Waco, TX: Stevens Publishing, pp. 26-28, February, 1996.

Freeman, Harry M., Editor. *Standard Handbook of Hazardous Waste Treatment and Disposal.* New York, NY: McGraw-Hill Book Company, 1988.

Kuman, P.B.A. Nanda, Niatcheslav Dushenkov, Harry Motto, and Ilya Raskin. "Phytoextraction: The use of Plants to Remove Heavy Metals from Soils," *Environmental Science & Technology,* Vol. 29, No. 5. American Chemical Society, 1995.

Sell, Nancy J. *Industrial Pollution Control: Issues and Techniques,* 2nd Edition. New York, NY: Van Nostrand Reinhold, 1992.

Chapter 7: Minimization of Waste and Pollution Prevention

Jacoby, Mitch. "ACS Receives Gore's Hammer Award," *Chemical & Engineering News,* Washington, D.C.: American Chemical Society, pp. 7-8, May 19, 1997

Potter, Jim, Editor. *Alternative Technology for Recycling and Treatment of Hazardous Wastes, The Third Biennial Report.* State of California Department of Health Services, Toxic Substances Control Division, Alternative Technology and Policy Development Section, July, 1986.

Raber, Linda. "Green Chemistry Honored," *Chemical & Engineering News.* Washington, D.C.: American Chemical Society, pp. 7-8, June 30, 1997.

Worobec, Mary Devine, and Girard Ordway. *Toxic Substances Controls Guide: Federal Regulation of Chemicals in the Environment.* Washington, D.C.: The Bureau of National Affairs, Inc., 1989.

Chapter 8: Life Cycle Design for General Manufacturing

Environmentally Conscious Design and Manufacturing: The Competitive Strategies of Industry Leaders. Highlights from the First International Congress on Environmentally Conscious Design and Manufacturing.

Product Recyclability Must Be Designed in Greenscore™, Assessment Tool for Achieving Environmental Excellence. National Center for Manufacturing Sciences, Environmentally Conscious Manufacturing Group.

Sarkis, Joseph, and Abdul Rasheed. "Greening the Manufacturing Function," *Business Horizons*, Vol. 38, No. 5.

U.S. Environmental Protection Agency. *Design for the Environment: Product Life Cycle Design Guidance Manual*. Government Institutes, Inc., U.S. Environmental Protection Agency, Office of Research and Development.

U.S. Environmental Protection Agency. *Life Cycle Design Framework and Demonstration Projects: Profiles of AT&T and Allied Signal*. Office of Research and Development, EPA/600/R-95/107, July 1995.

Chapter 9: Hazardous Waste Regulations for the Cleanup Industries

Corbitt, R.A. *Standard Handbook of Environmental Engineering*. New York, NY: McGraw-Hill, Inc., 1990.

Freeman, H.M. *Standard Handbook of Hazardous Waste Treatment and Disposal*. New York, NY: McGraw-Hill, Inc., 1989.

Grasso, D. *Hazardous Waste Site Remediation: Source Control*. Boca Raton, FL: CRC Press, Inc., 1993.

Chapter 10: The Petroleum Industry

Brouwer, L. E. J. (Foreword), *The Petroleum Handbook*, 5th Edition. Shell Centre, London SE1: Shell International Petroleum Company Limited, 1966.

Hoffman, H. L. Chapter 15, "Petroleum and Its Products," *Riegel's Handbook of Industrial Chemistry*, 9th Edition. James A. Kent, Editor. New York, NY: Van Nostrand Reinhold, 1992.

U. S. Environmental Protection Agency. *Crude Oil and Natural Gas Exploration and Production Wastes*. Exemption from RCRA Subtitle C Regulations. Office of Solid Waste (5305), U. S. Environmental Protection Agency, EPA 530-K-95-003, May 1995.

Chapter 11: The Chemical Production Industries

Cutting Chemical Waste. New York, NY: Inform, Inc., 1985.

CMA Waste Minimization Resource Manual. Washington, D.C.: Chemical Manufacturers Association.

Department of Energy. *Chemicals Industry Brochure*. Department of Energy, Office of Industrial Technologies. http://www.nrel.gov/oit/Industries-of-the-Future/chemical.html.

Higgins, Thomas. *Hazardous Waste Minimization Handbook*. Chelsea, MI: Lewis Publishers, 1989.

Industrial Process Design for Pollution Control. American Institute for Chemical Engineers, Vol. 4, 1972, Vol. 5, 1974, Vol. 6, 1974, and Vol. 7, 1975.

Plambeck, James A. *The Alkalie Industry*, 1995. (Jim.Plambeck@ualberta.ca).

Plambeck, James A. *The Acid Industry*, 1995. (Jim.Plambeck@ualberta.ca).

U.S. Environmental Protection Agency. *EPA Office of Compliance Sector Notebook Project – Profile of the Inorganic Chemical Industry*. Environmental Protection Agency, EPA/310-R-95-004 and EPA/310-R-95-012, September 1995.

Chapter 12: The Mining Industries

Given, Ivan A., Editor. *SME Mining Engineering Handbook*, Vol. 1 and 2, 1973.

Hartman, Howard L. *Introductory Mining Engineering*. New York, NY: John Wiley & Sons, 1987.

Hurlbut, Cornelius S. *Dana's Manual of Mineralogy*. 18th Edition. New York, NY: John Wiley & Sons, 1871.

Peele, Robert. *Mining Engineers' Handbook*. New York, NY: John Wiley & Sons, C1918.

Chapter 13: The Metal Production Industries

Gilchrist, J. D. *Extraction Metallurgy*, 3rd Edition. New York, NY: Pergamon Press, 1989.

Peele, Robert. *Mining Engineers' Handbook.* New York, NY: John Wiley & Sons, C1918.

van Zyl, Dirk J.A. Ian P.G. Hutchison, and Jean E. Kiel, Editors. *Introduction to Evaluation, Design and Operation of Precious Metal Heap Leaching Projects.* Littleton, CO: Society of Mining Engineering, Inc., 1988.

Wentz, Charles A. *Hazardous Waste Management.* New York, NY: McGraw-Hill, Inc., 1989.

Chapter 14: The Metalworking Industries

"Machining," *Metals Handbook*, 9th Edition, Volume 16. American Society for Metals, 1989.

Pollution Prevention Opportunities for Metal Fabricators. http://www.oia.org/p2metal.htm.

U.S. Environmental Protection Agency. *EPA Office of Compliance Sector Notebook Project – Profile of the Fabricated Metal Products Industry.* Environmental Protection Agency, EPA/310-R-95-007, September 1995.

U.S. Environmental Protection Agency. *Guides to Pollution Prevention: The Fabricated Metal Products Industry.* Environmental Protection Agency, EPA/625/7-90/006, July 1990.

U.S. Environmental Protection Agency. *Pollution Prevention in Metal Manufacturing: Saving Money through Pollution Prevention*, Environmental Protection Agency, OSW, October 1989.

U.S. Environmental Protection Agency. *Sustainable Industry: Promoting Strategic Environmental Protection in the Industrial Sector, Phase 1 Report*, Environmental Protection Agency, OERR, June 1994.

U. S. Environmental Protection Agency. *Waste Minimization for Metal Parts Cleaning.* Environmental Protection Agency, EPA/530/SW-89-049, August 1989.

Chapter 15: The Electroplating and Metal Finishing Industries

Bennett, Pat. *Assessment of the Metal Finishing and Plating Industry: Source Reduction Planning Efforts.* California Environmental protection Agency, Department of Toxic Substances Control, Office of Pollution Prevention and Technology Development, July 1996.

California Environmental Protection Agency. *Decorative Plating with Trivalent Chrome.* Fact Sheet, California Environmental Protection Agency: Department of Toxic Substances Control, Pollution Prevention, Public and Regulatory Assistance Program, 1992.

Kindschy, Jon W., David Ringwald, and Molly Carpenter. *Waste Minimization Assessment Procedures Module III: Waste Minimization in the Metal Finishing Industry.* California Department of Toxic Substance Control. Riverside, CA: University of California, May 1991.

U.S. Environmental Protection Agency. *Guides to Pollution Prevention: The Metal Finishing Industry.* Environmental Protection Agency, EPA/625/R-92/011, October 1992.

Chapter 16: The Paint and Surface Coating Industries

Chandler, K.A. and D.A. Bayliss. *Corrosion Protection of Steel Structures.* New York, NY: Elsevier Applied Sciences, 1985.

Chandler, R.H. *Epoxy Powder Coatings.* Braintree England: R.H. Chandler Ltd., 1973.

Journal of Protective Coatings and Linings. Pittsburgh, PA: Technology Publishing Company.

Lambourne, R. *Paint and Surface Coatings: Theory and Practice.* Chichester, England: Ellis Horwood, Ltd, 1993.

Metal Finishing Magazine. Hackensack, NJ: Metals and Plastics Publications.

Chapter 17: Integrated Circuits and Electronics Assembly Industries

"AT&T eliminates more CFCs with food ingredients," *Circuits Assembly*, June 1993.

Castaneda, Chris. "Eliminating Ozone-depleting Chemicals," *Circuits Assembly*, June 1993.

Tuck, John. "Getting the lead out," *Circuits Assembly*, December 1992.

Tuck, John. "A successor to solder?" *Circuits Assembly*, June 1993.

Wolf, Yazdani, and Yates. *Electronic Products Manufacture*. Source Reduction Research Partnership, June 1990.

Yazdani, Azita. "Source reduction of chlorinated solvents in the electronics industry." *Plating and Surface Finishing*, April 1993.

Chapter 18: The Printed Wiring Board Industries

U.S. Environmental Protection Agency. *Federal Environmental Regulations Affecting the Electronics Industry*, Design for the Environment Program Economics, Exposure and Technology Division Office of Pollution Prevention and Toxics, Environmental Protection Agency, EPA 744-B-95-001, September 1995.

U. S. Environmental Protection Agency. *Implementing Cleaner Technologies in the Printed Wiring Board Industry: Making Holes Conductive*. Design for the Environment Program Economics, Exposure and Technology Division Office of Pollution Prevention and Toxics, Environmental Protection Agency, EPA 744-R-97-001, February 1997.

U.S. Environmental Protection Agency. *Printed Wiring Board Industry and Use Cluster Profile*. Design for the Environment Program Economics, Exposure and Technology Division Office of Pollution Prevention and Toxics, U. S. Environmental Protection Agency, EPA 744-R-95-005, September 1995.

U.S. Environmental Protection Agency. *Printed Wiring Board Pollution Prevention and Control: Analysis of Survey Results*. Design for the Environment Program Economics, Exposure and Technology Division Office of Pollution Prevention and Toxics, Environmental Protection Agency, EPA 744-R-95-006, September 1995.

Chapter 19: The Pulp and Paper Industries

Biermann, C.J. *Handbook of Pulping and Papermaking*, 2nd Edition. New York, NY: Academic Press, 1996.

Sjöström, E. *Wood Chemistry: Fundamentals and Applications*, 2nd Edition. New York, NY: Academic Press, 1993.

U.S. Environmental Protection Agency. *Profile of the Pulp and Paper Industry*. Sector Notebook Project, Environmental Protection Agency, EPA/310-R-95-015, September, 1995.

Chapter 20: The Graphics and Printing Industries

DeSilver, Drew. "Waste Minimization in the Commercial Printing Industry." *EI Digest*. Minneapolis, MN: Environmental Information, Ltd., November 1994.

Great Printers Project, Council of Great Lakes Governors, Printing Industries of America, Environmental Defense Fund, Minneapolis, MN, July 1994.

Ramus, Catherine A. *Use Cluster Analysis of the Printing Industry (Draft Final)*. Regulatory Impacts Branch, Economics and Technology Division, Office of Pollution Prevention and Toxics, U.S. Environmental Protection Agency, May 26, 1992.

U.S. Environmental Protection Agency. *Federal Environmental Regulations Potentially Affecting the Commercial Printing Industry*. Design for the Environment Program, Economics, Exposure and Technology Division, Office of Pollution Prevention and Toxics, Environmental Protection Agency, March 1994.

Chapter 21: The Textile Industries

U.S. Environmental Protection Agency. *Profile of the Textile Industry*. EPA Compliance Sector Notebook, 1997.

Chapter 22: The Agricultural Industries

National Research Council. *Alternative Agriculture.* Washington DC: National Academy Press, 1989.

National Research Council. *Soil and Water Quality: An Agenda for Agriculture.* Washington, DC, National Academy Press, 1993.

NCSU Water Quality Group. *Evaluation of the Experimental Rural Clean Water Program.* Biological and Agricultural Engineering Department, North Carolina State University, Raleigh, NC, EPA-841/R-93-005.

Plucknett, Donald L. and Donald L. Winkelmann. "Technology for Sustainable Agriculture," *Scientific American*, September 1995.

U.S. Department of Agriculture. *National Handbook of Conservation Practices.* Natural Resources Conservation Service (formerly Soil Conservation Service), U.S. Department of Agriculture, 1994. http://www.ncg.nrcs.usda.gov/nhcp 2.html.

U.S. Department of Commerce. *Bureau of the Census Agricultural Brief – Large Farms Are Thriving in the United States.* U.S. Department of Commerce, AB/96-1, July 1996.

U.S. Environmental Protection Agency. *Journal of Soil and Water Conservation*, Vol. 45, No. 1. Environmental Protection Agency, EPA-841-S-95-004, Jan/Feb 1990.

U.S. Environmental Protection Agency. *Managing Nonpoint Source Pollution from Agriculture.* Environmental Protection Agency, EPA-841-F-96-004F, March 1996.

U. S. Environmental Protection Agency. *The Problem of Nonpoint Source Pollution.* Environmental Protection Agency, EPA-840-F-93-001b, July 1993.

Chapter 24: The Furniture Finishing Industries

Bralla, James G., Editor. *Handbook of Product Design for Manufacturing.* New York, NY: McGraw-Hill Book Company, 1986.

Bullard, S., B. Doherty, and P. Short. *The Mississippi Furniture Industry and its Use of Wood-Based Materials.* Research Report 13. Mississippi State University, December 1988.

Christianson, Rich. "Pennsylvania House Scores: A Finishing First," *Wood and Wood Products.* pp. 53-54, October 1991.

Hathaway, John. "Air Permitting Challenges," *Environmental Protection.* pp. 46-50, May 1994.

Kohl, Jerome. *Pollution Prevention: Managing and Recycling Solvents in the Furniture Industry.* NC Department of Environment, Health, and Natural Resources, May 1986.

Martin, R., K. Woods, and J. Schofield. "HAPhazard No More," *Industrial Wastewater.* pp. 17-26, May-June 1994.

Noyes, Robert, Editor. *Pollution Prevention Technology Handbook.* New Jersey: Noyes Publications, 1993.

"Pennsylvania House Expands Unicarb Use," *Furniture Design and Manufacturing.* pp. 48-49, November 1992.

Schrantz, Joe. "L.A. Basin VOC Rules Driving Industry Out," *Industrial Finishing.* February 1989.

Virginia Department of Environmental Quality. *Pollution Prevention in the Finishing of Wood Furniture: A Resource Manual and Guide.* Commonwealth of Virginia, Department of Environmental Quality, Waste Reduction Assistance Program, October 1993.

U.S. Environmental Protection Agency. *Control of Volatile Organic Compound Emissions from Wood Furniture Coating Operations.* Office of Air and Radiation, Environmental Protection Agency, October 1991.

U.S. Environmental Protection Agency. *Radiation Curable Coatings.* EPA Control Technology Center, North Carolina, EPA/600/2-91-035, July 1991.

Chapter 25: The Wood Preservation Industries

American Wood Preservers Association. *Book of Standards*. Granbury, TX: American Wood Preservers Association, 1996.

Eaton, R.A. and M.D.C. Hale. *Wood: Decay, Pests, and Protection*. New York, NY: Chapman and Hall, 1993.

Hunt, G.M. and G.A. Garratt. *Wood Preservation*. New York, NY: McGraw-Hill, 1967.

Nicholas, D.D., Editor. *Wood Deterioration and Its Prevention by Preservative Treatments*. Syracuse, NY: Syracuse University Press, 1973.

Richardson, B.A. *Wood Preservation*. Lancaster, England: The Construction Press, Ltd., 1978.

U.S. Department of Agriculture. *Wood Handbook: Wood as an Engineering Material*. Washington, D.C.: U.S.D.A. Forest Service Agricultural Handbook #72, 1974.

Wilkinson, J.G. *Industrial Timber Preservation*. London: Associated Business Press, 1979.

Zobel, R.A. and J.J. Morrell. *Wood Microbiology - Decay and It's Prevention*. San Diego, CA: Academic Press, Inc., 1992.

Chapter 26: The Medical Industries

Alderson, P. O. *The Specialty of Radiology Spotlight: Nuclear Medicine*, ACR Bulletin, August 1995. URL: http://www.acr.org/bulleting/8-95/10_spotlight_nm.html

Chapter 27: The Nuclear Industries

Benedict, M., T.H. Pigford, and H.W. Levi. *Nuclear Chemical Engineering*, 2nd Edition. New York, NY: McGraw Hill Book Company, 1981.

DOE/LLW-205, 1993. State-by-state assessment of low-level radioactive wastes received at commercial disposal sites. Idaho National Engineering Laboratory, EG&G Idaho Inc., National Low-level Waste Management Program, Idaho Falls, Idaho, September 1994.

Gershey E.L., R.C. Klien, E. Party, and A. Wilkerson. *Low-Level Radioactive Waste: From Cradle to Grave*, New York, NY: Van Nostrand Reinhold, 1990.

Hera, Christian. "Atoms for sustainable agriculture: Enriching the farmers yield," *IAEA Bulletin*, Vol. 37, No. 2. Vienna, Austria: International Atomic Energy Agency, 1995.

House W.B., and M.T. Ryan. "Barnwell to Have Improved Technology," *HPS Newsletter*, Vol. XXIV, No. 1, January 1996.

International Atomic Energy Agency. "Bituminization Processes Condition Radioactive Wastes," *Technical Reports Series*, No. 352. Vienna, Austria: International Atomic Energy Agency, 1993.

International Atomic Energy Agency. "Guidebook on Radioisotope Tracers in Industry," *Technical Reports Series*, No. 316. Vienna, Austria: International Atomic Energy Agency, 1990.

International Atomic Energy Agency. "Minimization of Radioactive Waste from Nuclear Power Plants and the Back End of the Nuclear Fuel Cycle," *Technical Reports Series*, No. 377. Vienna, Austria: International Atomic Energy Agency, 1995.

International Atomic Energy Agency. "Nuclear power reactors in operation and under construction at the end of 1995," PR96/8, April 19, 1996. URL:http//www.iaea.or.at/wordlatom/inforesource/pressrelease/prn896.

International Atomic Energy Agency. *Radioactive Waste Management, An IAEA Source Book*. Vienna, Austria: International Atomic Energy Agency, 1992.

Murray R.L. *Nuclear Energy, An Introduction to the Concepts, Systems, and Applications of Nuclear Process*. New York, NY: Peramon Press Inc., 1977.

Murray R.L. *Understanding Radioactive Waste*, 4th Edition. Columbus, OH: Battelle Press, 1994.

National Research Council, Nuclear Wastes. *Technologies for Separations and Transmutation*. Committee on Separations Technology and Transmutations Systems Board on Radioactive Waste Management, Commission on Geoscience, Environment, and Resources, National Research Council. Washington, D.C: National Academy Press, 1996.

Savannah River Technology Center. Savannah River Technology Center expertise in vitrification technology, June 28, 1996. URL:http://www.umr.edu /~cerengr/sem95/seminars.html.

Scientific Ecology Group. Brochure on Radwaste Processing and Management. Oak Ridge, TN: Scientific Ecology Group, Inc., July 1996.

Scientific Ecology Group. Brochure on Steam Reforming, Oak Ridge, TN: Scientific Ecology Group, Inc., July 1996.

Smyth, H.D. *Atomic Energy for Military Purposes.* Princeton University Press, 1946. Sixth Printing by Carey Press Co., New York.

Tang, Y.S. and J.H. Saling. *Radioactive Waste Management.* New York, NY: Hemisphere Publishing Corporation, 1990.

Title 10, Code of Federal Regulations, Energy. Parts 1-199. Washington, DC: U.S. Government Printing Office, Annual update.

Title 40, Code of Federal Regulations, Protection of Environment. Parts 1-799. Washington, DC: U.S. Government Printing Office, Annual update.

U.S. Atomic Energy Commission, *Nuclear Terms: A Brief Glossary,* 2nd Edition. USAEC Division of Technical Information Extension, Oak Ridge, TN, no date; C1968.

U.S. Department of Energy. *Waste Isolation Pilot Plant (WIPP),* DOE/EM-0036P(Revision 1). Office of Waste Management, U.S. Department of Energy, August 1994.

U.S. Ecology, "Permanent Disposal of NORM and NARM" brochure, July 1996.

University of Chicago, Office of Radiation Safety. *Radioactive Waste Management Guide.* Revised July 13, 1996. http://radiant.uchicago.edu/index.htm.

University of Iowa. *A Patients Guide to Nuclear Medicine.* URL:http://indy.radiology.uiowa.edu/ Patients. Accessed August 29, 1996.

Uranium Institute. *The Management of Radioactive Waste.* A report by an international group of experts. London, 1991.

Vitkus, T., W.L. Beck, and B. Freeman. *A Garage Sale Bargain: A leaking 2.2 GBq Source, Phase III - The Radiological Cleanup.* Health Physics Meeting, Seattle, July 25, 1996. Oak Ridge Institute of Science and Education and Tennessee Department of Environment and Conservation, 1996.

Chapter 28: Consumer-Related Waste Issues

California Integrated Waste Management Board. *Waste Reduction Ideas for Offices.* California Integrated Waste Management Board, http://www. ciwmb.ca.gov/mrt/wpw/wpbiz/fsoffice.htm, June 4, 1997.

Flynn, Alicia A. and Rory E. Kessler. *A Consumer Guide to Safer Alternatives to Hazardous Household Products,* April 1992.

Missouri Department of Natural Resources. *Preventing Pollution Begins with You.* Missouri Department of Natural Resources. http://es.inel.gov/techinfo/ facts/missouri/miss-p2.html, March 1995.

Ohio Environmental Protection Agency. *A Guide to Safe Management of Household Hazardous Waste.* Columbus, OH: Ohio EPA Public Interest Center, April 1990.

The World Is Full of Toxic Products: Your Home Shouldn't Be. San Diego, CA: Environmental Health Coalition, 1990.

U.S. Environmental Protection Agency. *Characterization of MSW in the U.S.: 1996 Update.* Environmental Protection Agency, Washington, DC. http:/ /www.epa.gov/epaoswer/non-hw...l/factbook/ internet/mswf/gen.htm#2.

Index

A

Abietic acid 358
Absorption 88
Acceptor ion 352
Acid 99
Acid cleaning 306
Acid deposition 32
Acid mine drainage (AMD)
    276, 298
Acid rain 32
Acidic pickling solution 335
Acidulant 439
Acrylic 449
Activated sludge 150
Activated sludge process 151
ACZA 462
Additive processing 364
Adhesive bonding 307
Adit 267
Administrative requirements 216
Adsorbent 251
Adsorption 88, 251
Aerobes 148
Aerobic respiration 143, 148

Agglomerate 60
Agglomeration 61
Air flotation 381
Air pollution control system (APCS)
    118
Air sparging 83, 162
Air stripping 83
Air toxics 32
Algae 153
Alkaline 103
Alkaline cleaning 306
Alkaline noncyanide copper 324
Alkylation 242
Alloys 304
Alpha particle 484
Alternative synthetic pathways for
    pollution prevention 190
AMD 276, 298
American Petroleum Institute (API)
    235
American Petroleum Institute gravity
    (°API) 235
Ammoniacal copper quat 462
Ammoniacal copper zinc arsenate
    (ACZA) 462

Anabolism 142
Anaerobic contact process 154
Anaerobic digester 154, 157
Anaerobic digestion 157
Anaerobic filter 151, 156
Anaerobic respiration 143
Anaerobic suspended growth
    150, 154
Anilox roller 389
Animal husbandry 421
Anions 97
Annealing 307
Anode 322, 369
Anoxic 157
Anoxic denitrification 151, 157
Antimony 293
AOC 216, 224
APCS 118
API 235
°API 235
Applicable or relevant and appropri-
    ate requirement 223
Aqueous cleaning 306, 357
Aqueous vapor degreasing (AVD)
    356

General Industry Safety Standards 28

General solubility rules 105

Geological repository 490

Glass 504, 511

Glassy slag 126

Global warming 32

Global Warming Treaty 245

Good housekeeping practices 451

Grade 267

Granular activated carbon 88

Gravimeter 237

Gravity, API (°API) 235

Gravure printing 389

Graywater 507

Green Chemistry Challenge 190, 191

Green Chemistry Program 186, 190

Green Lights Program 186, 188

Green liquor 378

Green manure 425

Green products 508

Green synthesis 259

Greenhouse 419

Grinding 286

Gypsum 267

H

Half-life 484

Halogen 97

Hanford 490

Hanford Engineering Works 485

HAPs 3, 5, 32, 394, 395, 444, 451

air toxics 32

Hardening 307

Hardwood 376, 456

Haulage drift 270

Hazard Communication Standard 28, 48

Hazardous air pollutants (HAPs) 3, 5, 32, 394, 395, 444, 451

air toxics 32

Hazardous and Solid Waste Amendments (HSWA) 11, 116, 171, 323

Hazardous Materials Transportation Act (HMTA) 9, 50, 51

environmental regulations 50

industrial impact 51

List of Hazardous Substances 51

marking 50

placard 50

RQs 51

warning label 50

Hazardous Solvent Substitution Data System (HSSDS) 197

Hazardous waste 215

Hazardous waste operations and emergency response (HAZWOPER) 28

HAZWOPER 28

HCFC 123 356

HCFC-141b 356

Heap leaching 295

Hearth 122

Heartwood 457

Heat exchanger 129

Heat set ink 388

Heat treating 302, 304, 307

Hemicellulose 376

Herbicide 420

Herbivore 422

Heterotroph 142

Hexavalent chrome 110, 324

Hexavalent chromium reduction 110

HHW 340

Hierarchy of hazardous waste management 11

High-level radioactive waste (HLW) 222, 490

High Volume/Low Pressure (HVLP) 337, 452

HLW 222, 490

HMTA 9, 50, 51

environmental regulations 50

industrial impact 51

List of Hazardous Substances 51

marking 50

placard 50

RQs 51

warning label 50

Hole barrel 364

Household hazardous waste (HHW) 340

HSSDS 197

HSWA 11, 116, 171, 323

HVLP 337, 452

Hydraulic gradient 270

Hydrocarbon 220, 232

Hydrochlorofluorocarbon (HCFC) 356

Hydrocracking 242

Hydrofluoric acid 242

Hydrogen bonding 60

Hydrogen ion 99

Hydrolysis 101

Hydrometallurgy 294

Hydronium ion 99

Hydrophobic 66

Hydrothermal fluid 267

Hydrotreating 242

Hydroxide ion 99

Hydroxide precipitation 104, 105

Hyperaccumulators 166

Hyperfiltration 78

I

IBSIN 170

IC 349, 362

Idaho Falls 490

IDW 215

Image area 389

Immobilization 90

Implementation 177

Impregnation 457

In situ bioremediation 160

In situ mining 270, 295

Incineration 116, 474, 478, 492

Incinerator chemistry 117

combustion 117

non-stoichiometric 117

starved oxygen 117

stoichiometric 117

three Ts 117, 125

TTUs 117

Incinerator regulations 133

Incinerators 7, 8, 474

bubbling fluidized-bed incinerator 124

circulating fluidized-bed incinerator 124

DREs 124

fixed hearth incinerator 122

fluidized-bed incinerator 124

liquid injection incinerator 119

multiple level hearth incinerator 122

operating permits 133

plasma thermal treatment unit 126

pyrolytic thermal treatment unit 128

reciprocating grate incinerator 125

regulations 133

rotary kiln 120

WFCOs 134

Inclined plate clarifier 64

Indicator paper 101

Indoor air pollutants 507

Induction hardening 308

Industrial impact

FIFRA 28

TSCA 26

Ingot 284, 350

Injection well 41

Injector nozzle 119

APR '01